Engineering Materials

This series provides topical information on innovative, structural and functional materials and composites with applications in optical, electrical, mechanical, civil, aeronautical, medical, bio- and nano-engineering. The individual volumes are complete, comprehensive monographs covering the structure, properties, manufacturing process and applications of these materials. This multidisciplinary series is devoted to professionals, students and all those interested in the latest developments in the Materials Science field, that look for a carefully selected collection of high quality review articles on their respective field of expertise.

Indexed at Compendex (2021) and Scopus (2022)

Bhasha Sharma · Vijay Chaudhary ·
Shashank Shekhar · Amit Kumar
Editors

Nanostructured Polymers

Synthesis and Performance

Springer

Editors
Bhasha Sharma
Deen Dayal Upadhyaya College
University of Delhi
New Delhi, Delhi, India

Vijay Chaudhary
Department of Mechanical Engineering
Amity School of Engineering
and Technology
Amity University Uttar Pradesh
Noida, Uttar Pradesh, India

Shashank Shekhar
Department of Chemistry
Sri Aurobindo College
University of Delhi
New Delhi, Delhi, India

Amit Kumar
Department of Chemistry
University of Delhi
New Delhi, Delhi, India

ISSN 1612-1317 ISSN 1868-1212 (electronic)
Engineering Materials
ISBN 978-981-95-3580-4 ISBN 978-981-95-3581-1 (eBook)
https://doi.org/10.1007/978-981-95-3581-1

This Springer imprint is published by the registered company Springer Nature Singapore Pte Ltd.
The registered company address is: 152 Beach Road, #21-01/04 Gateway East, Singapore 189721, Singapore

Preface

Nanopolymers have attracted considerable interest over the last few years due to their unique properties and behaviors resulting from their small size. As asserted by different authors, these nanoparticulate materials show potential for diversified functional applications ranging from biomedical and energy to sensing, labeling, and catalysis. Tailored structural features (i.e., size, shape, morphology, internal softness, and interior cross-linking) determine polymer nanoparticles' impact on the cargo loading capacity and controlled/sustained release, possibility of endocytosis, degradability, photostability, etc. The designed interfacial features, however (i.e., stimuli-responsive surfaces, wrinkling, surface porosity, shell-layer swellability, layer-by-layer surface functionalization, and surface charge), regulate nanopolymers' interfacial interactions, controlled assembly, movement and collision, and compatibility with the surroundings (e.g., solvent and biological environments). These features define nanopolymers' overall properties/functions of homogeneity, stability, interfacial tension, and minimization of the surface energy barrier. Lowering of the resultant outcomes is directly influenced by inhomogeneity in the structural and interfacial design through the structure–function relationship. Therefore, a key requirement is to produce well-defined polymer nanoparticles with controlled characteristics. Polymers are amorphous, flexible, and soft, and hence controlling their structural/interfacial features through the single-step process is a challenge. Overall, the structural and interfacial synthetic aspects of polymer nanoparticles such as the size, shape, morphology, core–shell design, layer-by-layer functionalization, and assembly provide very useful functions for a wide range of various nanotechnology-based applications. Additionally, the efficiency of nanopolymers as a reinforcing phase in various polymer nanocomposites, and the potential of nanopolymers as an ideal polymer matrix phase in nanocomposites, and their advanced applications is highlighted. This book will highlight the state-of-the-art features of nanopolymers with perspectives on their advanced applications.

This volume will serve as a valuable handbook for the development of novel class of materials-nanopolymers because it provides researchers, students, and entrepreneurs with all the material necessary to begin their projects in this field. Readers will find protocols to prepare polymer nanoparticles using different methods, since these are based on the variety of experiences that experts encounter in the field. In addition, complex topics such as the optimal characterization of polymer nanoparticles discussed, as well as practical guidelines on how to formulate polymer nanoparticles tailored for various applications, and how to modify the properties of nanoparticles to give them different functionalities will be presented. The book will discuss the translation of technology from research to practice, considering aspects related to industrialization of preparation and aspects of regulatory development, becoming an essential read for polymer and materials scientists, materials engineers, biochemists as well as environmental chemists.

New Delhi, India — Bhasha Sharma
Noida, India — Vijay Chaudhary
New Delhi, India — Shashank Shekhar
New Delhi, India — Amit Kumar

Contents

Fundamentals of Nanostructured Polymers

Nano Polymer Design: Fundamentals, Materials, and Strategies 3
Vijay Chaudhary, Manu Satya Prakash Pathariya, Bhasha Sharma, and Chanchal Ahlawat

Bioinspired and Biodegradable Nano Polymers: Fundamentals, Materials, and Green Strategies 33
Rukmani Sharma, V. K. Paliwal, Amit Kumar, Shashank Shekhar, Vijay Chaudhary, and Bhasha Sharma

Emerging Nano Polymer Technologies for Synthesis and Characterization of Polymer Nanoparticles 57
Ankit Tyagi, Anant Bhardwaj, Rohit Sahu, and Chanchal Ahlawat

Introduction of Nanostructure into Shape Memory Polymers from Materials to Applications 81
Ritesh Gupta, Gaurav Mittal, Surbhi Singh, Vijay Chaudhary, and Krishna Kumar

Engineered Nano Polymer-Based Hydrogel 103
Ameya Sharma, Divya Dheer, and Vivek Puri

Properties and Characteristics of Nanostructured Polymers

Surface Morphology, Structural, and Interfacial Features of Polymers at Nanoscale 131
Priti and Radha Sachan

Tunable Shape and Size of Nano Polymers and Their Effect on Properties 159
Monika Siddharthan, Dharani Rajasekaran, and Jolly Bhadra

Polysaccharide Nanoparticles-Based Functional Constructs 183
Surbhi Kharwar, Sheetal Jaiswal, Vinod Kumar, and Sunil Kumar

Nanoengineering of Polymer Particles for Traversing Bio, Nano Interactions 211
Saahil Hembrom, Soumya Pandey, and Neeta Kumari

Protein and Lipid Nanomaterial Based Functional Construct 235
Yamini Thakur, Anuradha Choudhary, Mamta Tripathi, Rama Sarojinee, Amit Kumar Sharma, and Reetu Sharma

Applications of Nanostructured Polymers

Nanopolymers for Energy Harvesting 261
Md. Arafat Hossain, Md. Wasikur Rahman, Victoria Padilla, Nazmul Islam, and Mohammed Jasim Uddin

Advanced Printing Techniques for Nanopolymers 299
Niranjan Patra, Pradeep Kumar Panda, and Ştefan Ţălu

Nano Polymer Composites for 3D and 4D Printing Applications 323
Chitte Prem Kumar, N. Sivanagaraju, D. Loknath, G. Naga Mallikarjun Rao, and M. R. K. Vakkalagadda

Nano Polymers Towards Advanced Energy Storage Devices 357
Barnika Chakraborty

Nanopolymers Applications in Medical Imaging, Diagnostics, Drug Delivery, Cell Activity Regulation, Sensors and Therapy 377
Ankita Rajput

Nano Polymers Applications in Food and Beverage Industries 405
Vijay Chaudhary, Bhasha Sharma, Chanchal Ahlawat, and Shashank Shekhar

Nanopolymer-Based Agro Nano-chemicals 427
Michael Ovbare Akharame, Aimuanmwosa Frank Eghomwanre, Fidelis Ifeakachukwu Okolafor, Bamidele Oladapo Fagbayigbo, and Ogheneochuko Utieyin Oputu

Nano Polymers for Clean Energy Solutions 461
Abdullah Mohammad Sarjish, Md. Wasikur Rahman, Victoria Padilla, Horacio Vasquez, and M. Jasim Uddin

Recent Advances and Prospects of Nano Polymers in Environmental Remediation 497
Surbhi Singh, Gaurav Mittal, Gaurav Saini, Ravindra Budania, and Ritesh Gupta

Fundamentals of Nanostructured Polymers

Nano Polymer Design: Fundamentals, Materials, and Strategies

Vijay Chaudhary, Manu Satya Prakash Pathariya, Bhasha Sharma, and Chanchal Ahlawat

Abstract The implementation of nanotechnology in polymer science has greatly advanced material engineering as the development of high-performance polymer nanocomposites with strengthened mechanical, thermal, and chemical properties is now possible. This chapter elaborately presents the essential concepts, components, and synthesis approaches concerning the design of nano-sized polymers. This section investigates the development of nanomaterials including their structural origins, states, pore sizes, toxicity levels, and the importance of nanoscale fillers in modifying polymer matrices. It also analyzes the importance of natural and synthetic polymer matrices, the interactions of these nanocomposites with their materials, and the consequences of the usage of these different composites in the aerospace, automotive, biomedical, and food packaging industries. Moreover, various synthesis methods like in situ polymerization, melt mixing, sol–gel processing, electrospinning, and solution blending processes are analyzed to assess their merits, drawbacks, and scope of application. This study highlights the growing importance of polymer nanocomposites in advanced materials development and engineering problem solving. There is greater emphasis on sustainable, lightweight, and high-performance materials which brings to focus the need to enhance polymer-nanomaterial interaction for innovative advancements in material science and industrial applications.

V. Chaudhary (✉) · M. S. P. Pathariya
Department of Mechanical Engineering, Amity School of Engineering and Technology, Amity University Uttar Pradesh, Noida 201313, India
e-mail: vijaychaudhary111@gmail.com

B. Sharma
Department of Chemistry, Deen Dayal Upadhyaya College, University of Delhi, New Delhi, India

C. Ahlawat
School of computer Science Engineering & Technology (SCSET), Bennett University, Greater Noida 201310, Uttar Pradesh, India
e-mail: chanchalchaudhary827@gmail.com

B. Sharma et al. (eds.), *Nanostructured Polymers*, Engineering Materials,
https://doi.org/10.1007/978-981-95-3581-1_1

1 Introduction

Modern materials science has entered a new phase with the advent of nanotechnology—the unique characteristics of nanoscale materials have offered new opportunities in many contexts. One of the most exciting developments in this field is the creation of nano-polymer composites—advanced materials produced by incorporating nanoscale fillers into polymers. These polymer composites possess exceptional mechanical, thermal, and chemical properties, making them essential in electronics, energy storage, aerospace, and medical devices. Therefore, nanotechnology and polymer science have become instrumental in modern industry in creating materials with specific details and solving challenges.

According to specialists, materials in the range of 1–100 nm have been classified as nanomaterials [1], and their unique characteristics make them important for various polymer nanocomposites. Further, nanomaterials have been proven to behave differently than bulk materials due to their high surface area-to-volume ratio. For example, their reactivity, optical properties, and even their surface chemistry is size dependent which is where phenomena such as quantum confinement and interfacial interactions come into play. Furthermore, with the progressive reduction in size, these phenomena become pronounced [2]. In addition to this, the use of nanoscale fillers including AgNPs, SiNPs, or AuNPs enhances the mechanical strength of polymers, thermal stability, biodegradability, and even gas barrier performance, which provides the ability to develop new features and functions [3, 4].

Natural polymers, or biopolymers, on the other hand, are receiving much attention due to their environmentally friendly features like being biodegradable and abundant. Various bio-based materials, such as starch, cellulose, chitosan, and alginate, which are extracted from plants, animals, or microorganisms, have received considerable attention for use in agriculture, packaging, and biomedical applications [5, 6]. These biopolymers, when integrated with nanomaterials, achieve greater plasticity and overcome limitations such as weak mechanical strength and quick deterioration, resulting in high-performance composites with versatile purposes [7, 8].

Applications of nano-polymer composites are also emerging in sensitive areas like agriculture and food security. With the rapid growth of the world's population that is expected to hit 9.7 billion mark by 2050 [9, 10], and one-third of the global food production being lost due to spoilage and waste within the food value chains, there is an urgent need for innovative approaches. With regards to agriculture, biopolymer-based hydrogels and nanocarriers can tremendously improve water retention, nutrient delivery, and pathogen control with lower depletion of environmental resources. Equally, biopolymer-nanoparticle composites offer the advantages of being both biodegradable and ecologically friendly when used in food packaging, which greatly reduces the ecological footprint of modern plastic packaging [11].

The design of polymer nanocomposites is governed by basic principles of science as well as the properties exhibited by the nanoscale fillers. The size and shape of nanofillers used as well as their interfacial interactions and dispersion critically influence the performance of these materials. These principles have facilitated the design of materials for specialized applications that include lightweight structural components for aerospace, flame-retardant coatings, and drug delivery systems.

This chapter investigates the principles, components, and approaches in the design of nano-polymers, specifically polymer nanocomposites as a distinctive category of advanced materials with polymers of the highest performance. It pursues the synthesis of the results of new developments and the results of multidisciplinary studies to understand the impact that these materials will have in developing sustainable technologies and solving problems of the changing world.

2 Historical Background

Examples of nanostructured materials can be found in nature which illustrates that the root of nanomaterials goes back to ancient times. Nanoscale features that formed through natural processes are evident in seashells, skeletal structures, and even in volcanic eruptions and meteorites. Another milestone is the discovery of fire, which produced nanoscale smoke particles. However, the deliberate use and synthesis of nanomaterials was not until much later. In 1857, Michael Faraday reported colloidal gold particles. The synthetic production of nanomaterials on an industrial scale initiated in the 1940s with the United States and Germany. During this time, Silica nanoparticles were used to replace carbon black in silicone rubber, which was subsequently used to manufacture fumed and precipitated nanomaterials. In addition, development of metallic nan powders for magnetic recording tapes took place in the 1960s and 1970s, which marked significant progress in the industry.

During the 1980s, several commercial research organizations and academic laboratories began to focus on polymer nanocomposites. Over this time frame, the foundational work by Blumstein in 1961 as well as the subsequent studies in 1965, the more primary research done by Theng in 1970 regarding what he described as a 'nanocomposite', and the promise shown by Hess and Parker in 1966 where they dispersed metallic cobalt nanoparticles of 100 nm polymer showed that nanocomposites were, at the very least, possible. The major breakthrough happened in 1988 when researchers from Toyota began to work on polymer layered silicate nanocomposites. Toyota used these for components of car parts, for example, timing belt covers, demonstrated their practical viability and set a precedent for broader exploration in this field [12, 13].

Nanocomposites as a development are not entirely new; they date back to more primitive activities. For instance, the ancient Egyptians incorporated straw into clay during brick production and the Mongols utilized composite materials in weaponry. The use of composites also served critical military purposes during World War II. Though, the serious foundation of contemporary nanotechnology started in 1981 with the invention of the scanning tunneling microscope which enabled scientists to see and reposition single atoms. This made it feasible to produce, characterize, and make use of elements in the nanometer scale, and later led to their incorporation into nanocomposites for many other uses [14].

The incorporation of nanomaterials in polymer composite has shown explosive improvements in thermal stability, mechanical strength, barrier properties, and the introduction of new functionalities. Researchers have improved the design and performance of the nanocomposites to meet the needs of a range of industries by optimizing the interactions of raw materials, processes, and nanostructures. These materials are now found use in the automotive, aerospace, packaging, and electronics industries, among others [15, 16].

The enormous historic shifts in the development of nanomaterials and nanocomposites marks the evolution from phenomenon useful in nature to a more purposefully created material with the ability to purposefully change forms. This progress strongly exemplifies the combination of ancient uses of matters like composite materials by the Egyptians and Mongols wit new epoch breaking developments in Nanotechnology within the last hundred years. The drastic changes in material properties such as Enhanced thermal stability, mechanical strength, barrier performance, as well as the development of new advanced functionalities has made it possible for humans to construct and modify materials on the nanoscale. These cross-disciplinary breakthroughs have positioned polymer nanocomposites as a pillar of many industries, including automotive, aerospace, packaging, and electronics that seek to solve and made the modern challenges of the world with new innovative strategies in material science.

3 Significance

As far as materials are concerned, nanomaterials taking the form of polymer nanocomposites are critical at this time. One of the most important aspects that distinguishes nanomaterials from bulk materials is their extremely elevated surface area. The volume of nanomaterials has about one billion times more particles than a similarly sized volume of micro-sized bulk [17]. Reactivity is an essential feature of these materials and based on their volume ratio, nanomaterials have a great deal more surface area than their bulk counterparts, leading to a higher level of surface activity. An example of this is the increased activity of nanomaterials as catalysts. Dependence of properties on size is another such feature that sets nanomaterials apart from other materials, such as the fact that nanoparticles with a size less than 10 nm show different colors when their size is altered. Also, their form and aspect ratio very much affect their behavior in combination with other materials, for example, polymers.

Including small quantities (0.1–5%) of nanomaterials to polymers improves their processes of creation when compared to previous systems while still enabling the lightweight characteristics of the original materials to remain intact. In polymer nanocomposites, improvements made observe an increase in mechanical strength, toughness, scratch resistance, thermal and chemical stability, and gas barrier properties. Furthermore, studies have documented improvements in retardancy during fire and biodegradability while still maintaining a lightweight and transparent composite.

Along with the advantages reported, polymer nanocomposites can also improve biocompatibility depending on the type of nanomaterials used, alongside introducing new properties such as magnetism, electricity, catalytic action, and optics without greatly increasing costs of production.

For these reasons, polymer nanocomposites possess a consistently better cost-to-performance ratio alongside nanomaterials when compared to conventional fillers or composite systems, making them very appealing for some applications commercially.

4 Classification of Nanomaterials

Nanomaterials can be classified into six groups based on structural configuration, based on state, based on dimensions, based on pore dimensions, based on origin, and based on toxicity [18, 19]. The various classifications of nanomaterials have been illustrated in Fig. 1.

4.1 Based on Dimensions

i. *0-D Nanomaterials*

Dendrimers, branched macromolecules, and various metallic or metal oxide nanoparticles including palladium, titanium dioxide, fullerenes, gold, tungsten, zinc oxide, vanadium oxide, magnesium oxide, molybdenum disulfide, and silicon carbide are

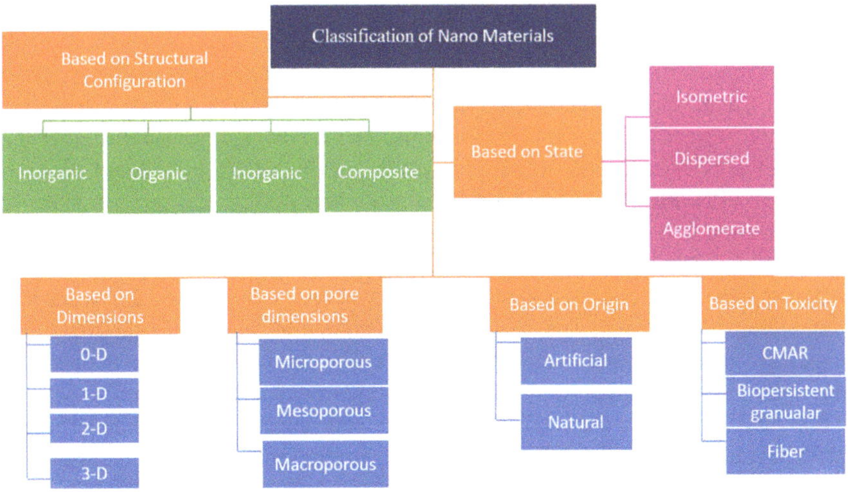

Fig. 1 The illustration above shows the classification of nanomaterials

all included in the category of zero-dimensional (0D) nanomaterials. They take on the form of distinct nanoparticles and range in size from 0 to 100 nm, qualifying as nanoscale [20, 21].

0D nanomaterials boast of properties that make them suitable, for use in a cell marker [22], emulsifier [23], or reinforcement filler [24]. One of the most popular 0D nanomaterials are quantum dots (QDs), which are commonly found in the semiconductor industry due to their size of less than 10 nm and their electrical properties. They can be electrons or electron voids, making these QDs important for electronics and enhancing the effectiveness of semiconductor devices [25].

ii. *1-D Nanomaterials*

One-dimensional (1D) nanomaterials have two out of their three dimensions (x, y) confined within the nanoscale range, while the other dimension is above 10 nm. This category includes structures such as nanofibers, nanotubes, nanohorns, nanorods, thin films, and nanowires. All of these structures have a needle-like shape. These materials can exist in various forms, including amorphous or crystal and single-crystalline or polycrystalline, as well as chemical pure or impure depending on the method of synthesis and intended application.

Furthermore, these materials exist either as free-standing nanostructure or incorporated into different matrices such as metallic, ceramic or polymeric materials. The composition of these matrices differs widely and may include metallic, ceramic or polymeric nanoparticles which makes them highly useful for many technological and industrial applications [26].

iii. *2-D Nanomaterials*

The "two-dimensional (2D) nanomaterials" class refers to materials that extend in two of its dimensions above a certain nanometric size, while one dimension is kept within a nanoscale limit, hence confining it to the nanoscale range. This includes in this group, amorphous or crystalline nanocoatings, thin films, nanoplates, and single or multilayered structures [10].

Physics and electronics employ ceramic and metallic coatings which are commonly present in the form of thin films. These coatings are used to alter the optical reflectivity of surfaces or construct electronic parts that are conductive or non-conductive [19]. Moreover, the development of nanocoatings designed to increase hardness, corrosion and abrasion resistance, or thermal and electrical insulation can also be achieved by the use of polymers, composites, and other materials [20].

Nanoplates can be obtained from smectic clay or can be produced using graphene, gold, silver, bismuth telluride, and bismuth selenide as starting materials. These nanoplates generally have a thickness of just a few nanometers, while their length and width vary considerably. For example, graphene nanoribbons can be as long as 600 nm, while nano clays are much shorter and range between 70 and 150 nm [21, 22]. Nanoplates are useful in improving the mechanical, thermal, and diffusion barrier properties.

iv. *3-D Nanomaterials*

Three-dimensional (3D) nanomaterials exhibit internal characteristics on the nanoscale, but do not bear external nanometer dimensions [23]. Consequently, the 3D materials Nanomaterials possess dimensions often larger than a hundred nm [8, 24]. This group includes different structures: nanotubes, carbon nanobuds, foams, fibers, polycrystals, pillars, fullerenes, and layer skeletons as well as honeycomb structures that can be oriented to have specific functional properties [24, 25]. Examples of 3D nanomaterials include nanocomposites and nanostructured materials.

Such materials where nanofillers are dispersed in a bulk matrix are referred to as nanocomposite. The matrix can be a polymer, ceramic, or metal, and the nanofillers are of 0, 1, or 2-dimensional nanomaterials. The nanocomposite can be in the form of films, fibers, or bulk structures. One of the natural examples of a nanocomposite is bone that consists of a collagen matrix and dispersed calcium hydroxyapatite nanocrystals.

Nanocomposite materials differ from traditional composite materials because the nanofillers have a rather volumetric surface area. Rigid polymer matrix nanocomposites are especially appreciated for their enormous tensile strength and stiffness, which makes them applicable for many purposes [26]. A great example of 3D nanomaterials applications are nanoscale stubby superhydrophobic coatings, block copolymer [28] and nanostructured alloys and metals.

4.2 Classification of Nanomaterials Based on Pore Dimensions

Nanomaterials have been classified into three major classes based on their pores diameters: microporous materials, mesoporous materials, and macroporous materials [27]. Such pores in these classifications serve as an important factor in knowing the dimensions of the molecules that can fit inside them and gives an indication of the diffusion and interaction characteristics. If the size of the guest molecules is less than the pore sizes, then during the diffusion process, it can be assumed that there will be minimal interaction between the molecules themselves and the pore walls while the interactions of the molecules with each other will be more favorable. This characteristic makes these porous nanomaterials especially useful in diffusion and adsorption systems.

i. *Microporous Materials*

With pore diameters of less than 2 nm, microporous materials have extremely narrow pores. Because of these limitations, the materials can only accommodate small molecules like linear molecular structures and gases. Their high molecular interaction capabilities and slow diffusion kinetics give them an advantage in applications that mandate selective molecular separation. Microporous materials include Na-Y zeolites and naturally occurring clay minerals. These materials are commonly used in membrane filtration, gas purification systems, and gas storage applications.

b. *Mesoporous Materials*

Because of their larger pore diameters of 2–50 nm, mesoporous materials can accept larger molecules unlike microporous materials. Some of the more common ones are MCM-41, MCM-48, SBA-15 and mesoporous carbons. They play key roles as nanoreactors in polymerization processes and as adsorption systems for liquids and vapor phase molecules. Their highly uniform porous structures and large surface areas make them very useful in a number of catalytic and separation applications.

iii. *Macroporous Materials*

Macroporous materials are characterzed by pores of diameter larger than 50 nm which allow the incorporation of larger molecules, such as small biological and polyaromatic compounds. Carbon microtubes, porous gels, and glass structures are some examples of macroporous materials. They are used extensively as matrices for functional molecule storage, as scaffolds for grafting functional groups (for example, catalytic centers), and as sensing materials for chemical detection.

To sum up, the pore size classification of nanomaterials is very important for defining their functional properties and range of use like in adsorption, diffusion, catalysis, and molecular storage. The incorporation of these materials in various scientific and industrial fields remains essential due to the relationship between pore size and molecular interactions [29].

4.3 Classification of Nanomaterials Based on Toxicity

Nanomaterials can be classified into three broad groups depending on their potential toxicity: fiber-like nanoparticles, bio-persistent granular nanoparticles, and CMAR (Carcinogenic, Mutagenic, Asthmagenic, Reproductive toxin) nanoparticles [28]. Each classification is based on certain structural characteristics, their persistence within a biological system, and their health-related risks.

i. *Fiber-Like Nanoparticles*

Few of these may include Fiber-Like Nanoparticles. Fiber-like nanoparticles resemble single rigid bio-persistent carbon nanotubes, fiber-like metal oxides, and other fibrous nanomaterials. Unlike asbestose however, these contain non-toxic properties. For occupational exposure, limits for bio-persistent carbon nanotubes and other rigid nanomaterials are reported to be around 10^4–10^5 fibers/m^3 [28]. These limits are set with the consideration of exposure to respiratory hazards due to fiber like nanomaterials.

ii. *Bio-persistent Granular Nanoparticles*

Bio-persistent granular nanoparticles are classified as nanoparticles possessing the capability to reside in biological systems without significant alteration. Suggested occupational exposure limits for these nanoparticles are 2×10^7 particles/m^3. This category contains materials such as gold, silver, cobalt, lanthanum, lead, iron, iron oxide, cerium oxide, antimony oxide, and tin oxide. A default value of 0.3 mg/m^3 has been recommended for insoluble nanoparticles that do not have specific occupational exposure limits in order to reduce health hazards due to inhalation and deposition in the respiratory tract [30].

iii. *CMAR (Carcinogenic, Mutagenic, Asthmagenic, Reproductive Toxin) Nanoparticles*

Materials that CMAR nanoparticles use as building blocks are known to be carcinogenic, mutagenic, asthmagenic and addictive reproductive toxins. These include some well-known materials such as nickel, quantum dots (QDs) containing cadmium, chromium VI, Beryllium, Arsenic, and Zinc chromate. Since these materials are toxic in nature, recommended occupational machines for both limits should range between 2×10^7 and 4×10^7 particles/m3. For insoluble CMAR nanoparticles without existing exposure limit guidelines, a recommended level of 0.003 mg/m^3 is suggested. For soluble CMAR nanoparticles, a precautionary level of 1.5 mg/m^3 is recommended to avoid future health issues from prolonged exposure risks [29].

In summation, the categorization of nanomaterials by toxicity is fundamental to evaluating occupational risks and safety control measures. These groupings form the basis of the regulatory framework intended to reduce the health risks from the use of nanoparticles in industry and research activities.

4.4 Classification of Nanomaterials Based on Origin

Nanomaterials can be further classified based on their origin, which can be natural or artificial.

i. *Natural*

Natural nanomaterials exist in different forms in nature and consist of biological, mineral, and colloidal forms. These include viruses, protein molecules, and minerals like clay, as well as milk and blood natural colloids, fog which is aerosol colloid and gelatin which is gel colloid. Also, mineralized biological structures such as shells, corals, bones, and biological surfaces such as insect wings, spider silk, lotus leaves, and gecko feet have nanoscale features. Moreover, natural nanomaterials are present in opals, volcanic ash, and ocean spray and all of them have distinct nanoscale properties that serve important functions in nature [23, 24].

ii. *Artificial*

Examples of artificially engineered nanomaterials that are built by mechanical and other engineering methods include carbon nanotubes and semiconductor nanoparticles like quantum dots (QDs). Based on their structural composition, nanomaterials can be classified into metal-based materials, dendrimers, or composite materials. These artifacts possess unique chemical and physical characteristics, thus making them applicable in many scientific and industrial disciplines [23, 24].

4.5 Classification of Nanomaterials Based on State

Nanoparticles can be grouped based on their shape, distribution, and uniformity. They may be agglomerated or dispersed and can also be categorized as inhomogeneous or isometric structures. Their morphological characteristics are determined by properties such as flatness, sphericity, and aspect ratio. NPs with a reduced aspect ratio are known to take on many forms such as cubic, helical, spherical, pillar-like, and even pyramidal structures. These structural attributes also affect their electromagnetic properties such as magnetism and surface charge. The agglomeration of nanoparticles in fluid environments relies on their functionalization and morphology, which in turn influence their hydrophilic or hydrophobic behaviour. Common types of NPs include nano hooks, nano stars, nanorods, nanotubes, nano cubes, nano helices, nano zigzags, nanobelts, nanopillars, and nanoplates, each possessing different and unique morphological characteristics [21, 31].

The incorporation of NPs into a nanocomposite often enhances the material's durability. Nanocomposites are made of a standard material matrix with included nanoparticles. Several metallic NMs such as zinc, alumina, iron, silver, copper, titanium, and silica are already in use because of their enhanced functionalities. Another type of NM is dendrimers, which are branched at their nanoscale.

Another means of classification of nanoparticles is by their chemical composition, which includes classes like mono-composites or nanocomposites. Carbon-based nanomaterials, otherwise termed as carbonaceous NMs, are constituents of carbon structures such as carbon nanotubes, graphene, and fullerenes. Like other materials, NMs also have different forms, including dispersed, colloidally suspended, or agglomerated states. It is worth mentioning that in the absence of functionalized surfaces, magnetic nanoparticles are prone to adsorb onto each other into agglomerates; this reinforces the need for surface modification to enhance controlled dispersion [32].

4.6 Classification of Nanomaterials Based on Structural Composition

The structural composition of nanomaterials allows their division into four broad groups: organic/dendrimers, inorganic, carbon-based, and composite nanomaterials [23–29]. Each group has unique properties and functions, making them essential in different applications in science and industry.

i. *Organic Nanomaterials*

At the nanoscale, organic materials are referred to as organic nanomaterials, which are of liposomes, dendrimers, micelles, and ferritin as per Fig. 2 [28, 31]. The non-toxic and biodegradable nanocapsulas which have reactive chain ends and can easily

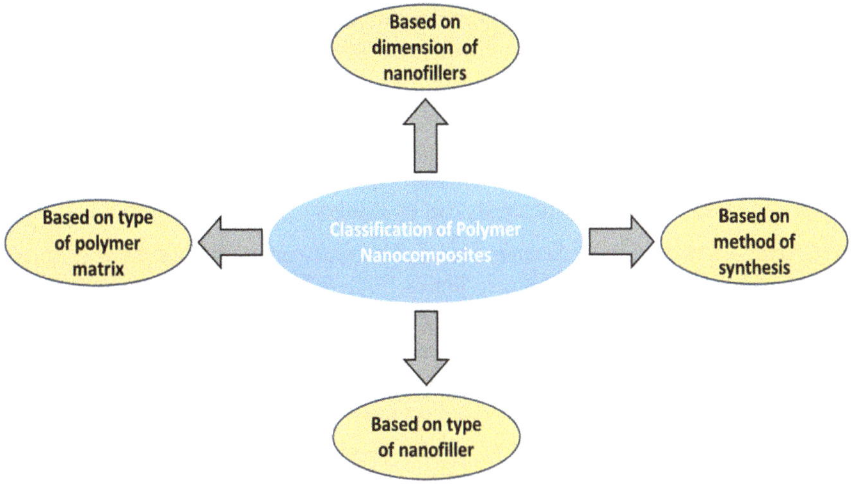

Fig. 2 The illustration above shows the classification of polymer nanocomposites

integrate into chemical reactions are a unique type of organic compounds. Of particular importance are dendrimers which have amazing characteristics and can be used for light harvesting, sensing at the nanoscale, and even in molecular recognition and in optoelectrochemical devices. Furthermore, their three-dimensional structure allows for the formation of internal cavities which can be target molecules, making them very useful in drug delivery systems [25, 26, 28–30].

ii. *Inorganic Nanomaterials*

Inorganic nanomaterials are distinguished by the absence of carbon atoms, as they are mostly made up of metal or metal oxide nanostructures.

Metal-Based Nanoparticles

Synthesis of metal nanoparticles can be achieved by either a destructive or a constructive process. The most commonly used metal nanoparticles are aluminum (Al), cadmium (Cd), cobalt (Co), copper (Cu), gold (Au), iron (Fe), lead (Pb), silver (Ag), and zinc (Zn). These nanoparticles have bulk quantum effects and a high surface volume ratio leading to strong optical, electrical, catalytic, thermal, and even antibacterial activities. As a result of these distinguished features, metal nanoparticles are frequently used in the field of optical and electronic research [29].

Metal Oxide Nanoparticles

These nanoparticles are commonly called metal oxide nanomaterials. Metal oxide nanoparticles that are frequently synthesized include silicon dioxide (SiO_2), titanium dioxide (TiO_2), zinc oxide (ZnO), and aluminum oxide (Al_2O_3). They are composed of positively charged metallic ions and negatively charged oxygen ions. Compared to their metal counterparts, these nanomaterials possess profoundly remarkable physicochemical characteristics, which makes them useful for catalysis, sensing, and even biomedical purposes [29].

Semiconductor Nanomaterials

Semiconductor nanomaterials are materials whose characteristics range between metals and insulators. These materials can be categorized into three types [21]:

1. *Concentrated Magnetic Semiconductor Nanomaterials*—In this class of materials, there are binary compounds such as EuTe (which is antiferromagnetic) that exhibit spontaneous magnetic ordering.
2. *Non-Magnetic Semiconductor Nanomaterials*—This class of nanomaterials do not have magnetic ions (other than the Fe et al.) and as such are widely used in information processing and communication technologies getters since the leverage the charge of electrons but rather are not used in mass data storage for disposable information technology.
3. *Diluted Magnetic Semiconductor Nanomaterials*—In this case. magnetic properties are achieved by the introduction of magnetic impurities into a host matrix. This includes substitution of diamagnetic host cations with magnetic transition metal (TM) cations. Therefore, the material can maintain semiconducting performance along with a magnetic function [32].

Ceramic Nanomaterials

Ceramic nanomaterials are self-supported inorganic solids made of carbides, carbonates, oxides, as well as phosphates, which are received by thermal treatment while sequentially cooling. These materials are progressively receiving great focus due to their promising uses in catalysis, photocatalysis, photodegradation of dyes, and medical imaginations. Besides that, ceramic nanoparticles are also important in drug delivery systems like in tumors, glaucoma, and even in bacterial infections [27, 28].

Lipid-Based Nanomaterials

Lipid-based nanomaterials are spherical nanoparticles with diameters ranging from 10 to 100 nm. These nanoparticles comprise of a solid lipid core that encapsulate the soluble lipophilic molecules. These nanomaterials are primarily used in biomedical applications such as drug delivery systems and RNA-based cancer therapies [27, 28].

iii. *Carbon-Based Nanomaterials*

Carbon nanomaterials are categorized wholly forming from carbon atoms that are divided into five main parts which are carbon black, carbon nanofibers, graphene, carbon nanotubes (CNTs), and fullerenes [23, 24].

- Fullerenes consist of carbon nanostructures in the shape of spheres or ellipsoids. These structures are known as Buckyballs. They are split into single-layered and multi-layered fullerenes. The single-layered style can reach an 8.2 nm diameter while the latter can range from 4–36 nm. All while having the ability to contain 28–1500 carbon atoms.

- Graphene is a two-dimensional form of carbon that is arranged in a hexagonal shape. It has a honey-comb lattice and a thickness of roughly 1 nm.
- Carbon nanotubes, also known as CNTs, are hollow cylinders of carbon. Their diameters can range significantly from 100 nm in multi-walled to 0.7 nm in single-walled, and their length can reach from micrometres to a couple of millimetres.
- Carbon Nanofibers are produced using graphene nanofossils and exhibit high strength and thermal stability.
- Carbon Black is an amorphous carbon material that is spherical, with a diameter between 20 and 70 nm [32, 33].

Carbon-based nanomaterials are highly sought after for their structural reinforcement applications, as they exhibit superior strength compared to steel. Furthermore, they possess high thermal conductivity along their length, while remaining non-conductive across their cylindrical structure [12, 25, 26, 28].

iv. *Composite Nanomaterials*

Composite nanomaterials are formed by the adding of nanoparticles to other nanoparticles or to larger scale materials and bulk-type materials. These composite nanomaterials provide products such as automobile parts or packaging materials with increased mechanical, thermal, and flame-retardant properties [32–35].

5 Classification of Polymer Nanocomposites

Polymer nanocomposites are a type of innovative material that has emerged because of the uniform distribution of nanofillers into a polymer matrix, which results in improved mechanical, thermal, electrical and barrier properties. Due to the wide range of applications in automotive, aerospace, biomedical and electronics, polymer nanocomposites have received considerable attention in recent years. They are classified to understand the structure–property relationships and to tune their functions for the intended application.

Polymer nanocomposites can be classified based on several key criteria.

5.1 Based on the Dimension of Nanofillers

Nanofillers are classified as 0D, 1D, 2D, or 3D depending upon the geometry of the fillers used in the polymer matrices. The dimensionality of the nanofillers has an impact on the mechanical reinforcement as well as functional needs of the composite.

5.2 Based on the Type of Nanofiller

The type of the nanofiller plays an essential part in deciding the characteristics of the polymer nanocomposites. Carbon-based materials, such as graphene or carbon nanotubes; metal, and metal oxide nanoparticles; nanoclays, or other hybrid nanomaterials are all examples of structural nanofillers used in the synthesis which provide various benefits to the polymer matrix.

5.3 Based on the Type of Polymer Matrix

Matrices in polymer-based nanocomposites are divided into thermoplastic, thermosetting polymers and elastomers. The polymer matrix chosen influences the processing, the mechanical properties of the composite, and the subsequent uses of the product.

5.4 Based on the Method of Synthesis

Multiple tactics for synthesis, such as solution blending, melt compounding, and in situ polymerization, affect the dispersal, stability and effectiveness of polymer nanocomposites. The criteria for choosing the synthesis method include polymer matrix compatibility, the type of nanofiller, etc.

These classification schemes are useful for categorization of polymer nanocomposites and help in the engineering design of materials to suit specific high-performance requirements. These classification bases will help in making more advanced polymer nanocomposites with greater functionality and efficiency.

5.5 Classification of Polymer Nanocomposites Based on Dimension of Nanofillers

Nanofillers are selected specifically for purposes at hand, while the structural classification of the filler material is key in determining the performance of the polymer nanocomposite. Nanofillers can be sub-divided into four major groups depending on their dimensions:

1. *Zero-Dimensional (0D) Nanofillers*

   These nanofillers have all their dimensions sized in the nanometer scale, with no single dimension expanding above 100 nm. These nanomaterials can be crystalline, amorphous, ceramic, or even metallic. A prime example of 0D fillers are particles—such nanomaterials are known to demonstrate extraordinary surface properties and quantum characteristics.

2. *One-Dimensional (1D) Nanofillers*

   The first dimensional nanofiller has nanoscale dimensions in one direction while two other dimensions are longer than the nanometric scale. The nanofillers have predominantly needle-shaped architecture encompasses nanowires, nanofibers, nanotubes, nanoplatelets, nanorods, nanoclays, and nanosheets.

3. *Two-Dimensional (2D) Nanofillers*

   Two-dimensional fillers are defined with limited scope of movement in two directions while their n-th direction is beyond nanoscale. For example, silicates, nanofibers, nanorods, and nanowhiskers are classified as 2D filler particles. It is noteworthy that carbon nanotubes claim the highest position among all 2D fillers and carbon nanofibers in terms of utilization and popularity because of their excellent mechanical, electrical, and thermal properties.

4. *Three-Dimensional (3D) Nanofillers*

   Three-dimensional fillers are unique in the sense that they have nanoscopic dimensions in all three directions, which makes them more intricate and stable. For example, zeolites, nanoclays, nanogranules, and equiaxed nanomaterials are 3D fillers, which are popular for providing structural strength, catalysis, and adsorption.

By selecting nanofillers based on their structural characteristics, researchers and engineers can develop polymer nanocomposites with tailored properties, optimizing their performance for specific applications.

5.6 Classification of Polymer Nanocomposites Based on Type of Polymer Matrix

Polymer matrix-based nanocomposites can be broadly categorized into the following types:

1. *Thermoplastic Matrix-Based Nanocomposites*
2. *Thermosetting Matrix-Based Nanocomposites*
3. *Elastomer Matrix-Based Nanocomposites*
4. *Natural and Biodegradable Polymer Matrix-Based Nanocomposites*

Thermoplastic Matrix-Based Nanocomposites

Thermoplastic polymers are those that can be reshaped by heating and cooling; During the cooling process, they remould themselves to the shape of the mould. These polymers lack covalent crosslinks, thus enabling them to be repeatedly softened and hardened via thermal processing. This flexibility coupled with the ability to be reprocessed and recycled makes thermoplastics widely accepted in different classes of industries.

Thermosetting Matrix-Based Nanocomposites

The addition polymerization of thermoplastics is different since there is irreversible hardening that happens due to the forming of a three-dimensional covalently bonded structure. Now that it has been set, it can no longer be reshaped or remoulded. This class includes polyesters, polyurethanes, vulcanized rubber, and epoxy resins [36]. However, Thermoset polymers have found applications where they are required to strengthen, resist, and endure to thermally elevated conditions.

Elastomer Matrix-Based Nanocomposites

Elastomers are elastic polymers that can be greatly deformed and subsequently returned to their original shape without any change to the internal structure. Such materials are viscoelastic in nature meaning that they exhibit both elasticities along with relatively high viscosity. Weak intermolecular forces, a low Young's modulus, and a high failure strain are characteristic features of elastomers. Examples of unsaturated rubbers include natural rubber, polybutadiene, and chloroprene while epichlorohydrin, silicone rubber, and polyacrylic rubber are examples of saturated rubbers [37].

Natural and Biodegradable Polymer Matrix-Based Nanocomposites

Natural and biodegradable polymers include materials that are found in nature and can be broken down by microorganisms over a period of time. Softness and flexibility are two attributes that these materials possess. Common examples include collagen, chitosan, starch, hyaluronic acid, and alginate, which find extensive uses in biomedicine, pharmacy, and environmental sustainability.

One of the reasons for the classification of polymer matrix-based nanocomposites is the proper selection of materials for applications. Each type has different structural and functional properties which affect the performance, durability, and processability of the nanocomposites.

5.7 Classification of Polymer Nanocomposites Based on Type of Nanofiller

Metal Sulfide/Polymer-Based Nanocomposites

Compared to the polymeric matrix, the incorporation of metal sulfide nanoparticles within the polymeric nanocomposite aids in improving thermal stability. This enhancement is the result of reduced material volatilization and low thermal efficiency movement along the polymer chains when metal sulfide nanofillers are located at intervals within the polymer structure. Cadmium sulfide (CdS) and zinc sulfide

(ZnS) are some of the major metal sulfides which desorb from the polymer matrix and result in intersegmental confinement and chain transfer imprinting. A variety of techniques are available for the fabrication of these nanocomposites, the most utilized are the hydrothermal and thermal decomposition techniques.

Metal Hydroxide/Polymer-Based Nanocomposites

Nanocomposites based on metal hydroxides and polymers exhibit flame retardant qualities. Such nanocomposites are produced which contain magnesium hydroxide ($Mg(OH)_2$) and aluminum hydroxide ($Al(OH)_3$) as nanofillers for application in wires, cables, fire resistant coatings etc. The polymer matrices for these composites are poly (vinyl chloride) (PVC), elastomers, phenolics, and unsaturated polyesters. Fireproofing is achieved through the exothermic decomposition of metal hydroxides which results in the emission of water vapor or carbon dioxide dilutes the flammable gas. In addition, magnesia (MgO) and alumina (Al_2O_3) are used as refractory powders which segregate MgO an insulating material and heat during processing into the composite. The further deposition of these fillers onto the surface of the composite creates a ceramic or glass barrier layer, additional protecting the composite. The general requirement for most additives and fillers is the ability to dry or extinguish fires is dependent on the endothermic decomposition process being favorable over a certain range.

Metal or Metal Oxide/Polymer-Based Nanocomposites

The melting point and flame resistance of a polymer increases with the addition of metal or metal oxide engraved nanofillers. These nanocomposites are fabricated using a variety of synthesis methods including hydrothermal method, sol–gel method, coprecipitation method, or sonochemical techniques. Silica, ferric oxide, titanium oxide, and copper oxide are examples of common nanofiller metal oxides. These materials should be selected carefully because they affect the mechanical, thermal, and chemical properties of the polymer composite.

Silicate/Polymer-Based Nanocomposites

The resistive nature of the clay materials enables silicate-based polymer nanocomposites to provide barrier against mass transport, this aid in suppression of heat buildup and improvement of flame resistance. Because of these qualities, these materials have an enormous potential in fireproof coatings.

5.8 Classification of Polymer Nanocomposites Based on the Method of Synthesis

Ex Situ Incorporation of Nanoparticles

This method implies to one step further processing, in which the polymer solution or the melted polymer incorporates nanoparticles that were not created within it.

The processes are often carried out separately for better control of properties of the composite material.

In Situ Synthesis of Nanoparticles within the Polymer Matrix

Following this technique, starting materials that can be transformed into nanostructures are placed in the voids of the polymer and then are transformed through a series of processes to nanoparticles. As an instance, the ions of metals can be reduced in the volume of the polymer matrix to create various forms of metallic nanoparticles so that the homogeneity within the polymer is preserved and the interaction with the polymer is increased.

Polymerization in the Presence of Pre-synthesized Nanoparticles

With this technique, the polymerization of monomers takes place in the presence of already synthesized nanoparticles. The mechanical and functional properties of the polymer nanocomposite are improved by the synthesized nanoparticles as they improve the overall composition of the product.

Simultaneous Synthesis of Polymer and Nanoparticles

In this process, both the polymer matrix and the nanoparticles are synthesized simultaneously from the precursor materials. This process facilitates the controlled growth of nanoparticles in the polymer so that the properties of the resulting nanocomposites are very well engineered.

6 Materials

6.1 Nanomaterials

The word 'nano' comes from the Greek language's 'nanos' which translates to either a dwarf or a short person. The International System of Units uses the prefix 'nano' to define a fractional part of a unit. Thus, a nanometer refers to one—billionth of a meter or a millionth of a millimeter. Likewise, a nanoliter is a billionth of a liter, and a nano-Kelvin is a billionth of a Kelvin. In the past ten years, the prefix "nano" has experienced a crazy rise in usage across multiple areas of knowledge and science papers [38, 39].

Their size is what fundamentally characterizes nanomaterials. It ranges from 1–100 nm, and the material has at least one dimension of the nanoscale range [40–42]. Also, according to the already established criteria, these materials must have a spherical surface area of more than 60 m^2/cm^3 [43]. Depending on the measure of their origin, size, structural configuration, pore diameters, and possible toxicity, they are separated into five primary types [3, 10, 11]. In addition, all matter at the nanoscale possesses different physical, chemical, and biological properties as compared to their larger counterparts. Most significantly, nanoparticulate matter is a state of matter different from solid, liquid, gas, and plasma. It has its own distinct form, and possesses unique optical, magnetic, and electrostatic properties.

Various procedures allow to produce nanomaterials, which are usually divided into two groups: 'top-down' and 'bottom-up' techniques. Top-down methods include, among others, lithography, mechanical milling, laser ablation, sputtering, and thermal decomposition. CVD, sol–gel processing, pyrolysis, and even biological processes are all bottom-up techniques [11, 12]. The discipline, which investigates the unique size-dependent properties of matter at nanoscale is called nanoscience and includes the study of nanomaterials [2, 11, 12].

Nanotechnology includes the design and application of nanomaterials which has received appreciable attention due to its potential applications in various industries. Some of the fields in which nanomaterials are used include agriculture, biomedicine, electronics, energy, pollution, food engineering, transportation, telecommunications, cosmetics, coatings, materials and mechanical engineering [2, 3, 6, 8, 9, 11, 12].

The value of nanomaterials is gradually gaining importance, as seen in the advancement of human civilization. For example, fights against global climate change and global warming have proven to be more effective with green technologies that incorporate nanomaterials rather than those using bulk materials [13–15]. Moreover, nanomaterials have played a pivotal role in diagnosing and controlling epidemic diseases on a global scale. The COVID-19 pandemic was an event where nanomaterials played an essential role in disease diagnosis and management [16–18]. So has the diagnosis and treatment of diseases such as monkeypox benefited from currently available methods that rely on nanomaterial-based technologies [19, 20]. Looking ahead, it is reasonable to expect that nanoscience, and nanotechnology will contribute significantly to augment the development of mankind. This, however, calls for intensive education on and awareness of nanomaterials [11, 12, 21, 22].

While over the last two decades there have been numerous investigations into nanoparticles and nanomaterials, nanomaterials are acknowledged for the unique attributes that most classify them separately from traditional materials. Their composition, such as size, shape, surface area and constitution, govern their classification, preparation methods and applications. Nanomaterials are able to exhibit novel chemical, physical and biological properties which make them easy to adapt for use in various fields. Along with these advantages, the versatility of new materials is boundless. It is because of this reason that nanomaterials have been able to find relevance and proven to be useful in medicine, renewable energy, environmental sustainability and advanced manufacturing.

6.2 Polymer Matrix

A wide variety of nano materials are added to other polymers to create polymer-based nanocomposites. These other polymers may include fibers, rubbers, and even plastics. For thermosetting polymers, both high molecular weight elastomers and low molecular weight resins serve as suitable matrics for the polymer nanocomposites

[44]. Moreover, polyurethanes are some of the most widely used thermoplastics and thermosetting polymers which make them easily accessible polymer based composite nanomaterial [45]. On top of that, polymer matrices can be formed from previously manufactured polymers, or they can be made in situ with the presence of nano materials. The addition of nano materials into polymer-based composites improves their durability, heat tolerance, and other functional capabilities.

In a nanocomposite, the polymeric matrix plays the role of a continuous phase while providing stability and facilitating strong interaction with the incorporated nanomaterials. At a more detailed level, these polymers can be defined as organic (mostly) and inorganic macromolecules formed by repeating monomer units that are covalently bonded together. These monomers, that are subsequently covalently bonded to form a macromolecule, are classified as additives, and the process of interlinking them is called polymerization. There are three basic forms of polymerization: addition, condensation, and rearrangement [45]. Given that polymers are developed from a repeating unit, there is a necessary degree of polymerization that one must meet to determine the overall weight of the polymer. Since polymers are designed with many repeating units, adding or removing a few of them will not greatly alter their overall characteristics.

7 Polymerization Techniques

7.1 Addition Polymerization

In addition, polymerization, monomers are linked sequentially through covalent bonds, forming long polymer chains without the elimination of byproducts. This process follows a chain-growth mechanism and is further classified into:

- Conventional Free Radical Polymerization
- *Controlled Radical Polymerization* (including atom transfer radical polymerization, group transfer polymerization, nitroxide-mediated polymerization, and reversible addition-fragmentation chain transfer)
- *Ionic Polymerization* (including cationic and anionic polymerization)

Polymers produced from addition polymerization include polyethylene (PE), polypropylene (PP), polystyrene (PS), poly(vinyl chloride) (PVC), poly(methyl methacrylate) (PMMA), polyacrylonitrile (PAN), polybutadiene, polyisoprene, polychloroprene, poly(tetrafluoroethylene) (PTFE), and poly(vinylidene fluoride) (PVDF). In addition to these, some copolymers are formed by linking different monomers together such as ethylene/propene copolymers, poly(styrene-butadiene), poly(acrylonitrile-butadiene), poly(acrylonitrile–butadiene–styrene) (ABS), poly(vinylidene fluoride-hexafluoropropylene-tetrafluoroethylene). Homopolymers have only one type of repeating unit, while copolymers have two or more types of repeating units and thus are more flexible with better mechanical and chemical properties.

7.2 Condensation Polymerization

In condensation polymerisation, the monomers react and join in step growth processes to form polymers, simultaneously removing small molecules like water (H_2O), hydrogen chloride (HCl), ammonia (NH_3), sodium chloride (NaCl), phenol (PhOH), and formaldehyde (HCHO). Examples of condensation polymerisation products are polyesters, polyethers, polycarbonates, polyamides, poly (ester amides), polyimides, poly (ether ketones), and polysulfones.

7.3 Rearrangement Polymerization

Rearrangement polymerization is a step-growth process that occurs without the elimination of byproducts, resulting in polymers such as polyurethanes, polyureas, and poly (urea urethanes).

Polymerization Techniques for Processing Nanocomposites

Polymerization techniques used for synthesizing polymer nanocomposites include:

- *Bulk (Mass) Polymerization*
- *Solution Polymerization*
- *Suspension Polymerization*
- *Emulsion Polymerization*

In addition to these, specialized techniques such as interfacial, template, electrochemical, and plasma polymerization are employed based on specific application requirements.

Structural Characteristics of Polymers

Polymers exist in various *structural configurations*, including:

- *Linear Polymers*—Possess a continuous, unbranched structure.
- *Low-Branched and High-Branched Polymers*—Exhibit branch-like molecular chains.
- *Hyperbranched Polymers*—Feature highly branched, three-dimensional macromolecular structures.

Hyperbranched polymers or dendritic polymers have a high number of functional surface groups, low solution and melt viscosities, and superior material compatibilities. These types of polymers can be easily processed and so they are regarded as the perfect matrices for polymer nanocomposites [46].

7.4 Processing and Versatility of Polymers

Polymers can be processed through (but are not limited to) melt mixing, rubbery stage mixing, solution, suspension, and even emulsion stage mixing. As long as reasonable care is observed, these processes can be repeated on thermoplastic polymers since their properties do not experience significant loss. However, these processes can only be done on thermosetting polymers before they are cured, sealed, and subjected to irreversible chemical cross-linking.

Polymers have their benefits and disadvantages as well, some of which include low mechanical strength, thermal stability, flame resistance, and limited biodegradability. On the flip side, in comparison to metals and ceramics, polymers have certain advantages, for instance:

- *Versatility in feedstock and structural modifications*
- *Mass-scale production at a relatively low cost*
- *Lightweight nature*
- *Ease of processing and fabrication*

Such benefits allow polymers to be used as matrix material for polymer nanocomposites, resulting in high-performance multifunctional materials suitable for many applications.

8 Synthesis Strategies

The central focus in the creation of polymer nanocomposites is strong interfacial bonding between the polymer matrix and the nanomaterials. Strengthening interfacial bonding is possible through uniform dispersion of the nanomaterials within the polymer matrix which increases the quantity of nano dimensional interphases in the final composite structure. Mechanical sharing and ultrasonic forces are two techniques often used for this purpose. However, one of the most important requirements for strengthened interfacial bonding is the presence of certain functional groups in the polymer matrix and the nanomaterials.

To conclude, the synthesis technique is one of the most important parameters that must be taken into consideration for optimizing the properties and performance of polymer nanocomposites. These methods include, but are not limited to, solution blending, in situ polymerization, and melt mixing for both lab and industrial purposes. Other methods like sol–gel processing, template synthesis, and microemulsion techniques can also be used depending on the particular materials and applications needed. The next section will give a detailed summary of different techniques for the fabrication of polymer nanocomposites.

8.1 Solution-Based Fabrication Technique for Polymer Nanocomposites

The process of preparing a polymer-nanomaterial mixture involves the use of a mechanical stirring combined with ultrasonication and treating the nanomaterial with a liquid medium. In the beginning, the nanomaterial is put in a polymer solution for it to swell, and after that, it is put in a homogenizer. At the same time, the polymer solution is put in a solvent that can be dispersed in the nanomaterial's medium. The final step in the process is to either evaporate or precipitate the solvent from the mixture, resulting in a polymer matrix that has the nanomaterial dispersed within.

The strength of the interaction between the polymer matrix and the nanomaterial's polymer chains determines how well the two components blend. To achieve homogeneous dispersion of nanomaterials in a polymer matrix, the interaction between the polymer and the nanomaterial has to be greater than the interaction between the nanomaterial and the solvent as well as the polymer and the solvent. This leads to enhanced polymer penetration into the nanomaterial structure and subsequent methacrylation, and the adsorption of chain molecules onto the nanomaterial surface resulting in a stable nanocomposite [47].

With the right degree of dispersion and interaction, it is possible to use this method to synthesize both intercalated and exfoliated nanocomposites. Entropic and enthalpic factors determine the thermodynamic stability of the final composite. The decrease in entropy is due to the restriction of the motion of polymer chains subjected to the rigid nanomaterials. However, this effect is offset by an increase of the appealing energy of the interlayer space as the motion of the nanomaterials increases. On the other hand, the Enthalpic loss is greatly increased because of the strong bond existing between the polymer matrix and the nanomaterials, which increases the stable value of the Gibbs free energy.

While these advances are beneficial, this method has its shortcomings such as increased cost of processing, risk of fire, health and environmental dangers due to the use of toxic organic solvents, and much lower interaction effectiveness. On the other hand, the solution method is advantageous as it is the easiest and most efficient since it makes use of preformed industrial polymers. Even the use of water instead of organic solvents can make the process environmentally friendly and economically feasible.

8.2 In Situ Polymerization Technique for Polymer Nanocomposites

The in-situ polymerization method employs polymerization of monomers in the presence of dispersed nanomaterials which ultimately leads to the formation of polymer nanocomposites. Initially, the nanomaterial is gently dispersed in a low viscosity monomer or a prepolymer. This allows the prepolymer molecules to adsorb onto the dispose surface of the nanomaterials. During the process of polymerisation, the chains of the polymers that are growing facilitate the dispersion of nanomaterials and the delamination and debundling. All of these actions work together to ensure a complete composite structure is formed [48].

Monomeric precursors or prepolymers tend to have very low viscosity when compared to the solution-based technique monomeric methods. Thus, in performed polymers, the viscosity is significantly higher. This points towards an observation that polymer-nanomaterial interactions are stronger, which enhances dispersion and interfacial adhesion within the polymer matrix. Due to the low viscosity of the monomers, strong interfacial interactions, and the uniform dispersion of nanomaterials, the exfoliation of nanocomposites is achieved much more effectively through the use of the in-situ polymerization method. This is why the in-situ method is preferred when fabricating various polymer nanocomposites, especially resin-based systems. It requires hardly any solvent and consumes very little energy.

In addition, the functional groups incorporated into the nanomaterials can take part in crosslinking reactions, especially in thermosetting nanocomposites, which may have an impact on the polymerization itself. This trait is particularly useful when emulsion polymerization is employed where the medium of reaction is water. In these circumstances, the polymerization of monomers is started by functionalized initiators, for instance, modified clay 2,2′-azobis(2-methyl propionamide) dihydrochloride. In such cases, functionalized clay is dispersed in water and serves two purposes an initiator (azo compound) and a nanomaterial (nano clay). The final polymer nanocomposite is obtained by coagulating the emulsion after classical emulsion polymerization.

In general, this approach is advantageous in systems where emulsions can be utilized in the same manner unrestrictedly. Altogether, however, the polymer counterpart in this case especially in situ is more popular, especially when it comes to laboratory scale making for the outstanding performance of the polymer nanocomposites resulting.

8.3 Melt-Mixing Technique for Polymer Nanocomposites

The melt-mixing method allows for the even distribution of nanomaterials in preformed polymer matrices with the use of conventional melt-processing equipment which leads to the fabrication of polymer nanocomposites. One reason this method is popular is because it removes the necessity for solvents, as well as being free of polymerization or crosslinking reactions, which makes this method best in industrial applications. Moreover, it is possible to use standard industrial polymer processing methods without the need for alterations or extra devices.

The process involves melting semicrystalline or the softening of amorphous polymers through the application of heat in mixing devices like extruders, Brabender plastographs, injection molding machines, and kneaders. After the polymer is adequately melted, the nanomaterials are added and dispersed within the polymer matrix by mechanical shear forces induced while mixing [49]. Even so, the high melt viscosity of the preformed polymers results in the resultant nanocomposites exhibiting poor intercalation which causes a lack of interaction between the polymer and nanomaterials.

Regardless of this difficulty, most industries find the melt-mixing technique appealing since it is less harmful compared to other methods. This is because they do not require dangerous solvents or any intricate chemical reactions. Hence, this technique is still popular and attractive for the process of creating polymer nanocomposites on a larger scale.

8.4 Sol–Gel Method for Polymer Nanocomposites

The sol–gel process is one of the most popular methods of producing nanomaterials and it is also used in the production of polymer nanocomposites. In this case, the nanoparticles are synthesized directly within the polymer. It entails the preparation of a water-based solution or gel which contains the precursor materials of the nanomaterial and the polymer matrix from which the material will be formed. The solution is then heated to a temperature that allows the nanomaterials to be formed and subsequently incorporated into the polymer matrix, initiating the nmatures of the nanocomposite structure [50]. Consequently, the new nanomaterials that are produced are incorporated into the polymer and thus the corresponding nanocomposite is produced.

This method is more effective for synthesizing colloidal nanocomposites because it can theoretically enable instant dispersion of the nanomaterials into the polymer matrix without any additional energy costs. This is, however, contrary to what the sol–gel technique can achieve. For example, in the case of clay/polymer nanocomposites, the fabrication of clay minerals is a high-temperature process, which may lead to the polymer matrix's thermal degradation. More so, the other extreme processing conditions will result in poor dispersion of nanomaterials due to severe nanomaterial agglomeration. Because of this, most researchers do not use the sol–gel method as frequently as other methods of nanocomposite fabrication.

8.5 Electrospinning Technique for Polymer Nanocomposite

Among the different techniques used in the fabrication of polymer nanocomposites-based nanofibers, the electrospinning technique is the most widely used. It is a common fiber forming technology that employs an automated hydraulic syringe and microtip needle, a high-voltage DC power supply, and grounded collector. During fiber formation, the process resembles that of the solution technique whereby a polymer solution or a polar polymer is molten and mixed with a well dispersed suspension of nanoparticles using ultrasonic agitation and mechanical shear. The resultant polymer-nanomaterial mixture obtained is then, as the name suggests, placed under an electrical charge.

In the first instance, when the polymer solution or melt tries to exit the needle tip, the polymer-nanomaterial mixture inserted in attempts to free itself from the obstruction. However, surface tension does not allow for the formation of a jet at first. It is imperative to note that to successfully generate nanofibers, the voltage needs to be sufficiently increased to overcome surface tension. Once the polymer jet has been stretched forming nano dimensional fibers, it action allows the jet to be released, forcing the needle tip into a jet like state. As these fibers travel to the collector plate, they need to spin in a direction that leads to solidifying and maintaining structural integrity.

The electrospun nanofibers do not differ in their morphology, uniformity, and dimensions, rather these complex structures are influenced by many factors including:

- *Viscosity, polarity, and concentration of the polymer solution or melt*
- *Hydraulic pressure of the syringe*
- *Flow rate and tip diameter*
- *Electric field strength*
- *Tip-to-collector distance*
- *Other spinning conditions*

Moreover, the configuration of the collector has an impact on the dimensions and morphology of the nanofibers. In order to obtain the desired electrospinning conditions, it is common to change different parameters as previously stated in other articles. These are:

- *Applied voltage:* 15–30 kV
- *Tip-to-collector distance:* 10–20 cm
- *Feed rate:* 1–15 mL/h
- *Relative humidity:* 25–35%
- *Temperature:* 25–60 °C

These optimized parameters help in obtaining high-quality electrospun nanofibers with the desired structural and functional properties.

9 Conclusion

As a transformative class of materials, polymer nanocomposites are potentially useful across a wide range of applications. The incorporation of nanoscale fillers into polymers has resulted in improvements in mechanical properties, thermal stability, chemical resistance, and biodegradability. In this chapter, I have examined the basics of nano-polymer design, especially material selection, classification, synthesis methods, and performance increases. The development of engineered nanomaterials from natural ones evidences their increasing significance for science and industry.

For optimization of nanomaterials for specific environments, their dimensional, origin and toxicity classification are crucial. Also, synthesis methods like solution blending, in situ polymerization, and electrospinning have made it possible to achieve uniform distribution and good adhesion of the polymers and nanofillers in the matrices. These techniques make it possible to formulate high performance polymer nanocomposite materials for aerospace, biomedical applications, energy storage, and solving environmental issues.

Large scale production, long-term stability, and the disposal of nanomaterials are significant concerns that must be addressed in the future of polymer nanocomposites. Future studies should focus on improving the compatibility of nanomaterials with polymer matrices, "greener" synthesis methods, and multifunctional adaptive nanocomposites. The combination of ever increasing and developing branches of nanotechnology and polymer science ensures that polymer nanocomposites will bear an essential burden in solving many engineering, environmental and sustainability issues for these materials will be one of the main factors in the realization of the next generation materials.

References

1. Jordan, J., Jacob, K.I., Tannenbaum, R., Sharaf, M.A., Jasiuk, I.: Experimental trends in polymer nanocomposites—a review. Mater. Sci. Eng. A **393**(1–2), 1–11 (2005). https://doi.org/10.1016/j.msea.2004.09.044
2. Xiao, Qi, L.: Surfactant-assisted, shape-controlled synthesis of gold nanocrystals. Nanoscale **3**(4), 1383 (2011). https://doi.org/10.1039/c0nr00814a
3. Karimi Sani, I., Geshlaghi, S.P., Pirsa, S., Asdagh, A.: Composite film based on potato starch/apple peel pectin/ZrO_2 nanoparticles/microencapsulated *Zataria multiflora* essential oil; investigation of physicochemical properties and use in quail meat packaging. Food Hydrocoll. **117**, 106719 (2021). https://doi.org/10.1016/j.foodhyd.2021.106719
4. Reddy, M.S.B., Ponnamma, D., Choudhary, R., Sadasivuni, K.K.: A comparative review of natural and synthetic biopolymer composite scaffolds. Polymers **13**(7), 1105 (2021). https://doi.org/10.3390/polym13071105
5. Ashfaq, M., Talreja, N., Chuahan, D., Srituravanich, W.: Polymeric nanocomposite-based agriculture delivery system: Emerging technology for agriculture. In: Genetic Engineering—A Glimpse of Techniques and Applications. IntechOpen, London, UK (2019). https://doi.org/10.5772/intechopen.89702
6. Panchal, S.S., Vasava, D.V.: Fabricating approaches for synthesis of miktoarm star-shaped polymers having tailored biodegradability. Int. J. Polym. Mater. Polym. Biomater. **71**, 1407–1424 (2022). https://doi.org/10.1080/00914037.2021.1981319
7. Tomadoni, B., Casalongué, C., Alvarez, V.A.: Biopolymer-based hydrogels for agriculture applications: swelling behavior and slow release of agrochemicals. In: Gutiérrez, T.J. (ed.) Polymers for Agri-Food Applications. Springer, New York, pp. 99–125 (2019). https://doi.org/10.1007/978-3-030-19416-1_7
8. Sikder, A., Pearce, A.K., Parkinson, S., Napier, R.M., O'Reilly, R.K.: Recent trends in advanced polymer materials in agriculture-related applications. ACS Appl. Polym. Mater. **3**, 1203–1217 (2021). https://doi.org/10.1021/acsapm.0c00982
9. Lahlali, R., et al.: Biological control of plant pathogens: a global perspective. Microorganisms **10**, 596 (2022). https://doi.org/10.3390/microorganisms10030596

10. Godfray, H.C.J.: The challenge of feeding 9–10 billion people equitably and sustainably. J. Agric. Sci. **152**, 2–8 (2014). https://doi.org/10.1017/S0021859613000774
11. Karimi Sani, I., et al.: Value-added utilization of fruit and vegetable processing by-products for the manufacture of biodegradable food packaging films. Food Chem. **405**, 134964 (2023). https://doi.org/10.1016/j.foodchem.2022.134964
12. Komarneni, S., Roy, R., Li, Q.H.: Microwave-hydrothermal synthesis of ceramic powders. Mater. Res. Bull. **27**, 1397 (1992). https://doi.org/10.1016/0025-5408(92)90004-J
13. Noordermeer, J.W.M., Dierkes, W.K.: Carbon black reinforced elastomers. In: Kobayashi, S., Müllen, K. (eds.) Encyclopedia of Polymeric Nanomaterials. Springer, Berlin, Heidelberg (2015). https://doi.org/10.1007/978-3-642-36199-9_287-1
14. Blumstein, A.: Polymerization of adsorbed monolayers. I. Preparation of the clay–polymer complex. J. Polym. Sci. Part A: Gen. Pap. **3**(7), 2653–2664 (1965). https://doi.org/10.1002/pol.1965.100030720
15. Karak, N.: Fundamentals of nanomaterials and polymer nanocomposites. In: Nanomaterials and Polymer Nanocomposites, pp. 1–45 (2019). https://doi.org/10.1016/b978-0-12-814615-6.00001-1
16. Wypij, M., Trzcińska-Wencel, J., Golińska, P., Avila-Quezada, G.D., Ingle, A.P., Rai, M.: The strategic applications of natural polymer nanocomposites in food packaging and agriculture: chances, challenges, and consumers' perception. Front. Chem. **10**, 1106230 (2023). https://doi.org/10.3389/fchem.2022.1106230
17. Karak, N.: Vegetable Oil-Based Polymers: Properties, Processing and Applications. Woodhead Publishing Ltd., Oxford, UK (2012)
18. Khan, S., Hossain, M.K.: Nanoparticle-based polymer composites. In: Nanoparticle-Based Polymer Composites, London, p. 15. Woodhead Publishing, UK (2022)
19. Cho, G., Park, Y., Hong, Y.K., Ha, D.H.: Ion exchange: an advanced synthetic method for complex nanoparticles. Nano Converg. **6**, 17 (2019). https://doi.org/10.1186/s40580-019-0187-0
20. Dolez, P.I.: Nanomaterials definitions, classifications, and applications. In: Nanoengineering: Global Approaches to Health and Safety. Elsevier, pp. 3–40 (2015)
21. Buzea, C., Pacheco, I.: Nanomaterials and their classification. In: Advanced Structured Materials. Springer, pp. 3–45 (2017)
22. Salata, O.V.: Applications of nanoparticles in biology and medicine. J. Nanobiotechnol. **2**, 3 (2004)
23. Luo, J., Zhou, Q., Sun, J., Liu, R., Liu, X.: Micelle-assisted synthesis of PANI nanoparticles and application as particulate emulsifier. Colloid Polym. Sci. **292**, 653–660 (2014)
24. Gao, F.: Advances in Polymer Nanocomposites: Types and Applications. Woodhead Publishing, Cambridge, UK (2012)
25. Smith, A.M., Nie, S.: Semiconductor nanocrystals: structure, properties, and band gap engineering. Acc. Chem. Res. **43**, 190–200 (2010)
26. Alkaç, I., Cerci, B., Timuralp, C., Şen, F.: Nanomaterials and their classification. In: Nanomaterials for Direct Alcohol Fuel Cells (2021). https://doi.org/10.1016/B978-0-12-821713-9.00011-1
27. O'Reilly, R.K., Hawker, C.J., Wooley, K.L.: Cross-linked block copolymer micelles: functional nanostructures of great potential and versatility. Chem. Soc. Rev. **35**, 1068–1083 (2006)
28. Dolez, P. (ed.): Nanoengineering: Global Approaches to Health and Safety Issues. Elsevier, Amsterdam, Netherlands (2015)
29. Mekuye, B., Abera, B.: Nanomaterials: an overview of synthesis, classification, characterization, and applications. Nano Select. **4**, 486–501 (2023). https://doi.org/10.1002/nano.202300038
30. Sharma, A., Bhalla, P., Kaith, B.S.: Polymer nanocomposite matrices: classification, synthesis methods, and applications. https://doi.org/10.1007/978-3-030-10614-0_51-1
31. Camargo, P.H.C., Satyanarayana, K.G., Wypych, F.: Nanocomposites: synthesis, structure, properties and new application opportunities. Mater. Res. **12**, 1–39 (2009)
32. Glezer, A.M.: Structural classification of nanomaterials. Russ. Metall. **2011**, 263–269 (2011)

33. Paul, D.R., Robeson, L.M.: Polymer nanotechnology: nanocomposites. Polymer **49**(15), 3187–3204 (2008). https://doi.org/10.1016/j.polymer.2008.04.017
34. Pham, J.Q., Mitchell, C.A., Bahr, J.L., Tour, J.M., Krishnamurti, R., Green, P.F.: Article title. J. Polym. Sci. Part B Polym. Phys. **41**, 3339–3345 (2003)
35. Okamoto, M.: Polymer nanocomposites. Eng. **4**, 457–479 (2023). https://doi.org/10.3390/eng4010028
36. Kausar, A., Ahmad, I., Zhao, T., Aldaghri, O., Ibnaouf, K.H., Eisa, M.H.: Multifunctional polymeric nanocomposites for sensing applications—design, features, and technical advancements. Crystals **13**, 1144 (2023). https://doi.org/10.3390/cryst13071144
37. Akpan, E.I., Shen, X., Wetzel, B., Friedrich, K.: Design and synthesis of polymer nanocomposites. In: Pielichowski, K., Majka, T.M. (eds.) Polymer Composites with Functionalized Nanoparticles, Micro and Nano Technologies. Elsevier, pp. 47–83 (2019). https://doi.org/10.1016/B978-0-12-814064-2.00002-0
38. Gobena, S.T., Woldeyonnes, A.D.: A review of synthesis methods, and characterization techniques of polymer nanocomposites for diverse applications. Discov. Mater. **4**, 52 (2024). https://doi.org/10.1007/s43939-024-00119-0
39. Li, T., Ding, G., Han, S.T., Zhou, Y.: Introduction of polymer nanocomposites (2021). https://doi.org/10.1002/9783527826490.ch1
40. Clifton, S., Thimmappa, B.H.S., Selvam, R., Shivamurthy, B.: Polymer nanocomposites for high-velocity impact applications—a review. Compos. Commun. **17**, 72–86 (2020). https://doi.org/10.1016/j.coco.2019.11.013
41. Senthil, M.S., Selvan, M., Santhanam, K., Arumugam, K., Chandraprabu, V., SampathKumar, L.: Effect of nanomaterials on tribological and mechanical properties of polymer nanocomposite materials. Adv. Mater. Sci. Eng. **2022**, 1–16 (2022). https://doi.org/10.1155/2022/2165855
42. Tyagi, M.: Polymer nanocomposites and their applications in the electronics industry. Int. J. Electron. Electr. Eng. **7**(6), 603–608 (2014)
43. Jamil, H., Faizan, M., Adeel, M., Jesionowski, T., Boczkaj, G., Balčiūnaitė, A.: Recent advances in polymer nanocomposites: unveiling the frontier of shape memory and self-healing properties—a comprehensive review. Molecules **29**, 1267 (2024). https://doi.org/10.3390/molecules29061267
44. Hu, H., Zhang, F., Luo, S., Chang, W., Yue, J., Wang, C.-H.: Recent advances in rational design of polymer nanocomposite dielectrics for energy storage. Nano Energy **74**, 104844 (2020). https://doi.org/10.1016/j.nanoen.2020.104844
45. Pandey, P.P.: Preparation and characterization of polymer nanocomposites. Soft Nanosci. Lett. **10**, 1–15 (2020). https://doi.org/10.4236/snl.2020.101001
46. Li, S., Lin, M.M., Toprak, M.S., Kim, D.K., Muhammed, M.: Nanocomposites of polymer and inorganic nanoparticles for optical and magnetic applications. Nano Rev. **1**(1), 5214 (2010). https://doi.org/10.3402/nano.v1i0.5214
47. Gajević, S., Krstić, J., Miladinović, S., Blagojević, J., Stojanovic, B.: Polymer matrix nanocomposites from the ecological aspect in the automotive industry (2022)
48. Coiai, S., Passaglia, E., Pucci, A., Giacomo, R.: Nanocomposites based on thermoplastic polymers and functional nanofiller for sensor applications. Materials **8**, 3377–3427 (2015). https://doi.org/10.3390/ma8063377
49. Dantas de Oliveira, A., Augusto Gonçalves Beatrice, C.: Polymer nanocomposites with different types of nanofiller. Nanocompos. Recent Evol. (2019). https://doi.org/10.5772/intechopen.81329
50. Choudhary, V., Gupt, A.: Polymer/carbon nanotube nanocomposites. Carbon Nanotubes—Polym. Nanocompos. (2011). https://doi.org/10.5772/18423
51. Schadler, L.S., Brinson, L.C., Sawyer, W.G.: Polymer nanocomposites: a small part of the story. JOM **59**, 53–60 (2007). https://doi.org/10.1007/s11837-007-0040-5

Bioinspired and Biodegradable Nano Polymers: Fundamentals, Materials, and Green Strategies

Rukmani Sharma, V. K. Paliwal, Amit Kumar, Shashank Shekhar, Vijay Chaudhary, and Bhasha Sharma

Abstract The emergence of bio-inspired and biodegradable nano polymers has found wide applications in the fields of biomedicine that include drug delivery, regenerative medicine and tissue engineering, sustainable packaging, biosensors and in agricultural solutions with sustainable approaches as they are biodegradable by themselves. This chapter focuses on the scope of bioinspired biopolymers. It also discusses the classification, design and fabrication of bioinspired nano polymers. The greener approach of biodegradable nano polymer is studied for sustainability. Some characterisation techniques were described which help in the study of bioinspired nano polymer. Applications of bio-inspired and biodegradable polymer with challenges and future perspectives is also presented.

Keywords Nanopolymer · Biodegradable polymers · Green materials · Biopolymer

---

R. Sharma (✉)
Department of Chemistry, GTBIT, IP University Delhi, Delhi, India
e-mail: rukmanis13@gmail.com

V. K. Paliwal
Department of Physics, Dayal Singh College, University of Delhi, Delhi, India

A. Kumar
Department of Chemistry, University of Delhi, Delhi, India

S. Shekhar
Department of Chemistry, Sri Aurobindo College, University of Delhi, Delhi, India

V. Chaudhary
Department of Mechanical Engineering, Amity School of Engineering and Technology, Amity University, Noida, Uttar Pradesh, India
e-mail: vijaychaudhary111@gmail.com

B. Sharma
Department of Chemistry, Deen Dayal Upadhyaya College, University of Delhi, Delhi, India

B. Sharma et al. (eds.), *Nanostructured Polymers*, Engineering Materials,
https://doi.org/10.1007/978-981-95-3581-1_2

1 Introduction to Bioinspired and Biodegradable Nanopolymers

Nanotechnology is the a of innovation and research which enables a fast development in the field of medicine, electronics, sensors and many other areas [1]. Although designing a nano-polymers of desired size is challenging and it requires continuous development to make nano-polymer of precise size for the synthesis and use in various applications. Nano polymers comprise unique properties due to their small size as surface area increases which makes them more potent than the traditional polymers [2]. Over the years, bio-inspired and biodegradable nano-polymers have received a considerable amount of interest not only due to their small size but also as they degrade with time in aerobic or anaerobic conditions [3]. Polymeric materials derived from biomolecules or synthetic monomers are made of monomeric units that result in the polymer which can be reduced to the nano-polymer for the specific applications. In biomedicine, these nanopolymers act as carriers for targeted drug delivery, enabling localized therapeutic action.. So, with the biodegradable nano-polymers, optimal results can be obtained by minimizing toxicity and environmental hazards as compared to the non-biodegradable carriers [4]. Bioinspired nanopolymers represent a class of materials that is manufactured or intentionally engineered to replicate the functionality, structure or mechanism observed in the natural biological systems [5, 6]. In chemistry, this is a novel trajectory by which materials with unique characteristics can be developed to address the broader aspect of scientific, environmental and technical obstacle in various multidimensional sectors like bioengineering and biotechnology [7, 8], medicine and healthcare [9–11], environment [10, 12], material science [13], energy [14] and many more [15–17]. Bioinspired adhesives derived from gecko foot structures exhibit remarkable adhesion properties and have inspired researchers to develop revolutionary adhesive materials [18]. Similarly, biological entities like DNA [19, 20], protein [21, 22], cells [23] can can construct shapes and structure of DNA at nano level [24, 25]. Biodegradable nano-polymers (BNP) are microscopic particles of polymer typically less than100 nm in diameter that degrade under aerobic or anaerobic conditions [26, 27]. These materials decompose into environmentally benign byproducts such as carbon dioxide, water, and biomass, in contrast to conventional non-biodegradable plastics They are widely used in applications like tissue engineering [28, 29], drug delivery [4, 30, 31] and packaging etc. [32, 33]. Nanopolymers offer several advantages in comparison to conventional polymers like enhanced performance, biocompatibility, sustainability and self-assembly characterization [34]. The utilization of nano polymers spans diverse disciplines such as environment, biomedicine, optics, and electronics etc.

Enhanced Performance

Bio-inspired nano-polymers exhibit enhanced performance due to their distinct structural and functional properties. One of the potential applications of these nano-polymers is their ability to replicate the adhesive properties of gecko feet [35]. Another example includes shark-skin-inspired nanomaterials, which possess drag-reducing properties and have the potential to enhance the energy efficiency of microfluidic devices [36].

Biocompatibility

One of the main advantages of bioinspired nanopolymers is their excellent biocompatibility. The nano polymers that are inspired by or derived from biological systems possesses inherent compatibility with living tissues, making them suitable for biomedical applications [37]. In drug delivery systems, these nanopolymers act as carriers that transport therapeutic agents to targeted sites within the body, thereby minimizing the risk of adverse immune responses and improving treatment efficacy [38–39].

Self-assembly

It is a characteristic observed in numerous biological systems, where in complex structures are formed spontaneously through molecular interactions . Bioinspired materiasl inherit thisintriguing propertry. Factors such as temperature, pH, and ionic strength can be controlled to regulate this process. The same ability can be used for constructing complex structures with minimal external intervention.

Sustainability

The synthesis and design of bioinspired nano polymers include green chemistry and biomimicry, which promote sustainability [40]. By using a greener approach, advancements in fabrication can be achieved that are also environmentally sustainable. With the use of greener approachs, many nano polymers can be synthesized using gentler reagents in comparison to toxic reagents or byproducts. Some nano

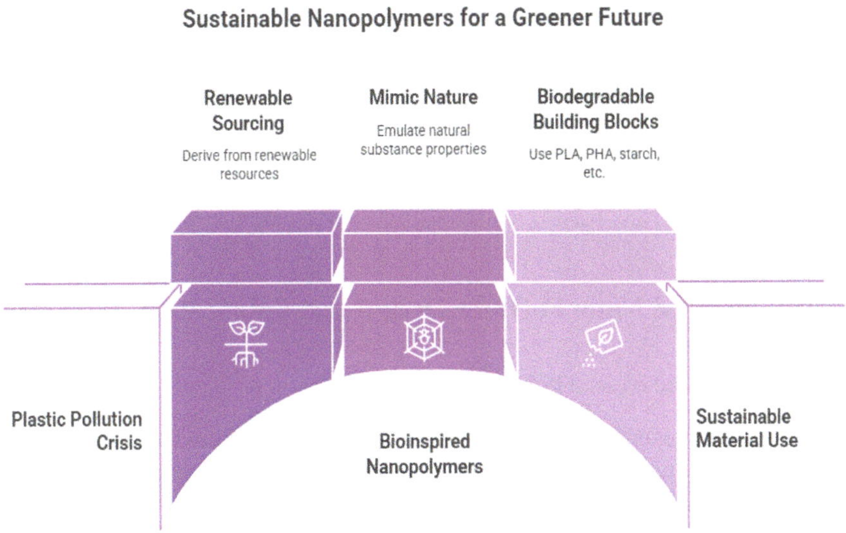

Fig. 1 Introduction to bioinspired nanopolymers

polymers can degrade in the environment over time, which reduces the threat to the environment.

2 Classification of Bio-nanopolymers

Bio nano polymers can be classified into three types based on the origin of the biopolymers , namely natural, synthetic and hybrid.

- Natural Nanopolymers (Proteins, Polysaccharides, Nucleic Acids)

Polymers that are derived from natural sources like flora and fauna have nanoscale dimensions (1–100 nm). They are renewable, bio-compatible and biodegradable, making them versatile in various applications, including biomedicine [41], material science [42] and environmental remediation [43]. Some examples of natural nano polymers are proteins, polysaccharides and nucleic acids.

- Synthetic Bioinspired Nanopolymers (Polyesters, Polyamides, Polyurethanes)

Polymers that mimic the natural polymers but are man-made, are synthetic bio inspired polymers having dimensions (1–100 nm). They are synthesized in laboratory using synthetic or natural polymers. They are biodegradable and biocompatible, making them a sustainable alternative for conventional plastics. Examples of synthetic bioinspired nano polymers are polyester, polyamides and polyurethane.

- Hybrid Nano polymers (Composites of Natural and Synthetic Components)

This class incorporates the natural and synthetic components at the nanoscale, which enables synthesis of innovative polymers that enhance properties of individual constituents. These materials provide enhanced or completely new properties that are not present in the individual constituents. They can enhance durability, mechanical properties, specific property tuning and are also cost effective. These nanopolymers can be fabricated using different approaches like building block approach, In-situ formation, grafting method, and self-assembly etc. They find applications in construction, aerospace, automotive and sports equipment (Fig. 2).

3 Bio-inspired Nano Polymer Design and Fabrication

- Principles of Bioinspired Polymerization

These are nano polymers that are designed to mimic the structure found in nature at nanoscale. They are prepared by learning from biological systems such as DNA and proteins.

- Self-Assembly and Supramolecular Chemistry in Biopolymer Nanostructures

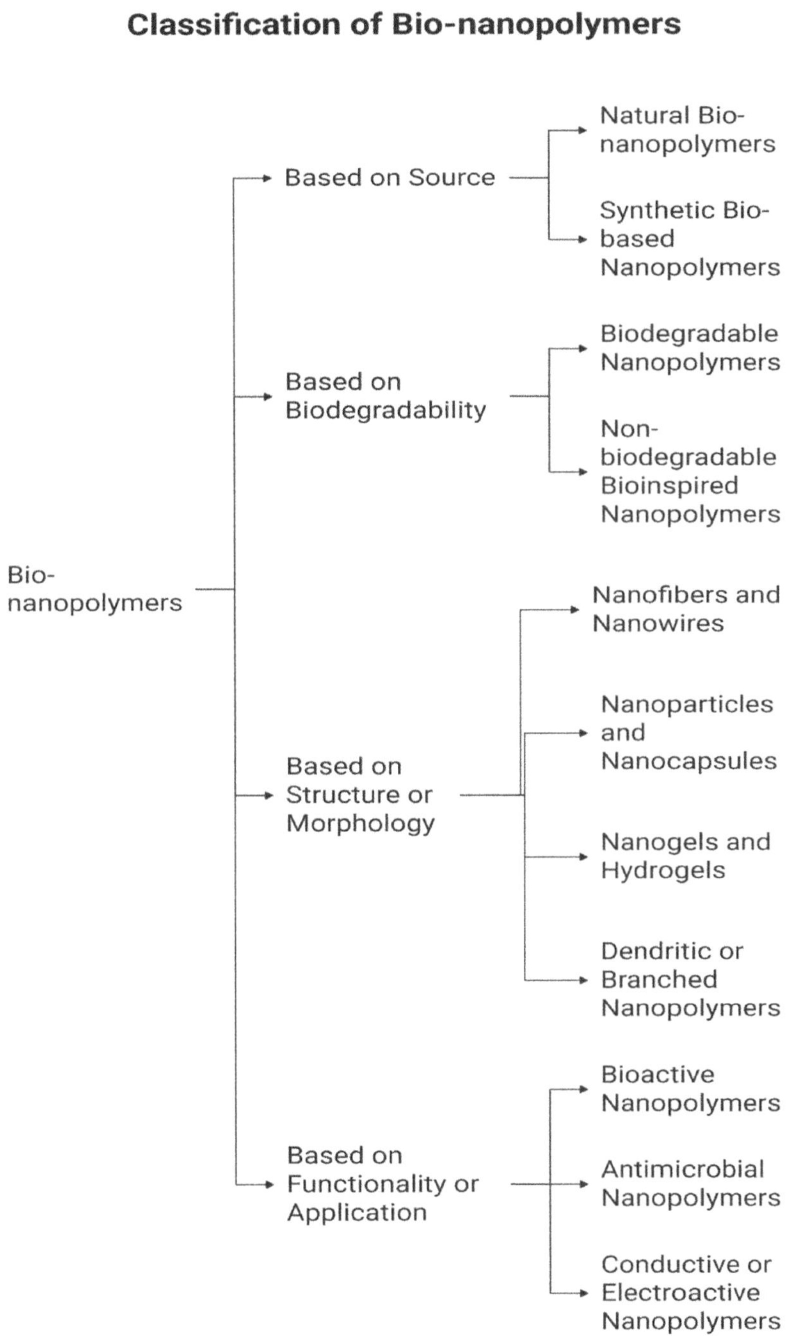

Fig. 2 Classification of bio-nano polymers

Self-assembly and supramolecular chemistry are foundational principles in the creation of biopolymer nanostructures. These mechanisms mimic biological processes to produce highly organized and functional materials without external manipulation [44]. In recent years, they have become essential tools in the design of nanoscale devices for applications in drug delivery, tissue engineering, biosensing, and nanomedicine.

Self-assembly is the spontaneous organization of molecules into structured and functional arrangements through non-covalent interactions such as hydrogen bonding, electrostatic interactions, van der Waals forces, π-π stacking, and hydrophobic effects [45]. Natural examples include DNA double helix, lipid bilayers, and protein folding. Supramolecular chemistry focuses on chemistry beyond the molecule, how molecules interact and recognize each other via non-covalent bonds to form larger, organized systems. The advantages of self-assembly include a bottom-up approach, which allows the creation of complex structures starting from simpler building blocks [46]. These systems are dynamic and reversible and they exhibit a high potential for structural complexity.

- Nanostructure Formation via Peptide and Protein Engineering

Nanostructure formation via peptide and protein engineering is a cutting-edge field at the intersection of nanotechnology, synthetic biology, and materials science. It involves designing and utilizing peptides and proteins to self-assemble into nanoscale architectures with defined structures and functions [47]. This approach has wide-ranging applications in drug delivery, biosensing, tissue engineering, and nanoelectronics. Peptides and proteins are natural polymers composed of amino acids that can fold into complex 3D shapes. By harnessing or engineering their sequence and structure, researchers can control how these molecules interact and assemble into nanostructures such as fibers, tubes, sheets, cages, and more. These systems follow mechanism of self-assembly and directed assembly.

Peptides and proteins can self-assemble via non-covalent interactions, hydrogen bonding, hydrophobic interactions, electrostatic forces, π–π stacking and van der Waals forces. β-sheet structures lead to the formation of nanofibers and tapes, α-helices form coiled-coil bundles, and loop and turn motifs result in cages or rings [48]. Directed assembly involves designing sequences that fold and assemble predictably and are inspired by natural proteins like collagen, silk, elastin, etc.

They follow the following design strategy: de novo protein design, modular assembly, bioinspired and biomimetic approaches. De novo protein design provides computational tools to predict folding and interactions [49]. It allows the design of entirely new protein structures with programmable features. Modular assembly is designed by repeating peptide motifs or domains that self-associate, such as coiled-coil peptides forming nanorods or nanotubes. Bio-inspired and biomimetic approaches mimic natural self-assembling systems such as amyloid fibrils or virus capsids [50].

- Natural Template-Assisted Nanopolymer Synthesis

Natural template-assisted nanopolymer synthesis refers to a green and bio-inspired approach to creating nanostructured polymers using natural materials (biological or bio-derived) as templates to direct the shape, size, and morphology of the resulting polymer nanostructures [51]. In conventional nanopolymer synthesis, synthetic templates, surfactants, or harsh chemical conditions are often used to control the architecture of polymer nanomaterials. In contrast, natural templates offer an eco-friendly, biocompatible, and sustainable alternative.

Natural templates are biological structures or materials that serve as scaffolds or molds to guide the formation of nanostructures. These include cellulose fibre, proteins, DNA [52], bacterial biofilms etc. These templates often have well-defined nano- or micro-architectures that can be replicated or coated with polymers. The process usually involves template preparation, which includes extraction or isolation of the natural material. Polymerization involves the initiation of polymer growth on or around the template surface using techniques like in situ polymerization, graft polymerization, or electropolymerization. The template may be removed (e.g., calcination and enzymatic degradation) to leave behind a nanostructured polymer replica. Alternatively, it may be retained for hybrid bio-nanocomposites.

4 Green Strategies for Biodegradable Nanopolymers

When green chemistry blends with nanotechnology, a hostile approach can achieve sustainability in future.

- Sustainable Synthesis Techniques (Enzymatic Polymerization, Green Catalysis)

Sustainable synthesis of bionanopolymers, biodegradable polymers at the nanoscale used in fields like biomedicine, packaging, and agriculture has become a central research area. Among eco-friendly methods, enzymatic polymerization and green catalysis stand out due to their mild reaction conditions, reduced environmental impact, and potential for high selectivity.

Enzymatic polymerization uses enzymes (biocatalysts) to synthesize polymers under environmentally benign conditions (ambient temperature, neutral pH, aqueous or green solvents). Enzymes used for enzymatic polymerization are lipase, peroxides and lactases. These reactions can proceed under mild conditions and are energy efficient. They are highly stereo- and regio- selective, and metal catalysts are not needed, thereby avoid contamination. Through this synthesis, molecular weight and polymer architecture can be controlled. An example includes CALB-catalyzed ring-opening polymerization of ε-caprolactone to produce biodegradable polycaprolactone nanoparticles for drug delivery. Biopolymers prepared using this technique include polyesters such as poly (ε-caprolactone), poly (lactic acid); polysaccharides through enzymatic grafting on chitosan, cellulose; and polypeptides and proteins via------ enzyme-mediated ring-opening polymerization of N-carboxyanhydrides (NCAs).

Green catalysis involves using environmentally benign catalysts—often metal-free, bio- derived, or recyclable—to carry out polymerization reactions efficiently and safely. Preferred catalysts include are organocatalyst such as thioureas, phosphines, N-heterocyclic carbenes, ionic liquids; used as both solvents and catalysts; biobased catalyst derived from amino acids, sugar, and heterogeneous catalysts such assilica-supported, zeolite-based systems. Examples include ring opening polymerization of L-lactide using an organocatalyst (e.g., DBU) to produce polylactic acid nanostructures with controlled molecular weight and biodegradability. RAFT and ATRP use green solvents and low toxicity initiators, while click chemistry is used for cross-linking or surface functionalization, Ring opening polymerization of lactones or lactides using organocatalysts is widely applied. The advantages of these approaches include reusability of catalysts, reduced waste and toxicity, enhanced control over polymer architecture, and compatibility with bio renewable monomers (Table 1).

- Eco-friendly processing methods for bio-nano polymers are gaining significant attention due to their potential to reduce environmental impact, energy consumption, and reliance on toxic chemicals. Among these, most promising approaches are solvent-free methods and supercritical fluid technologies, these processing methods such as solvent-free extrusion and supercritical fluid technologies offer sustainable routes to fabricate high-performance bio-nanopolymers. These methods not only reduce environmental impact but also align with the principles of green chemistry and circular bioeconomy. They can be prepared using different approaches such as solvent free processing and supercritical fluid processing Solvent free processing involves synthesizing or processing bio-nanopolymers without using traditional organic solvents, which are often volatile, flammable, and toxic. Common techniques are melt processing/extrusion in which nano fillers and bionanopolymers are blended at high temperature. These techniques are mainly suitable for thermoplastic biopolymers like PLA, PHA, starch-based polymers, while nano fillers like cellulose, nano clay are dispersed directly. Another approach is reactive extrusion, where polymerization and compounding occur simultaneously in an extruder. It enables in-situ grafting or functionalization of nanofiller, and separate solvent is not needed. The final method is solid state shear pulverisation (SSSP), in which mechanical shear at low temperatures is used to mix polymers and nanoparticle [53]. Polymer structure can be preserved while fine dispersion occurs. This process can eliminate toxic solvent disposal, is energy efficient for thermoplastics, scalable and compatible with industrial equipment.

Table 1 Types of catalyst used for techniques with products

Technique	Catalyst type	Typical products	Benefits
Enzymatic polymerization	Enzymes (e.g., CALB)	Polyesters, polysaccharides	Mild conditions, high selectivity
Green catalysis	Organocatalysts, bio-catalysts	PLA, PCL, hybrid nano polymers	Low toxicity, tunable polymer structures

Supercritical fluid processing uses fluids above their critical temperature and pressure to process polymers in a green medium, most commonly supercritical CO_2 ($scCO_2$) [54]. These methods help in the dispersion of nanoparticles, foaming and porous structures, extraction and purification, and supercritical anti solvent (SAS) processes. They have found their usefulness as they are non- toxic, non-flammable, and recyclable. These processes operate at relatively low temperatures (ideal for heat-sensitive biopolymers). They also enhance polymer- nanoparticle interactions and homogeneity (Fig. 3).

- Renewable Resource-Based Nano polymers (Plant-Based, Algae-Derived, Bacterial Polymers)v.:?

Renewable resource-based nanopolymers, a which are derived from plants, algae, and bacteria, are increasingly explored in the development of bio-nanopolymers due to their sustainability, biodegradability, and reduced environmental impact. They originaate from different sources , such as plant based nanopolymers, algae derived nanopolymers, and bacteria derived nanopolymers. Plant based nanopolymers are

Fig. 3 Green strategies for bio-nanopolymers

sourced from cellulose, starch, lignin, pectin, natural rubber, protein. Their nanostructured forms include cellulose nanocrystals (CNCs), cellulose nanofibers (CNFs), starch nanocrystals, and zein nanoparticles [55]. They find applications in nanocomposites, packaging (biodegradable films), drug delivery systems, tissue engineering scaffolds, and biosensors.

Algae derived nanopolymers are sourced from alginate (from brown algae), carrageenan (from red algae), Agar, ulvan (from green algae), and fucoidan [56]. Their nanostructure forms include alginate nanoparticles andagar-based hydrogels at the nanoscale. They find applications in the encapsulation of drugs and nutrients, antiviral and antimicrobial delivery, hydrogels for wound healing, food packaging and stabilization, and tissue regeneration scaffolds.

Bacterial derived nanopolymers are sourced from polyhydroxyalkanoates (PHAs) (e.g., polyhydroxybutyrate—PHB), bacterial cellulose (BC), exopolysaccharides (e.g., xanthan gum, levan), and polylactic acid (PLA) (produced via bacterial fermentation of sugars) [57]. Their nanostructure forms include nanocellulose from bacterial cellulose, PHB nanoparticles, and PLA nanofibers via electrospinning. They find applications in bioplastics, medical implants and scaffolds, targeted drug delivery, controlled-release fertilizers, and environmental remediation (e.g., heavy metal adsorbents). Advantages of renewable bio-nanopolymers include biodegradablility and compostability, low toxicity, derivation from abundant non-petroleum resources, tunable mechanical and chemical properties, and compatibility with green synthesis techniques..

5 Characterization Techniques for Bio-nanopolymers

- Structural Characterization (FTIR, NMR, XRD)

FTIR is a crucial tool for analyzing the chemical structure of bio-nanopolymers. It analyses functional groups, molecular interactions and structural changes due to nanomaterials incorporated or chemical modification. It indicates interaction between polymer matrix and nanoparticles, successful incorporation of nanomaterials and cross linkers. It monitors degradation, crystallinity and chemical modification over time.

Depending on the biopolymer used (e.g., chitosan, cellulose, starch, PLA, PHA), common peaks include (Table 2).

Nuclear Magnetic Resonance (NMR) spectroscopy is a powerful and non-destructive technique for the structural characterization of bio-nanopolymers, which are hybrid materials composed of natural or synthetic polymers and nanoscale biological or inorganic components. NMR provides detailed molecular-level information about structure, dynamics, and interactions. Several types of NMR techniques are employed depending on the material's physical state (solution or solid) and the desired information.

NMR of biopolymers are given below

Table 2 Wavenumber with the functional group of different nano polymeric groups

Wavenumber (cm^{-1})	Functional group/vibration	Interpretation
~3400	O–H/N–H stretching	Hydrogen bonding, hydroxyl, amine groups
~2920/2850	C–H stretching (asymmetric/symmetric)	Aliphatic chains, CH_2 groups
~1740	C=O stretching (carbonyl)	Ester, carboxylic acid groups (e.g., PLA, PHA)
~1650–1600	N–H bending/Amide I (proteins, chitosan)	Presence of peptide bonds or amine groups
~1420–1370	CH_3 bending/symmetric COO^-	Methyl groups, carboxylates
~1250–1050	C–O–C/C–O stretching	Polysaccharides (cellulose, starch), and ester groups
~800–600	Metal–O vibrations (if nanoparticles present)	Indicative of metal oxide nanoparticles (e.g., ZnO, TiO_2)

Chitosan–silver nanoparticle composites: 1H NMR shows broadening of signals due to nanoparticle interactions. SSNMR reveals changes in the polysaccharide backbone.

Cellulose nanocrystals in PLA matrix: 13C SSNMR shows chemical shifts indicating matrix– filler interactions and altered crystallinity.

Protein-polymer nanogels: 2D NMR (HSQC) used to monitor protein folding and chemical environment in the nanostructure (Table 3).

X-ray diffraction (XRD) is a powerful technique for structural characterization of materials, including bionanopolymers. It provides information about crystallinity,

Table 3 NMR technique with characteristics of nano-polymers

Structural feature	NMR technique	Characteristics
Chemical composition	1H, 13C NMR	Identification of monomers, functional groups
Nanoparticle–polymer interaction	SSNMR, 2D NMR	Evidence of binding (e.g., chemical shift changes, new peaks)
Chain dynamics	T1, T2 relaxation times	Mobility of polymer chains near nanoparticle interfaces
Crystallinity	13C CP/MAS	Degree of crystallinity and polymorphism
Phase separation	2D and SSNMR	Observation of microphase-separated domains
Crosslinking or degradation	1H/13C NMR	Chemical changes due to crosslinking, hydrolysis

phase composition, crystallite size, and lattice parameters. XRD identifies crystalline and amorphous regions in the polymer matrix. It also provides information about semi-crystalline bio-nanopolymers (e.g. PLA, PCL, PHB) which show characteristic peaks. In composites, peaks from added nanoparticles (e.g., ZnO, TiO_2, clay, AgNPs) appear in the XRD pattern. Sucessful dispersion can be confirmed using this method. Some XRD results are discussed below.

Pure bio-nanopolymer (e.g., Chitosan, Starch): Broad hump at low 2θ (amorphous) or few weak peaks (semi-crystalline).

Nanoparticle-loaded composite: Sharp peaks corresponding to nanoparticle (e.g., ZnO at ~31.8°, 34.5°, 36.3° for wurtzite phase).

Clay nanocomposites (e.g., MMT): Shift in basal spacing peak (001) indicates intercalation/exfoliation.

- Morphological Analysis (SEM, TEM, AFM)

To investigate the surface morphology, nanoparticle dispersion, and microstructural features of the synthesized bionanopolymer, the surface of the pure biopolymer appears smooth and homogeneous, indicating a uniform polymeric phase. With nanoparticle incorporation, roughness increases, suggesting effective embedding and interaction of nanofillers with the biopolymer matrix. The dispersion of nanoparticles can also be observed in SEM, as at lower nanoparticle loadings (e.g., 1–2 wt%), the nanoparticles appear evenly dispersed throughout the matrix while at higher loadings (>5 wt%), signs of agglomeration or clustering are visible, thereby affecting mechanical or barrier properties. SEM images show good interfacial adhesion between the biopolymer and nanoparticles, as evidenced by minimal gaps or voids at the filler-matrix boundary. In some cases, pull-outs or microvoids are present, indicating potential weak interaction zones. Cross-sectional SEM may reveal porous or lamellar structures, especially if the material is freeze-dried or cast from solution. Pore sizes and distributions can be estimated using image analysis software (e.g., Image J).

- Thermal and Mechanical Properties (TGA, DSC, DMA)

To analyze thermal and mechanical properties of a bio-nanopolymer, three key techniques are commonly used: Thermogravimetric Analysis (TGA), Differential Scanning Calorimetry (DSC), and Dynamic Mechanical Analysis (DMA) (Table 4).

Thermogravimetric Analysis (TGA) is used to measure the change in mass of a material when heated, providing thermal stability and composition data. It predicts decomposition temperature i.e. onset and peak degradation temperature, and provide thermal stability. It gives an idea about the moisture loss and volatile content present in the compound. When the technique is applied to bio-nanopolymers, they show enhanced thermal stability due to nanoparticle reinforcement (e.g., nanoclays, cellulose nanocrystals). Multiple degradation steps may be observed if blends or multiple components are present (e.g., polymer matrix + biofiller).

Differential Scanning Calorimetry (DSC) is used to measure heat flow associated with transitions in a polymer (e.g., glass transition, melting, crystallization). It provides the glass transition temperature (Tg), which indicates the onset of molecular

Table 4 Different characterization techniques

Technique	Works on	
SEM (Scanning Electron Microscopy)	Surface morphology	Dispersion of nanoparticles, surface roughness, fiber optics
TEM (Transmission Electron Microscopy)	Internal structure at nanometer scale	Nanoparticle distribution, interfaces, crystallites
AFM (Atomic Force Microscopy)	Topographical mapping	Surface roughness and mechanical mapping
XRD (X-ray Diffraction)	Crystallinity and phase analysis	Degree of crystallinity and crystalline structure
FTIR (Fourier Transform Infrared Spectroscopy)	Chemical bonds and interactions	Indirectly indicates morphological changes (e.g., due between matrix and nanofillers)
DSC (Differential Scanning Calorimetry)	Thermal behavior	Can indicate structural/ morphological transitions (e.g. transition)

mobility in amorphous regions. It also provides the melting temperature (Tm), which reflects crystalline phase melting. Crystallization temperature (Tc) and enthalpy can be used to calculate the degree of crystallinity. Nanofillers often shift Tg/Tm or alter crystallinity due to nucleation effects. Bio-nanopolymers may show a slight increase in Tg due to restricted polymer chain mobility. There are changes in melting/ crystallization behavior due to nanoparticle interaction. Sometimes, cold crystallization may occur if rapid cooling prevents full crystallization.

Dynamic Mechanical Analysis (DMA) measures viscoelastic properties of materials as a function of temperature, time, frequency, or stress. It examines the storage modulus (E'), which provides elastic behavior or stiffness. The loss modulus (E'') represents viscous behavior or energy dissipation. Tan δ (E''/E') represents damping behavior, and its peak indicates Tg. Nanoparticles typically increase stiffness and raise the modulus across the temperature range. Bio-nanopartickes possess higher storage modulus, which indicates improved stiffness. A shift in the tan δ peak shift suggests Tg modification. There is greater thermomechanical stability at elevated temperatures (Table 5).

Table 5 Different thermal techniques employed for analysis

Technique	Measured property	Relevance to bionanopolymers
TGA	Mass loss versus temperature	Degradation behavior, thermal stability
DSC	Heat flow versus temperature	Tg, Tm, crystallinity, phase transitions
DMA	Modulus versus temperature	Stiffness, damping, viscoelastic transitions

- Biodegradability and Environmental Impact Assessment

Biodegradability refers to the ability of a material to be broken down by microorganisms into natural substances like water, carbon dioxide (or methane in anaerobic conditions), biomass, and mineral salts. The factors that influence biodegradability include polymer matrix composition such as polylactic acid [PLA], polyhydroxyalkanoates [PHAs], starch-based polymers; types of nano fillers like cellulose, nanocrystal, chitin nano fibre, nano clays and carbon-based nano materials. Environmental conditions such as temperature, pH, moisture, oxygen level, and microbial community etc also influence biodegradability. Surface area and morphology of the nanopolymer also play an important role.

Standard testing methods are: ASTM D5338—Aerobic biodegradation under controlled composting conditions; ISO 14855—Determination of ultimate aerobic biodegradability and OECD 301 series—ready biodegradability test. The pathway for degradation has three stages. The first is biodeterioration, i.e., surface weakening by abiotic and biotic factors. The second is depolymerization, i.e., breakdown into oligomers and monomers. The third is assimilation, i.e., microbial uptake and mineralization.

Environmental Impact Assessment has three aspects The first is Life Cycle Assessment (LCA), which evaluates the environmental impacts associated with all stages of a bio-nanopolymer's life. Here, raw material extraction is formed by nanopolymers synthesis, polymer-nanofiller blending, usage and end of treatment. The second is ecotoxicity concerns, as some nano fillers such as metal oxide nanoparticles, and carbon nanotubes may possess ecotoxicity risks. The risks involved include accumulation in soil and water, bioavailability to microorganisms, plants, and aquatic life, and disruption of microbial ecosystem necessary for biodegradation. The last aspect is end of life scenarios, which include compositing where fully biodegradable systems are present,soil degradation, aquatic degradation and incineration or landfill. The challenges faced include lack of uniformity in testing protocols for nanocomposite biodegradability, poor understanding of long-term fate of nanoparticles post- degradation, and limited standardized assessment methods. To overcoming these challenges, some recommendations are to use natural or bio-based nanofillers (e.g., cellulose nanocrystals) to minimize ecotoxicity; design fully biodegradable systems that decompose under common environmental conditions; perform comprehensive LCA and eco-toxicological studies before commercialization; and ensure compliance with regulatory frameworks (e.g., REACH, EPA, ISO).

6 Applications of Bio-nanopolymers

Bionanopolymers hold promise as sustainable alternatives to conventional plastics, offering enhanced properties along with potential for biodegradability. However, to ensure environmental safety, rigorous biodegradability and environmental impact assessments must be conducted. Future development should focus on green nanotechnology principles, safe-by-design approaches, and robust life cycle assessments (Fig. 4).

- Biomedical Applications (Drug Delivery, Tissue Engineering, Biosensors)

Bioinspired and BNPs have emerged as powerful tools in biomedical applications due to their unique properties—biocompatibility, controlled degradability, molecular tunability, and their ability to mimic natural biological structures. Their nanoscale size allows for interactions with biological systems at the cellular and molecular levels, making them ideal for several cutting-edge medical technologies. In drug delivery, they control and sustain the release of the drug. The drug is delivered to the targeted site which also reduces the side effects caused by the drug. They also enhance the solubility and bioavailability of the drug [58]. Nano polymers like poly (lactic-co-glycolic acid) (PLGA), chitosan, and PEG-based nanocarriers are widely used for encapsulating chemotherapeutic agents. Ligands such as antibodies or peptides can be attached for tumor- targeting (e.g., folic acid-targeted PLGA nanoparticles for breast cancer). Biodegradable polymers protect drugs from degradation in the digestive system and allow for oral delivery of proteins like insulin. mart nanopolymers respond to environmental triggers like pH, temperature, or enzymes. For example, pH-sensitive chitosan nanoparticles release drugs preferentially in acidic tumor environments [59].

For tissue engineering, scaffold biomimicry of extracellular matrix is the key feature. Biodegradability matches the regeneration rate of tissues. Nano polymer promotes cell adhesion, proliferation and differentiation in tissue engineering. They find its application in bone tissue engineering as nanofibrous scaffolds from polylactic acid (PLA) or collagen/ chitosan blends enhance osteoblast adhesion and proliferation [60]. Incorporation of hydroxyapatite nanoparticles further improves osteoconductivity. They also support natural tissue repair by using nanopolymers supporting neural stem cell growth and guide neurite extension. Conductive polymers like polypyrrole combined with biodegradable scaffolds promote nerve regeneration. Electrospun nanofiber mats of silk fibroin or PCL (polycaprolactone) loaded with antibacterial agents accelerate healing and prevent infections, thereby helping in skin and wound healing.

Biosensors have a high surface-to-volume ratio for enhanced sensitivity, which is attracting the industries toward them. They provide a functionalized surface for specific molecular recognition, which enhances the sensitivity of the biosensor [61]. These sensors are rapidly finding their application in glucose sensing in diabetic patients. Nanopolymer-based enzymatic sensors (e.g., PANI or chitosan with glucose oxidase) for continuously used for glucose monitoring. They can also detect the

Application Area	Functionality	Examples	Advantages
Biomedical and Healthcare	Drug Delivery System	Chitosan nanoparticles for oral insulin delivery, PLA-PEG nano carriers for cancer drug, Zein protein-based nano capsules for hydrophobic drugs	Improves drug stability, bioavailability and reduces side effects.
Biomedical and Healthcare	Tissue Engineering and Regenerative Medicine	Collagen or gelatin -based nano fibre scaffolds for bone and skin regeneration , Nanocellulose hydrogels for cartilage regeneration	Promotes cell adhesion, proliferation and differentiation.
Biomedical and Healthcare	Wound Healing and Antimicrobial Dressings	Chitosan-silver nanoparticle dressing, PLA nanofibres loaded with curcumin	Accelerates healing and prevents infection.
Biomedical and Healthcare	Biosensors and Diagnosis Devices	Silk fibroin nnanofilms in glucose sensors, conductive PHA or chitosan composites in electrochemical sensors	Biocompatible and suitable for wearable devices.
Food Packaging and Safety	Biodegradable Active Packaging	starch-PVA films with essential oil-loaded nano capsules, PLA films embedded with ZnO or TiO2 nanoparticles	Extended Shelf life and reduces plastic waste.
Food Packaging and Safety	Intelligent Packaging	Chitosan films with anthocyanin based pH sensors	Improves food safety and traceability
Agriculture and Agrochemicals	Controlled Release of Fertilizers and Pesticides	Lignin-based nanocarrieers for urea or herbicides, gelatine nanoparticles for slow-release micronutrients	Minimises environmental contamination and improves nutrient uptake.
Agriculture and Agrochemicals	Plant Growth, Promotion and Seed Coating	Chitosan nanoparticles enhancing seed germination, penicillin coated nanoparticles delivering plant hormones	Boost yield and resistance
Environmental Application	Water Purification	Chiotosan /magnetite nano composites for arsenic removal, Cellulose nanofiber filters for bacterial removal	Eco-friendly, reusable and efficient.
Environmental Application	Biomediation	Alginate-based nano carriers for microbial remediations of oil spills	Supports green clean-up strategies.

Fig. 4 Applications of bio-nanopolymers

pathogens in the living organism as nanopolymers functionalized with aptamers or antibodies detect bacteria or viruses with high sensitivity. The important use of biosensors is in detection of cancer, as PLGA- or dendrimer-based biosensors detect proteins (e.g., PSA, HER2) in blood samples, enabling early diagnosis for cancer patients.

- Food Packaging and Edible Coating

Bio-inspired or BNP packaging offers eco-friendly alternatives to the packaging that leads to long-term environmental pollution and health risk. This approach enhances food safety and extend the shelf life of packed food. It also reduces ecological impact by using advanced packaging technologies. The natural polymers which can be used for packaging include chitosan that has antimicrobial properties and film forming capabilities. Another is starch, which is very abundant, biodegradable but needs reinforcement. Alginate is another reinforcement that has gel forming capacity and provides a barrier to oxygen. Gelatin is also biocompatible and flexible for packaging. Some of the nanomaterials for packaging include nanocellulose, which is a better reinforcement and has better barrier enhancement. Nanochitosin had antimicrobial activity and is a better option for packaging. Nano clay can also be used as packaging material because gas barrier properties can be enhanced by using nanomaterials. Active packaging incorporates substances that interact with the food, such as antimicrobials and antioxidants [62]. For antimicrobial packaging, chitosan-infused essential oils are used, while for antioxidant properties, starch film is incorporated with green tea extract. Another form of packaging is biodegradable films and trays, which can be disposed of after using. They are used for fruits, vegetables, meat, and bakery products. Smart packaging is the most advanced type of packaging where the food is monitored with the change of pH-sensitive colour change.

Edible coatings are the coatings in which a very thin layer of coating is applied to the food so that moisture can be controlled, provide a gas barrier to the food and prevent it from fouling. They act as active carriers that protect food from antioxidant and probiotics. They can be used to protect food like sliced fruits, which turn brown when exposed to the air, by applying a very thin, fine layer of edible coating to the sliced fruits. Cheese can be prevented from developing mould on the surface if a thin layer of edible coating is applied to it. Packaging with bioinspired nanopolymers or BNPs is advantageous for the environment due to its compostability and low ecological footprint [63]. The packaging is user friendly, as it is derived from natural sources and has a safer perception.

- Environmental Remediation and Water Purification

Environmental pollution that contaminates water by heavy metals, dyes, pharmaceuticals and microplastics is a global issues. Traditional remediation methods often involve toxic chemicals or generate secondary pollutants. Bio-inspired and biodegradable nanopolymers provide an eco-friendly alternative by mimicking nature's strategies and breaking down harmlessly after use. They can be employed in soil cleanup, where chitosin nanoparticles absorb heavy metals like $Pb2^{+}$ and $Cd2^{+}$. Bio-inspired hydrophobic nanofiber mats, modeled after lotus leaves, are used for oil spill recovery. Biodegradable nanocomposites with catalytic function imitate pesticide degradation [64]. They are sustainable, biodegradable, selective removal, non-toxic and allow scalable production.

- Agricultural and Smart Coatings

Nano capsules or nanogels release nutrients or pesticides slowly over time. With the use of agricultural coatings, leaching loss is reduced. Some examples include chitosin nano carriers for nitrogen fertilizer delivery. Seed coatings may include nanopolymer-based coatings to improve germination, drought resistance, and pest protection [65]. Biopolymer coatings are embedded with micronutrients or beneficial microbes. Smart hydrogels that retain water and release it under specific moisture conditions are used for soil conditioning. This enhances resilience in arid zones [66].

Smart coatings are coatings for stimuli-responsive coatings that response to changes in temperature, pH, moisture or light. They release agrochemicals or change properties when triggered, such as pH-sensitive biodegradable nanogels for pesticide release only in alkaline soils. They can embed nanosensors in coatings to monitor soil pH, humidity, and nutrient level. This could enable precision agriculture with real-time date feedback. Biodegradable nanopolymers used in the smart coatings include chitosin, cellulose nanocrystal, PLA and starch based nonomaterials, as they have extensive properties such as antimicrobial, biodegradable, film-forming, compostability, being abundant and inexpensive [67]. They provide a reduction in micro pollution, and biodegradation returns safe by-products to the soil (CO_2, water, biomass).

7 Challenges and Future Perspectives

- Scalability and Commercialization Barriers

Scalability and commercialization of BNMs, which are gaining attention in fields like biomedicine, packaging, agriculture, and environmental remediation, face several significant barriers. These can be broadly categorized into technical, economic, regulatory, and market-related challenges, such as technical barriers, economic cost barrier, and regularity and safety barriers. These can be overcome by process optimisation, cost reduction strategies, standardisation, public private partnerships, education and advocacy.

- Stability, Performance, and Functionalization Challenges

When working with BNMs, especially in applications like drug delivery, tissue engineering, or packaging, researchers face a range of interconnected challenges. These typically fall into three major categories: stability, performance, and functionalization.

Stability Challenges can be classified as thermal stability, chemical and hydraulic stability, and shelf life and storage. Performance challenges include mechanical properties, controlled design rate, barrier and permeability properties, reproducibility and batch variability. Functional challenges include surface modification, nanoparticle loading and release, targeting and biocompatibility, and scalability of functional techniques. These can be addressed by using copolymers or blends such as PLA-PEG, by adding nano fillers like nano clay and graphene oxide, by inducing smart design of polymer architecture, implementing greener approaches and by incorporating advanced fabrication methods like 3-D printing and emulsion polymerisation for better control.

- Future Trends in Biodegradable Nanopolymer Research

Biodegradable nanopolymers are an exciting and rapidly evolving field at the intersection of nanotechnology, polymer science, and sustainability. Future trends in this research area reflect growing global interest in environmentally friendly materials, biomedical applications, and advanced manufacturing techniques. The most prominent emerging trend is green and sustainable synthesis method in which eco-freindly synthesis techniques are used that minimize the toxic solvents and energy consumption. One example is the use of enzymatic polymerization or microbiological biosynthesis. Nano polymer-based scaffolds and drug delivery systems have mainly attracted researchers to biomedical and tissue engineering innovations, and 3-D printing scaffolds using nano polymer inks for regenerative medicines can be a future outlook. Researchers are also trying to expand the large-scale applications in agricultural and environmental remediation by using controlled release of fertilizers using BNPs or water purification via nano polymer based fillers or absorbents. Biopolymer- nanoparticle hybrids are one of the future trends, as they combine biodegradable polymers with functional nanoparticles such as metallic, magnetic, and carbon-based materials. These are gaining popularity as it improves mechanical strength, drug loading capacity, or antimicrobial activity. They can also tune electric and optical properties.

References

1. Pramanik, P.K.D., Solanki, A., Debnath, A., Nayyar, A., El-Sappagh, S., Kwak, K.S.: Advancing modern healthcare with nanotechnology, nanobiosensors, and internet of nano things: taxonomies, applications, architecture, and challenges. IEEE Access **8**, 65230–65266 (2020)
2. Njuguna, J., Ansari, F., Sachse, S., Rodriguez, V.M., Siqqique, S., Zhu, H.: Nanomaterials, nanofillers, and nanocomposites: types and properties. In: Health and Environmental Safety of Nanomaterials. Woodhead Publishing, pp. 3–37 (2021)
3. Bher, A., Mayekar, P.C., Auras, R.A., Schvezov, C.E.: Biodegradation of biodegradable polymers in mesophilic aerobic environments. Int. J. Mol. Sci. **23**(20), 12165 (2022)
4. Idrees, H., Zaidi, S.Z.J., Sabir, A., Khan, R.U., Zhang, X., Hassan, S.: A review of biodegradable natural polymer-based nanoparticles for drug delivery applications. Nanomaterials **10**(10), 1970 (2020)

5. Chakraborty, S., Bera, D., Roy, L., Ghosh, C.K.: Biomimetic and bioinspired nanostructures: recent developments and applications. Bioinspired Green Synth. Nanostruct. A Sustain. Approach 353–404 (2023)
6. Barhoum, A., García-Betancourt, M.L., Jeevanandam, J., Hussien, E.A., Mekkawy, S., Mostafa, M., Omran, M.M., Abdalla, M.S., Bechelany, M.: Review on natural, incidental, bioinspired, and engineered nanomaterials: history, definitions, classifications, synthesis, properties, market, toxicities, risks, and regulations. Nanomaterials **12**(2), 177 (2022)
7. Fatima, G., Magomedova, A., Parvez, S.: Biotechnology and Sustainable Development. Shineeks Publishers (2024)
8. Sofian, B.A., Aiman, A.D., Sun, X., Gupta, V.K., Berenjian, A., Xia, A., Ma, Z., Show, P.K.: Advances, synergy, and perspectives of machine learning and biobased polymers for energy, fuels, and biochemicals for a sustainable future. Energy Fuels **38**(3), 1593–1617 (2024)
9. Mitchell, S., Martín, A.J., Guillén-Gosálbez, G., Pérez-Ramírez, J.: The future of chemical sciences is sustainable. Angew. Chem. Int. Ed. **63**(26), e202318676 (2024)
10. Tawiah, B., Ofori, E.A., George, S.C.: Nanotechnology in societal development. In: Nanotechnology in Societal Development, pp. 1–64. Springer Nature Singapore, Singapore (2024)
11. Singh, A.V., Chandrasekar, V., Prabhu, V.M., Bhadra, J., Laux, P., Bhardwaj, P., Al-Ansari, A.A., Aboumarzouk, O.M., Luch, A., Dakua, S.P.: Sustainable bioinspired materials for regenerative medicine: balancing toxicology, environmental impact, and ethical considerations. Biomed. Mater. **19**(6), 060501 (2024)
12. Ozin, G.A., Loh, J.Y.Y.: Energy Materials Discovery: Enabling a Sustainable Future. Royal Society of Chemistry (2022)
13. Shahzad, K., Mardare, A.I., Hassel, A.W.: Accelerating materials discovery: combinatorial synthesis, high-throughput characterization, and computational advances. Sci. Technol. Adv. Mater. Methods **4**(1), 2292486 (2024)
14. Zhu, X.: AI and Robotic Technology in Materials and Chemistry Research. John Wiley & Sons (2024)
15. Kumar, V., Kumar, M.: Temporal-spatial-energy resolved advance multidimensional techniques to probe photovoltaic materials from atomistic viewpoint for next-generation energy solutions. Energy Environ. Sci. **14**(9), 4760–4802 (2021)
16. Khan, K., Tareen, A.K., Aslam, M., Wang, R., Zhang, Y., Mahmood, A., Ouyang, Z., Zhang, H., Guo, Z.: Recent developments in emerging two-dimensional materials and their applications. J. Mater. Chem. C **8**(2), 387–440 (2020)
17. Khan, Y., Sadia, H., Shah, S.A.Z., Khan, M.N., Shah, A.A., Ullah, N., Ullah, M.F., et al.: Classification, synthetic, and characterization approaches to nanoparticles, and their applications in various fields of nanotechnology: a review. Catalysts **12**(11), 1386 (2022)
18. Sikdar, S., Rahman, M.H., Siddaiah, A., Menezes, P.L.: Gecko-inspired adhesive mechanisms and adhesives for robots—a review. Robotics **11**(6), 143 (2022)
19. Keller, A., Linko, V.: Challenges and perspectives of DNA nanostructures in biomedicine. Angew. Chem. Int. Ed. **59**(37), 15818–15833 (2020)
20. Zhang, J.-H., Zhou, Y.-G.: Nano-impact electrochemistry: analysis of single bioentities. TrAC, Trends Anal. Chem. **123**, 115768 (2020)
21. Bilardo, R., Traldi, F., Vdovchenko, A., Resmini, M.: Influence of surface chemistry and morphology of nanoparticles on protein corona formation. Wiley Interdisc. Rev. Nanomed. Nanobiotechnol. **14**(4), e1788 (2022)
22. Kihara, S., Ghosh, S., McDougall, D.R., Whitten, A.E., Mata, J.P., Köper, I., McGillivray, D.J.: Structure of soft and hard protein corona around polystyrene nanoplastics—particle size and protein types. Biointerphases **15**(5) (2020)
23. Rana, K.L., Kour, D., Yadav, N., Yadav, A.N.: Endophytic microbes in nanotechnology: current development, and potential biotechnology applications. In: Microbial Endophytes, pp. 231–262. Woodhead Publishing (2020)
24. Athanasiadou, D., Carneiro, K.M.M.: DNA nanostructures as templates for biomineralization. Nat. Rev. Chem. **5**(2), 93–108 (2021)

25. Li, F., Li, J., Dong, B., Wang, F., Fan, C., Zuo, X.: DNA nanotechnology-empowered nanoscopic imaging of biomolecules. Chem. Soc. Rev. **50**(9), 5650–5667 (2021)
26. Rizwan, K., Rasheed, T., Bilal, M.: Nano-biodegradation of polymers. In: Biodegradation and Biodeterioration at the Nanoscale, pp. 213–238. Elsevier (2022)
27. Glaskova-Kuzmina, T., Starkova, O., Gaidukovs, S., Platnieks, O., Gaidukova, G.: Durability of biodegradable polymer nanocomposites. Polymers **13**(19), 3375 (2021)
28. Mohammadi Nasr, S., Rabiee, N., Hajebi, S., Ahmadi, S., Fatahi, Y., Hosseini, M., Bagherzadeh, M., et al.: Biodegradable nanopolymers in cardiac tissue engineering: from concept towards nanomedicine. Int. J. Nanomed. 4205–4224 (2020)
29. Modrak, M., Trebuňová, M., Balogova, A.F., Hudák, R., Živčák, J.: Biodegradable materials for tissue engineering: development, classification and current applications. J. Funct. Biomater. **14**(3), 159 (2023)
30. Afzal, O., Altamimi, A.S.A., Nadeem, M.S., Alzarea, S.I., Almalki, W.H., Tariq, A., Mubeen, B., et al.: Nanoparticles in drug delivery: from history to therapeutic applications. Nanomaterials **12**(24), 4494 (2022)
31. Su, S., Kang, P.M.: Systemic review of biodegradable nanomaterials in nanomedicine. Nanomaterials **10**(4), 656 (2020)
32. Jafarzadeh, S., Jafari, S.M.: Impact of metal nanoparticles on the mechanical, barrier, optical and thermal properties of biodegradable food packaging materials. Crit. Rev. Food Sci. Nutr. **61**(16), 2640–2658 (2021)
33. Fahmy, H.M., Eldin, R.E.S., Serea, E.S.A., Gomaa, N.M., AboElmagd, G.M., Salem, S.A., Elsayed, Z.A., Edrees, A., Shams-Eldin, E., Shalan, A.E.: Advances in nanotechnology and antibacterial properties of biodegradable food packaging materials. RSC Adv. **10**(35), 20467–20484 (2020)
34. Musa, A.A., Bello, A., Adams, S.M., Onwualu, A.P., Anye, V.C., Bello, K.A., Obianyo, I.I.: Nano-enhanced polymer composite materials: a review of current advancements and challenges. Polymers **17**(7), 893 (2025)
35. Hassan, M.M., Mahajan, A.: Gecko-inspired adhesives: mechanical principles, synthetic fabrications, and applications. J. Adhes. 1–49 (2025)
36. Zheng, Z., Gu, X., Yang, S., Wang, Y., Zhang, Y., Han, Q., Cao, P.: Underwater drag reduction applications and fabrication of bio-inspired surfaces: a review. Biomimetics **10**(7), 470 (2025)
37. Yeh, Y.-C., Huang, T.-H., Yang, S.-C., Chen, C.-C., Fang, J.-Y.: Nano-based drug delivery or targeting to eradicate bacteria for infection mitigation: a review of recent advances. Front. Chem. **8**, 286 (2020)
38. Chen, C., Ng, D.Y.W., Weil, T.: Polymer bioconjugates: modern design concepts toward precision hybrid materials. Prog. Polym. Sci. **105** (2020)
39. Manzari, M.T., Shamay, Y., Kiguchi, H., Rosen, N., Scaltriti, M., Heller, D.: Targeted drug delivery strategies for precision medicines. Nat. Rev. Mater. **6**(4), 351–370 (2021)
40. Mundekkad, D., Mallya, A.R.: Biomimicry at the nanoscale—a review of nanomaterials inspired by nature. Nano Trends 100119 (2025)
41. Pesode, P., Barve, S., Wankhede, S.V., Ahmad, A.: Sustainable materials and technologies for biomedical applications. Adv. Mater. Sci. Eng. **2023**(1), 6682892 (2023)
42. Wang, H.-M., Yuan, T.-Q., Song, G.-Y., Sun, R.-C.: Advanced and versatile lignin-derived biodegradable composite film materials toward a sustainable world. Green Chem. **23**(11), 3790–3817 (2021)
43. Darie-Niţă, R.N., Frąckowiak, S.: An overview of potential applications of environmentally friendly hybrid polymeric materials. Polymers **17**(2), 252 (2025)
44. Ariga, K., Jia, X., Song, J., Hill, J.P., Leong, D.T., Jia, Y., Li, J.: Nanoarchitectonics beyond self-assembly: challenges to create bio-like hierarchic organization. Angewandte Chemie Int. Edn. **59**(36), 15424–15446 (2020)
45. Talreja, S., Tiwari, S.: Supramolecular chemistry: unveiling the fascinating world of non-covalent interactions and complex assemblies. J. Pharm. Pharmacol. Res **7**, 133–139 (2023)
46. Gaspar, V.M., Lavrador, P., Borges, J., Oliveira, M.B., Mano, J.F.: Advanced bottom-up engineering of living architectures. Adv. Mater. **32**(6), 1903975 (2020)

47. Levin, A., Hakala, T.A., Schnaider, L., Bernardes, G.L., Gazit, E., Knowles, T.P.: Biomimetic peptide self-assembly for functional materials. Nat. Rev. Chem. **4**(11), 615–634 (2020)
48. Miller, J.G.: Design and characterization of natural and naturally-inspired helical protein assemblies. Ph.D. Dissertation, Emory University (2023)
49. Kortemme, T.: De novo protein design—from new structures to programmable functions. Cell **187**(3), 526–544 (2024)
50. Yuca, E., Khan, A., Hacıosmanoğlu, N., Şeker, U.O.S.: Design and construction of protein and peptide-based self-assembled nanostructures. In: Design, Principle and Application of Self-Assembled Nanobiomaterials in Biology and Medicine, pp. 193–204. Academic Press (2022)
51. Liang, W., Zhou, C., Liu, X., Xie, Q., Xia, L., Li, Q., Lin, H., et al.: Synthetic nanopillars for stimulating osteoblast activity and osteointegration in bone-related disorders. Int. J. Nanomed. 2205–2223 (2025)
52. TL, S., Jagajjanani Rao, K., Korumilli, T.: Natural biogenic templates for nanomaterial synthesis: advances, applications, and environmental perspectives. ACS Biomater. Sci. Eng. **11**(3), 1291–1316 (2025)
53. Will, T.A.: Solid-State Shear Pulverization: A Comprehensive Study on Polymer Processing Technology from a Systematic Innovation Perspective (2023)
54. Nadeem, F.: Review on polymer processing techniques using supercritical carbon dioxide (scCO2). Pak. J. Chem **11**(1–4), 01–09 (2021)
55. Ghavidel, N., Fatehi, P.: Recent developments in the formulation and use of polymers and particles of plant-based origin for emulsion stabilizations. Chemsuschem **14**(22), 4850–4877 (2021)
56. Zhang, S., Qamar, S.A., Junaid, M., Munir, B., Badar, Q., Bilal, M.: Algal polysaccharides-based nanoparticles for targeted drug delivery applications. Starch-Stärke **74**(7–8), 2200014 (2022)
57. Blanco, F.G., Hernandez, N., Rivero-Buceta, V., Maestro, B., Sanz, J.M., Mato, A., Hernandez-Arriaga, A.M., Auxiliadora Prieto, M.: From residues to added-value bacterial biopolymers as nanomaterials for biomedical applications. Nanomaterials **11**(6), 1492 (2021)
58. Hua, S.: Advances in oral drug delivery for regional targeting in the gastrointestinal tract-influence of physiological, pathophysiological and pharmaceutical factors. Front. Pharmacol. **11**, 524 (2020)
59. Piroonpan, T., Rimdusit, P., Taechutrakul, S., Pasanphan, W.: PH-responsive water-soluble chitosan amphiphilic core–shell nanoparticles: radiation-assisted green synthesis and drug-controlled release studies. Pharmaceutics **15**(3), 847 (2023)
60. Huang, T., Zeng, Y., Li, C., Zhou, Z., Xu, J., Wang, L., Wang, K.: Application and development of electrospun nanofiber scaffolds for bone tissue engineering. ACS Biomater. Sci. Eng. **10**(7), 4114–4144 (2025)
61. Gideon, O., Samuel, H.S., Okino, I.A.: Biocompatible materials for next-generation biosensors. Discov. Chem. **1**(1), 34 (2024)
62. Deshmukh, R.K., Gaikwad, K.K.: Natural antimicrobial and antioxidant compounds for active food packaging applications. Biomass Convers. Biorefin. **14**(4), 4419–4440 (2024)
63. Zhong, J., McClements, D.J., He, K., Zhang, Z., Zhang, R., Qiu, C., Long, J., et al.: Innovative applications of bio-inspired technology in bio-based food packaging. Crit. Rev. Food Sci. Nutr. 1–14 (2025)
64. Rose, A., Rana, S., Pisupati, H., Vinodkumar, V., Nani, D., Munjal, S.: Nanotechnology in pesticide degradation: mechanisms, applications, and environmental implications. Pure Appl. Chem. **0** (2025)
65. Durgadevi, P., Girigoswami, K., Girigoswami, A.: Biodegradable nanomaterials in boosting seed vigor and germination: seed coating towards sustainability. Discov. Appl. Sci. **7**(7), 695 (2025)
66. Agbna, G.H., Zaidi, S.J.: Hydrogel performance in boosting plant resilience to water stress—a review. Gels **11**(4), 276 (2025)

67. Khezerlou, A., Tavassoli, M., Sani, M.A., Mohammadi, K., Ehsani, A., McClements, D.J.: Application of nanotechnology to improve the performance of biodegradable biopolymer-based packaging materials. Polymers **13**(24), 4399 (2021)

Emerging Nano Polymer Technologies for Synthesis and Characterization of Polymer Nanoparticles

Ankit Tyagi, Anant Bhardwaj, Rohit Sahu, and Chanchal Ahlawat

Abstract Recent study offers a thorough examination of their synthesis methods, salient features, and real-world applications. It examines many methods for reinforcing polymer matrixes and emphasizes how these methods enhance the thermal, mechanical, barrier qualities and electrical, of the resultant materials. Thanks to innovative production methods, these high-tech composites with improved qualities have found useful applications in many fields, such as biomedical, wastewater treatment, petroleum, automotive, and aerospace. Thermomechanical, optoelectronic, and magnetic features, as well as a high density, enhanced strength, huge surface area, and greater elastic modulus make these materials attractive for innovative uses. The evolution of polymer composite synthesis and production is explored in depth in this book. Biomedical engineering, aerospace, the automobile industry, and energy storage are just a few of the several additional sectors that use these cutting-edge materials. The field's prospects and problems are brought to light via discussions of possible advances and planned research initiatives. By revealing heretofore untapped regions and merging current information, this study increases our understanding of and development in the manufacture of polymer composites, hence addressing substantial industrial and technical needs.

Keywords Carbon composite · Nanoparticle · Polymer composite · Synthesis · Applications

A. Tyagi · A. Bhardwaj
Department of Mechanical Engineering, Poornima College of Engineering, Jaipur, Rajasthan, India

R. Sahu (✉)
Department of Mechanical Engineering, G.L Bajaj Institute of Technology and Management, Greater Noida, Uttar Pradesh, India
e-mail: rohitsahu87@gmail.com

C. Ahlawat
School of Computer Science Engineering & Technology, Bennett University, Greater Noida, Uttar Pradesh, India
e-mail: chanchalchaudhary827@gmail.com

B. Sharma et al. (eds.), *Nanostructured Polymers*, Engineering Materials,
https://doi.org/10.1007/978-981-95-3581-1_3

1 Introduction

Polymer composite technology has grown, carbon-based composites have received substantial attention owing to their better structural, chemical, and physical qualities over its component components. These composites may find usage in a variety of sectors, including wastewater treatment, petroleum, automotive, aerospace, and construction, owing to their increased properties [1–3]. Carbon-based polymer-based composites and foams have emerged as a viable material class with a number of industrial uses [4]. These cutting-edge materials combine the benefits of carbon-based reinforcements, as carbon fibres, CNTs, and graphene, with the flexibility and processing simplicity of polymers [5].

Significant gains in mechanical strength and barrier qualities have been made by integrating these carbon compounds into polymer matrices, which makes them ideal for a variety of applications. Microwave absorption, reinforcement in NFRC, solar cells, EMI shielding, adsorbents, corrosion resistance, field-emission devices, chemical sensors, and hydrogen storage are among the documented possible applications [6–28]. Both CNTs' adherence to polymer surfaces and their dispersion within polymer matrices rely on their nanoscale properties. Contrary to what is expected theoretically for individual CNTs, actual data reveal that the production of bundles and agglomerates may adversely affect the mechanical characteristics of the resultant composites. Foams are materials containing distributed gas bubbles in a liquid or solid substrate. They are frequently linked with being inexpensive, portable, and energy-efficient. In addition to these qualities, foams offer substantial advantages including soundproofing, thermal insulation, energy absorption, and unique mechanical properties [29]. These days, polymer foam is the most widely used sort of foam, utilised in packaging, medical equipment, upholstery, building materials, and automobile components. The worldwide polymer foam market is predicted to develop rapidly, reaching roughly $90.7 billion in 2020 and an estimated $114.8 billion by 2025 [30]. Polyurethane (PU) foam, one of the most prominent thermosetting foams, is well-known for its good mechanical qualities, simplicity of manufacture, and possibility for customization [31–34]. The industry has also experienced a surge in thermoplastic foams, including PP foams for automotive applications, TPE foams for sportswear, and PS foams for thermal insulation [35–41].

Recent advancements in material science and nanotechnology have hastened the development of synthesis techniques and the widespread usage of polymer composites. These materials are acknowledged as advanced functional composites owing to their huge surface area, high elastic modulus, thermal stability, magnetic characteristics, flame retardancy, enhanced density, and optoelectronic capabilities. Nanotechnology research has placed polymer nanocomposites at the forefront of innovation. Nanofillers function as high-performance reinforcing agents in bioresins by creating a synergistic link, increasing the mechanical properties of bio-composites [42, 43]. In terms of mechanical strength, fire resistance, heat distortion temperature, and thermal stability, these nano-biocomposites perform better than standard polymer composites. Nanoclay may be disseminated into bioresins by either intercalation or exfoliation; exfoliation delivers greater increases in mechanical and physical characteristics. Exfoliated nanoclay structures give superior reinforcing than intercalated morphologies, according to a number of research on nanoclay-modified bioresins [44, 45]. PLA, chitosan, soy protein, PU, starch, PCL, and other synthetic and bio-based polymeric components have all been employed to generate nanoclay-reinforced composites [46–53].

When we talk about polymer, Polymers are large macromolecules composed of repeating structural units known as monomers. They play a crucial role in modern life, serving as the foundation of materials used in various industries such as packaging, construction, medicine, and electronics. These materials can be naturally occurring, like proteins and DNA, or synthetically created, such as plastics and synthetic rubbers. Their adaptability, strength, and cost efficiency have made them indispensable across multiple sectors. The origin, structure, polymerization process, and physical characteristics of polymers are used to classify them in various ways. According to their origin, they are both synthetic and natural. They are cross-linked, branching, and linear in structure. They are addition and condensation polymers, depending on the polymerization method. They are also categorized as thermoplastic and thermosetting according to their physical characteristics. The many categories of polymers include:

1.1 Based on Origin

- Proteins, cellulose, starch, and natural rubber are examples of natural polymers, which are found in nature. In living things, they are essential for both structural support and biological processes.
- Man-made materials such as polyethylene, polystyrene, and polyvinyl chloride (PVC) are examples of synthetic polymers. They are designed to serve particular functions.

1.2 Based on Structure

- Linear Polymers: These polymers are strong and stiff due to their straight-chain structure. Polyvinyl chloride (PVC) and high-density polyethylene (HDPE) are two examples.
- •Branched Polymers: The density and flexibility of these polymers are influenced by their side chains. One example is low-density polyethylene (LDPE).
- Cross-linked Polymers: These are extremely strong and stable polymer chains joined by covalent bonds. Epoxy resins and vulcanized rubber are two examples.

1.3 Based on Polymerization Process

- Addition polymers are created when monomers with double or triple bonds react with one another without producing any byproducts. Polyethylene and polypropylene are two examples.
- Condensation Polymers: These are created when tiny molecules like methanol or water are released during a condensation reaction. Polyester and nylon are two examples.

1.4 Based on Physical Properties

- Thermoplastics: These polymers are recyclable and moldable because they soften when heated and solidify when cooled. Polypropylene and polystyrene are two examples.
- Thermosetting Polymers: These become robust and heat-resistant when heated through irreversible chemical changes. Epoxy and bakelite resins are two examples.

2 Properties of Polymers

Polymers exhibit a variety of characteristics that determine their applications. These properties depend on the polymer type, molecular weight, and intermolecular forces.

2.1 Mechanical Properties

Polymers can be rigid or flexible based on their molecular structure. High-density polymers are generally strong and resistant to wear, whereas low-density ones are lightweight and more flexible.

2.2 Thermal Properties

Thermoplastic polymers soften at elevated temperatures and can be reshaped, whereas thermosetting polymers retain their structure due to cross-linking. Some, like PTFE (Teflon), have high heat resistance.

2.3 *Chemical Resistance*

Many polymers are highly resistant to chemicals, making them ideal for packaging, piping, and protective coatings. For example, polyethylene withstands acids and bases, while fluoropolymers resist harsh chemicals.

2.4 *Electrical Properties*

Certain polymers, such as rubber and polyethylene, are excellent insulators, making them essential for electrical applications. Conductive polymers, like polyaniline, are used in sensors and flexible electronics.

3 Applications of Polymers

Polymers are present in nearly every aspect of daily life and industry. Their applications range from basic consumer products to advanced technologies.

3.1 *Packaging*

Plastics such as polyethylene, polypropylene, and polystyrene dominate the packaging sector due to their lightweight, durability, and moisture resistance. They are used for containers, bottles, and films.

3.2 *Construction*

Polymers contribute significantly to construction materials. PVC is used for pipes and fittings, while polystyrene serves as insulation. Acrylic polymers are utilized in durable paints and coatings.

3.3 *Automotive Industry*

Vehicle weight can be decreased with the use of lightweight polymers, improving fuel economy. Seat cushions are made of polyurethane, whilst bumpers and dashboards are made of polypropylene and polycarbonate.

3.4 Medicine and Healthcare

In medical applications such as drug delivery systems, biodegradable sutures, and prosthetics, polymers are essential. Pharmaceutical formulations use materials like polyvinyl alcohol (PVA) and polyethylene glycol (PEG).

3.5 Electronics

Conductive polymers are increasingly being used in OLED displays, flexible electronics, and sensors. Polymers with high dielectric strength, such as polytetrafluoroethylene (PTFE), are used for cable insulation.

4 Environmental Impact of Polymers

Notwithstanding their benefits, synthetic polymers provide serious environmental risks, especially in relation to pollution and plastic waste.

4.1 Plastic Pollution

Because plastics take centuries to break down, they accumulate in landfills and contaminate the ocean. Oceans have been shown to contain micro plastics, which damage marine ecosystems and get into the food chain.

4.2 Recycling and Waste Management

Mechanical and chemical methods are being used to improve polymer recycling. As sustainable substitutes, biodegradable polymers like polylactic acid (PLA) and polyhydroxyalkanoates (PHA) are being investigated.

4.3 Sustainable Polymer Innovations

The goal of research is to create bio-based polymers from renewable resources like cellulose and maize starch. The creation of environmentally friendly polymers is also being aided by developments in green chemistry.

5 Future of Polymers

Polymers continue to change as a result of continuous technical breakthroughs, spurring innovation across many industries. The future is being shaped by the creation of nanocomposites, self-healing materials, and smart polymers.

5.1 Smart Polymers

These materials are useful in medication delivery systems and adaptable coatings because they react to external stimuli including temperature, pH, and light.

5.2 Self-healing Polymers

Self-healing materials can fix damage on their own, increasing product longevity and lowering maintenance expenses. Applications for this technology in coatings and automobiles are very beneficial.

5.3 Nanocomposites

Applications in electronics, medical devices, and aircraft result from the addition of nanoparticles to polymers, which improves their mechanical, thermal, and barrier qualities.

Because of their strength, affordability, and versatility, polymers have revolutionized a number of sectors. Although their effects on the environment are still a worry, advancements in recyclable materials and recycling techniques are assisting in resolving these problems. We should expect much more innovative uses of polymer science as it develops further, which will shape materials and technology in the future.

In reference to nanoparticles, Usually between 1 and 100 nm in size, nanoparticles are minuscule materials that differ from their bulk counterparts in specific physical, chemical, and biological characteristics. They have a higher surface-area-to-volume ratio because of their nanoscale size, which makes them extremely reactive and adaptable. With uses in the fields of medicine, electronics, energy, and environmental sciences, these properties have put nanoparticles at the front of scientific advancement. Numerous industries have seen revolutionary breakthroughs as a result of the study and use of nanoparticles, which has expanded the subject of nanotechnology to new heights.

6 Classification of Nanoparticles

Depending on their composition, structure, and functional characteristics, nanoparticles can be categorized; choosing the right nanoparticles for a given application requires an understanding of these classifications.

6.1 Based on Composition

Due to their exceptional conductivity and antibacterial qualities, metallic nanoparticles—which are made of metals like gold, silver, and platinum—have a wide range of uses in electronics, medicine, and catalysis.

Ceramic nanoparticles: These nanoparticles, which include carbides, nitrides, and oxides, are valuable in medication delivery systems, coatings, and sensors because of their exceptional stability and heat resistance.

Polymeric Nanoparticles: These organic polymer-based nanoparticles are widely used in medicines, especially in gene therapy and controlled drug release.

Carbon-Based Nanoparticles: These nanoparticles, which include fullerenes, graphene, and carbon nanotubes, are prized for their remarkable mechanical, electrical, and thermal characteristics.

6.2 Based on Structure

Core–Shell Nanoparticles: These nanoparticles are more stable and functional in industrial and biomedical applications because their core substance is encased in a distinct shell material.

Quantum Dots: These semiconductor nanoparticles are very useful for imaging, biosensing, and display technologies because of their size-dependent optical characteristics.

Dendrimers: Because of their carefully regulated molecular structure, these highly branched polymeric nanoparticles are used in targeted drug delivery and nanomedicine.

7 Properties of Nanoparticles

The unique properties of nanoparticles contribute to their extensive applications across various fields. These characteristics include:

7.1 *Optical Properties*

Nanoparticles, particularly quantum dots, exhibit size-dependent fluorescence, which is extensively used in imaging, biosensing, and display technologies.

7.2 *Mechanical Strength*

Composite materials are perfect for structural reinforcement in building and aerospace applications because nanoparticles increase their mechanical strength and endurance.

7.3 *Thermal Stability*

Excellent heat resistance is a feature of many nanoparticles, which is useful for applications like fireproof coatings and insulation in high-temperature settings.

7.4 *Electrical Conductivity*

Advanced electrical and electronic applications, like flexible circuits and high-efficiency batteries, are made possible by carbon nanotubes, silver nanoparticles, and other conductive nanoparticles.

8 Applications of Nanoparticles

The diverse properties of nanoparticles enable their widespread use in various industries and scientific disciplines. Below are some of their most significant applications.

8.1 *Medicine and Healthcare*

Targeted Drug Delivery: By guaranteeing targeted delivery to particular cells or tissues and reducing negative effects, nanoparticles improve therapeutic efficacy.

Cancer Treatment: In photothermal therapy, gold and silver nanoparticles are used to specifically kill malignant cells.

Medical Imaging: Quantum dots and magnetic nanoparticles enhance imaging methods like fluorescence and magnetic resonance imaging (MRI) for precise disease diagnosis.

Antibacterial Coatings: To stop bacterial infections, silver and copper nanoparticles are included into textiles and medical equipment.

Gene Therapy: The potential of polymeric and lipid-based nanoparticles to transport genetic material to cure inherited illnesses is being investigated.

8.2 Electronics

Nanoelectronics: Faster and more effective microchips have been created as a result of the use of nanoparticles to reduce the size of electronic components.

Flexible Displays: Nanoparticles of graphene and silver improve the functionality of lightweight, flexible display displays.

Wearable Sensors: The creation of high-sensitivity biosensors for wearable health monitoring systems depends heavily on nanoparticles.

Quantum Computing: Quantum dots for quantum computing applications are being created using sophisticated nanoparticles, opening the door to ground-breaking data processing capabilities.

High-Performance Batteries: Lithium-ion and solid-state batteries are incorporating nanoparticles to enhance conductivity and energy storage.

8.3 Environmental Applications

Water Purification: Water quality is improved by nanoparticles' assistance in eliminating contaminants and heavy metals.

Air Filtration: Fine contaminants are captured by nano-based air filters, improving the quality of the air in cities and industries.

Renewable Energy: The efficiency of next-generation battery technology, solar panels, and hydrogen storage systems is enhanced by nanoparticles.

Waste Management: Nanoparticles help in the breakdown of hazardous materials and facilitate eco-friendly waste disposal.

Carbon Capture: Nanoparticles facilitate environmentally friendly waste disposal and aid in the breakdown of hazardous materials.

9 Environmental and Health Concerns

Nanoparticles have enormous advantages, yet they can also be harmful to the environment and human health. For their proper use, these issues must be resolved.

Toxicity: Certain nanoparticles have the potential to accumulate in biological systems and have cytotoxic effects, which could be harmful to human health.

Environmental Persistence: Concerns regarding the long-term ecological effects of nanoparticles are raised by their environmental stability.

Regulatory Challenges: To reduce these concerns, ongoing research attempts to develop standards for the safe manufacture, use, and disposal of nanoparticles.

Bioaccumulation: Some nanoparticles have the ability to linger in ecosystems, possibly making their way into food chains and having unidentified biological impacts.

Workplace Exposure: Strict safety regulations must be put in place by industries handling nanoparticles in order to reduce workplace risks.

It is anticipated that science and technology will undergo revolutionary transformations as nanoparticles continue to evolve. Among the encouraging developments are:

Smart Drug Delivery Systems: Personalized medicine will be transformed by developments in drug carriers based on nanoparticles, which will provide targeted therapeutic interventions.

Advanced Nanomaterials: The next generation of nanoparticle-infused materials will enhance the durability and efficiency of aerospace, automotive, and construction applications.

Sustainable Nanotechnology: Research is underway to develop biodegradable and eco-friendly nanoparticles, reducing their environmental footprint.

Nanorobotics: The integration of nanoparticles into nanorobots will enable precise medical interventions, such as targeted drug delivery and non-invasive surgeries.

10 Self-healing Materials

Nanoparticles are being added to coatings and materials that have the ability to mend themselves on their own, increasing their durability across a range of sectors.

The use of nanoparticles for lightweight, high-strength materials in spacecraft and space habitats is being studied by NASA and other space agencies.

At the forefront of contemporary technology, nanoparticles provide ground-breaking answers in the fields of environmental sciences, electronics, and medicine. Although their special qualities allow for a wide range of uses, strict restrictions and

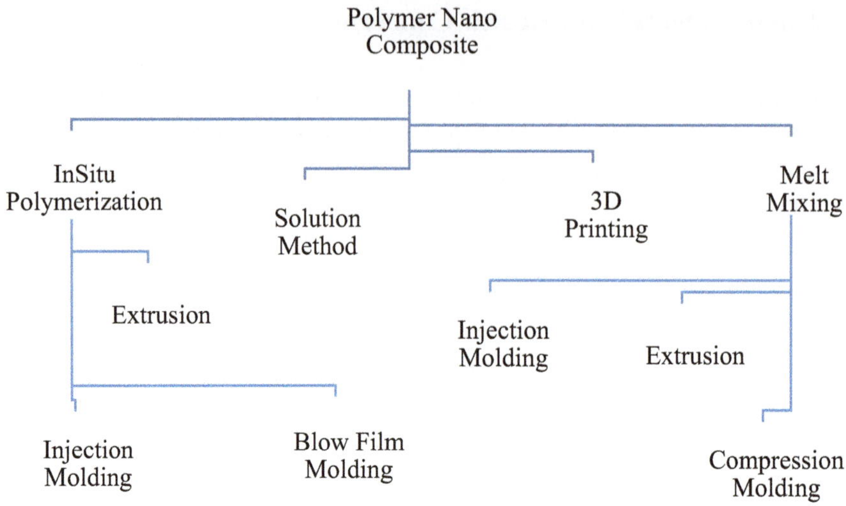

Fig. 1 Synthesis techniques for polymer nanocomposites

responsible innovation are necessary to meet safety and environmental concerns. Nanoparticles have the potential to revolutionize technological developments as study continues, influencing both science and industry in the future. They remain one of the most fascinating areas of research due to their adaptability and promise, with several ground-breaking applications anticipated in the ensuing decades.

This review study presents a complete overview of carbon-based polymer composites. It looks at their synthesis processes, characteristics, and uses in an aim to be a useful resource for researchers, scientists, and engineers. The knowledge gained from this examination will be beneficial in generating new materials and expanding the industry. Figure 1 shows various synthesis techniques of polymer nano composites.

11 Synthesis Techniques for Polymer Nanocomposites

Carbon-based composites may be integrated into polymer matrices using a number of synthetic processes to generate materials with specified characteristics. Some of the criteria that determine the choosing of a given technique include the ideal material quality, processing parameters, and the compatibility of carbon additions with the polymer matrix. Much research has been done on the enhancement of polymer matrices by carbon nanotubes (CNTs) in order to generate CNT-reinforced polymer composites. Because of their nanoscale dispersion, which produces an unusually wide surface area inside nanocomposites, CNTs give benefits over typical carbon-based micron fillers.

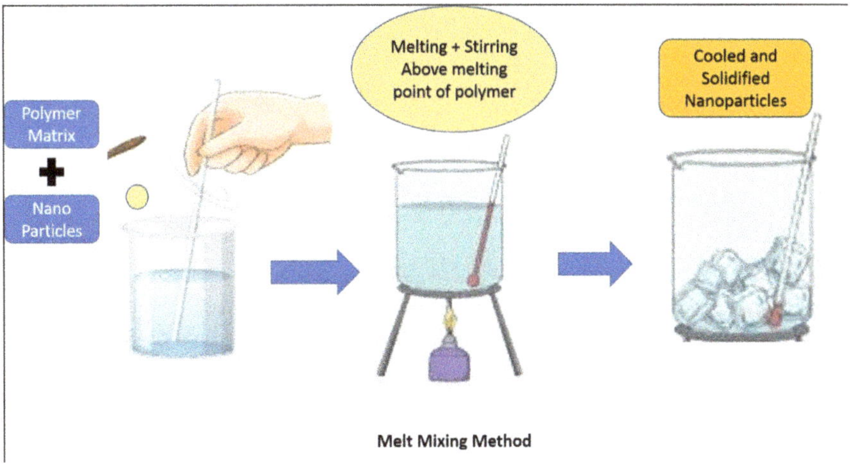

Fig. 2 Melt mixing method

11.1 Melt Mixing

Molten polymer matrices and carbon-based chemicals are directly blended; this process is called “melt mixing.” This technique, which is often used to thermoplastic polymers, entails heating the polymer matrix to its melting point before utilizing to add and distribute carbon components. Melt mixing ensures excellent dispersion of carbon compounds, but due to the high processing temperatures, its use may be limited to polymers with sufficient thermal stability. To disintegrate CNT bundles and disperse them uniformly throughout the polymer matrix, this method typically comprises combining CNTs with a polymer matrix and applying shear forces using Banbury mixer [54–56]. Figure 2 describes the melt mixing method pictorially.

11.2 In Situ Polymerization

In situ polymerisation can polymerise with carbon fillers because it forms a polymer matrix when carbon-based components are present. Before beginning the polymerisation process, carbon elements may be added directly in the monomer solution. Applying functional groups compatible with the monomer to the surfaces of carbon nanotubes (CNTs) may enhance their dispersion, which is a major benefit of in situ polymerization [57–59]. The method is depicted in Fig. 3.

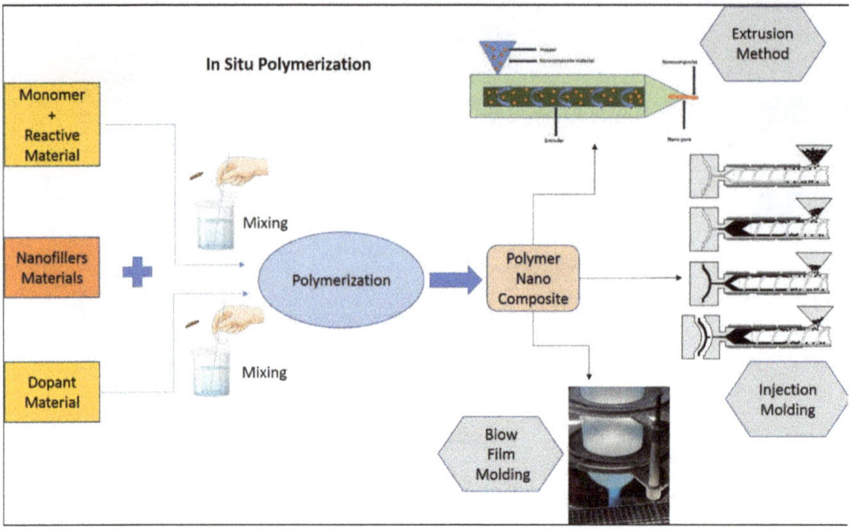

Fig. 3 In situ polymerization

12 Application of Polymer Nanocomposites

Because of their special combination of qualities, carbon-integrated polymer composites are very adaptable materials with a wide range of uses in many sectors. The following sections go over the main areas in which these materials may be used in aerospace and aviation, energy storage, automotive industry, military, biomedical engineering, marine industry [60–70]. Figure 4 shows application of Polymer Nanocomposites.

13 Future Perspective and Challenges of Polymer Nanocomposites

Researchers can enhance interfacial bonding, techniques hold a great lot of potential to boost material performance, which may lead to the identification of new applications and economic prospects. However, one of the main hurdles is the high cost of carbon-based components, especially for large-scale manufacturing. The economic feasibility of these materials hinges on the development of technologies for the manufacturing of carbon fillers and their insertion into polymer matrices. Scaling up the manufacture of carbon-integrated foams and composites is another tough problem. It is crucial to have efficient manufacturing processes that can dependably generate enormous volumes of high-quality items. Robust interfacial bonding and homogeneous dispersion across polymer matrices are needed for these materials to attain

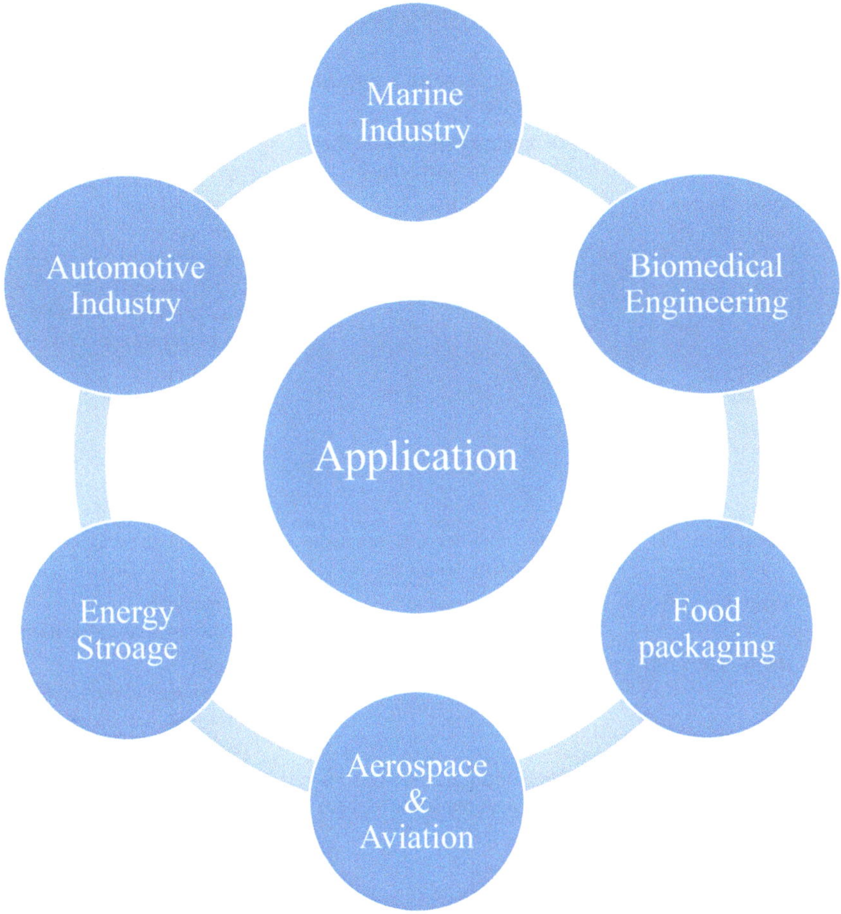

Fig. 4 Application of polymer nanocomposites

their full potential [71]. When compared to composites manufactured with separate carbon fillers, these hybrid composites may have synergistic effects that result in better mechanical, thermal, and electrical characteristics [72]. These hybrid systems may lead to ground-breaking applications in a variety of sectors as this area of study develops. Another tactic to improve the characteristics and performance of these materials is to functionalise carbon fillers with certain chemical groups or additions.

The main issues that need to be resolved are achieving ideal dispersion, controlling agglomeration, and enhancing interfacial interactions. Concerns have also been expressed about the effects of carbon-integrated polymer composites and foams on the environment, specifically with respect to waste management and the energy-intensive character of their production methods [73]. Table 1 shows the latest applications and synthesis method for nanoparticles.

Table 1 Latest applications of nanocomposites along with technology used in synthesizing nanoparticles

Sr. No	Authors	Technology used for synthesizing nanoparticles	Application
1	Doris et al. [74] (2024)	• Template-assisted method • Polymer-melt technique	• Potential use in optoelectronic devices
2	Ravichandran and Santhosh [75]	Melt intercalation, sol–gel processes In situ polymerization	Enhancement of • thermal, • mechanical, • barrier properties in polymer nanocomposites for industrial applications
3	Uddin [76]	Nanoclays	Applications in • aerospace, • automotive, • construction, • biomedical, • food packaging industries
4	Wang et al. [77]	Interface engineering composite solid electrolytes	All-solid-state batteries
5	Yashtulov [78]	Synthesis polymeric nanocomposites	• biomedical • environmental applications
6	Batlle et al. [79]	Iron-oxide-based magnetic nanoparticles	• health, • environmental remediation, • ultra-high-density magnetic recording
7	George et al. [80]	Carbon dots integration (Pr3)+ gahnite nanophosphor (doped)	• Thermal sensing, • data fortification, • fingerprint visualization analysis
8	September et al. [81]	silica + silicon from sugarcane bagasse ash	• sustainable material development
9	Pagar et al. [82]	CdO nanoparticles + Citrus sinensis peel extract	• Degradation of Rhodamine B dye in wastewater treatment
10	Sarifujjaman et al. [83]	copper oxide nanoparticles using leaf extracts of Dillenia Indica + Mikania Micrantha	• Potential applications as photocatalysts • antibacterial agents
11	Mallikarjunaswamy et al. [84]	lanthanum oxide nanoparticles	• Photocatalytic • electrochemical applications • energy sectors
12	Saridewi et al. [85]	Sansevieria trifasciata extract + $ZnFe_2O_4$ nanoparticles	• photocatalyst • dye degradation in wastewater

(continued)

Table 1 (continued)

Sr. No	Authors	Technology used for synthesizing nanoparticles	Application
13	Bopape et al. [86]	Gold + silver-incorporated carbon-based zinc oxide nanocomposites	• Photo-degradation of textile dyes Pharmaceuticals
14	Gidey et al. [87]	Rice husk + mesoporous silica materials	Used in • catalysis, • adsorption, • drug delivery systems

Nanoparticles are minute materials, typically ranging between 1 and 100 nm in size, that exhibit unique physical, chemical, and biological properties distinct from their bulk counterparts. Due to their nanoscale dimensions, they have an increased surface-area-to-volume ratio, making them highly reactive and versatile. These characteristics have positioned nanoparticles at the forefront of scientific innovation, with applications spanning medicine, electronics, energy, and environmental sciences. The study and application of nanoparticles have led to groundbreaking advancements in various industries, propelling the field of nanotechnology into new dimensions.

The exploration of space presents numerous challenges, including extreme temperatures, radiation exposure, micrometeoroid impacts, and the need for lightweight, durable materials. Nanocomposites, which are materials reinforced with nanoparticles, play a crucial role in overcoming these challenges. Their remarkable mechanical strength, thermal stability, and resistance to radiation make them indispensable for spacecraft, satellites, and space missions.

- Lightweight and High-Strength Materials

One of the most significant advantages of nanocomposites in space technology is their ability to reduce weight while maintaining superior strength. Traditional materials used in spacecraft, such as aluminum and titanium alloys, are being replaced or reinforced with nanocomposites made of carbon nanotubes, graphene, and ceramic nanoparticles. These materials enhance structural integrity, allowing spacecraft to withstand high stress while reducing overall launch weight and fuel consumption.

- Thermal Protection Systems

Spacecraft experience extreme temperature variations, from the frigid cold of deep space to the intense heat upon re-entry into Earth's atmosphere. Nanocomposite materials, particularly those infused with carbon nanotubes and ceramic nanoparticles, provide excellent thermal insulation and heat resistance. These materials are used in thermal protection systems (TPS) for spacecraft, ensuring that astronauts and onboard equipment remain safe from thermal fluctuations.

- Radiation Shielding

Exposure to cosmic radiation and solar energetic particles poses a serious threat to both astronauts and spacecraft components. Traditional radiation shielding materials, such as lead and aluminum, add significant weight to space vehicles. Nanocomposites incorporating boron nitride nanotubes, graphene, and polymer-based nanomaterials offer lightweight alternatives with enhanced radiation absorption capabilities. These materials help protect astronauts and sensitive electronic systems from harmful radiation.

- Self-Healing Materials

The harsh space environment increases the likelihood of structural damage due to micrometeoroid impacts and material degradation. Self-healing nanocomposites, embedded with microcapsules containing healing agents or nanoparticles that react upon damage, provide a revolutionary solution. These materials can autonomously repair minor cracks and punctures, prolonging the lifespan of spacecraft and reducing maintenance requirements.

- Enhanced Spacecraft Electronics

Nanocomposites are also making significant contributions to the miniaturization and enhancement of spacecraft electronics. Conductive nanomaterials, such as silver and graphene nanoparticles, improve the efficiency of electronic circuits, sensors, and antennas. These advancements enable the development of more compact, high-performance electronic systems crucial for deep-space missions and satellite communications.

- Advanced Propulsion Systems

Efficient propulsion is essential for interplanetary travel and deep-space exploration. Nanocomposite materials are used in ion thrusters, solar sails, and next-generation propulsion systems to improve efficiency and durability. Carbon-based nanomaterials, such as graphene and carbon nanotubes, contribute to lightweight, high-strength components for propulsion systems, enhancing the thrust-to-weight ratio of spacecraft.

- Space Habitats and Life Support Systems

Future space exploration will require the establishment of long-term habitats on the Moon, Mars, and beyond. Nanocomposites are being integrated into habitat structures, radiation shields, and life support systems to ensure safety and sustainability. These materials contribute to efficient air and water filtration systems, lightweight structural components, and enhanced durability for extraterrestrial colonies.

As space exploration advances, the demand for innovative nanocomposite materials will continue to grow. Researchers are actively developing multifunctional nanocomposites that combine various properties, such as thermal resistance, self-healing capabilities, and radiation shielding, into a single material. The integration of artificial intelligence and nanotechnology is also expected to play a crucial role in designing adaptive materials for space applications.

Nanocomposites have revolutionized space technology by providing lightweight, high-strength, and multifunctional materials essential for spacecraft, propulsion systems, and extraterrestrial habitats. Their unique properties, including thermal resistance, radiation shielding, and self-healing capabilities, make them indispensable for future space exploration. As research continues, the development of advanced nanocomposites will play a pivotal role in enabling deep-space missions and establishing sustainable space colonies.

14 Conclusion

Polymer composites' remarkable qualities and adaptability have made them essential to many sectors. In the aerospace industry, their high strength-to-weight ratio improves fuel efficiency and aircraft performance. By using these materials to lighten vehicles, the automobile sector can increase fuel efficiency and lower pollution. Polymer composites, which provide durability and resilience to environmental variables, are used in civil engineering to strengthen structures. They are used in equipment that improves athlete performance by making it stronger and lighter, which benefits the sports business. Additionally, because of their biocompatibility and adaptability, polymer composites are used in biomedical applications for implants and prosthetics. Polymer composites' ongoing development and application in a variety of industries highlights their importance in contemporary innovation and technology.

A family of materials known for its exceptional qualities and broad range of applications are carbon-integrated polymer composites. Numerous production techniques are used, each with unique advantages. These technologies include 3D printing, electrospinning, melt and solution mixing, and in situ polymerization. Thermomechanical, optoelectronic, and magnetic features, as well as a high density, enhanced strength, huge surface area, and greater elastic modulus make these materials attractive for innovative uses. The evolution of polymer composite synthesis and production is explored in depth in this article. Biomedical engineering, aerospace, the automobile industry, and energy storage are just a few of the several additional sectors that use these cutting-edge materials. The field's prospects and problems are brought to light via discussions of possible advances and planned research initiatives. Future developments in hybrid systems, functionalization of carbon materials. However, issues with interfacial compatibility, scalability, cost-effective manufacturing, and environmental sustainability must be resolved before widespread use is feasible.

References

1. Okamoto, M.: Recent advances in polymer/layered silicate nanocomposites: an overview from science to technology. Mater. Sci. Technol. **22**, 756–779 (2006)

2. Rhim, J.-W., Hong, S.-I., Ha, C.-S.: Tensile, water vapor barrier and antimicrobial properties of PLA/nanoclay composite films. LWT Food Sci. Technol. (Lebensmittel-Wissenschaft-Technol.) **42**, 612–617 (2009)
3. Jamshidian, M., Tehrany, E.A., Imran, M., Akhtar, M.J., Cleymand, F., Desobry, S.: Structural, mechanical and barrier properties of active PLA–antioxidant films. J. Food Eng. **110**, 380–389 (2012)
4. Banerjee, R., Gebrekrstos, A., Orasugh, J.T., Ray, S.S.: Nanocarbon-containing polymer composite foams: a review of systems for applications in electromagnetic interference shielding, energy storage, and piezoresistive sensors. Ind. Eng. Chem. Res. **62**(18), 6807–6842 (2023)
5. Zhao, C., Mark, L.H., Kim, S., Chang, E., Park, C.B., Lee, P.C.: Recent progress in micro-/nano-fibrillar reinforced polymeric composite foams. Polym. Eng. Sci. **61**(4), 926–941 (2021)
6. Wu, Z., Yang, Z., Pei, K., Qian, X., Jin, C., Che, R.: Dandelion-like carbon nanotube assembly embedded with closely separated Co nanoparticles for high-performance microwave absorption materials. Nanoscale **12**(18), 10149–10157 (2020)
7. Janudin, N., Abdullah, N., Wan Yunus, W.M.Z., Yasin, F.M., Yaacob, M.H., Mohamad Saidi, N., Mohd Kasim, N.A.: Effect of functionalized carbon nanotubes in the detection of benzene at room temperature. J. Nanotechnol. **2018** (2018)
8. Mananghaya, M., Yu, D., Santos, G.N., Rodulfo, E.: Scandium and titanium containing single-walled carbon nanotubes for hydrogen storage: a thermodynamic and first principle calculation. Sci. Rep. **6**(1), 27370 (2016)
9. Zhou, E., Xi, J., Guo, Y., Liu, Y., Xu, Z., Peng, L., Gao, W., Ying, J., Chen, Z., Gao, C.: Synergistic effect of graphene and carbon nanotube for high-performance electromagnetic interference shielding films. Carbon **133**, 316–322 (2018)
10. Souto, L.F.C., Soares, B.G.: Polyaniline/carbon nanotube hybrids modified with ionic liquids as anticorrosive additive in epoxy coatings. Prog. Org. Coat. **143**, 105598 (2020)
11. Medupin, R.O., Abubakre, O.K., Abdulkareem, A.S., Muriana, R.A., Abdulrahman, A.S.: Carbon nanotube reinforced natural rubber nanocomposite for anthropomorphic prosthetic foot purpose. Sci. Rep. **9**(1), 20146 (2019)
12. Nurazzi, N.M., Harussani, M.M., Zulaikha, N.S., Norhana, A.H., Syakir, M.I., Norli, A.: Composites based on conductive polymer with carbon nanotubes in DMMP gas sensors—an overview. Polimery **66**(2), 85–97 (2021)
13. Maity, D., Rajavel, K., Kumar, R.T.R.: Polyvinyl alcohol wrapped multiwall carbon nanotube (MWCNTs) network on fabrics for wearable room temperature ethanol sensor. Sens. Actuators B **261**, 297–306 (2018)
14. Feng, D., Xu, D., Wang, Q., Liu, P.: Highly stretchable electromagnetic interference (EMI) shielding segregated polyurethane/carbon nanotube composites fabricated by microwave selective sintering. J. Mater. Chem. C **7**(26), 7938–7946 (2019)
15. Chen, M., Zhao, G., Shao, L.L., Yuan, Z.Y., Jing, Q.S., Huang, K.J., Huang, Z.Y., Zhao, X.H., Zou, G.D.: Controlled synthesis of nickel encapsulated into nitrogen-doped carbon nanotubes with covalent bonded interfaces: the structural and electronic modulation strategy for an efficient electrocatalyst in dye-sensitized solar cells. Chem. Mater. **29**(22), 9680–9694 (2017)
16. Yahya, M.S., Ismail, M.: Improvement of hydrogen storage properties of MgH2 catalyzed by K2NbF7 and multiwall carbon nanotube. J. Phys. Chem. C **122**(21), 11222–11233 (2018)
17. Song, Y., Li, J., Wu, Q., Yi, C., Wu, H., Chen, Z., Ou-Yang, W.: Study of film thickness effect on carbon nanotube based field emission devices. J. Alloy. Compd. **816**, 152648 (2020)
18. Kumar, R., Ansari, M.O., Barakat, M.A.: DBSA doped polyaniline/multi-walled carbon nanotubes composite for high efficiency removal of Cr (VI) from aqueous solution. Chem. Eng. J. **228**, 748–755 (2013)
19. Park, S., Gupta, A.P., Yeo, S.J., Jung, J., Paik, S.H., Mativenga, M., Kim, S.H., Shin, J.H., Ahn, J.S., Ryu, J.: Carbon nanotube field emitters synthesized on metal alloy substrate by PECVD for customized compact field emission devices to be used in X-ray source applications. Nanomaterials **8**(6), 378 (2018)

20. Abidin, M.S.Z., Herceg, T., Greenhalgh, E.S., Shaffer, M., Bismarck, A.: Enhanced fracture toughness of hierarchical carbon nanotube reinforced carbon fibre epoxy composites with engineered matrix microstructure. Compos. Sci. Technol. **170**, 85–92 (2019)
21. Chen, M., Wang, G.C., Shao, L.L., Yuan, Z.Y., Qian, X., Jing, Q.S., Huang, Z.Y., Xu, D.L., Yang, S.X.: Strategic design of vacancy-enriched Fe1–x S nanoparticles anchored on Fe3C-encapsulated and N-doped carbon nanotube hybrids for high-efficiency triiodide reduction in dye-sensitized solar cells. ACS Appl. Mater. Interfaces **10**(37), 31208–31224 (2018)
22. Guo, F., Kang, T., Liu, Z., Tong, B., Guo, L., Wang, Y., Liu, C., Chen, X., Zhao, Y., Shen, Y., Lu, W.: Advanced lithium metal–carbon nanotube composite anode for high-performance lithium–oxygen batteries. Nano Lett. **19**(9), 6377–6384 (2019)
23. Chen, M., Jing, Q.S., Sun, H.B., Xu, J.Q., Yuan, Z.Y., Ren, J.T., Ding, A.X., Huang, Z.Y., Dong, M.Y.: Engineering the core–shell-structured NCNTs-Ni2Si@ porous Si composite with robust Ni–Si interfacial bonding for high-performance li-ion batteries. Langmuir **35**(19), 6321–6332 (2019)
24. Chen, M., Wang, G.C., Yang, W.Q., Yuan, Z.Y., Qian, X., Xu, J.Q., Huang, Z.Y., Ding, A.X.: Enhanced synergetic catalytic effect of Mo2C/NCNTs@ Co heterostructures in dye-sensitized solar cells: fine-tuned energy level alignment and efficient charge transfer behavior. ACS Appl. Mater. Interfaces **11**(45), 42156–42171 (2019)
25. Hassan, A.G., Yajid, M.M., Saud, S.N., Bakar, T.A., Arshad, A., Mazlan, N.: Effects of varying electrodeposition voltages on surface morphology and corrosion behavior of multi-walled carbon nanotube coated on porous Ti-30 at.%-Ta shape memory alloys. Surf. Coat. Technol. **401**, 126257 (2020)
26. Mo, Z., Yang, R., Lu, D., Yang, L., Hu, Q., Li, H., Zhu, H., Tang, Z., Gui, X.: Lightweight, three-dimensional carbon Nanotube@ TiO_2 sponge with enhanced microwave absorption performance. Carbon **144**, 433–439 (2019)
27. Chen, M., Shao, L.L., Lv, X.W., Wang, G.C., Yang, W.Q., Yuan, Z.Y., Qian, X., Han, Y.Y., Ding, A.X.: In situ growth of Ni-encapsulated and N-doped carbon nanotubes on N-doped ordered mesoporous carbon for high-efficiency triiodide reduction in dye-sensitized solar cells. Chem. Eng. J. **390**, 124633 (2020)
28. Shubhadarshinee, L., Mohapatra, P., Behera, S., Jali, B.R., Mohapatra, P., Barick, A.K.: Review on synthesis and characterization of metal nanoparticles doped carbon nanofillers based nanohybrids reinforced polyaniline nanocomposites. Polym. Plast. Technol. Mater. 1–25 (2024)
29. Singh, N.P., Gupta, V.K., Singh, A.P.: Graphene and carbon nanotube reinforced epoxy nanocomposites: a review. Polymer **180**, 121724 (2019)
30. Silva, A.L., Bordado, J.C.: Recent developments in polyurethane catalysis: catalytic mechanisms review. Catal. Rev. **46**(1), 31–51 (2004)
31. Polymer foam market by resin type (PU, PS, PO, Phenolic), foam type (rigid, flexible), end-use industry (building & construction, packaging, automotive, furniture & bedding, footwear, sports & recreational), and region—global forecast to 2025 (2020)
32. Yang, C., Fischer, L., Maranda, S., Worlitschek, J.: Rigid polyurethane foams incorporated with phase change materials: a state-of-the-art review and future research pathways. Energy Build. **87**, 25–36 (2015). https://doi.org/10.1016/j.enbuild.2014.10.075
33. Zia, K.M., Bhatti, H.N., Bhatti, I.A.: Methods for polyurethane and polyurethane composites, recycling and recovery: a review. React. Funct. Polym. **67**(8), 675–692 (2007)
34. Kausar, A.: Polyurethane composite foams in high-performance applications: a review. Polym.-Plast. Technol. Eng. **57**(4), 346–369 (2018)
35. Bhasney, S.M., Kumar, A., Katiyar, V.: Microcrystalline cellulose, polylactic acid and polypropylene biocomposites and its morphological, mechanical, thermal and rheological properties. Compos. B Eng. **184**, 107717 (2020)
36. Han, C.D., Ma, C.Y.: Rheological properties of mixtures of molten polymer and fluorocarbon blowing agent. I. mixtures of low-density polyethylene and fluorocarbon blowing agent. J. Appl. Polym. Sci. **28**(2), 831–850 (1983)

37. Jiang, R., Liu, T., Xu, Z., Park, C.B., Zhao, L.: Improving the continuous microcellular extrusion foaming ability with supercritical CO_2 of thermoplastic polyether ester elastomer through in-situ fibrillation of polytetrafluoroethylene. Polymers **11**(12), 1983 (2019)
38. Wang, G., Zhao, G., Zhang, L., Mu, Y., Park, C.B.: Lightweight and tough nanocellular PP/PTFE nanocomposite foams with defect-free surfaces obtained using in situ nanofibrillation and nanocellular injection molding. Chem. Eng. J. **350**, 1–11 (2018)
39. Zhao, J., Zhao, Q., Wang, C., Guo, B., Park, C.B., Wang, G.: High thermal insulation and compressive strength polypropylene foams fabricated by high-pressure foam injection molding and mold opening of nano-fibrillar composites. Mater. Des. **131**, 1–11 (2017)
40. Williams, J.M., Wrobleski, D.A.: Spatial distribution of the phases in water-in-oil emulsions. Open and closed microcellular foams from cross-linked polystyrene. Langmuir **4**(3), 656–662 (1988)
41. Zhao, J., Qiao, Y., Wang, G., Wang, C., Park, C.B.: Lightweight and tough PP/talc composite foam with bimodal nanoporous structure achieved by microcellular injection molding. Mater. Des. **195**, 109051 (2020)
42. Lau, K.A., Gu, C., Hui, D.: A critical review on nanotube and nanotube/nanoclay related polymer composite materials. Compos. Part B-Eng. **37**, 425–436 (2006)
43. Huang, X., Netravali, A.N.: Characterization of nano-clay reinforced phytagel-modified soy protein concentrate resin. Biomacromol **7**, 2783–2789 (2006)
44. Zhao, R., Torley, P., Halley, P.J.: Emerging biodegradable materials: starch- and protein-based bio-nanocomposites. J. Mater. Sci. **43**, 3058–3071 (2008)
45. Alexandre, M., Dubois, P.: Polymer-layered silicate nanocomposites: preparation, properties and uses of a new class of materials. Mater. Sci. Eng. R. Rep. **28**, 1–63 (2000)
46. Aouada, F.A., Mattoso, L.H.C., Longo, E.: New strategies in the preparation of exfoliated thermoplastic starch–montmorillonite nanocomposites. Ind. Crop Prod. **34**, 1502–1508 (2011)
47. Chen, P., Zhang, L.: Interaction and properties of highly exfoliated soy protein/montmorillonite nanocomposites. Biomacromol **7**, 1700–1706 (2006)
48. Kushwaha, P.K., Kumar, R.: Reinforcing effect of nanoclay in bamboo-reinforced thermosetting resin composites. Polym. Plast. Technol. Eng. **50**, 127–135 (2011)
49. Zhang, J., Gupta, R.K., Wilkie, C.A.: Controlled silylation of montmorillonite and its polyethylene nanocomposites. Polymer **47**, 4537–4543 (2006)
50. Balachandra, A.M., Corporation, T., Co, M., Darsanasiri, N., Corporation, M.: Fabrication of carbon nanotube/polymer nanocomposite sheets and its mechanical performance. Adv. Nanosci. Nanotechnol. **2** (2018)
51. Spitalsky, Z., Tasis, D., Papagelis, K., Galiotis, C.: Carbon nanotube–polymer composites: chemistry, processing, mechanical and electrical properties. Prog. Polym. Sci. **35**(3), 357–401 (2010)
52. Basheer, B.V., George, J.J., Siengchin, S., Parameswaranpillai, J.: Polymer grafted carbon nanotubes—synthesis, properties, and applications: a review. Nano-Struct. Nano-Objects **22**, 100429 (2020)
53. Pei, X., Hu, L., Liu, W., Hao, J.: Synthesis of water-soluble carbon nanotubes via surface initiated redox polymerization and their tribological properties as water-based lubricant additive. Eur. Polym. J. **44**(8), 2458–2464 (2008)
54. Sahoo, N.G., Rana, S., Cho, J.W., Li, L., Chan, S.H.: Polymer nanocomposites based on functionalized carbon nanotubes. Prog. Polym. Sci. **35**(7), 837–867 (2010)
55. Rane, A.V., Kanny, K., Abitha, V.K., Thomas, S.: Methods for Synthesis of Nanoparticles and Fabrication of Nanocomposites. Woodhead Publishing, pp. 121–139 (2018)
56. Venkataraman, A., Amadi, E.V., Chen, Y., Papadopoulos, C.: Carbon nanotube assembly and integration for applications. Nanoscale Res. Lett. **14**, 1–47 (2019)
57. Martin, D.J., Osman, A.F., Andriani, Y., Edwards, G.A.: Thermoplastic polyurethane (TPU)-based polymer nanocomposites. In: Advances in Polymer Nanocomposites, Woodhead Publishing, pp. 321–350 (2012)
58. Jin Yoo, H., Jung, Y.C., Sahoo, N.G., Cho, J.W.: Polyurethane-carbon nanotube nanocomposites prepared by in-situ polymerization with electroactive shape memory. J. Macromol. Sci. Part B **45**(4), 441–451 (2006)

59. Sahoo, N.G., Jung, Y.C., Yoo, H.J., Cho, J.W.: Influence of carbon nanotubes and polypyrrole on the thermal, mechanical and electroactive shape-memory properties of polyurethane nanocomposites. Compos. Sci. Technol. **67**(9), 1920–1929 (2007)
60. Mangalgiri, P.D.: Composite materials for aerospace applications. Bull. Mater. Sci. **22**, 657–664 (1999)
61. Fogorasi, M.S., Barbu, I.: The potential of natural fibres for automotive sector-review. In: IOP Conference Series: Materials Science and Engineering, vol. 252, no. 1. IOP Publishing, p. 012044 (2017)
62. Rana, S., Fangueiro, R.: Advanced Composite Materials for Aerospace Engineering: Processing, Properties and Applications. Woodhead Publishing (2016)
63. Ahmad, H., Markina, A.A., Porotnikov, M.V., Ahmad, F.: A review of carbon fiber materials in automotive industry. In: IOP Conference Series: Materials Science and Engineering, vol. 971, no. 3, p. 032011. IOP Publishing (2020)
64. Du Yuan, Y.G.Z., Jia, C.: Carbon nanotube-polymer composites for energy storage applications. Carbon Nanotubes: Curr. Prog. Polym. Comp. 439 (2016)
65. Ramli, N., Mazlan, N., Ando, Y., Leman, Z., Abdan, K., Aziz,m A.A., Sairy, N.A.: Natural fiber for green technology in automotive industry: a brief review. In: IOP Conference Series: Materials Science and Engineering, vol. 368, p. 012012. IOP Publishing (2018)
66. Liu, X.M., dong Huang, Z., Woon Oh, S., Zhang, B., Ma, P.C., Yuen, M.M., Kim, J.K.: Carbon nanotube (CNT)-based composites as electrode material for rechargeable Li-ion batteries: a review. Compos. Sci. Technol. **72**(2), 121–144 (2012)
67. Friedrich, K., Almajid, A.A.: Manufacturing aspects of advanced polymer composites for automotive applications. Appl. Compos. Mater. **20**, 107–128 (2013)
68. Díez-FPascual, A.M.: State of the art in the antibacterial and antiviral applications of carbon-based polymeric nanocomposites. Int. J. Mol. Sci. **22**(19), 10511 (2021)
69. Prasanth, R., Shankar, R., Gupta, N., Ahn, J.H.: Advances in lithium-ion battery technology based on functionalized carbon nanotubes for electrochemical energy storage. In: Handbook of Polymer Nanocomposites. Processing, Performance and Application: Volume B: Carbon Nanotube Based Polymer Composites, pp. 447–478 (2015)
70. Udayakumar, M., T´oth, P., Wiinikka, H., et al.: Hierarchical porous carbon foam electrodes fabricated from waste polyurethane elastomer template for electric double-layer capacitors. Sci. Rep. **12**, 11786 (2022)
71. Nasrollahzadeh, M., Sajjadi, M., Iravani, S., Varma, R.S.: Carbon-based sustainable nanomaterials for water treatment: state-of-art and future perspectives. Chemosphere **263**, 128005 (2021)
72. Iwata, T.: Biodegradable and bio-based polymers: future prospects of eco-friendly plastics. Angew. Chem. Int. Ed. **54**(11), 3210–3215 (2015)
73. Wu, N., Hu, Q., Wei, R., Mai, X., Naik, N., Pan, D., Guo, Z., Shi, Z.: Review on the electromagnetic interference shielding properties of carbon based materials and their novel composites: recent progress, challenges and prospects. Carbon **176**, 88–105 (2021)
74. Doris, M., Sulaiman, K., Supangat, A.: Template-assisted method and polymer-melt technique for fabrication of PCPDTBT nanostructures. J. Nanomater. **15**(2), 123–134 (2024)
75. Ravichandran, G., Santhosh, N.: Melt intercalation, sol-gel processes, and in situ polymerization for enhancing properties of polymer nanocomposites. Mater. Sci. Adv. **10**(4), 567–580 (2024)
76. Uddin, F.: Synthesis methods for nanoclays and their industrial applications. Adv. Eng. Mater. **22**(3), 345–356 (2024)
77. Wang, S., La Monaca, A., Demopoulos, G.P.: Interface engineering and composite solid electrolytes in all-solid-state batteries. Energy Storage Mater. **18**(1), 45–59 (2025)
78. Yashtulov, N.A., Zhang, W., Melnikov, P.V.: Synthesis strategies for polymeric nanocomposites in biomedical and environmental applications. J. Polym. Sci. **56**(7), 789–802 (2024)
79. Batlle, X., Moya, C., Torroellla, M.E., Iglesias, O., Rodriguez, A.F., Labarta, A.: Iron-oxide-based magnetic nanoparticles: synthesis and applications. Magn. Mater. J. **12**(5), 234–250 (2024)

80. George, A., Sreedhara, R., Navya, N.: Integration of carbon dots in Pr3+ doped gahnite nanophosphor for multifunctional applications. Optoelectron. Adv. Mater. **9**(3), 112–125 (2024)
81. September, L.A., Kheswa, N., Seroka, N.S., Khotseng, L.: Green synthesis of silica and silicon from sugarcane bagasse ash for environmental applications. Sustain. Mater. Sci. **5**(2), 98–110 (2023)
82. Pagar, K., Gadore, V., Mishra, S.R., Basnet, P.: Bio-Inspired Fabrication of CdO nanoparticles using citrus Sinensis peel extract for photocatalytic applications. Catal. Today **30**(4), 456–467 (2024)
83. Sarifujjaman, M., Pal, P.K., Saha, P., Rahman, S.M., Islam, M.E.: Leaf extract-mediated synthesis of copper oxide nanoparticles for photocatalytic and antibacterial applications. Green Chem. Lett. Rev. **17**(1), 23–35 (2024)
84. Mallikarjunaswamy, C., Soundarya, T.L., Al-Kahtani, A.A., Al-Odayni, A.B., Nagaraju, G.: Green synthesis of lanthanum oxide nanoparticles: photocatalytic and electrochemical applications. J. Mater. Sci. **59**(8), 3456–3470 (2024)
85. Saridewi, N., Utami, D.J., Zulys, A., Nurbayti, S.: Utilization of Sansevieria Trifasciata extract for green synthesis of $ZnFe_2O_4$ nanoparticles. J. Environ. Chem. Eng. **12**(3), 567–578 (2024)
86. Bopape, D.A., Motaung, D.E., Hintsho-Mbita, N.C.: Biosynthesis of gold- and silver-incorporated carbon-based zinc oxide nanocomposites for environmental applications. Nanomaterials **14**(1), 112–125 (2024)
87. Gidey, A.G., Pillantakath, A.R.K., Witoon, T., Lim, J.-W., Banat, F.: Conversion of rice husk waste into mesoporous silica materials for catalysis and adsorption. Chemosphere **290**, 133–145 (2023)

Introduction of Nanostructure into Shape Memory Polymers from Materials to Applications

Ritesh Gupta, Gaurav Mittal, Surbhi Singh, Vijay Chaudhary, and Krishna Kumar

Abstract This chapter explores the integration of nanostructures into Shape Memory Polymers (SMPs), a class of smart materials renowned for their ability to recover predefined shapes after deformation. SMPs, driven by molecular architecture, exhibit the shape memory effect (SME) through programming, temporary shape fixation, and recovery. The incorporation of nanostructures, such as carbon nanotubes, graphene, and nanoparticles, has significantly enhanced SMPs' mechanical, thermal, and electrical properties, addressing limitations like slow recovery kinetics and weak mechanical strength. These advancements enable multi-stimuli responsiveness, making SMPs suitable for diverse applications in biomedical engineering, aerospace, robotics, and flexible electronics. The chapter delves into fabrication techniques, including nanocomposite blending and advanced nanofabrication strategies, while highlighting challenges such as nanoparticle dispersion and scalability. Performance enhancements, including improved tensile strength, toughness, and fatigue resistance, are discussed, alongside applications like self-expanding stents, deployable aerospace structures, and soft robotics. Despite progress, issues like durability, reproducibility, and cost remain. Emerging trends, such as AI-driven material design and biomimetic nanostructures, offer promising avenues for future research. Overall, nanostructured SMPs represent a transformative class of materials with vast potential across various industries.

R. Gupta (✉) · G. Mittal · K. Kumar
Department of Metallurgical and Materials Engineering, Malaviya National Institute of Technology Jaipur, Jaipur, Rajasthan, India
e-mail: ritesh.gupta641@gmail.com

S. Singh
Shodh Lab, Suresh Gyan Vihar University, Jaipur, Rajasthan, India

Materials Research Centre, Malaviya National Institute of Technology Jaipur, Jaipur, Rajasthan, India

V. Chaudhary
Department of Mechanical Engineering, Amity School of Engineering and Technology, Amity University, Noida, Uttar Pradesh, India
e-mail: vijaychaudhary111@gmail.com

B. Sharma et al. (eds.), *Nanostructured Polymers*, Engineering Materials,
https://doi.org/10.1007/978-981-95-3581-1_4

1 Introduction

Shape Memory Polymers (SMPs) represent a groundbreaking class of smart materials that have garnered significant attention due to their ability to recover a predefined shape after undergoing deformation. This unique property, known as the shape memory effect (SME), is driven by the material's molecular architecture, which consists of two key components: a crosslinked network (fixed phase) that retains the permanent shape and a reversible phase (e.g., glassy or crystalline domains) that enables temporary shape fixation and recovery [1, 2]. The SME process involves three main steps: programming, temporary shape fixation, and recovery. During programming, the SMP is deformed above its transition temperature T_{trans}, where the polymer chains become flexible. Upon cooling below T_{trans}, the temporary shape is fixed as the chains lose mobility. Finally, reheating the material above T_{trans} triggers the recovery of the original shape as the polymer chains regain mobility and the stored elastic energy is released [3, 4].

Compared to other smart materials, such as Shape Memory Alloys (SMAs), SMPs offer several advantages, including higher strain recovery (up to 400%), lower density, and tunable stiffness [5]. While SMAs are limited by their high cost, heavy weight, and relatively low strain recovery (typically 8–10%), SMPs provide a lightweight and cost-effective alternative with superior flexibility in chemical composition and fabrication methods [6]. This versatility allows SMPs to be tailored for a wide range of stimuli, including heat, light, pH, electrical fields, and solvents, making them suitable for diverse applications [7]. For example, thermally activated SMPs are widely used in biomedical devices, while light-activated SMPs are employed in remote-controlled actuators [2, 8]. The ability to customize SMPs for specific stimuli and applications has positioned them as a cornerstone of modern smart material research.

The integration of nanostructures into SMPs has revolutionized their performance, addressing key limitations such as slow recovery kinetics, weak mechanical strength, and limited functionality. Nanoscale additives, such as carbon nanotubes (CNTs), graphene, and nanoparticles, have been shown to significantly enhance the thermal and electrical conductivity of SMPs, enabling rapid and remote actuation via Joule heating or magnetic induction [9, 10]. For instance, CNT-reinforced SMP composites achieve near-instantaneous shape recovery under electrical stimulation, outperforming conventional thermally activated systems [11]. The incorporation of Fe_3O_4 nanoparticles allows SMPs to respond to alternating magnetic fields, enabling selective actuation of distinct segments within a single material [12]. This multi-stimuli responsiveness is a critical advancement, as it expands the potential applications of SMPs in complex environments where multiple triggers may be present.

Nanofibers and porous microstructures, fabricated via techniques such as electrospinning or 4D printing, provide high surface-to-volume ratios, improving responsiveness and enabling applications in drug delivery and tissue engineering [13, 14]. For example, electrospun SMP nanofibers have been used as dynamic scaffolds for cell growth, where the fibers' shape memory properties can be programmed to change morphology during cell culture, influencing cell behavior and differentiation [15]. Similarly, 4D-printed SMPs with embedded nanostructures achieve complex, reversible morphing for applications in soft robotics and biomedical devices [16]. The ability to precisely control the microstructure of SMPs at the nanoscale has opened new avenues for designing materials with tailored mechanical, thermal, and electrical properties.

The introduction of nanostructures also enhances the mechanical properties of SMPs, such as tensile strength, toughness, and durability. For example, the addition of graphene oxide to SMP foams has been shown to improve their compressive strength and shape recovery ratio, making them suitable for high-performance applications in aerospace and engineering [17]. Furthermore, nanostructured SMPs exhibit improved fatigue resistance, allowing them to undergo multiple shape memory cycles without significant degradation in performance [14, 17]. These advancements underscore the critical role of nanotechnology in unlocking multifunctionality and precision in SMP design, paving the way for next-generation smart materials.

The unique properties of SMPs, particularly when enhanced with nanostructures, have enabled their use in a wide range of high-impact applications across various fields. In biomedical engineering, SMPs are employed in dynamic scaffolds for tissue regeneration, where their shape memory properties can be programmed to change morphology during cell culture, influencing cell behavior and differentiation [3, 16]. SMP-based nanofibers and microparticles are also used in targeted drug delivery systems, where the material's ability to change shape in response to physiological stimuli (e.g., temperature or pH) enables controlled release of therapeutic agents [7, 9]. For example, SMP microparticles can be programmed to switch between spherical and non-spherical shapes, allowing them to navigate through the bloodstream and release drugs at specific sites [12].

In aerospace, lightweight SMP foams are used for deployable satellite components, such as self-expanding solar sails and antennas, which reduce payload volume during launch and deploy autonomously in space [3]. The incorporation of nanostructures, such as carbon nanotubes or graphene, enhances the mechanical and thermal properties of these foams, making them suitable for the harsh conditions of space [1]. Similarly, SMPs are being explored for use in morphing skins for aircraft, where their ability to change shape in response to environmental conditions can improve aerodynamic efficiency [4, 16].

In the field of robotics, 4D-printed SMPs with embedded nanostructures are used to create soft actuators and grippers that can perform complex, reversible morphing tasks [15]. These materials are particularly well-suited for applications in minimally invasive surgery, where their flexibility and responsiveness enable precise manipulation of tissues and organs [8]. Additionally, SMPs are being developed for use in wearable electronics, where their ability to conform to the human body and respond to external stimuli makes them ideal for flexible sensors and artificial muscles [4, 17].

The increasing demand for high-performance, multi-functional SMP composites is driven by the need for materials that can meet the complex requirements of modern technologies. As industries continue to prioritize miniaturization, multifunctionality, and sustainability, nanostructured SMPs represent a cornerstone for next-generation smart materials. Their ability to combine lightweight, flexibility, and responsiveness with enhanced mechanical and thermal properties makes them indispensable for applications in biomedical engineering, aerospace, robotics, and beyond.

2 Fundamentals of Shape Memory Polymers

2.1 Working Mechanism of SMPs

Shape Memory Polymers (SMPs) exhibit a unique ability to recover their original shape from a deformed state when exposed to specific external stimuli. This behavior, known as the shape memory effect (SME), is governed by the material's molecular architecture, which consists of a crosslinked network (fixed phase) and a reversible phase (e.g., glassy or crystalline domains) [1, 2]. The SME process involves three key steps: programming, temporary shape fixation, and recovery. During programming, the SMP is deformed above its transition temperature T_{trans}, where the polymer chains become flexible. Upon cooling below T_{trans}, the temporary shape is fixed as the chains lose mobility. Finally, reheating the material above T_{trans} triggers the recovery of the original shape as the polymer chains regain mobility and the stored elastic energy is released [3, 4].

Thermal-responsive SMPs are the most widely studied, with shape recovery driven by changes in temperature relative to the glass transition temperature T_g or melting temperature T_g of the polymer [5]. For example, polyurethane-based SMPs exhibit shape recovery when heated above their T_g, while semicrystalline polymers like polycaprolactone (PCL) rely on melting and recrystallization for shape memory behavior [6]. Light-responsive SMPs incorporate photoresponsive moieties, such as azobenzene or cinnamic acid, which undergo reversible photoisomerization upon exposure to specific wavelengths of light, enabling remote and precise control of shape recovery [7]. Electrical and magnetic-responsive SMPs are typically reinforced with conductive fillers like carbon nanotubes (CNTs) or magnetic nanoparticles (e.g., Fe_3O_4, allowing shape recovery to be triggered by Joule heating or magnetic induction, respectively [8, 9].

Multi-stimuli responsive SMPs represent a significant advancement, enabling shape recovery in response to multiple triggers, such as heat, light, and magnetic fields. For instance, SMP composites with embedded CNTs and Fe_3O_4 nanoparticles can be activated by both electrical and magnetic stimuli, offering greater flexibility in applications [10]. Dual- and triple-shape memory effects further expand the functionality of SMPs, allowing sequential recovery of multiple temporary shapes in response to different stimuli or temperature ranges [11]. The underlying mechanism of these effects relies on the presence of multiple reversible phases with distinct transition temperatures, enabling complex shape transformations [12].

The role of polymer networks and phase transition mechanisms is critical in determining the performance of SMPs. The fixed phase, typically formed by chemical or physical crosslinks, provides the structural integrity needed to retain the permanent shape, while the reversible phase enables temporary shape fixation and recovery [13]. The interplay between these phases, along with the material's viscoelastic properties, governs the shape memory behavior, including recovery speed, strain recovery ratio, and cycle durability [14].

2.2 Key Material Properties of SMPs

The performance of SMPs is largely determined by their material properties, including glass transition temperature (T_g), melting point (T_m), viscoelasticity, elasticity, and strain recovery. (T_g) and (T_m) are critical parameters that define the temperature range over which shape recovery occurs. For amorphous polymers, (T_g) marks the transition from a rigid glassy state to a rubbery state, while for semicrystalline polymers, (T_m) governs the melting and recrystallization processes that drive shape recovery [15]. The ability to tailor (T_g) and (T_m) through chemical modification or blending with other polymers allows SMPs to be customized for specific applications, such as biomedical devices operating at body temperature or aerospace components requiring high-temperature stability [1, 16].

Viscoelasticity and elasticity are key factors influencing the deformation and recovery behavior of SMPs. Viscoelasticity, characterized by a combination of viscous and elastic responses, determines the material's ability to store and release energy during deformation and recovery [17]. Elasticity, on the other hand, governs the material's ability to return to its original shape after deformation. High elasticity is essential for achieving high strain recovery ratios, which can exceed 400% in some SMPs [6, 18]. The stress–strain behavior of SMPs is also critical, as it determines the material's ability to withstand mechanical loads during deformation and recovery. SMPs typically exhibit a nonlinear stress–strain relationship, with significant strain hardening at high deformations [19].

Durability under multiple deformations is another important consideration for SMPs, particularly in applications requiring repeated shape memory cycles. Fatigue resistance, which refers to the material's ability to maintain its shape memory properties over multiple cycles, is influenced by factors such as crosslink density, polymer chain mobility, and the presence of reinforcing fillers [14, 20]. For example, SMPs with high crosslink densities exhibit greater fatigue resistance but may have reduced strain recovery due to restricted chain mobility [21].

2.3 Limitations of Conventional SMPs

Despite their unique properties, conventional SMPs face several limitations that restrict their performance and applicability. One of the most significant challenges is slow response times, particularly in thermally activated SMPs, where shape recovery can take several minutes or even hours depending on the heating rate and material thickness [11, 22]. This limitation is particularly problematic in applications requiring rapid actuation, such as robotics or biomedical devices.

Another major limitation is the low thermal and electrical conductivity of conventional SMPs, which restricts their use in electronic applications. While the incorporation of conductive fillers like CNTs or graphene can enhance conductivity, achieving uniform dispersion and maintaining mechanical properties remains a challenge [23]. Mechanical weaknesses, such as brittleness or low tensile strength, further limit the applicability of SMPs in high-stress environments. For example, some SMPs exhibit poor resistance to crack propagation, leading to premature failure under cyclic loading [24].

The limited number of shape recovery cycles is another critical issue, as repeated deformation and recovery can lead to degradation of the polymer network and loss of shape memory properties [25]. This is particularly problematic in applications requiring long-term durability, such as aerospace or automotive components. Additionally, conventional SMPs often exhibit poor environmental stability, with properties degrading under exposure to UV radiation, moisture, or chemical agents [3, 26].

Efforts to address these limitations have focused on the development of advanced SMP composites and nanostructured materials. For example, the incorporation of nanofillers like graphene oxide or clay nanoparticles has been shown to improve mechanical strength, thermal conductivity, and fatigue resistance [27]. Similarly, the use of multi-stimuli responsive SMPs enables faster and more versatile actuation, expanding the range of potential applications [28].

3 Types of Nanostructures Used in SMPs

3.1 Carbon-Based Nanostructures

Carbon-based nanostructures, such as carbon nanotubes (CNTs), graphene, and carbon black, have been widely incorporated into Shape Memory Polymers (SMPs) to enhance their mechanical, electrical, and thermal properties. Carbon nanotubes (CNTs) are particularly notable for their ability to improve mechanical strength and electrical conductivity. When dispersed uniformly within an SMP matrix, CNTs form a conductive network that enables rapid Joule heating, significantly reducing shape recovery times [1, 2]. For example, CNT-reinforced SMP composites have demonstrated near-instantaneous shape recovery under electrical stimulation, making them ideal for applications in soft robotics and wearable electronics [3]. Additionally, CNTs enhance the tensile strength and toughness of SMPs, enabling them to withstand higher mechanical loads without failure [4].

Graphene and graphene oxide (GO) are another class of carbon-based nanostructures that have gained attention for their ability to improve the flexibility and actuation efficiency of SMPs. Graphene's high surface area and exceptional mechanical properties make it an excellent filler for enhancing the stiffness and durability of SMPs [5, 29]. Graphene oxide, with its oxygen-containing functional groups, improves interfacial adhesion with the polymer matrix, leading to better dispersion and enhanced mechanical performance [6]. Moreover, GO-based SMPs exhibit improved thermal conductivity, enabling faster and more uniform heat transfer during shape recovery [7].

Carbon black, a low-cost carbon-based nanomaterial, is often used in electroactive SMPs to provide electrical conductivity and improve mechanical properties. While its performance is generally inferior to CNTs and graphene, carbon black is more economical and easier to process, making it suitable for large-scale applications [8]. For instance, carbon black-reinforced SMPs have been used in low-cost sensors and actuators, where moderate conductivity and mechanical reinforcement are sufficient [9, 30].

3.2 Metallic and Ceramic Nanoparticles

Metallic and ceramic nanoparticles are another important class of nanostructures used to enhance the functionality of SMPs. Gold (Au) and silver (Ag) nanoparticles are commonly employed to induce optical and electrical responses in SMPs. These nanoparticles exhibit localized surface plasmon resonance (LSPR), which can be exploited for light-triggered shape recovery [10]. For example, Au nanoparticle-embedded SMPs have been used in photonic devices, where their ability to absorb specific wavelengths of light enables precise control over shape recovery [11]. Silver nanoparticles, on the other hand, are valued for their high electrical conductivity, making them suitable for electroactive SMPs [12].

Iron oxide (Fe_3O_4) nanoparticles are widely used to impart magnetic responsiveness to SMPs. When exposed to an alternating magnetic field, Fe_3O_4 nanoparticles generate heat through magnetic induction, enabling remote and contactless actuation of SMPs [13]. This property is particularly useful in biomedical applications, such as minimally invasive surgery, where magnetic fields can be used to trigger shape recovery in SMP-based devices [14, 31].

Ceramic nanoparticles, such as barium titanate ($BaTiO_3$), silica (SiO_2), and titania (TiO_2), are used to improve the thermal stability and mechanical toughness of SMPs. $BaTiO_3$ nanoparticles, known for their piezoelectric properties, enhance the electromechanical performance of SMPs, making them suitable for sensors and energy harvesting devices [15]. Silica and titania nanoparticles improve the thermal stability and wear resistance of SMPs, enabling their use in high-temperature applications [16]. For example, SiO_2-reinforced SMPs have been employed in aerospace components, where their ability to withstand extreme temperatures and mechanical stress is critical [17].

3.3 Polymeric Nanostructures

Polymeric nanostructures, such as core–shell nanospheres, electrospun nanofibers, and dendritic polymers, offer unique opportunities for enhancing the functionality of SMPs. Core–shell nanospheres, consisting of a polymeric core surrounded by a shell of another material, are used to create multi-functional SMPs with tailored properties. For instance, SMPs with core–shell nanospheres containing a thermally responsive core and a mechanically robust shell exhibit improved shape recovery and durability [18, 32]. These nanostructures are particularly useful in biomedical applications, where controlled drug release and mechanical stability are required [19].

Electrospun nanofibers are another important class of polymeric nanostructures used to enhance the tensile strength and flexibility of SMPs. Electrospinning allows for the fabrication of nanofibers with high surface-to-volume ratios, which improve the interfacial adhesion between the fibers and the polymer matrix [20]. SMPs reinforced with electrospun nanofibers exhibit superior mechanical properties, such as higher tensile strength and strain recovery, making them suitable for applications in tissue engineering and wearable electronics [21, 33]. For example, polyurethane-based SMP nanofibers have been used as dynamic scaffolds for cell growth, where their shape memory properties enable controlled changes in morphology during cell culture [22].

Dendritic and hyperbranched polymer nanostructures are also used to enhance the performance of SMPs. These highly branched polymers provide a large number of functional groups, which can be used to tailor the mechanical, thermal, and electrical properties of SMPs [23]. For instance, hyperbranched polymers with terminal epoxy groups have been used to improve the crosslink density and mechanical strength of SMPs [24, 34]. Dendritic polymers, with their unique three-dimensional architecture, enhance the toughness and fatigue resistance of SMPs, making them suitable for applications requiring long-term durability [25].

4 Fabrication and Processing Techniques of Nanostructured SMPs

4.1 Nanocomposite Blending Approaches

The fabrication of nanostructured Shape Memory Polymers (SMPs) often begins with nanocomposite blending approaches, which aim to uniformly disperse nanofillers within the polymer matrix. Solution mixing is one of the most common methods, where nanoparticles are dispersed in a solvent along with the polymer, followed by solvent evaporation to form a composite film or bulk material [1]. This method is particularly effective for achieving uniform dispersion of carbon-based nanomaterials like graphene oxide (GO) and carbon nanotubes (CNTs) [2]. However, the choice of solvent and mixing conditions can significantly influence the final properties of the composite, as poor dispersion can lead to agglomeration and reduced performance [3, 35].

In-situ polymerization is another widely used technique, where nanoparticles are dispersed in a monomer solution, and polymerization is initiated to form the composite material. This method ensures strong interfacial adhesion between the nanoparticles and the polymer matrix, enhancing mechanical and thermal properties [4]. For example, in-situ polymerization has been used to fabricate SMPs with embedded Fe_3O_4 nanoparticles, enabling magnetic responsiveness and improved mechanical strength [5]. Melt blending, on the other hand, involves mixing nanoparticles directly into the polymer melt using techniques like extrusion or injection molding. This method is suitable for large-scale production but often faces challenges in achieving uniform nanoparticle dispersion due to high viscosity and shear forces [6].

Solvent-casting and hot-pressing techniques are also employed to fabricate nanostructured SMPs. Solvent-casting involves dissolving the polymer and dispersing nanoparticles in a solvent, followed by casting into a mold and evaporating the solvent to form a solid film [7, 36]. This method is simple and cost-effective but may result in residual solvent, which can affect the material's properties. Hot-pressing involves compressing a mixture of polymer and nanoparticles at elevated temperatures to form a dense composite. This technique is particularly useful for creating SMPs with high mechanical strength and thermal stability [8].

4.2 Advanced Nanofabrication Strategies

Advanced nanofabrication strategies, such as electrospinning, 3D printing, and layer-by-layer (LbL) assembly, have emerged as powerful tools for creating nanostructured SMPs with precise control over their architecture and properties. Electrospinning is a versatile technique for producing nanofibers with high surface-to-volume ratios, which can be used to reinforce SMPs [9]. For example, electrospun polyurethane nanofibers have been incorporated into SMP matrices to enhance tensile strength and shape recovery performance [10]. The ability to tailor fiber diameter and alignment during electrospinning allows for the creation of SMPs with anisotropic properties, making them suitable for applications in tissue engineering and wearable electronics [11, 37].

3D printing, particularly fused deposition modeling (FDM) and stereolithography (SLA), has revolutionized the fabrication of SMPs by enabling the creation of complex, multi-material structures with spatially varying properties [12, 38]. 4D printing, an extension of 3D printing, adds the dimension of time by incorporating shape memory functionality into printed structures. For instance, 4D-printed SMPs with embedded CNTs have been used to create self-folding structures that respond to electrical stimuli [13]. Layer-by-layer (LbL) assembly is another advanced technique that involves the sequential deposition of polymer and nanoparticle layers to create hierarchical structures with tailored properties. This method has been used to fabricate SMPs with enhanced mechanical and thermal properties, as well as multi-stimuli responsiveness [14].

Self-assembly methods, such as micelle formation and block copolymer templating, are also employed to achieve structured nanofiller dispersion within SMPs. These methods rely on the spontaneous organization of nanoparticles or polymer chains into ordered structures, which can enhance the material's mechanical and functional properties [15]. For example, self-assembled Fe_3O_4 nanoparticle clusters have been used to create SMPs with improved magnetic responsiveness and mechanical strength [16, 39].

4.3 Challenges in Processing

Despite the significant advancements in fabrication techniques, several challenges remain in the processing of nanostructured SMPs. One of the primary challenges is achieving uniform nanoparticle dispersion within the polymer matrix. Agglomeration of nanoparticles can lead to localized stress concentrations, reducing the material's mechanical properties and shape memory performance [17]. Various strategies, such as surface functionalization of nanoparticles and the use of surfactants, have been employed to improve dispersion, but these methods often require complex processing steps and may introduce additional costs [18, 40].

Another major challenge is ensuring compatibility between nanoparticles and the polymer matrix. Poor interfacial adhesion can result in weak mechanical properties and reduced shape memory performance [19]. Surface modification of nanoparticles, such as grafting polymer chains or functional groups, has been shown to improve compatibility and enhance the overall performance of SMPs [20, 41]. However, these modifications can alter the intrinsic properties of the nanoparticles, requiring careful optimization.

Scaling up the production of nanostructured SMPs for industrial applications is another significant challenge. Many advanced fabrication techniques, such as electrospinning and 3D printing, are limited by low production rates and high costs, making them unsuitable for large-scale manufacturing [21]. Melt blending and solvent-casting techniques, while more scalable, often face challenges in achieving the desired level of nanoparticle dispersion and interfacial adhesion [22]. Additionally, the incorporation of nanoparticles can increase the viscosity of the polymer melt, complicating processing and requiring specialized equipment [23].

Efforts to address these challenges have focused on developing new processing techniques and optimizing existing methods. For example, the use of twin-screw extrusion has been shown to improve nanoparticle dispersion and reduce processing times in melt blending [24, 42]. Similarly, advances in 3D printing technology, such as multi-material printing and high-speed printing, are enabling the production of complex SMP structures at larger scales [25]. Despite these advancements, further research is needed to overcome the remaining challenges and fully realize the potential of nanostructured SMPs in industrial applications.

5 Performance Enhancement in SMPs via Nanostructuring

5.1 *Mechanical Property Improvements*

The integration of nanostructures into Shape Memory Polymers (SMPs) has revolutionized their mechanical properties, significantly enhancing tensile strength, toughness, and fatigue resistance. These improvements are primarily achieved through the incorporation of nanofillers such as carbon nanotubes (CNTs), graphene, silica nanoparticles, and other reinforcing agents, which interact with the polymer matrix to create a more robust and durable material. The mechanical performance of SMPs is critical for their application in fields such as aerospace, biomedical engineering, and robotics, where materials must withstand high stresses and repeated deformations without failure [1, 2, 43].

One of the most notable improvements is in tensile strength, which is the maximum stress a material can withstand while being stretched or pulled before breaking. Nanofillers like CNTs and graphene act as reinforcing agents by forming a network within the polymer matrix that distributes stress more evenly. This network prevents localized stress concentrations, which are often the starting points for material failure. For example, CNTs have been shown to increase the tensile strength of polyurethane-based SMPs by up to 50%, making them suitable for high-load applications such as actuators and structural components [3, 44]. Similarly, graphene oxide (GO) enhances tensile strength by creating strong interfacial interactions with the polymer chains, effectively bridging gaps and preventing crack propagation [18].

Toughness, which refers to a material's ability to absorb energy and deform plastically without fracturing, is another critical property improved by nanostructuring. Nanofillers such as silica nanoparticles and core–shell structures enhance toughness by acting as energy-dissipating agents. When stress is applied, these nanoparticles deflect cracks and absorb energy through mechanisms such as interfacial sliding and crack bridging [15]. For instance, SMPs reinforced with silica nanoparticles exhibit significantly higher toughness compared to unfilled polymers, making them ideal for applications requiring impact resistance, such as protective coatings and automotive components [6].

Fatigue resistance, or the ability of a material to withstand repeated loading and unloading cycles without failure, is also greatly enhanced by the addition of nanofillers. Fatigue failure is a common issue in conventional SMPs, particularly in applications such as actuators and biomedical devices, where materials are subjected to cyclic stresses. Nanofillers like CNTs and graphene improve fatigue resistance by creating a more uniform stress distribution and preventing the initiation and propagation of micro-cracks [7, 45]. For example, SMPs reinforced with CNTs have demonstrated superior fatigue performance, withstanding thousands of shape memory cycles without significant degradation in mechanical properties [8].

The role of nanofillers in preventing micro-cracking is particularly important in SMPs, as cracks can propagate rapidly under cyclic loading, leading to premature failure. Nanoparticles act as barriers to crack propagation, effectively increasing the material's resistance to fracture. This is achieved through mechanisms such as crack deflection, where cracks are forced to change direction when they encounter a nanoparticle, and crack bridging, where nanoparticles span the crack and prevent it from widening [25]. For example, SMPs reinforced with silica nanoparticles exhibit significantly higher resistance to micro-cracking compared to unfilled polymers, making them suitable for long-term applications in harsh environments [10].

5.1.1 Electrical and Thermal Properties

The electrical and thermal properties of SMPs are critical for their performance in applications such as electroactive actuators, sensors, and thermally responsive devices. The incorporation of nanostructures, particularly conductive and thermally conductive nanofillers, has led to significant improvements in these properties, enabling the development of SMPs with enhanced functionality and responsiveness [11, 46].

Electrical conductivity is a key property for electroactive SMPs, which rely on electrical stimulation to trigger shape recovery. The addition of conductive nanofillers such as CNTs, graphene, and silver nanoparticles significantly improves the electrical conductivity of SMPs, enabling rapid and efficient Joule heating. This property is particularly useful in applications requiring fast and precise actuation, such as soft robotics and wearable electronics [12]. For example, CNT-reinforced SMP composites exhibit near-instantaneous shape recovery under electrical stimulation, with recovery times as short as a few seconds. This rapid response is achieved through the formation of a conductive network within the polymer matrix, which allows for uniform and efficient heat generation [13].

Thermal conductivity is another critical property enhanced through nanostructuring. The addition of thermally conductive nanofillers such as boron nitride (BN), aluminum oxide (Al_2O_3), and graphene improves heat transfer within the SMP matrix, leading to faster and more uniform shape recovery. This is particularly important in applications where rapid actuation is required, such as in aerospace deployable structures or biomedical devices [14]. For instance, SMPs reinforced with BN nanoparticles exhibit significantly reduced recovery times compared to their unfilled counterparts, making them suitable for time-sensitive applications [15].

The optimization of thermal and electrical properties through nanostructuring also enables the development of multi-functional SMPs that can respond to multiple stimuli. For example, SMPs with embedded Fe_3O_4 nanoparticles exhibit both magnetic and thermal responsiveness, allowing for remote actuation through alternating magnetic fields [16, 47]. Similarly, the incorporation of gold (Au) nanoparticles enables light-triggered shape recovery, expanding the range of potential applications in photonics and optoelectronics [17]. These multi-responsive properties open up new possibilities for SMPs in complex environments where multiple triggers may be present.

5.2 Shape Recovery and Actuation Speed

One of the most significant advantages of nanostructured SMPs is their ability to achieve faster and more controlled shape recovery. The addition of nanofillers enhances the material's responsiveness to external stimuli, such as heat, light, and magnetic fields, enabling rapid and precise actuation [18]. For example, SMPs reinforced with CNTs exhibit near-instantaneous shape recovery under electrical stimulation, with recovery times as short as a few seconds [19, 48]. This rapid response is critical for applications in robotics and wearable electronics, where quick and reliable actuation is essential [20].

Multi-responsive SMPs, which can respond to multiple stimuli, represent a significant advancement in the field. For instance, SMPs with embedded Fe_3O_4 nanoparticles can be activated by both magnetic fields and heat, providing greater flexibility in applications [21, 49]. Similarly, light-responsive SMPs, which incorporate photoresponsive moieties such as azobenzene, enable precise control over shape recovery through exposure to specific wavelengths of light [22]. These multi-responsive properties open up new possibilities for SMPs in complex environments where multiple triggers may be present.

The ability to control the actuation speed and recovery behavior of SMPs through nanostructuring has also led to the development of advanced applications in biomedical engineering and aerospace. For example, SMP-based stents with embedded nanoparticles can be programmed to expand at body temperature, providing a minimally invasive solution for treating vascular diseases [23]. In aerospace, nanostructured SMPs are used in deployable structures, such as solar sails and antennas, where rapid and reliable shape recovery is critical for mission success [24, 50, 51].

6 Applications of Nanostructured SMPs

6.1 Biomedical Applications

Nanostructured Shape Memory Polymers (SMPs) have revolutionized the field of biomedical engineering, offering innovative solutions for minimally invasive surgeries, drug delivery, and tissue regeneration. One of the most prominent applications is in self-expanding stents, which are used to treat vascular diseases such as atherosclerosis. These stents are fabricated from SMPs that can be compressed into a small diameter for insertion into the body and then expand to their original shape upon exposure to body temperature [7]. The incorporation of nanostructures, such as carbon nanotubes (CNTs) or silica nanoparticles, enhances the mechanical properties of these stents, ensuring they can withstand the dynamic environment of blood vessels without fracturing [52]. Additionally, the use of SMPs allows for precise control over the expansion process, reducing the risk of complications such as restenosis [53].

Another significant application of nanostructured SMPs is in shape-changing drug carriers. These carriers can be programmed to change shape in response to specific stimuli, such as temperature, pH, or light, enabling targeted drug delivery. For example, SMP-based microparticles can be designed to release therapeutic agents at specific sites within the body, such as tumor tissues or inflamed areas [34]. The incorporation of nanostructures, such as gold nanoparticles or Fe_3O_4 nanoparticles, allows for remote actuation using external stimuli like light or magnetic fields, providing a non-invasive method for controlled drug release [5, 54]. This approach minimizes side effects and improves the efficacy of treatments, particularly in cancer therapy [17].

SMP-based tissue scaffolds are another groundbreaking application in regenerative medicine. These scaffolds provide a temporary structure for cell growth and tissue regeneration, gradually degrading as new tissue forms. The incorporation of nanostructures, such as electrospun nanofibers or graphene oxide, enhances the mechanical and biological properties of these scaffolds, promoting cell adhesion, proliferation, and differentiation [7]. For instance, SMP scaffolds with embedded nanofibers have been used to regenerate bone tissue, where the shape memory effect allows the scaffold to adapt to the defect site and provide mechanical support during healing [8]. Furthermore, the ability to control the porosity and surface topography of these scaffolds at the nanoscale enables precise regulation of cell behavior, leading to more effective tissue regeneration [9, 55].

6.2 Aerospace and Automotive Applications

In the aerospace industry, nanostructured SMPs are being used to develop deployable structures that can be compactly stored during launch and then expanded in space. These structures, such as solar sails, antennas, and reflectors, are critical for space missions, where weight and volume constraints are paramount [20]. SMPs offer a lightweight and reliable solution for these applications, as they can be programmed to change shape upon exposure to specific stimuli, such as heat or light. The incorporation of nanostructures, such as CNTs or graphene, enhances the mechanical and thermal properties of these materials, ensuring they can withstand the harsh conditions of space [11]. For example, SMP-based solar sails reinforced with CNTs exhibit superior strength and durability, enabling them to maintain their shape and function over extended periods [19].

Another important application in aerospace is the development of morphing structures, which can change shape in response to environmental conditions to improve aerodynamic efficiency. For instance, SMP-based wing flaps can be designed to adjust their shape during flight, reducing drag and improving fuel efficiency [13, 56]. The use of nanostructured SMPs allows for precise control over the shape-changing process, enabling rapid and reliable actuation. Additionally, the incorporation of thermally conductive nanofillers, such as boron nitride (BN), improves heat transfer within the material, ensuring uniform shape recovery even in extreme temperatures [14].

In the automotive industry, nanostructured SMPs are being used to develop self-healing coatings that can repair damage caused by scratches or impacts. These coatings are fabricated from SMPs that can recover their original shape upon exposure to heat or light, effectively "healing" the damaged area [8, 15]. The incorporation of nanostructures, such as silica nanoparticles or graphene, enhances the mechanical properties of these coatings, ensuring they can withstand the wear and tear of daily use [7, 16]. For example, SMP-based coatings reinforced with silica nanoparticles exhibit superior scratch resistance and durability, making them ideal for protecting automotive surfaces [57].

Another promising application in the automotive sector is the development of smart bumpers that can absorb impact energy and recover their shape after a collision. These bumpers are fabricated from SMPs that can undergo large deformations without permanent damage, providing enhanced safety and reducing repair costs [18]. The incorporation of nanostructures, such as CNTs or Fe_3O_4 nanoparticles, enhances the mechanical and thermal properties of these materials, ensuring they can withstand high-impact forces and recover their shape quickly [19]. Additionally, the use of SMPs allows for the integration of sensors and actuators, enabling real-time monitoring and control of the bumper's performance [20].

6.3 *Flexible Electronics and Wearables*

Nanostructured Shape Memory Polymers (SMPs) have opened up exciting possibilities in the field of flexible electronics and wearables, where materials must be lightweight, stretchable, and capable of adapting to dynamic environments. One of the most promising applications is in shape-adaptive circuits, which can conform to irregular surfaces and maintain functionality under mechanical deformation. SMPs reinforced with conductive nanofillers, such as carbon nanotubes (CNTs) or silver nanoparticles, enable the creation of circuits that can stretch, bend, and recover their original shape without losing electrical conductivity [1]. For example, SMP-based circuits have been integrated into wearable devices, such as smartwatches and fitness trackers, where they can withstand repeated stretching and bending during daily use [2].

Stretchable displays are another groundbreaking application of nanostructured SMPs in flexible electronics. These displays are fabricated using SMPs with embedded conductive nanomaterials, such as graphene or CNTs, which allow the display to stretch and recover without compromising image quality [3]. The shape memory effect of SMPs ensures that the display can return to its original form after deformation, making it ideal for applications in foldable smartphones, rollable TVs, and wearable displays [4]. Additionally, the incorporation of nanostructures enhances the mechanical durability and thermal stability of these displays, ensuring long-term performance in demanding environments [58].

Smart clothing is another area where nanostructured SMPs are making a significant impact. SMP-based conductive pathways can be integrated into fabrics to create clothing that can adapt to the wearer's movements and environmental conditions. For instance, SMPs with embedded CNTs or graphene can be used to create stretchable and conductive threads that are woven into fabrics, enabling the development of smart garments with integrated sensors, heating elements, or communication devices [6]. These garments can monitor physiological signals, such as heart rate and body temperature, and provide real-time feedback to the wearer [7]. The shape memory effect of SMPs ensures that the conductive pathways maintain their functionality even after repeated stretching and washing, making them suitable for everyday use [8, 59].

6.4 Soft Robotics and Actuators

In the field of soft robotics, nanostructured SMPs are being used to develop bio-inspired robotic grippers and adaptive locomotion systems that can mimic the flexibility and dexterity of natural organisms. SMP-based grippers can change shape in response to external stimuli, such as heat or light, enabling them to grasp and manipulate objects of various shapes and sizes [9]. The incorporation of nanostructures, such as Fe_3O_4 nanoparticles or CNTs, enhances the mechanical and thermal properties of these grippers, allowing for precise control over their actuation [10]. For example, SMP grippers with embedded Fe_3O_4 nanoparticles can be remotely controlled using magnetic fields, making them ideal for applications in minimally invasive surgery and hazardous environments [11].

Adaptive locomotion systems, such as soft robots that can crawl, swim, or climb, are another exciting application of nanostructured SMPs. These robots are fabricated using SMPs with embedded conductive or magnetic nanomaterials, which enable them to change shape and move in response to external stimuli [12]. For instance, SMP-based soft robots with embedded CNTs can be actuated using electrical signals, allowing them to perform complex movements with high precision [13]. The shape memory effect of SMPs ensures that these robots can return to their original form after deformation, enabling them to perform repetitive tasks without losing functionality [14].

Nano-enhanced actuators with high precision are also being developed using nanostructured SMPs. These actuators can generate precise and controlled movements in response to external stimuli, such as heat, light, or magnetic fields [15]. The incorporation of nanostructures, such as gold nanoparticles or graphene, enhances the responsiveness and durability of these actuators, making them suitable for applications in microelectromechanical systems (MEMS) and biomedical devices [16]. For example, SMP-based actuators with embedded gold nanoparticles can be actuated using light, enabling precise control over their movement in applications such as microsurgery and drug delivery [17].

7 Challenges and Future Perspectives

7.1 Current Limitations and Unresolved Issues

Despite the remarkable advancements in nanostructured Shape Memory Polymers (SMPs), several challenges remain that limit their widespread adoption and application. One of the primary limitations is durability, particularly in applications requiring repeated shape memory cycles. Over time, the repeated deformation and recovery of SMPs can lead to material degradation, such as micro-cracking, loss of mechanical strength, and reduced shape recovery efficiency [11]. For example, SMPs used in biomedical stents or soft robotics may experience fatigue failure after prolonged use, compromising their performance and safety [12, 60]. The incorporation of nanostructures, such as carbon nanotubes (CNTs) or silica nanoparticles, has improved fatigue resistance, but further research is needed to develop SMPs that can withstand millions of cycles without degradation [32].

Processing challenges also pose significant hurdles in the fabrication of nanostructured SMPs. Achieving uniform dispersion of nanofillers within the polymer matrix is critical for ensuring consistent mechanical and functional properties. However, nanoparticles tend to agglomerate due to their high surface energy, leading to inhomogeneous material properties and reduced performance [4, 61]. Techniques such as surface functionalization and in-situ polymerization have been employed to improve dispersion, but these methods often require complex processing steps and may increase production costs [15]. Additionally, the scalability of advanced fabrication techniques, such as electrospinning and 3D printing, remains a challenge, as these methods are often limited by low production rates and high costs [62].

Reproducibility is another critical issue in the development of nanostructured SMPs. Variations in nanoparticle size, shape, and distribution can lead to inconsistent material properties, making it difficult to produce SMPs with predictable performance [63]. Standardization of fabrication processes and quality control measures is essential to ensure reproducibility and reliability in industrial applications. Furthermore, the lack of standardized testing protocols for evaluating the shape memory performance of SMPs complicates the comparison of different materials and limits their adoption in regulated industries, such as aerospace and healthcare [8].

Cost and scalability issues also hinder the commercialization of nanostructured SMPs. The production of high-quality nanofillers, such as CNTs and graphene, is expensive, and their incorporation into SMPs often requires specialized equipment and processing techniques [19]. These factors contribute to the high cost of nanostructured SMPs, making them less competitive compared to conventional materials. Scaling up production while maintaining material quality and performance is another significant challenge, particularly for applications requiring large volumes of SMPs, such as automotive components or construction materials [10].

7.2 Emerging Trends in Nanostructured SMPs

The field of nanostructured SMPs is rapidly evolving, with several emerging trends that promise to address current limitations and unlock new possibilities. One of the most exciting trends is the use of artificial intelligence (AI) and machine learning (ML) to accelerate material design and optimization. AI-driven approaches can analyze vast datasets of material properties and processing parameters to identify optimal compositions and fabrication methods for SMPs with specific performance requirements [11]. For example, AI algorithms have been used to predict the shape memory behavior of SMPs based on their molecular structure and nanofiller content, enabling the rapid development of new materials with tailored properties [12].

Biomimetic nanostructures are another emerging trend, where SMPs are designed to mimic the structure and function of natural materials. For instance, SMPs with hierarchical nanostructures inspired by plant leaves or animal skins have been developed to achieve superior mechanical properties and multi-stimuli responsiveness [63]. These biomimetic materials offer new opportunities for applications in soft robotics, wearable electronics, and biomedical devices, where flexibility, durability, and adaptability are critical [4].

Next-generation shape-shifting materials are also being developed, with a focus on achieving more complex and reversible shape transformations. For example, SMPs with multi-shape memory effects can transition between multiple temporary shapes in response to different stimuli, enabling more sophisticated functionality in applications such as deployable structures and adaptive textiles [25]. Additionally, the integration of SMPs with other smart materials, such as liquid crystal elastomers or piezoelectric materials, is opening up new possibilities for creating hybrid materials with enhanced performance and functionality [13].

7.3 Commercialization Potential

The commercialization of nanostructured SMPs is gaining momentum, driven by increasing demand for smart materials in various industries. Market trends indicate growing interest in SMPs for applications such as biomedical devices, aerospace components, and consumer electronics, where their unique properties offer significant advantages over conventional materials [17, 19]. For example, the global market for SMPs in biomedical applications is expected to grow significantly, driven by the increasing adoption of minimally invasive surgical techniques and the development of advanced drug delivery systems [38].

However, several challenges must be addressed to fully realize the commercialization potential of nanostructured SMPs. Regulatory challenges, particularly in the healthcare and aerospace sectors, require rigorous testing and certification to ensure the safety and reliability of SMP-based products [42]. Additionally, the high cost of production and limited scalability of advanced fabrication techniques pose barriers to widespread adoption [64].

Future industrial developments are likely to focus on overcoming these challenges through innovations in material design, processing techniques, and quality control. For example, the development of low-cost nanofillers and scalable fabrication methods could reduce production costs and enable the mass production of SMPs for large-scale applications [21]. Furthermore, collaborations between academia, industry, and regulatory bodies will be essential to establish standardized testing protocols and accelerate the commercialization of SMP-based products [65].

8 Conclusion

This chapter has provided a comprehensive overview of nanostructured Shape Memory Polymers (SMPs), highlighting their unique properties, fabrication techniques, and diverse applications. The integration of nanostructures, such as carbon nanotubes, graphene, and metallic nanoparticles, has significantly enhanced the mechanical, thermal, and electrical properties of SMPs, enabling their use in cutting-edge fields such as biomedical engineering, aerospace, flexible electronics, and soft robotics. The ability of nanostructured SMPs to respond to multiple stimuli, including heat, light, and magnetic fields, has opened up new possibilities for creating smart materials with unprecedented functionality and adaptability.

Despite these advancements, several challenges remain, including issues related to durability, processing scalability, and reproducibility. Addressing these challenges will require continued innovation in material design, fabrication techniques, and quality control. Future research should focus on developing low-cost, scalable production methods, improving the fatigue resistance of SMPs, and exploring new applications in emerging fields such as AI-driven material design and biomimetic systems.

In conclusion, nanostructured SMPs represent a transformative class of materials with the potential to revolutionize industries ranging from healthcare to aerospace. Their unique combination of properties, coupled with ongoing advancements in nanotechnology, positions them as a cornerstone of future smart material development. As research progresses, nanostructured SMPs are poised to play a pivotal role in shaping the next generation of adaptive, responsive, and intelligent materials.

References

1. Leng, X., Lan, Y., Liu, Y., Du, S.: Prog. Mater. Sci. **56**, 1077–1135 (2011)
2. Lendlein, A., Jiang, H., Ju¨nger, O., Langer, R.: Nature, **434**, 879 (2005)
3. Xie, T.: Nature, **464**, 267–270 (2010)
4. Huang, W.M., Yang, B., Zhao, Y., Ding, Z.: J. Mater. Chem. **20**, 3367–3381 (2010)
5. Meng, Q., Hu, J.: Compos. A **40**, 1661–1672 (2009)
6. Liu, Y., Lv, H., Lan, X., Leng, J., Du, S.: Compos. Sci. Technol. **69**, 2064–2068 (2009)
7. Zhang, F., Zhou, T., Liu, Y., Leng, J.: Sci. Rep. **5**, 11152 (2015)
8. Lu, H., Leng, J., Du, S.: Soft Matter **9**, 3851–3858 (2013)
9. Bao, M., Lou, X., Zhou, Q., Dong, W., Yuan, H., Zhang, Y., Appl, A.C.S.: Mater. Interfaces **6**, 2611–2621 (2014)
10. Ge, Q., Sakhaei, A.H., Lee, H., Dunn, C.K., Fang, N.X., Dunn, M.L.: Sci. Rep. **6**, 31110 (2016)
11. Li, W., Liu, Y., Leng, J.: J. Mater. Chem. A **3**, 24532–24539 (2015)
12. Zhang, F., Xia, Y., Liu, Y., Leng, J.: Nanoscale Horiz. **5**, 1155–1173 (2020)
13. Wischke, C., Schossig, M., Lendlein, A.: Small **10**, 83–87 (2014)
14. Santo, L., Quadrini, F., Mascetti, G., Dolce, F., Zolesi, V.: Acta Astronaut. **91**, 333–340 (2013)
15. Ding, Z., Yuan, C., Peng, X., Wang, T., Qi, H.J., Dunn, M.L.: Sci. Adv. **3**, e1602890 (2017)
16. Zhang, F., Xia, Y., Wang, L., Liu, L., Liu, Y., Leng, J., Appl, A.C.S.: Mater. Interfaces **10**, 35526–35532 (2018)
17. Liu, S., Li, A., He, S., Xuan, P.: Compos. A **78**, 35–41 (2015)
18. Fan, D., Li, M., Qiu, J., Xing, H., Jiang, Z., Tang, T., Appl, A.C.S.: Mater. Interfaces **10**, 22669–22677 (2018)
19. Tseng, L.-F., Mather, P.T., Henderson, J.H.: Acta Biomater. **9**, 8790–8801 (2013)
20. Gupta, R., Mittal, G., Kumar, K., Pandel, U.: Funct. Compos. Struct. **6**, 025007 (2024)
21. Yang, Z., Huck, W.T., Clarke, S.M., Tajbakhsh, A.R., Terentjev, E.M.: Nat. Mater. **4**, 486–490 (2005)
22. Gupta, R., Mittal, G., Nhaichaniya, G.K., Kumar, K., Pandel, U.: J. Appl. Polym. Sci. **141**, e56233 (2024)
23. Pretsch, T., Ecker, M., Schildhauer, M., Maskos, M.: J. Mater. Chem. **22**, 7757–7766 (2012)
24. Li, P., Han, Y., Wang, W., Liu, Y., Jin, P., Leng, J.: Sci. Rep. **7**, 44333 (2017)
25. Gong, T., Zhao, K., Liu, X., Lu, L., Liu, D., Zhou, S.: Small **12**, 5769–5778 (2016)
26. Friess, F., Nöchel, U., Lendlein, A., Wischke, C.: Adv. Healthcare Mater. **3**, 1986–1990 (2014)
27. Xiao, R., Guo, J., Safranski, D.L., Nguyen, T.D.: Soft Matter **11**, 3977–3985 (2015)
28. Zhao, Y., Huang, W.M., Fu, Y.Q.: J. Micromech. Microeng. **21**, 067007 (2011)
29. Gupta, R., Mittal, G., Kumar, K., Pandel, U.: Model. Simul. Mater. Sci. Eng. **32**, 055021 (2024)
30. Li, C., Jiao, Y., Lv, X., Wu, S., Chen, C., Zhang, Y., Li, J., Hu, Y., Wu, D., Chu, J., Appl, A.C.S.: Mater. Interfaces **12**, 13464–13472 (2020)
31. Boyle, A.J., Wierzbicki, M.A., Herting, S., Weems, A.C., Nathan, A., Hwang, W., Maitland, D.J.: Med. Eng. Phys. **49**, 56–62 (2017)
32. Dong, Y., Fu, Y., Ni, Q.-Q.: J. Appl. Polym. Sci. **132**, 42599 (2015)
33. Yang, C., Boorugu, M., Dopp, A., Ren, J., Martin, R., Han, D., Choi, W., Lee, H.: Mater. Horiz. **6**, 1244–1250 (2019)

34. Hendrikson, W.J., Rouwkema, J., van Blitterswijk, C.A., Moroni, L.: RSC Adv. **5**, 54510–54516 (2015)
35. Kolesov, I., Dolynchuk, O., Jehnichen, D., Reuter, U., Stamm, M., Radusch, H.-J.: Macromolecules **48**, 4438–4450 (2015)
36. Pandini, S., Riccò, T., Borboni, A., Bodini, I., Vetturi, D., Cambiaghi, D., Toselli, M., Paderni, K., Messori, M., Pilati, F., Chiellini, F., Bartoli, C.: J. Mater. Eng. Perform. **23**, 2545–2552 (2014)
37. Tippets, C.A., Li, Q., Fu, Y., Donev, E.U., Zhou, J., Turner, S.A., Jackson, A.-M.S., Ashby, V.S., Sheiko, S.S., Lopez, R., Appl, A.C.S.: Mater. Interfaces **7**, 14288–14293 (2015)
38. Lu, H., Gou, J., Leng, J., Du, S.: Appl. Phys. Lett. **98**, 174105 (2011)
39. Gupta, R., Deshmukh, M., Mittal, G.: Smart Mater. Struct. **33**, 125032 (2024)
40. Du, H., Sóng, Z., Wang, J., Liang, Z., Shen, Y., You, F.: Sens. Actuators, A, **228**, 1–8 (2015)
41. Tan, L., Gan, L., Hu, J., Zhu, Y., Han, J.: Compos. A **76**, 115–123 (2015)
42. Yu, K., Liu, Y., Leng, J.: RSC Adv. **4**, 2961–2968 (2014)
43. Kumar, U.N., Kratz, K., Heuchel, M., Behl, M., Lendlein, A.: Adv. Mater. **23**, 4157–4162 (2011)
44. Mohr, R., Kratz, K., Weigel, T., Lucka-Gabor, M., Moneke, M., Lendlein, A.: Proc. Natl. Acad. Sci. U. S. A. **103**, 3540 (2006)
45. Zhang, F.H., Zhang, Z.C., Luo, C.J., Lin, I.T., Liu, Y., Leng, J., Smoukov, S.K.: J. Mater. Chem. C **3**, 11290–11293 (2015)
46. Razzaq, M.Y., Behl, M., Lendlein, A.: Adv. Funct. Mater. **22**, 184–191 (2012)
47. Huang, W.M., Yang, B., An, L., Li, C., Chan, Y.S.: Appl. Phys. Lett. **86**, 114105 (2005)
48. Fang, Y., Ni, Y., Leo, S.-Y., Taylor, C., Basile, V., Jiang, P.: Nat. Commun. **6**, 7416 (2015)
49. Li, G., Fei, G., Xia, H., Han, J., Zhao, Y.: J. Mater. Chem. **22**, 7692–7696 (2012)
50. Fang, Y., Ni, Y., Choi, B., Leo, S.Y., Gao, J., Ge, B., Taylor, C., Basile, V., Jiang, P.: Adv. Mater. **27**, 3696–3704 (2015)
51. Deng, S., Wu, J., Dickey, M.D., Zhao, Q., Xie, T.: Adv. Mater. **31**, 1903970 (2019)
52. Zhao, Q., Qi, H.J., Xie, T.: Prog. Polym. Sci. **49–50**, 79–120 (2015)
53. Gupta, R., Mittal, G., Kumar, K., Pandel, U.: In: International conference on fundamental and industrial research in materials, pp. 151–164. Springer Nature Singapore, Singapore (2023)
54. Miao, S., Zhu, W., Castro, N.J., Nowicki, M., Zhou, X., Cui, H., Fisher, J.P., Zhang, L.G.: Sci. Rep. **6**, 27226 (2016)
55. Wei, H., Zhang, Q., Yao, Y., Liu, L., Liu, Y., Leng, J., Appl, A.C.S.: Mater. Interfaces **9**, 876–883 (2017)
56. Galabura, Y., Soliani, A.P., Giammarco, J., Zdyrko, B., Luzinov, I.: Soft Matter **10**, 2567–2573 (2014)
57. Zhang, D., Petersen, K.M., Grunlan, M.A., Appl, A.C.S.: Mater. Interfaces **5**, 186–191 (2013)
58. Fang, Y., Ni, Y., Leo, S.Y., Wang, B., Basile, V., Taylor, C., Jiang, P., Appl, A.C.S.: Mater. Interfaces **7**, 23650–23659 (2015)
59. FengLong, J., Yong, Z., JinLian, H., Yan, L., Lap-Yan, Y., GuangDou, Y.: Smart Mater. Struct. **15**, 1547 (2006)
60. Zhang, F., Zhang, Z., Liu, Y., Cheng, W., Huang, Y., Leng, J.: Compos. A **76**, 54–61 (2015)
61. Gong, T., Li, W., Chen, H., Wang, L., Shao, S., Zhou, S.: Acta Biomater. **8**, 1248–1259 (2012)
62. Tan, L., Hu, J., Huang, H., Han, J., Hu, H.: Int. J. Biol. Macromol. **79**, 469–476 (2015)
63. Fu, G.-D., Xu, L.-Q., Yao, F., Li, G.-L., Kang, E.-T., Appl, A.C.S.: Mater. Interfaces **1**, 2424–2427 (2009)
64. Zhang, F., Zhang, Z., Liu, Y., Lu, H., Leng, J.: Smart Mater. Struct., **22**, (2013)
65. Zhuo, H.T., Hu, J.L., Chen, S.J.: J. Mater. Sci. **46**, 3464–3469 (2011)

Engineered Nano Polymer-Based Hydrogel

Ameya Sharma, Divya Dheer, and Vivek Puri

Abstract Engineered nano polymer-based hydrogels are a transformative class of flexible materials that combine the distinctive properties of hydrogels with the improved functionalities provided by nanomaterials. The development, synthesis, and diverse applications of these advanced hybrid systems are the focus of this comprehensive review. The report commences with an introduction to hydrogels and the importance of nano-engineering. It then explores the fundamental physicochemical properties, swelling behavior, and crosslinking mechanisms that are essential for their functionality. A variety of synthesis and fabrication techniques, such as chemical, physical, and nanotechnological methods, are assessed for their ability to customize hydrogel performance. The detailed discussion of the incorporation of nanomaterials, including metallic nanoparticles, carbon-based structures (e.g., graphene, carbon nanotubes), polymeric nanostructures, and nanoclays, emphasizes their impact on bioactivity, responsiveness, and mechanical strength. Extensive reviews of application domains are conducted to emphasize their potential in biomedicine (drug delivery, wound healing, tissue engineering), agriculture, and environmental remediation. Challenges such as long-term stability, regulatory constraints, and scalability persist, despite optimistic advancements. The chapter concludes by identifying emerging trends, such as the integration of hydrogels with wearable or implantable devices and smart and stimuli-responsive systems, which emphasize their potential future roles in nanomedicine and material science. The purpose of this work is to impart a comprehensive comprehension of hydrogels based on nano polymers and to motivate additional research and innovation in the field.

A. Sharma (✉) · D. Dheer · V. Puri
Chitkara School of Pharmacy, Chitkara University, Baddi 174103, Himachal Pradesh, India
e-mail: ameya.sharma@chitkarauniversity.edu.in

B. Sharma et al. (eds.), *Nanostructured Polymers*, Engineering Materials,
https://doi.org/10.1007/978-981-95-3581-1_5

1 Introduction

Hydrogels are hydrophilic polymer networks that are three-dimensional and possess a unique ability to incorporate and retain significant amounts of water or biological fluids within their structure without dissolving [1]. The presence of hydrophilic functional groups—including hydroxyl, carboxyl, amino, and sulfonic groups—distributed throughout the polymer chains is the primary cause of this distinctive capability [2]. Hydrogels are capable of maintaining their structural integrity by physically or chemically crosslinking polymer chains, despite their high-water content. This allows them to maintain a solid or gel-like form even in the presence of mechanical stress [3]. Hydrogels are physically similar to natural biological tissues, such as epidermis, cartilage, and mucosa, due to their soft and elastic nature. Hydrogels are highly appealing materials for a diverse array of biomedical and pharmaceutical applications, including biosensing, wound healing, controlled drug delivery, and regenerative medicine, due to their tissue-mimicking behaviour [4].

The polymeric building elements of hydrogels can be derived from both natural and synthetic sources, each of which has its own unique advantages and drawbacks. Natural polymers, including alginate, chitosan, gelatin, hyaluronic acid, and collagen, are often biodegradable, biocompatible, and promote the development of cellular interactions. They are derived from biological sources [5]. These characteristics render natural hydrogels highly advantageous for in vivo applications, particularly in the fields of tissue engineering and cell encapsulation. Nevertheless, they frequently experience inadequate mechanical strength, limited tunability of physicochemical properties, and batch-to-batch variability. In contrast, synthetic polymers, such as polyacrylamide (PAAm), polyethylene glycol (PEG), polyvinyl alcohol (PVA), and poly(N-isopropylacrylamide) (PNIPAm), provide a high degree of control over molecular architecture, mechanical strength, and degradation kinetics. Although these materials can be precisely engineered to comply with the requirements of particular applications, they may necessitate further chemical modification to enhance their biocompatibility and mitigate toxicity [6].

2 Importance of Nano-Engineering in Hydrogels

Through the integration of nanotechnology into the design, synthesis, and functionalization of hydrogels, nanoengineering has substantially altered the field of hydrogel research. Materials exhibit distinctive chemical, physical, and biological characteristics that are distinct from those of their bulk counterparts at the nanoscale. The incorporation of nanostructures or nanomaterials into hydrogels leads to enhanced bioactivity, controlled responsiveness, and improved mechanical strength [7]. For instance, nanocomposite hydrogels are composed of a hydrogel matrix that is embedded with nanoparticles, including silica, graphene oxide, carbon nanotubes, or metallic nanoparticles. These inclusions have the potential to provide mechanical reinforcement, antimicrobial activity, or electrical conductivity. Furthermore, nano-engineered hydrogels can provide precise drug delivery through stimulus-responsive release, in which nanostructures function as carriers or gates. Additionally, the advancement of hydrogels based on nano-polymers has facilitated the exploration of novel opportunities in the field of regenerative medicine [8]. These hydrogels have the ability to closely resemble the extracellular matrix (ECM), thereby promoting tissue regeneration and guiding cell behavior. For example, hydrogels that incorporate nanofibers are well-suited for the engineering of bone, cartilage, and nerve tissue, as they can regulate cell adhesion and proliferation. In essence, the functional complexity of hydrogels is enhanced by the convergence of nanotechnology and hydrogel science, which broadens their applicability in high-performance biomedical and industrial systems [9].

3 Fundamentals of Nano Polymer-Based Hydrogels

Nano polymer-based hydrogels are an intricate category of flexible materials that integrate the water-retaining capabilities of traditional hydrogels with the distinctive physicochemical properties of nanostructured materials. Typically, these hybrid systems are comprised of or embedded with polymers and structures at the nanoscale, with dimensions ranging from 1 to 100 nm. The hydrogel matrix is endowed with increased mechanical strength, increased surface area, and functional versatility through the incorporation of nanomaterials, including metallic nanoparticles, silica, graphene derivatives, or nanofibers. These properties are essential for biomedical and engineering applications [10]. The objective of hydrogel nano-engineering is to enhance their structural stability, compatibility with biological systems, and responsiveness to environmental stimuli, thereby rendering them suitable for applications in biosensing, drug delivery, wound healing, and tissue engineering [11].

In accordance with their internal structure, composition, and the nature of their crosslinking mechanisms, nano polymer-based hydrogels can be systematically classified. They are classified into three primary categories based on their structural characteristics. Nanocomposite hydrogels, the initially developed type, are composed of dispersed nanoparticles, including silver, gold, silica, and graphene oxide, that are incorporated into the hydrogel matrix. This process improves the hydrogel's mechanical resilience, conductivity, and antimicrobial activity [12]. The second form of hydrogel, nanofiber-reinforced hydrogels, employs electrospun or self-assembled nanofibers to fabricate a scaffold-like reinforcement that closely resembles the extracellular matrix (ECM). This reinforcement supports cell attachment and tissue regeneration. The third category encompasses vesicular or micellar hydrogels, which are produced through the self-assembly of amphiphilic block copolymers. These structures are particularly advantageous for the encapsulation of hydrophobic pharmaceuticals or bioactive molecules, as they provide compartmentalization within the hydrogel [13].

Nano polymer-based hydrogels are typically classified into three categories based on their composition: synthetic, natural, and hybrid. Materials such as poly (N-isopropylacrylamide), polyethylene glycol (PEG), and poly(acrylic acid) are synthetic polymer-based hydrogels that enable precision control over mechanical and chemical characteristics. However, they frequently necessitate functionalization to ensure biological compatibility [14]. Although natural polymer-based hydrogels, including those derived from chitosan, alginate, and gelatin, exhibit exceptional biocompatibility and biodegradability, they may not possess the necessary mechanical strength for applications requiring load bearing capability. Hybrid hydrogels have been developed by amalgamating synthetic and natural polymers in order to circumvent these constraints. This method synergistically combines the biological affinity of natural polymers with the tunability of synthetic polymers, resulting in the development of materials that are both biointeractive and robust [15].

The underlying structure as well as functional behavior of hydrogels are significantly influenced by the crosslinking mechanism. Crosslinking can be classified as either physical or chemical. Hydrogels that are physically crosslinked depend on non-covalent interactions, including ionic interactions, hydrogen bonding, and polymer chain entanglement. For example, hydrogen bonding is frequently employed because of its reversibility, which renders it appropriate for drug-delivery systems and injectables. Stable yet reversible gels are formed by ionic interactions, such as those between alginate and divalent cations like calcium, which are ideal for encapsulating biological agents [16]. Crystallization zones that function as natural crosslinking sites are induced by physical entanglement through freeze–thaw cycles, particularly in poly(vinyl alcohol) (PVA) systems, without the need for external chemicals. Conversely, chemically crosslinked hydrogels utilize covalent bonds to establish networks that tend to be more durable and stable. Monomers and crosslinkers are activated by thermal or chemical initiators to form robust hydrogel structures, which can be synthesized using techniques such as free radical polymerization [17]. Clicking chemistry, which enables substantially specific and efficient covalent bonding under benign conditions, and photo-crosslinking, which activates light-sensitive groups by UV or visible light to initiate gelation, are more sophisticated methods. These chemical methods are particularly advantageous for the development of hydrogels that are spatially controlled or patterned, which are utilized in 3D bioprinting and biofabrication [18].

The physicochemical properties of nano polymer-based hydrogels are a significant factor in determining their performance and suitability for specific applications. One of the most critical properties is the hydrogel's capacity to absorb and retain water, which is evident in its swelling capacity. This behavior is influenced by environmental conditions, ionic strength, crosslinking density, and polymer composition. In general, high edema is advantageous for applications such as drug delivery or wound dressings, as it enables diffusion due to the water content [19]. This behavior can be substantially altered by the nanomaterials present in the hydrogel. Hydrophilic nanofillers, such as modified cellulose nanocrystals or specific metal oxides, can improve water absorption, whereas hydrophobic nanoparticles, such as carbon nanotubes, may decrease edema but contribute to increased mechanical strength and stability.

Mechanical strength is an additional critical characteristic that frequently restricts the practical application of conventional hydrogels. The elasticity, compressive strength, and resilience of hydrogels are significantly enhanced by nano-reinforcement strategies, such as the incorporation of graphene oxide or carbon-based nanostructures [20]. This renders nano polymer-based hydrogels particularly well-suited for load-bearing applications, such as the engineering of bone or cartilage tissue. The rate at which nutrients, waste products, and therapeutic agents diffuse through the hydrogel matrix is determined by permeability, which is equally important, particularly in biomedical contexts. It is essential to regulate permeability in order to create hydrogels that have selective filtration capabilities or controlled drug release profiles [21].

The fundamental prerequisites for in vivo use are biocompatibility and biodegradability. The nature of the base polymers and the degradation products that are produced over time are the determining factors for these properties. Synthetic polymers frequently necessitate meticulous selection or modification to guarantee safe degradation, whereas natural polymers are inherently biodegradable and offer superior biocompatibility [22]. Lastly, the stimuli-responsive behavior of nano polymer-based hydrogels is one of their most innovative characteristics. These hydrogels can be designed to respond to a diverse array of external stimuli, such as pH, temperature, light, magnetic fields, or the presence of specific enzymes, by causing modifications in their mechanical properties, solubility, or swelling. This responsiveness is of great value in applications such as biosensors, smart wound dressings, and on-demand drug delivery systems [23].

4 Synthesis and Fabrication Techniques

Each of the numerous chemical and physical strategies employed in the synthesis and the production of nano polymer-based hydrogels is intended to adapt the structural, mechanical, as well as functional properties of the resulting hydrogels. The hydrogel's morphology, porosity, degradation behavior, and compatibility with encapsulated drugs or cells are all directly influenced by the synthesis method selected. Consequently, the selection of a method is of the utmost importance. In general, fabrication techniques are categorized into three main categories: chemical methods, physical approaches, and advanced nanotechnological strategies. Each of these categories provides unique benefits for specific applications [24].

4.1 Chemical Methods

One of the most frequently employed chemical synthesis techniques is free radical polymerization, a scalable and robust method that enables the formation of covalently crosslinked hydrogel networks. Reactive monomers, such as acrylamide or N-isopropylacrylamide, crosslinking agents like N,N′-methylenebisacrylamide (MBA), and initiators like ammonium persulfate (APS) are frequently employed in conjunction with catalysts like tetramethylethylenediamine (TEMED) in this process [25]. Depending on the desired characteristics, this method can be implemented in a variety formulation, ranging including bulk, solution, emulsion, and suspension polymerization. Each formulation has an impact on properties such as particle size distribution, mechanical integrity, and internal porosity. Free radical polymerization is an acceptable choice for both laboratory and industrial production due to its high yield, process scalability, and tunability of polymer structure. Nevertheless, this method is restricted by the potential cytotoxicity of residual monomers or initiators, which requires extensive purification procedures, particularly in biomedical applications [26, 27].

Graft copolymerization is an additional valuable technique that allows for the chemical modification of natural polymers by affixing synthetic polymer chains to their backbone. The mechanical stability, solubility, and functional efficacy of natural hydrogels are enhanced by this hybridization approach. For instance, the addition of polyethylene glycol (PEG) to chitosan not only improves its water solubility but also confers antifouling and non-immunogenic properties, thereby expanding its applications in tissue engineering and drug delivery. This procedure capitalizes on the processability of synthetic polymers while preserving the biocompatibility of the base material [28].

The bio-orthogonality, specificity, and efficacy of emerging 'click chemistry' techniques and Michael addition reactions have also gained prominence under mild, aqueous conditions. Click chemistry, particularly the azide-alkyne cycloaddition reaction, enables the construction of highly controlled networks and site-specific functionalization. These reactions are optimal for the development of biofunctional hydrogels that are intended for therapeutic and diagnostic applications due to their rapidity, minimal residues, and compatibility with sensitive biomolecules. Michael addition reactions, which frequently involve thiol and maleimide functional groups, also provide comparable advantages, particularly in the development of hydrogels that are injectable and in situ-forming [29, 30].

4.2 Physical Methods

Nano polymer-based hydrogels are constructed without the use of covalent crosslinking agents through the application of a variety of physical fabrication methods. Self-assembly is one such method, in which amphiphilic block copolymers spontaneously organize into micelles, vesicles, or nanofibrillar networks in aqueous environments. The balance of hydrophobic and hydrophilic interactions is the cause of this self-organization, which results in physically crosslinked hydrogels that are frequently responsive to environmental changes and reversible. These hydrogels are highly desirable for minimally invasive therapeutics and drug delivery, particularly when they are designed to be stimuli-responsive and injectable [31].

The freeze–thaw cycle is another frequently employed physical method that is frequently employed in the synthesis of poly(vinyl alcohol) (PVA) hydrogels. This method entails the repeated process of freezing and thawing, which results in the formation of crystallite regions within the polymer matrix that serve as physical crosslinking sites and induce phase separation. This technique is particularly well-suited for biomedical applications due to its simplicity, non-toxicity, and absence of chemical crosslinkers. Nevertheless, the hydrogels that result from this process are relatively weak in terms of mechanical strength, which limits their applicability in applications that necessitate structural support or extended mechanical stability [32].

Ionic gelation is an additional physical crosslinking strategy that is particularly pertinent for hydrogels that are based on polyelectrolytes. This method involves the formation of crosslinked structures by naturally occurring polymers, such as alginate, in response to multivalent cations such as calcium (Ca^{2+}), barium (Ba^{2+}), or strontium (Sr^{2+}). Encapsulating cells, proteins, or bioactive molecules without compromising their functionality is advantageous due to the fact that the gelation process is instantaneous and can take place underneath ambient conditions. Ionic gelation is a preferred choice for applications in regenerative medicine, drug delivery, and cell encapsulation due to its biocompatibility, simplicity, and benign reaction conditions [33].

4.3 Advanced Nanotechnological Approaches

The fabrication of nano polymer-based hydrogels has been transformed by the integration of advanced nanotechnological approaches, which has enabled unprecedented control over architecture, functionality, and application-specific performance. One such method is electrospinning, which results in nanofiber mats with a high surface area and an ECM-like topography. Composite scaffolds with improved cell adhesion, proliferation, and mechanical reinforcement can be produced by incorporating these nanofibers into hydrogel matrices. In contrast, 3D bioprinting allows for the precise deposition of photopolymerizable hydrogels in a layer-by-layer fashion, resulting in the development of intricate, cell-laden constructs for tissue engineering. Gradient structures or multiple cell types can be incorporated into these printed hydrogels to facilitate functional tissue regeneration, and they can be customized to replicate patient-specific anatomy [34].

Layer-by-Layer (LbL) assembly is another innovative approach that enables the nanoscale deposition of alternating polymer or nanoparticle layers on hydrogel substrates. This method is particularly advantageous for the functionalization of hydrogel surfaces with biomolecules, growth factors, or therapeutic agents, as it provides precise spatial and temporal control over the release kinetics. The LbL method also facilitates the development of multi-responsive surfaces that can continually engage the biological environment [35].

Another frontier in hydrogel fabrication is nanoparticle-driven crosslinking. In this context, metallic nanoparticles, including gold (AuNPs), silver (AgNPs), or silica particles, perform dual functions as functional additives and crosslinkers. These nanoparticles have the potential to improve mechanical strength, introduce antimicrobial properties, and respond to external stimuli, including light, magnetic fields, and redox conditions. Their uniform dispersion within the hydrogel matrix guarantees their consistent mechanical and functional performance throughout the material [36].

Lastly, stimuli-responsive nanogels, which are nanoscale hydrogel particles that respond to specific physiological triggers, can be incorporated into bulk hydrogels to introduce intelligent, programmable behavior. These nanogels are highly suitable for targeted drug delivery and biosensing applications due to their ability to respond to pH shifts, temperature changes, enzymatic activity, or oxidative environments. These responsive nanogels enhance the efficacy of the hydrogel in real-time biological systems by enabling the on-demand release of therapeutic agents or dynamic adaptation to environmental changes when incorporated into a hydrogel matrix [37].

5 Types of Nanomaterials Used

Polymer-based hydrogels have emerged as flexible and promising materials in biomedical, pharmacological, and engineering applications owing to their remarkable biocompatibility, elevated water content, and controlled physical properties [38]. The incorporation of nanoparticles into these hydrogels has significantly improved their structural, mechanical, and functional properties, resulting in the development of nanocomposite hydrogels with increased attributes [39]. A diverse array of nanomaterials, such as nanoparticles, nanofibers, nanotubes, nanosheets, and quantum dots, has been thoroughly investigated to enhance the mechanical strength, drug transport efficacy, responsiveness, and bioactivity of polymer-based hydrogels [40, 41]. Metallic and metal oxide nanoparticles, including silver, gold, zinc oxide, and titanium dioxide, are widely incorporated into hydrogels due to their antibacterial, catalytic, and optical properties. Additionally, carbon-based nanoparticles such as carbon nanotubes, graphene oxide, and fullerenes provide substantial electrical conductivity, mechanical enhancement, and bioactivity to the hydrogel matrix [42, 43]. Polymeric nanostructures such as dendrimers, micelles, and nanogels are utilized to permit controlled release of drugs and targeted delivery [44]. Moreover, ceramic and silica nanoparticles have been utilized to enhance the thermal stability and mechanical integrity of hydrogels, whereas nanoclay and nanocellulose enhance their structural strength and swelling capacity [45]. The collaborative interaction between nanoparticles and polymer networks enables the development of multifunctional hydrogels designed for specific purposes, including tissue engineering, wound healing, biosensing, and intelligent drug delivery systems [46]. Comprehending the various types and functions of nanomaterials utilized in polymer-based hydrogels is essential for progressing next-generation hydrogel technologies with improved performance and improved application potential [47].

5.1 Metallic and Metal Oxide Nanoparticles

Metallic and metal oxide nanoparticles have drawn significant interest in the advancement of polymer-based hydrogels owing to their distinctive physicochemical characteristics, including elevated surface area, antibacterial efficacy, catalytic activities, and optical attributes. Incorporating these nanoparticles into hydrogel matrix enhances the mechanical strength, stability, bioactivity, and responsiveness of the hydrogels, rendering them extremely appropriate for biomedical and pharmaceutical applications. Silver (Ag) and gold (Au) nanoparticles are the most often employed metallic nanoparticles [48, 49]. Silver nanoparticles (AgNPs) are recognized for their comprehensive antimicrobial potential and have been widely integrated into hydrogels for wound healing, infection management, and antibacterial coatings [50]. AgNP-embedded chitosan or polyvinyl alcohol (PVA) hydrogels exhibit remarkable antibacterial activity and excellent biocompatibility, rendering them suitable for wound dressing applications. Gold nanoparticles (AuNPs) are esteemed for their biocompatibility, surface plasmon resonance, and capacity to enhance photothermal therapy [51]. Gold nanoparticles integrated into polymer hydrogels can be utilized for biosensing, delivery of drugs, and cancer treatment. Metal oxide nanoparticles, including zinc oxide (ZnO), titanium dioxide (TiO_2), iron oxide (Fe_3O_4), and cerium oxide (CeO_2), are extensively utilized in polymer-based hydrogels [52]. ZnO nanoparticles exhibit antibacterial characteristics and UV protection, rendering ZnO-hydrogel composites appropriate for skincare and wound healing applications. Titanium dioxide nanoparticles enhance photocatalytic efficacy and have been utilized in hydrogels for environmental purposes such as water purification [53, 54]. Magnetic Fe_3O_4 nanoparticles confer magnetic responsiveness, advantageous for targeted drug administration or automated hydrogel systems. Consequently, the incorporation of metallic and metal oxide nanoparticles into hydrogels provides a means to develop sophisticated, multifunctional materials for various applications [55]. Zaoui et al. [56] synthesized chitosan-encapsulated hydrogel composite beads incorporating activated carbon@metal oxide (CaO, Al_2O_3, CaO/Al_2O_3) derived from olive stone biomass, characterized through SEM, BET, TGA, Zeta potential, XRD, and FTIR; the porous biosorbents exhibited substantial adsorption capacities for Orange G (Q_{max} = 323.625 mg/g) and methylene blue (Q_{max} = 134.589 mg/g) in accordance with Langmuir and pseudo-second order kinetics, showing remarkable reusability [56]. Peng et al. [57] showed an innovative method for the fabrication of acrylamide hydrogels utilizing eutectic gallium-indium (EGaIn) liquid metal nanoparticles as UV-activated photoinitiators, resulting in transparent hydrogels with adjustable mechanical properties, thereby highlighting the potential of liquid metal in the design of multifunctional hydrogels [57]. Devi et al. [59] developed antibacterial chitosan-casein nanocomposite patches containing CuO-NiO nanoparticles synthesized using plant-mediated methods, exhibiting antibacterial properties, hemocompatibility, and accelerated diabetic wound healing by promoting cell survival and migration [58]. Using 3D printing, Yang et al. [58] developed ZIFC nanozyme-loaded hybrid hydrogels for bone tissue engineering. These hydrogels showed major osteogenic repair through the synergistic catalysis of nitrogen-doped carbon nanostructures, broad-spectrum ROS scavenging, antioxidant defense activation, and inflammation suppression, providing a promising antioxidant scaffold for clinical bone regeneration [59].

5.2 Carbon-Based Nanomaterials (Graphene, CNTs)

Carbon-based nanomaterials, including graphene and carbon nanotubes (CNTs), have proven to be highly efficient nanofillers for improving the structural, mechanical, electrical, and biological properties of polymer-based hydrogels [60]. Their distinctive attributes, including as elevated surface area, electrical conductivity, mechanical robustness, and biocompatibility, render them exemplary choices for diverse biological, pharmaceutical, and technical applications. Graphene, a monolayer of carbon atoms organized in a two-dimensional hexagonal lattice, together with its derivative, graphene oxide (GO), is extensively integrated into hydrogel matrices [61, 62]. Graphene oxide, characterized by its many oxygen-containing functional groups, has superior dispersibility in aqueous environments and robust interactions with polymer chains, hence improving the mechanical strength, flexibility, and swelling properties of hydrogels [63]. Moreover, GO provides electrical conductivity and antibacterial properties, rendering GO-reinforced hydrogels appropriate for use in tissue engineering, biosensors, and drug delivery devices. Graphene oxide integrated into polyvinyl alcohol (PVA) or chitosan hydrogels has exhibited enhanced mechanical stability and biological efficacy, especially in wound healing and scaffold fabrication for tissue regeneration. Single-walled carbon nanotubes (SWCNTs) and multi-walled carbon nanotubes (MWCNTs) are cylindrical nanostructures that exhibit remarkable electrical conductivity, thermal stability, and tensile strength [64, 65]. The incorporation of carbon nanotubes into polymer-based hydrogels markedly improves their mechanical strength, electrical characteristics, and responsiveness to stimuli. CNT-based hydrogels have been investigated for uses in flexible sensors, smart drug delivery systems, and electrically conductive scaffolds for nerve regeneration. CNT-reinforced polyethylene glycol (PEG) and alginate hydrogels showed potential in wearable sensors and bioelectronic systems. Carbon-based nanomaterials are essential in the advancement of next-generation functional hydrogels, offering improved performance and expanded application possibilities [66, 67]. Yang et al. [68] developed a multifunctional nano-enzymatic hydrogel (TCBP) by integrating taxifolin-loaded CuO nanoparticle-modified biochar into a PVA/GO/dextran matrix, providing rapid hemostasis, antibacterial properties, anti-inflammatory effects, hypoxia mitigation, and angiogenesis for the treatment of diabetic wounds [68]. Jiang et al. [69] developed a 3D-structured GO/CNT/Fe_3O_4/PVA hydrogel solar evaporator (GHCFeP) that attained significant evaporation rates (up to 2.30 kg•m^{-2}•h^{-1}) and 92.4% photothermal efficiency, utilizing a honeycomb layout to minimize inactive evaporation zones [69]. Noruzi et al. [70] developed a magnetically sensitive GO/casein/LDH/alginate nanobiocomposite that showed low hemolysis (<3.84%), antibacterial properties, good cell viability, and effective specific absorption rate (67.04 W•g^{-1}) for hyperthermia applications [70]. Cao et al. [71] developed highly reinforced PVA/GO-tannic acid hydrogels highlighting dense multi-crosslinked networks that exhibit markedly improved tensile strength (14.38 MPa) and fracture toughness for load-bearing biomedical applications [71]. Das et al. [72] exhibited GO-chitosan-PVA hydrogels that effectively removed Congo red dye (88.17% at pH 2) in accordance with pseudo-second-order kinetics, suggesting their potential as environmentally friendly adsorbents [72]. Chen et al. [73] presented CNT-wrapped polyacrylamide/alginate hydrogels (CPSA) including convoluted microchannels for optimized photothermal evaporation (1.97 kg•m^{-2}•h^{-1}), methylene blue degradation, and superior EMI shielding (81.1 dB) [73]. Liu et al. [74] formulated agar/polyacrylamide/CNT hydrogels that attained 98.6% light absorption and an evaporation rate of 3.19 kg•m^{-2}•h^{-1}, showing WHO-compliant desalination efficacy and sustained operational stability [74]. Ryu et al. [75] showed AgNP-MWCNT-loaded PF127 hydrogels exhibiting temperature-responsive gelation, improved antibacterial and anti-biofilm properties, antioxidant activity, and exceptional in vitro wound healing (99% scratch closure), while reducing AgNP toxicity [75]. Han et al. [76] developed ionic liquid-modified dual-network hydrogels incorporating CNTs and biomacromolecules for the development of stretchable, sticky, and biocompatible wearable sensors designed to detect joint mobility and subtle biological signals [76].

5.3 Polymeric Nanoparticles

Polymeric nanoparticles are nanoscale entities made from natural or synthetic polymers, commonly included into hydrogel systems to improve their functioning, stability, and controlled release properties. Advantages of these nanoparticles include their high drug-loading capacity, biocompatibility, biodegradability, and adaptability in surface characteristics for targeted delivery [77, 78]. Incorporating them into polymer-based hydrogels results in nanocomposite systems exhibiting enhanced mechanical strength, prolonged drug release profiles, and stimuli-responsive characteristics [79]. Commonly used polymeric nanoparticles comprise nanogels, dendrimers, micelles, and polymeric nanospheres. Nanogels are crosslinked polymer networks at the nanoscale capable of encapsulating drugs, biomolecules, or therapeutic agents, providing controlled and targeted delivery [80]. Hydrogels loaded with nanogels offer dual functionality of localized retention and prolonged release, making them appropriate for wound healing, tissue engineering, and drug delivery applications [81]. Hussein et al. [82] used tannic acid-gelatin hydrogels as in situ templates for bio-synthesized silver nanoparticles (10–15 nm) exhibiting exceptional catalytic activity in A3-coupling reactions and stable reusability across six cycles [82]. Nowfal et al. [83] discovered Arabic gum-chitosan hydrogels containing Au NPs (10–20 nm), exhibiting significant catalytic efficiency in Sonogashira coupling with prolonged activity over eight cycles [83]. Mezzasalma et al. [84] introduced the generalized Weibull rule to effectively describe hydrogel mesh size in various networks, enhancing predictive control in drug delivery, hyperthermia, and polymer transport research [84]. Schutz et al. [85] developed multi-responsive $CoFe_2O_4$@Au\@polymer hydrogels that utilize NIR-triggered photothermal water release, integrating adjustable magneto-plasmonic characteristics and thermosensitivity for enhanced delivery systems [85]. Athavale et al. [86] developed coconut oil-based nanocarrier hydrogels incorporating BMS-SLNs for atopic dermatitis, showing extended drug release, enhanced skin penetration, and cytocompatibility [86]. Rauf et al. [87] developed aceclofenac-loaded carbopol hydrogels incorporating permeation enhancers for rheumatoid arthritis, resulting in sustained drug release, elevated permeation flux, and in vivo anti-arthritic effectiveness [87]. Turk et al. [88] reported a selective, in situ production approach for gold nanoparticles within 3D-printed pHEMA hydrogels, utilizing residual olefinic groups for the targeted inclusion of nanoparticles in multifunctional devices [88]. Malik et al. [89] exhibited PVA-P(AMPS)-g-C_3N_4 hydrogels incorporating ZnO nanoparticles of diverse morphologies, with quasi-spherical nanoparticles achieving superior tensile strength, swelling capacity, antibacterial properties, and over 80% wound healing in vivo, making them appropriate for advanced burn dressings [89].

5.4 *Nanoclays and Other Inorganic Fillers*

Nanoclays and other inorganic fillers have drawn substantial interest in the formulation of polymer-based hydrogels owing to their capacity to enhance mechanical strength, thermal stability, swelling characteristics, and barrier properties of hydrogel networks. Nanoclays are layered silicate substances characterized by a large surface area and aspect ratio, providing exceptional reinforcing properties even at minimal concentrations [90]. Prevalent varieties of nanoclays utilized in hydrogel compositions comprise montmorillonite (MMT), halloysite nanotubes (HNTs), and laponite. The hydrogels dimensional stability, load-bearing capabilities, and structural integrity are all improved when nanoclays are distributed throughout a polymer matrix and interact chemically or physically with the polymer chains. Montmorillonite (MMT)-reinforced hydrogels have been employed in biomedical applications, including wound dressings and delivery of drugs, owing to their enhanced mechanical qualities and regulated swelling characteristics [91, 92]. Halloysite nanotube-infused hydrogels facilitate prolonged drug release and improved mechanical integrity, providing them appropriate for tissue engineering and regenerative medicine. Alongside nanoclays, various inorganic fillers such as silica nanoparticles, calcium phosphate, hydroxyapatite, and bioactive glass are integrated into hydrogels to enhance bioactivity, osteoconductivity, and mineralization capacity, especially for bone tissue engineering [45]. These fillers enhance cell adhesion, proliferation, and differentiation, rendering nanocomposite hydrogels optimal for advanced biological applications. In order to address diabetic bone defects, Li et al. [93] formulated a 3D-printed GelMA-Nanoclay hydrogel scaffold that was loaded with metformin. This scaffold showed improved osteogenesis, angiogenesis, and immunomodulation both in vitro and in vivo [93]. Wu et al. [94] developed injectable ChABC-immobilized Laponite® nanoclay hydrogels for spinal cord injury, resulting in decreased glial scarring, neuronal regeneration, and restoration of motor function [94]. Zhao et al. [95] developed a multifunctional kaolin-OMCC-NHS/PAA hydrogel exhibiting enhanced hemostatic efficacy, substantial tissue adhesion, mechanical robustness, and antibacterial properties for the management of emergency haemorrhage [95]. Liu et al. [96] developed a dECM-GelMA-SA-Nanoclay hydrogel by 3D bioprinting for colorectal cancer modeling, promoting tumor spheroid formation and drug resistance investigations, thereby offering a functional in vitro platform for anticancer research [96]. Far et al. [97] synthesized CS-g-GEL/BNC nanocomposite hydrogels showing enhanced thermal stability, rapid adsorption kinetics, and a high removal efficiency (93.85%, 453.87 mg/g) for Congo red dye, adhering to the Langmuir isotherm and pseudo-first-order kinetics, indicating potential for the treatment of dye-contaminated wastewater [97].

6 Applications of Engineered Nano Polymer-Based Hydrogels (ENPHs)

Engineered Nano Polymer-based Hydrogels (ENPHs) have appeared as a powerful class of smart materials with broad applications across various disciplines owing to their unique physicochemical properties, high water content, tunable mechanical strength, and nano-scale functionalization [98, 99]. These materials have shown immense potential in biomedical, agricultural, and environmental fields, offering targeted solutions to complex challenges as described below.

7 Biomedical Applications

7.1 Drug Delivery

ENPHs have a tremendous potential in the development of drug delivery systems [99]. Their hydrophilic network and nano-polymer architecture helps in increased drug loading, zero-order release, and targeted delivery [100]. For instance, chitosan, cyclodextrins and acrylic acid have been engineered for pH-sensitive hydrogels in colon-targeted delivery [101]. Similarly, alginate based multiresponsive hydrogels have been explored by Li et al. in 2025 for on-demand drug release in respond to specific physiological conditions [102]. Injectable thermoresponsive hydrogels have been investigated for treating central nervous system disorders including Alzhiemer's and Parkinson's disease [103]. Engineered nanocellulose-based hydrogels have been discovered as smart materials because of biocompatibility, biodegradability, recyclability and tunable surface properties [104]. Additionally, ENPHs are playing a significant role in cancer treatment owing to nanoparticles incorporated within hydrogels can increase drug retention at tumor sites via the enhanced permeability and retention (EPR) effect [105, 106]. Further, chemical modification with ligands or antibodies enables active transport and reduced toxicity [107, 108]. Bai et al. in 2024 explored hyaluronic acid and carboxymethyl chitosan for doxorubicin-loaded hydrogel based nanocarriers in triple negative breast cancer [109]. Wu et al. in 2023 designed a new liposomes packed alginate hydrogel system resulting in increased bioavailability of insulin orally [110].

7.2 *Tissue Engineering*

ENPHs impart a three-dimensional structure mimicking the extracellular matrix (ECM), making them suitable for tissue engineering [98]. Their potential to include bioactive agents, cell adhesion, and degrade in a controlled fashion allows for the regeneration of ruptured tissues such as bone, cartilage, skin, and nerve tissues [111]. Phan et al. in 2023 constructed alginate based injectable hydrogels made up of layered double hydroxides for sustained release of recombinant human erythropoietin [112]. Interestingly, nano-hydroxyapatite can be formulated into ENPH network for improved dental implant applications [113]. Apart from bone tissue regeneration, conductive polymers including collagen, polypyrrole or polyaniline-based hydrogel has reported to promote neural stem cell proliferation and differentiation in the presence of electrical stimulation [114, 115].

7.3 *Wound Healing*

ENPHs have also exhibited tremendous potential in wound healing owing to moisture-retention, antimicrobial, and bioactive effect [116]. These hydrogels can be used as dressings maintaining a moist environment ideal for healing, while delivering therapeutic agents such as antibiotics, growth factors, or anti-inflammatory drugs directly to the wounded site [117]. Biopolymers such as chitosan, alginate, and gelatin, when nano-engineered into hydrogels, help in enhancing hemostasis, tissue regeneration, and re-epithelialization [118, 119]. Mokhtari et al. in 2024 fabricated sodium alginate/gelatin and natural extract based hydrogel loaded with dopamine for enhanced antioxidant and antibacterial effect [120]. Hao et al. in 2025 investigated hydrogel loaded with silver nanoparticles bearing photothermal property observed better healing of diabetic wounds [121]. Further, smart hydrogels have the ability to respond in the wound microenvironment (e.g., pH changes due to bacterial infection) can provide on-demand drug release [122].

7.4 Agricultural Applications

The agricultural sector is rapidly adopting ENPHs to address limitations related to water scarcity, nutrient loss, and pesticide overuse [123]. These hydrogels bear superabsorbent nature and controlled release capacity, thus contributing to more sustainable and efficient farming practices [124]. In this direction, ENPHs can encapsulate fertilizers and pesticides slowly releasing them in the presence of different conditions of soil such as moisture content, pH [125]. This reduces nutrient leaching, minimizes environmental pollution, and ensures a controlled release of fertilizers and pesticides to plants [126]. The engineered hydrogels have the tendency of water retention and thus provides soil conditioning [127]. These formulations act as aqueous reservoirs resulting in absorption thus retaining significant amounts of water, which can be gradually released to plant roots, enhancing drought resistance [128]. This water-retentive property helps in improving soil structure, minimizing irrigation frequency, and promotes resource conservation [129]. Ghobashy and group in 2023 synthesized hydroxyl ethyl cellulose/acrylamide based ultra-absorbent hydrogel using radiation method and demonstrated its use in soil conditioning [130]. In a different study, coating seeds engulfed bioactive ENPHs allows the delivery of micronutrients or growth stimulants directly during germination [131]. These coated formulations protect seeds from pathogens or abiotic stress, thus increasing seedling vigor and nascent plant development [132]. The nano-scale functionalized smart hydrogels allows precise modulation of the coating thickness and composition [133].

7.5 Environmental Applications

ENPHs has greatly been investigated for environmental remediation because of their ability to interact with and remove a variety of pollutants from aqueous environment [134]. Their increased surface area, porosity, and chemical tunability make them ideal for adsorption, filtration, and catalytic degradation processes [135]. One of the major applications of engineered hydrogels is in water treatment [136], particularly for the adsorption of heavy metals such as lead, cadmium, and mercury [137]. Functional groups such as carboxyl, amine, or thiol moieties can be grafted onto hydrogel matrices to selectively bind and remove metal ions from water [138]. For instance, nanosized chitosan-based hydrogels have high affinity for chromium thus eliminates it owing to electrostatic interactions and chelation. The hydrogel form allows for easy separation after treatment, and the treatment can be repeated after regeneration of the hydrogel [139].

The engineered hydrogels embedded with photocatalytic nanoparticles such as TiO_2, ZnO have a tendency to degrade organic pollutants such as dyes, pesticides, and pharmaceutical residues under light irradiation [140]. These hydrogels act as carriers, stabilizes nanoparticles thus enhances their surface area for improved degradability. Interestingly, hydrophobic ENPHs have been explored to adsorb oil and organic solvents from contaminated water bodies [141]. These hydrogels can float on water surfaces and selectively absorb hydrocarbons, offering a biodegradable and efficient cleanup solution due to their high porosity and less density [142]. In fact, novel developments in aerogel forms of nano-polymer-based hydrogels have shown utilization in air filtration, including the adsorption of volatile organic compounds, greenhouse gases, and particulate matter [143–145].

8 Challenges and Limitations

The incorporation of nanoparticles into polymeric hydrogels shows significant potential to develop innovative materials with enhanced mechanical, biological, and functional characteristics for biomedical, pharmaceutical, and engineering applications. Numerous nanomaterials, including metallic and metal oxide nanoparticles, carbon-based nanomaterials, polymeric nanoparticles, nanoclays, and inorganic fillers, have been extensively employed to improve the efficacy of hydrogels. Metallic nanoparticles, including silver (Ag) and gold (Au), are integrated to impart antimicrobial, optical, and catalytic attributes, whereas metal oxides such as zinc oxide (ZnO), titanium dioxide (TiO_2), and iron oxide (Fe_3O_4) enhance mechanical strength, magnetic responsiveness, and antimicrobial efficacy. Carbon-based nanomaterials, such as graphene, graphene oxide (GO), and carbon nanotubes (CNTs), enhance the electrical conductivity, mechanical strength, and bioactivity of hydrogels, making them suitable for applications in tissue engineering, biosensing, and drug delivery. Polymeric nanoparticles, including nanogels, dendrimers, and micelles, are utilized to enable controlled drug release, targeted distribution, and enhanced stability of hydrogel systems. Moreover, nanoclays such as montmorillonite (MMT) and halloysite nanotubes (HNTs), along with inorganic fillers like silica, hydroxyapatite, and bioactive glass, markedly improve the mechanical strength, swelling characteristics, and bioactivity of hydrogels, particularly in applications related to wound healing and bone tissue engineering. Despite these advantages, multiple challenges and restrictions impede the extensive translation of nanocomposite hydrogels. A significant challenge is to scale-up and manufacturing, as achieving reproducible synthesis of nanomaterials with consistent size, shape, and properties at an industrial scale is difficult. Ensuring a uniform distribution of nanoparticles within the hydrogel matrix during large-scale production poses challenges, resulting in variability in material performance. Furthermore, the fabrication procedures frequently necessitate meticulous regulation of reaction conditions, hence increasing complexity and expenses in manufacturing. A further notable issue is to regulatory and safety challenges, particularly in biomedical applications. The emergence of nanomaterials prompts concerns regarding their long-term biocompatibility, cytotoxicity, immunogenicity, and possible accumulation in the body. The regulatory frameworks for nanocomposite hydrogels are still developing, and there is an absence of standardized testing techniques to thoroughly evaluate their safety, limiting clinical translation. Moreover, the long-term stability and degradation of nanocomposite hydrogels pose significant constraints. Certain nanoparticles may leach or disintegrate unpredictably under physiological applications affecting the mechanical stability and usefulness of the hydrogel. In tissue engineering and implantable devices, the hydrogel's breakdown rate must align with tissue regeneration processes; yet, attaining regulated degradation while preserving structural integrity poses a significant problem. The long-term behavior of inorganic fillers or metallic nanoparticles in biological systems remains unclear, prompting environmental and health concerns. Therefore, despite significant advancements in the development of nanomaterial-reinforced polymer-based hydrogels, it is essential to solve scale-up, regulatory, safety, and stability concerns to fully realize their potential in clinical and industrial applications.

9 Future Perspectives

The engineered nano polymer-based hydrogels are made to revolutionize pharmaceuticals through their ability to mimic biological systems in the presence of external stimuli, further integrating with emerging technologies. These formulations are expected to take center stage in the development of smart, responsive, and multi-functional systems in the era of interdisciplinary research. Future advancements will focus on maximizing performance, precision, and scalability, enabling these systems to transition from laboratory research to widespread clinical and industrial use. One of the most promising smart hydrogel based systems are capable of responding dynamically to environmental or biological cues such as pH, temperature, light, enzymes, or electric fields finding their major applicability in personalized and real-time therapeutic interventions. The integration of nanoscale switches, aptamers, or CRISPR-based components may further evolve the intelligence of these systems. The convergence of these hydrogels with wearable electronics and implantable biomedical devices observes a significant leap toward regular health monitoring and on-demand therapy. Hydrogels have high biocompatibility and flexibility making them ideal interfaces for developing biosensors, drug reservoirs, and responsive actuators. These systems can be aligned with cutting-edge innovations such as gene therapy, immunomodulation, and cellular delivery systems in the field of nanotechnology. Their ability to encapsulate and protect delicate biological molecules such as RNA, DNA, and proteins makes them suitable vectors for targeted delivery.

10 Conclusion

A versatile platform that bridges the barrier between traditional hydrogels and high-performance nanomaterials is provided by engineered nano polymer-based hydrogels, which represents a significant advancement in material science. By incorporating nanostructures, these hybrid hydrogels exhibit improved bioactivity, responsiveness, and mechanical properties, rendering them highly appropriate for a diverse array of applications, including biomedicine, agriculture, as well as environmental remediation. Future research must address the critical areas of scalability, long-term stability, and regulatory sanction, despite the promising progress. The next iteration of innovations is anticipated to be driven by the ongoing investigation of smart, stimuli-responsive designs and their integration with wearable or implantable technologies. The transformative potential of nano-engineered hydrogels is emphasized in this review, which also serves as a foundation for further scientific exploration and technological development in this dynamic field.

References

1. Nanda, D., Behera, D., Pattnaik, S.S., Behera, A.K.: Advances in natural polymer-based hydrogels: Synthesis, applications, and future directions in biomedical and environmental fields. Discov. Polym. **2**(1), 6 (2025)
2. Kakapour, A., Nouri Khorasani, S., Khalili, S., Hafezi, M., Sattari-Najafabadi, M., Najarzadegan, M., Saleki, S., Bagheri-Khoulenjani, S.: An injectable IPN nanocomposite hydrogel embedding nano silica for tissue engineering application. Macromol. Mater. Eng. **310**(1), 2400242 (2025)
3. Sekar, V., Santhanam, A.: In situ forming nanocomposite hydrogel scaffold reinforced with biogenic gold nanoparticles for enhanced cell proliferation. Inorg. Chem. Commun. **1**(174), 113959 (2025)
4. Cháirez-Ramírez, M.H., Díaz-Rivas, J.O., Contreras-Ramírez, J.I., González-Laredo, R.F., Gallegos-Infante, J.A.: Green polymer-based materials as promising therapeutical agents for non-communicable diseases. Polym. Int. **74**(6), 483–493 (2025)
5. Amponsah, O., Nopuo, P.S., Manga, F.A., Catli, N.B., Labus, K.: Future-oriented biomaterials based on natural polymer resources: characteristics, application innovations, and development trends. Int. J. Mol. Sci. **26**(12), 5518 (2025)
6. Li, C., Feng, C., Zhang, L., Zhang, L., Wang, L.: Direct ink writing of polymer-based materials—A review. Polym. Eng. Sci. **65**(2), 431–454 (2025)
7. Puri, V., Sharma, A., Kumar, P., Singh, I.: Thiolation of biopolymers for developing drug delivery systems with enhanced mechanical and mucoadhesive properties: A review. Polymers **12**(8), 1803 (2020)
8. Assie-Souleille, S., Seguier, L., Gauchard, D., Drobecq, I., Franc, B., Malaquin, L., Foncy, J.: Stereolithography 3D printing method for multi-material hydrogel 2D photo-patterning in a microfluidic chip. Micro Nano Eng. **2**, 100301 (2025)
9. Shi, D., Ji, D., Li, X., Guan, Z., Li, M., Bae, J.: Firefly-inspired 3d printable multi-stimuli-responsive fluorescent hydrogels with adaptive luminescence. ACS Nano., (2025)
10. Praveen, H.S., Naveen, N.R., Goudanavar, P.: Engineering biomaterial scaffolds for melanoma: Innovations in 3D printing and nanoparticle-hydrogel scaffolds for enhanced drug delivery. Biomed. Mater. & Devices. **23**, 1–9 (2025)

11. Sharma, A., Puri, V., Kumar, P., Singh, I.: Biopolymeric, nanopatterned, fibrous carriers for wound healing applications. Curr. Pharm. Des. **26**(38), 4894–4908 (2020)
12. Narayana, S., Gowda, B.J., Hani, U., Ahmed, M.G., Asiri, Z.A., Paul, K.: Smart poly (N-isopropylacrylamide)-based hydrogels: a tour D'horizon of biomedical applications. Gels. **11**(3), 207 (2025)
13. Xie, L., Wang, X., Bai, Y., Zou, X., Liu, X.: Fast-developing dynamic radiative thermal management: full-scale fundamentals, switching methods, applications, and challenges. Nano-Micro Letters. **17**(1), 146 (2025)
14. Kaur, M., Sharma, A., Puri, V., Aggarwal, G., Maman, P., Huanbutta, K., Nagpal, M., Sangnim, T.: Chitosan-based polymer blends for drug delivery systems. Polymers **15**(9), 2028 (2023)
15. Chen, Y., Yu, W., Zhu, Q., Huang, Y., Xia, F., Li, F.: Polymer/hydrogel integrated solid-state nanochannel composite membranes. ACS Nano, (2025)
16. Hossain, K.R., Lu, Y., Hu, D., Ji, Z., Yao, X., Wu, T., Zhao, H., Wang, X.: Polymer-based lubricating in 3d printing. Macromol. Chem. Phys. e00047 (2025)
17. Hamidi, M., Tabatabaei, M.S. (eds.) Novel drug delivery systems: Fundamentals and applications. John Wiley & Sons, (2025)
18. Li, Y., Tan, S., Zhang, X., Li, Z., Cai, J., Liu, Y.: Design strategies and emerging applications of conductive hydrogels in wearable sensing. Gels. **11**(4), 258 (2025)
19. Parvin, N., Joo, S.W., Mandal, T.K.: Injectable biopolymer-based hydrogels: a next-generation platform for minimally invasive therapeutics. Gels. **11**(6), 383 (2025)
20. Tamnna, Sharma, T., Mittal, R., Kaushik, P.: Advances in polymer-based scaffolding for enhanced wound healing: a comprehensive review. Int. J. Polym. Mater. Polym. Biomater. **74**(8), 684–702 (2025)
21. Gonzalez, V.: Future applications of polymer-based two-dimensional nanocomposite. In: Polymers and two-dimensional nanocomposites, pp. 657–681. Woodhead Publishing, (2025)
22. Mo, Y., Zhou, T., Li, W., Niu, Y., Sheu, C.: Advances in nanohybrid hydrogels for wound healing: from functional mechanisms to translational prospects. Gels. **11**(7), 483 (2025)
23. Chakraborty, G., Das, P.: Bio-absorbable polymer-based scaffolds as drug carriers: fabrication, characterization, and drug release. In: Bioabsorbable polymers for drug carriers and fixation devices, pp. 31–59. Apple Academic Press, (2025)
24. Cai, C., Li, W., Zhang, X., Cheng, B., Chen, S., Zhang, Y.: Natural polymer–based hydrogel dressings for wound healing. Adv. Wound Care **14**(6), 295–322 (2025)
25. Beyki, M.S., Mashhadi, E., Rashidi, J., Khavar, A.H., Safaei-Ghomi, J.: Nature-inspired chemical methods. In: Green biomaterials in tissue engineering, pp. 1–53. American Chemical Society, (2025)
26. Shukla, A., Syaifie, P.H., Jaya, S., Rochman, N.T., Jauhar, M.M., Mardliyati, E.: A recent study of natural hydrogels: improving mechanical properties for biomedical applications. Biomedical Mat. (2025)
27. Şener Raman, T., Claus, C., Abel, B., Schulze, A.: Study of the influence of structure-chemical properties of electron beam-polymerized PEGDA/Gelatin hybrid hydrogels on the uptake and release dynamics of different photosensitizer molecules. Biomacromolecules. (2025)
28. Matini, A., Naghib, S.M.: Microwave-assisted natural gums for drug delivery systems: Recent progresses and advances over emerging biopolymers and technologies. Curr. Med. Chem. **32**(13), 2547–2571 (2025)
29. Papagiannopoulos, A., Sentoukas, T., Triantafyllopoulou, E., Pispas, S., Pippa, N.: Designing polysaccharide graft copolymers for drug delivery applications. In: Graft copolymers of polysaccharides for drug delivery systems, pp. 1–33. Apple Academic Press, (2025)
30. Tan, C., Si, G., Zou, C., Chen, C.: Functional polyolefins and composites. Angew. Chem. Int. Ed. **64**(12), e202424529 (2025)
31. Zhang, S., Li, H., Zhang, B., Ai, S., Shan, Y., Ding, S.: Non-covalent in situ self-assembly of fruit peel waste into eco-friendly pectocellulosic bioplastics with high strength, flexibility and processability properties. Chem. Eng. J. **15**(504), 158697 (2025)
32. Chen, X., Wang, Y., Wang, A., Xu, L., Du, B., Guo, D., Zhao, H., Sha, L.: Demethylated lignin-reinforced multifunctional polyvinyl alcohol-based composites materials enabled by natural physical intertwining effect. Chem. Eng. J. **1**(509), 161192 (2025)

33. Uysal, B., Madduma-Bandarage, U.S., Jayasinghe, H.G., Madihally, S.: 3d-printed hydrogels from natural polymers for biomedical applications: Conventional fabrication methods, current developments, advantages, and challenges. Gels. **11**(3), 192 (2025)
34. Sharma, A., Mittal, A., Puri, V., Kumar, P., Singh, I.: Curcumin-loaded, alginate–gelatin composite fibers for wound healing applications. 3 Biotech. **10**(11), 464 (2020)
35. Hamzat, Y., Aljabali, A.A., El-Tanani, M., Tambuwala, M.M.: From nanomaterials to well-defined structures: exploring layer-by-layer assembly techniques. Curr. Nanosci. **21**(3), 404–422 (2025)
36. Hu, Y., Zhou, J., Gao, Y., Fan, Y., Chen, B., Su, J., Li, H.: Multifunctional nanocomposite hydrogels: An effective approach to promote diabetic wound healing. Biomed. Mater. **20**(3), 032006 (2025)
37. Rezanejade Bardajee, G., Mahmoodian, H., Shafiei, N., Amiri, B.: Development of a multi-stimuli-responsive magnetic nanogel–hydrogel nanocomposite for prolonged and controlled doxorubicin release. Bioconjugate Chem. (2025)
38. Thang, N.H., Chien, T.B., Cuong, D.X.: Polymer-based hydrogels applied in drug delivery: An overview. Gels. **9**(7), 523 (2023)
39. Huang, S., Hong, X., Zhao, M., Liu, N., Liu, H., Zhao, J., Shao, L., Xue, W., Zhang, H., Zhu, P., Guo, R.: Nanocomposite hydrogels for biomedical applications. Bioeng. & Transl. Med. **7**(3), e10315 (2022)
40. Mousavi, S.M., Hashemi, S.A., Kalashgrani, M.Y., Omidifar, N., Bahrani, S., Vijayakameswara Rao, N., Babapoor, A., Gholami, A., Chiang, W.H.: Bioactive graphene quantum dots based polymer composite for biomedical applications. Polymers **14**(3), 617 (2022)
41. Cheng, T., Xiang, Y., He, X., Pang, J., Zhu, W., Luo, L., Cao, Y., Pei, R.: Nanostructured conductive polymers: synthesis and application in biomedicine. J. Mater. Chem. B., (2025)
42. Edwin, E.R., Sakthi Sanjana, D., Michael, B., Kathiresan, S., Elumalai, K., Jayaprakash, N.: Metallic nanoparticle-functionalized graphene oxide nanocomposites for wound healing: A comprehensive review. Biomed. Mater. & Devices. **11**, 1–23 (2025)
43. Sangnim, T., Puri, V., Dheer, D., Venkatesh, D.N., Huanbutta, K., Sharma, A.: Nanomaterials in the wound healing process: new insights and advancements. Pharmaceutics. **16**(3), 300 (2024)
44. Gulati, S., Ansari, N., Moriya, Y., Joshi, K., Prasad, D., Sajwan, G., Shukla, S., Kumar, S., Varma, R.S.: Nanobiopolymers in cancer therapeutics: advancing targeted drug delivery through sustainable and controlled release mechanisms. J. Mater. Chem. B., (2024)
45. Tipa, C., Cidade, M.T., Borges, J.P., Costa, L.C., Silva, J.C., Soares, P.I.: Clay-based nanocomposite hydrogels for biomedical applications: a review. Nanomaterials **12**(19), 3308 (2022)
46. Jacob, S., Nair, A.B., Shah, J., Sreeharsha, N., Gupta, S., Shinu, P.: Emerging role of hydrogels in drug delivery systems, tissue engineering and wound management. Pharmaceutics. **13**(3), 357 (2021)
47. Ghosh, S., Yadav, P., Das, B.: Materials technology and its advancements involving nanotechnology, hydrogels, and its impact assessment on various aspects of improving the health-care system. In: Industrial microbiology and biotechnology: An insight into current trends, pp. 405–435. Springer Nature Singapore, Singapore
48. Alshangiti, D.M., El-Damhougy, T.K., Zaher, A., Madani, M.: Revolutionizing biomedicine: advancements, applications, and prospects of nanocomposite macromolecular carbohydrate-based hydrogel biomaterials: a review. RSC Adv. **13**(50), 35251–35291 (2023)
49. Dzulkharnien, N.S., Rohani, R.: A review on current designation of metallic nanocomposite hydrogel in biomedical applications. Nanomaterials **12**(10), 1629 (2022)
50. Jangid, H., Singh, S., Kashyap, P., Singh, A., Kumar, G.: Advancing biomedical applications: an in-depth analysis of silver nanoparticles in antimicrobial, anticancer, and wound healing roles. Front. Pharmacol. **8**(15), 1438227 (2024)
51. Zhao, J., Gao, C., Guo, W., Zhang, B., Ren, S., Wu, S., Guo, J., Qu, W.: Conductive hydrogels as an "innovative healer" for the treatment of diabetic wounds. Mater. Chem. Front. **8**(18), 2944–2977 (2024)

52. Kadhim, R.J., Jabir, M.S., Sulaiman, G.M., Nayef, U.M., Mohammed, H.A., Abomughaid, M.M.: The advancing of zinc oxide and gold nanoparticles as a promising therapeutic approach for future biomedical applications. Plasmonics **25**, 1–22 (2025)
53. Wafi, A., Khan, M.M.: Green synthesized ZnO and ZnO-based composites for wound healing applications. Bioprocess Biosyst. Eng. **48**(4), 521–542 (2025)
54. Wu, T., Sawut, A., Simayi, R., Gong, X., Zhang, X., Jiang, M., Zhu, Z.: Green synthesis and environmental applications of alginate/polyacrylamide/titanium dioxide composite hydrogel. J. Appl. Polym. Sci. **140**(37), e54394 (2023)
55. Nordin, A.H., Ahmad, Z., Husna, S.M., Ilyas, R.A., Azemi, A.K., Ismail, N., Nordin, M.L., Ngadi, N., Siti, N.H., Nabgan, W., Norfarhana, A.S.: The state of the art of natural polymer functionalized Fe3O4 magnetic nanoparticle composites for drug delivery applications: A review. Gels. **9**(2), 121 (2023)
56. Zaoui, F., Elhadj-Daouadji, B., Alaoui, C., Saleh, N.I., Gherdaoui, C.E., Beldjilali, M., Chipiri, T., Ibrahim, F., Sebba, F.Z., Bounaceur, B., Ma, H.: Adsorption behaviour of cationic and anionic dyes on new chitosan-activated carbon@ metal oxide hydrogels beads: Effect of the metal nature and comparative study. Int. J. Biol. Macromol. **14**, 144186 (2025)
57. Peng, L., Ding, J., Liu, M., Yang, X., Sui, G.: Liquid metal nanoparticles as photo-initiators for preparation of transparent hydrogel with adjustable mechanical properties. J. Colloid Interface Sci. **15**(672), 415–422 (2024)
58. Yang, Y., Zeng, Q., Zhao, C., Shi, J., Wang, W., Liang, Y., Li, C., Guan, Q., Chen, B., Li, W.: Metal-free antioxidant nanozyme incorporating bioactive hydrogel as an antioxidant scaffold for accelerating bone reconstruction. Biomaterials **1**(320), 123285 (2025)
59. Devi, S.G., Kanagalakshmi, M., Subasini, S., Pius, A.: Binary metal oxide nanoparticle-infused chitosan/casein patches for enhanced diabetic wound healing: An in vitro investigation. Inorg. Chem. Commun. **27**, 114963 (2025)
60. Özgenç, E., Çoruh, M.K., Keleş Özgenç, E., Dönmez, A.H., Töre, G.Y.: The properties and functional effects of carbon-based nanofillers in environmental applications. In: Handbook of nanofillers, pp. 1–37. Springer Nature Singapore, Singapore
61. Saba, T., Saad, K.S., Rashid, A.B.: Precise surface engineering: Leveraging chemical vapor deposition for enhanced biocompatibility and durability in biomedical implants. Heliyon. **10**(18), (2024)
62. Gostaviceanu, A., Gavrilaş, S., Copolovici, L., Copolovici, D.M.: Graphene-oxide peptide-containing materials for biomedical applications. Int. J. Mol. Sci. **25**(18), 10174 (2024)
63. Qu, X., Xu, N., Zhao, L., Xu, Z.: High-performance flexible zinc–air batteries enabled by carboxymethyl cellulose/graphene oxide composite hydrogel electrolytes. New J. Chem. **49**(18), 7376–7383 (2025)
64. Veeramani, S., Thulasimuthu, E., Mayakrishnan, V., Ilangovan, R.: Development and characterization of ZnO-, CuO-, and CuO–ZnO-incorporated porous sodium alginate–polyvinyl alcohol–gelatin hybrid bio-nanocomposite scaffolds for wound healing applications. Polym. Bull. **20**, 1–23 (2025)
65. Vasluianu, R.I., Dima, A.M., Bobu, L., Murariu, A., Stamatin, O., Baciu, E.R., Luca, E.O.: Dentistry insights: single-walled and multi-walled carbon nanotubes, carbon dots, and the rise of hybrid materials. J. Funct. Biomater. **16**(3), 110 (2025)
66. Xiang, S., Ménard-Moyon, C., Bianco, A.: Carbon and 2D nanomaterial smart hydrogels for therapeutic applications. Nanotechnol. Rev. **14**(1), 20250156 (2025)
67. Bhavsar, A., Pati, F., Chakraborty, P.: Supramolecular conductive hydrogels for tissue engineering applications. ChemBioChem **26**(1), e202400733 (2025)
68. Yang, J., Ding, C., Ding, Q., Zhang, S., Chai, G., Wei, H., Li, D., Sui, Q., Liu, W., Liang, D., Zhao, T.: Polyvinyl alcohol/graphene oxide/dextran hydrogel doped copper oxide modified biochar with nano-enzymatic activity and multiple synergistic effects for accelerated diabetic wound healing. Int. J. Biol. Macromol. **22**, 144480 (2025)
69. Jiang, Y., Gong, Y., Guo, C., Xiang, X.: Carbon nanotube-nano-Fe3O4 composite graphene hydrogel with optimized 3D structure for high-performance solar evaporation. Desalination **1**(608), 118840 (2025)

70. Noruzi, E.B., Vasigh, S.A., Eivazzadeh-Keihan, R., Aliabadi, H.A., Bani, M.S., Shaabani, B.: Chemical and physical modification of graphene oxide nano-sheets using casein, Zn-Al layered double hydroxide, alginate hydrogel, and magnetic nanoparticles for biomedical applications. Int. J. Biol. Macromol. **1**(269), 132047 (2024)
71. Cao, J., Zhao, X., Ye, L.: Super-strong and anti-tearing poly (vinyl alcohol)/graphene oxide nano-composite hydrogels fabricated by formation of multiple crosslinking bonding network structure. J. Ind. Eng. Chem. **25**(112), 366–378 (2022)
72. Das, L., Das, P., Bhowal, A., Bhattachariee, C.: Synthesis of hybrid hydrogel nano-polymer composite using Graphene oxide, Chitosan and PVA and its application in waste water treatment. Environ. Technol. Innov. **1**(18), 100664 (2020)
73. Chen, J., Li, H., Chen, B., Li, X., Yao, D., Gao, X., Lu, C., Pang, X.: Carbon nanotube-wrapped particles in sodium alginate hydrogels for enhanced solar evaporation and EMI shielding through tortuous channel. Int. J. Biol. Macromol. **24**, 144501 (2025)
74. Liu, G., Ma, Q., Zhang, X.: Agar-polyacrylamide dual network hydrogel-carbon nanotube composites with long-term stability for high efficient solar water purification. Compos. Commun. **1**(53), 102248 (2025)
75. Ryu, S.J., Jang, Y., Hong, I.K., Sathiyaseelan, A., Baek, J.S.: Efficacy of antibacterial and antibiofilm of silver nanoparticles decorated multi-walled carbon nanotubes loaded PF127 hydrogel. J. Drug Deliv. Sci. Technol. **1**(105), 106640 (2025)
76. Han, Y., Li, Y., Liu, Y., Alsubaie, A.S., El-Bahy, S.M., Qiu, H., Jiang, D., Wu, Z., Ren, J., El-Bahy, Z.M., Jiang, B.: Carbon nanotube reinforced ionic liquid dual network conductive hydrogels: Leveraging the potential of biomacromolecule sodium alginate for flexible strain sensors. Int. J. Biol. Macromol. **1**(282), 137123 (2024)
77. Idumah, C.I., Nwuzor, I.C., Odera, S.R., Timothy, U.J., Ngenegbo, U., Tanjung, F.A.: Recent advances in polymeric hydrogel nanoarchitectures for drug delivery applications. Int. J. Polym. Mater. Polym. Biomater. **73**(1), 1–32 (2024)
78. Geszke-Moritz, M., Moritz, M.: Biodegradable polymeric nanoparticle-based drug delivery systems: comprehensive overview, perspectives and challenges. Polymers **16**(17), 2536 (2024)
79. Asadi, K., Samiraninezhad, N., Akbarizadeh, A.R., Amini, A., Gholami, A.: Stimuli-responsive hydrogel based on natural polymers for breast cancer. Front. Chem. **18**(12), 1325204 (2024)
80. Dehchani, A.J., Jafari, A., Shahi, F.: Nanogels in biomedical engineering: revolutionizing drug delivery, tissue engineering, and bioimaging. Polym. Adv. Technol. **35**(10), e6595 (2024)
81. Delgado-Pujol, E.J., Martínez, G., Casado-Jurado, D., Vázquez, J., León-Barberena, J., Rodríguez-Lucena, D., Torres, Y., Alcudia, A., Begines, B.: Hydrogels and nanogels: pioneering the future of advanced drug delivery systems. Pharmaceutics. **17**(2), 215 (2025)
82. Hussein, A.M., Saleh, L.H., Diab, M.A., Muniyandy, E., Abdullaev, A., Latipova, M., Madaminov, B., Issa, S.K., Hasan, M.A., Islam, S.: Synthesis and application of tannic acid–gelatin hydrogel polymers capped silver nanoparticles as efficient and reusable nanocatalyst for A3 coupling reactions. J. Organomet. Chem. **6**, 123766 (2025)
83. Nowfal, S.H., Rusho, M.A., Kamolova, N., Pérez-Larios, A., Soto-Robles, C., Muniyandy, E., Smerat, A., García-Áviles, O., Arab, A.I., Islam, S.: Design and decorated gold nanoparticles over the Arabic gum-chitosan hydrogel polymers as a novel catalyst for Sonogashira coupling reactions. J. Organomet. Chem. **1**, 123762 (2025)
84. Mezzasalma, S.A., Abrami, M., Grassi, G., Grassi, M.: Towards a universal size distribution in a polymer network. Implications for drug delivery and plasmonic nanoparticle transport phenomena in polysaccharide and synthetic hydrogels. International J. Biol. Macromol., 144741 (2025)
85. Schütz, P., Lemich, S.B., Weißpflog, M., Körner, P., Abetz, V., Hankiewicz, B.: Multi-responsive hydrogels based on magneto-plasmonic nanoparticles in a thermo-responsive polymer matrix. Nano Trends. **23**, 100113 (2025)
86. Athavale, T.M., Nalliah, M., Obuobi, S., Grimstad, Ø., Škalko-Basnet, N.: Sustained release of betamethasone from solid lipid nanoparticles loaded polymeric hydrogels with natural emollient: One step closer to effective topical therapy for atopic dermatitis. J. Pharm. Sci. **114**(8), 103876 (2025)

87. Rauf, S., Ahmed, N., Fatima, S.K., Khan, D., ur Rehman, A., ur Rehman, A.: Aceclofenac delivery through polymeric nanoparticles loaded with transdermal hydrogel against rheumatoid arthritis. J. Drug Deliv. Sci. Technol. **102**, 106419 (2024)
88. El Turk, S., Chekkaramkodi, D., Ali, M., Sajini, A.A., Butt, H.: Programmed polymeric integration of gold nanoparticles into multi-material 3D printed hydrogels. Mater. Des. **1**(250), 113650 (2025)
89. Malik, H., Amir, F., Jahan, Z., Liaqat, U., Andleeb, S., Bandyopadhyay, S., Niazi, M.B.: Study of shape of zinc oxide nanoparticles on the in-vitro and in-vivo performance of polymeric hydrogels for wound dressing. Int. J. Pharm. **15**(674), 125482 (2025)
90. Uddin, M.N., Hossain, M.T., Mahmud, N., Alam, S., Jobaer, M., Mahedi, S.I., Ali, A.: Research and applications of nanoclays: A review. SPE Polymers. **5**(4), 507–535 (2024)
91. Hwang, H.S., Lee, C.S.: Nanoclay-composite hydrogels for bone tissue engineering. Gels. **10**(8), 513 (2024)
92. Krajišnik, D., Usković-Marković, S., Daković, A.: Chitosan–clay mineral nanocomposites with antibacterial activity for biomedical application: advantages and future perspectives. Int. J. Mol. Sci. **25**(19), 10377 (2024)
93. Li, H., Mao, B., Zhong, J., Li, X., Sang, H.: Localized delivery of metformin via 3D printed GelMA-Nanoclay hydrogel scaffold for enhanced treatment of diabetic bone defects. J. Orthop. Transl. **1**(47), 249–260 (2024)
94. Wu, T., Li, Y., Wu, Z., Wang, Z., Li, Y., Jian, K., He, C., Zhang, C., Shi, L., Dai, J.: Enzyme-immobilized nanoclay hydrogel simultaneously reduces inflammation and scar deposition to treat spinal cord injury. Chem. Eng. J. **15**(484), 149642 (2024)
95. Zhao, T., Ren, R., Qiao, S., Tang, X., Chi, Z., Jiang, F., Liu, C.: Multi-crosslinking nanoclay/oxidized cellulose hydrogel bandage with robust mechanical strength, antibacterial and adhesive properties for emergency hemostasis. J. Colloid Interface Sci. **1**(683), 828–844 (2025)
96. Liu, X., Shu, Y., Zhu, J., Fang, H., Su, Y., Ma, H., Li, B., Xu, J., Cheng, Y.Y., Pan, B., Song, K.: A 3D bioprinted potential colorectal tumor model based on decellularized matrix/gelatin methacryloyl/nanoclay/sodium alginate hydrogel. Int. J. Biol. Macromol. **1**(293), 139346 (2025)
97. Far, B.F., Naimi-Jamal, M.R., Jahanbakhshi, M., Khalafvandi, S.A., Alian, M., Jahromi, D.R.: Decontamination of Congo red dye from aqueous solution using nanoclay/chitosan-graft-gelatin nanocomposite hydrogel. J. Mol. Liq. **1**(395), 123839 (2024)
98. Cernencu, A.I., Dinu, A.I., Stancu, I.C., Lungu, A., Iovu, H.: Nanoengineered biomimetic hydrogels: A major advancement to fabricate 3D-printed constructs for regenerative medicine. Biotechnol. Bioeng. **119**(3), 762–783 (2022)
99. Quazi, M.Z., Park, N.: Nanohydrogels: Advanced polymeric nanomaterials in the era of nanotechnology for robust functionalization and cumulative applications. Int. J. Mol. Sci. **23**(4), 1943 (2022)
100. Thirumalaivasan, N., Kanagaraj, K., Nangan, S., Pothu, R., Rajendra, S.P., Karuppiah, P., et al.: Bioactive Hydrogels (Bio-HyGs): Emerging trends in drug delivery and wound healing applications. Polym. Adv. Technol. **36**(4), e70132 (2025)
101. Suhail, M., Shao, Y.-F., Vu, Q.L., Wu, P.-C.: Designing of pH-sensitive hydrogels for colon targeted drug delivery; characterization and in vitro evaluation. Gels. **8**(3), 155 (2022)
102. Li, B.-Y., Lin, T.-Y., Lai, Y.-J., Chiu, T.-H., Yeh, Y.-C.: Engineering multiresponsive Alginate/PNIPAM/carbon nanotube nanocomposite hydrogels as on-demand drug delivery platforms. Small **21**(12), 2407420 (2025)
103. Bellotti, E., Schilling, A.L., Little, S.R., Decuzzi, P.: Injectable thermoresponsive hydrogels as drug delivery system for the treatment of central nervous system disorders: A review. J. Control. Release **329**, 16–35 (2021)
104. Liu, S., Qamar, S.A., Qamar, M., Basharat, K., Bilal, M.: Engineered nanocellulose-based hydrogels for smart drug delivery applications. Int. J. Biol. Macromol. **181**, 275–290 (2021)
105. Kim, K.: Hybrid systems of gels and nanoparticles for cancer therapy: advances in multifunctional therapeutic platforms. Gels. **11**(3), 170 (2025)

106. Subhan, M.A., Yalamarty, S.S.K., Filipczak, N., Parveen, F., Torchilin, V.P.: Recent advances in tumor targeting via EPR effect for cancer treatment. J. Pers. Medicine. **11**(6), 571 (2021)
107. Singh, D., Dheer, D., Samykutty, A., Shankar, R.: Antibody drug conjugates in gastrointestinal cancer: From lab to clinical development. J. Control. Release **340**, 1–34 (2021)
108. Dheer, D., Gupta, R., Singh, D., Magotra, A., Singh, G., Gupta, P.N., et al.: Hyaluronic acid-tacrolimus bioconjugate: synthesis, characterization, and pharmacokinetic investigation of an acid-responsive macromolecular prodrug. ACS Appl. Bio Mater. **2**(11), 4728–4736 (2019)
109. Bai, X., Xu, Y., Liu, Y.: Therapeutic effects of DOX-loaded hydrogel MOF nanocarriers on triple negative breast cancer and derivative design via reinforcement learning. Sci. Rep. **14**(1), 23946 (2024)
110. Wu, H., Nan, J., Yang, L., Park, H.J., Li, J.: Insulin-loaded liposomes packaged in alginate hydrogels promote the oral bioavailability of insulin. J. Control. Release **353**, 51–62 (2023)
111. Mo, X., Zhang, D., Liu, K., Zhao, X., Li, X., Wang, W.: Nano-hydroxyapatite composite scaffolds loaded with bioactive factors and drugs for bone tissue engineering. Int. J. Mol. Sci. **24**(2), 1291 (2023)
112. Phan, V.H.G., Duong, H.-S., Le, Q.-G.T., Janarthanan, G., Vijayavenkataraman, S., Nguyen, H.-N.H. et al.: Nanoengineered injectable hydrogels derived from layered double hydroxides and alginate for sustained release of protein therapeutics in tissue engineering applications. J. Nanobiotechnology. **21**(1), 405 (2023)
113. Islam, M.A., Hossain, N., Hossain, S., Khan, F., Hossain, S., Arup, M.M.R., et al.: Advances of hydroxyapatite nanoparticles in dental implant applications. Int. Dent. J. **75**(3), 2272–2313 (2025)
114. Xu, X., Wang, L., Jing, J., Zhan, J., Xu, C., Xie, W., et al.: Conductive collagen-based hydrogel combined with electrical stimulation to promote neural stem cell proliferation and differentiation. Front. Bioeng. Biotechnology. **10** (2022)
115. Omidian, H., Chowdhury, S.D., Cubeddu, L.X.: Hydrogels for neural regeneration: exploring new horizons. Materials. **17**(14), 3472 (2024)
116. Wang, M., Deng, Z., Guo, Y., Xu, P.: Engineering functional natural polymer-based nanocomposite hydrogels for wound healing. Nanoscale Advances. **5**(1), 27–45 (2023)
117. Dam, P., Celik, M., Ustun, M., Saha, S., Saha, C., Kacar, E.A., et al.: Wound healing strategies based on nanoparticles incorporated in hydrogel wound patches. RSC Adv. **13**(31), 21345–21364 (2023)
118. Parvin, N., Joo, S.W., Mandal, T.K.: Injectable biopolymer-based hydrogels: a next-generation platform for minimally invasive therapeutics. Gels. **11**(6), 383 (2025)
119. Zarrintaj, P., Seidi, F., Youssefi Azarfam, M., Khodadadi Yazdi, M., Erfani, A., Barani, M., et al.: Biopolymer-based composites for tissue engineering applications: A basis for future opportunities. Compos. B Eng. **258**, 110701 (2023)
120. Nezhad-Mokhtari, P., Hamishehkar, H., Farahpour, M.R., Mehdipour, A., Rahbarghazi, R., Milani, M., et al.: Engineered bioadhesive Self-Healing nanocomposite hydrogel to fight infection and accelerate cutaneous wound healing. Chem. Eng. J. **489**, 150992 (2024)
121. Hao, M., Wang, Y., Wang, Y., Su, S., Zhou, Y., Wei, S.: A carrier-free hydrogel loading silver nanoparticles with photothermal effect to heal refractory diabetic wound by robust bacteria elimination and repair-promotion. Chem. Eng. J. **519**, 165250 (2025)
122. Jaiswal, R., Sherje, A.P.: Recent advances in biopolymer-based smart hydrogel for wound healing. J. Drug Deliv. Sci. Technol. **99**, 105990 (2024)
123. Alcalde-Garcia, F., Prasher, S., Kaliaguine, S., Tavares, J.R., Dumont, M.-J.: Desorption strategies and reusability of biopolymeric adsorbents and semisynthetic derivatives in hydrogel and hydrogel composites used in adsorption processes. ACS Engineering Au. **3**(6), 443–460 (2023)
124. Vinzant, K., Rashid, M., Khodakovskaya, M.V.: Advanced applications of sustainable and biological nano-polymers in agricultural production. Front. Plant. Science. **13**, 2022 (2023)
125. Mikhailidi, A., Ungureanu, E., Tofanica, B.-M., Ungureanu, O.C., Fortună, M.E., Belosinschi, D. et al.: Agriculture 4.0: Polymer hydrogels as delivery agents of active ingredients. Gels. **10**(6), 368 (2024)

126. Dhiman, A., Sharma, A.K., Agrawal, G.: Polymer based engineered materials for sustainable agriculture. ACS Agric. Sci. & Technology. **2**(4), 693–711 (2022)
127. Rizwan, M., Rubina Gilani, S., Iqbal Durani, A., Naseem, S.: Materials diversity of hydrogel: Synthesis, polymerization process and soil conditioning properties in agricultural field. J. Adv. Res. **33**, 15–40 (2021)
128. Patra, S.K., Poddar, R., Brestic, M., Acharjee, P.U., Bhattacharya, P., Sengupta, S., et al.: Prospects of hydrogels in agriculture for enhancing crop and water productivity under water deficit condition. Int. J. Polym. Science. **2022**(1), 4914836 (2022)
129. Mohamady Ghobashy, M.: Chapter 12—the application of natural polymer-based hydrogels for agriculture. In: Chen, Y. (ed.) Hydrogels based on natural polymers, pp. 329–356. Elsevier (2020)
130. Ghobashy, M.M., Amin, M.A., Ismail, M.A., Nowwar, A.I., El-diehy, M.A., Gayed, H.M.: Radiation cross-linked ultra-absorbent hydrogel to rationalize irrigation water and fertilizer for maize planting in drought conditions. Int. J. Biol. Macromol. **252**, 126467 (2023)
131. Skrzypczak, D., Jarzembowski, Ł, Izydorczyk, G., Mikula, K., Hoppe, V., Mielko, K.A., et al.: Hydrogel alginate seed coating as an innovative method for delivering nutrients at the early stages of plant growth. Polymers **13**(23), 4233 (2021)
132. Chen, F.-B., Feng, Y.-C., Huo, S.-P.: Seed coating with micronutrients improves germination, growth, yield and microelement nutrients of maize (Zea mays L.). Biotechnic & Histochemistry. **98**(4), 230–242 (2023)
133. Park, J., Guan, W., Yu, G.: Smart hydrogels for sustainable agriculture. EcoMat. **7**(4), e70011 (2025)
134. Martin, S.S., Zhang, L., Cen, X., Liu, P., Zhou, J., Tan, Z., et al.: Smart nanocomposite hydrogel for controlled bioactive delivery in environmental remediation. J. Environ. Chem. Eng. **13**(4), 117367 (2025)
135. Song, M., Wang, J., He, J., Kan, D., Chen, K., Lu, J.: Synthesis of hydrogels and their progress in environmental remediation and antimicrobial application. Gels. **9**(1), 16 (2023)
136. Baby, R., Hussein, M.Z., Abdullah, A.H., Zainal, Z.: Nanomaterials for the treatment of heavy metal contaminated water. Polymers **14**(3), 583 (2022)
137. Ethaib, S., Al-Qutaifia, S., Al-Ansari, N., Zubaidi, S.L.: Function of nanomaterials in removing heavy metals for water and wastewater remediation: A review. Environments **9**(10), 123 (2022)
138. Karnwal, A., Malik, T.: Nano-revolution in heavy metal removal: engineered nanomaterials for cleaner water. Front. Environ. Sci. **12**, (2024)
139. Li, X., Li, S., Peng, C., Wang, Y., Li, Y., Yang, S., et al.: Chitosan-based composite featuring dual cross-linking networks for the removal of aqueous Cr(VI). Carbohyd. Polym. **348**, 122859 (2025)
140. Visan, A.I., Negut, I.: Environmental and wastewater treatment applications of stimulus-responsive hydrogels. Gels. **11**(1), 72 (2025)
141. Marouch, S., Benbellat, N., Duran, A., Yilmaz, E.: Nanoclay- and TiO2 nanoparticle-modified poly(N-vinyl pyrrolidone) hydrogels: A multifunctional material for application in photocatalytic degradation and adsorption-based removal of organic contaminants. ACS Omega **7**(39), 35256–35268 (2022)
142. Kumar, R., Rauwel, P., Rauwel, E.: Nanoadsorbants for the removal of heavy metals from contaminated water: current scenario and future directions. Processes. **9**(8), 1379 (2021)
143. Sepahvand, S., Kargarzadeh, H., Jonoobi, M., Ashori, A., Ismaeilimoghadam, S., Varghese, R.T., et al.: Recent developments in nanocellulose-based aerogels as air filters: A review. Int. J. Biol. Macromol. **246**, 125721 (2023)
144. Gao, B., Feng, X., Zhang, Y., Zhou, Z., Wei, J., Qiao, R., et al.: Graphene-based aerogels in water and air treatment: A review. Chem. Eng. J. **484**, 149604 (2024)
145. Chemere, E.A-O.X., Mhlabeni, T.L., Mhike, W., Mavhungu, M.L., Shongwe, M.B.: A comprehensive review of types, synthesis strategies, advanced designing and applications of aerogels. R. Soc. Open. Sci. **12**(5), 241975 (2025). https://doi.org/10.1098/rsoseCollection

Properties and Characteristics of Nanostructured Polymers

Surface Morphology, Structural, and Interfacial Features of Polymers at Nanoscale

Priti and Radha Sachan

Abstract This chapter offers a comprehensive overview of methods and concepts used to study and interpret polymer behavior at the nanoscale. It starts with an introduction to nanoscale surface morphology, highlighting how structural and topographical features influence the physical, mechanical, and chemical properties of polymers. Advanced imaging techniques such as Atomic Force Microscopy (AFM), Scanning Electron Microscopy (SEM), and Transmission Electron Microscopy (TEM) are covered for examining nanoscale structures, surface roughness, and wettability. X-ray–based methods like Small-Angle X-ray Scattering (SAXS), Wide-Angle X-ray Scattering (WAXS), and X-ray Diffraction (XRD) are also included for their ability to reveal crystallinity, molecular orientation, and nanoscale organization. The impact of processing parameters and additives on polymer morphology is discussed, with special focus on blends, copolymers, and phase separation in block copolymers. Interfacial aspects such as polymer–substrate interactions, polymer–polymer boundaries, and interfiber contacts in nanofibers are analyzed to explain their roles in adhesion, stability, and molecular packing. Additional tools like X-ray Photoelectron Spectroscopy (XPS) and Dynamic Light Scattering (DLS) are highlighted for investigating chemical composition and nanoscale particle size distributions. By integrating these techniques and perspectives, this chapter establishes a comprehensive framework for linking nanoscale structure and interfacial properties to the overall behavior of polymer systems.

Priti
Department of Paint Technology, Mahamaya Polytechnic of Information Technology, Chail Kaushambi, Uttar Pradesh, India

R. Sachan (✉)
Department of Paint Technology, School of Chemical Technology, Harcourt Butler Technical University, Kanpur, Uttar Pradesh, India
e-mail: radha.sachan@gmail.com

B. Sharma et al. (eds.), *Nanostructured Polymers*, Engineering Materials,
https://doi.org/10.1007/978-981-95-3581-1_6

1 Introduction

The surface morphology of polymers at the nanoscale is a critical characteristic that significantly influences their mechanical strength [1], optical behavior [2], thermal stability [3], adhesion, and interfacial interactions [4]. It refers to the nanoscale structural features, including roughness, patterning, porosity, and phase separation, which define the surface characteristics of a polymer material. The surface morphology is primarily determined by polymer composition [5], molecular architecture [6], fabrication techniques [7, 8] processing conditions, and environmental factors such as temperature and humidity [9]. Understanding and controlling these nanoscale features are essential for optimizing polymer performance in diverse applications, including coatings, nanocomposites, biomaterials, flexible electronics, and filtration membranes. Polymers at the nanoscale can exhibit a variety of surface morphologies, ranging from smooth, uniform surfaces to highly structured and patterned textures. Surface roughness and topography play a significant role in defining properties such as adhesion, friction, and wettability. For instance, a rougher surface can enhance adhesion in coatings and adhesives, whereas a controlled nanostructured surface can impart superhydrophobicity or antifouling properties [10–12].

At the nanoscale, the structural features of polymers exhibit unique properties that differ significantly from their bulk counterparts. One of the key features is the increased surface area-to-volume ratio, which can enhance interactions with surrounding environments, such as other materials or solvents [13]. Polymers at this scale often exhibit enhanced mechanical properties, like increased strength or flexibility, due to the alignment of molecular chains and reduced defects [14]. The molecular structure, such as the degree of polymerization and the arrangement of monomers, can influence properties like crystallinity, viscosity, and thermal stability [15]. Furthermore, at the nanoscale, polymers can form self-assembled nanostructures, such as nanoparticles or nanofibers, which can exhibit properties like tunable porosity, increased reactivity, or electrical conductivity [16]. The behavior of polymer chains at the nanoscale is also influenced by factors like entanglement density, chain mobility, and local nanoscale interactions, which play a critical role in defining their overall performance in applications like drug delivery, nanocomposites, and sensors [17].

At the nanoscale, the interfacial features of polymers are crucial in determining their interactions with other materials, such as fillers, substrates, or solvents. One of the most notable characteristics is the significant influence of the polymer's surface properties, which can be drastically different from the bulk properties due to the increased surface-to-volume ratio. These properties include surface energy, roughness, and chemical composition, which can affect the adhesion, wetting behavior, and mechanical bonding between the polymer and surrounding materials [18].

Polymers at the nanoscale can exhibit altered surface functionality, as their molecular chains may be more exposed to the environment. This exposure allows for enhanced interactions with neighboring surfaces, such as stronger van der Waals forces, hydrogen bonding, or electrostatic interactions. The interface between the polymer and other materials often plays a pivotal role in the overall performance of nanocomposites, where the dispersion and distribution of nanoparticles within the polymer matrix can significantly influence the material's mechanical, electrical, and thermal properties [19].

Additionally, at the nanoscale, polymers can form organized structures at the interface, such as layered or ordered nanostructures, which can improve the material's stability and performance in applications like sensors, coatings, and biomedical devices. These interfacial features also influence phenomena like polymer chain alignment, crystallinity, and the formation of nanoscopic domains that affect properties like conductivity, permeability, and selectivity in various nanotechnological applications [20].

By tailoring the surface morphology of polymers at the nanoscale opens new opportunities for designing advanced materials with customized properties and enhanced functionalities. By precisely controlling nanoscale roughness, porosity, and patterning, researchers can develop polymers with improved biocompatibility, self-cleaning surfaces, controlled drug release capabilities, and enhanced mechanical performance. These innovations are driving the advancement of nanotechnology, biomedical engineering, and smart material development, paving the way for next-generation polymer applications in healthcare, energy, and sustainable materials [21].

This chapter explores the fundamental aspects of polymer surfaces and interfaces at the nanoscale, focusing on morphology, structural properties, and interfacial interactions.

2 Surface Morphology of Polymer at Nanoscale

Surface morphology refers to the topographical and structural features observed on the polymer surface. At the nanoscale, these features include roughness, patterns, and hierarchical structures, which influence properties such as wettability, adhesion, and biocompatibility. Since surface morphology of polymers such as both amorphous and crystalline regions play an important role in defining physical, chemical and mechanical properties of polymer at nanoscale. It is important to know about factors influencing surface morphology. The surface morphology of polymers is influenced by various factors, including, 1. Polymer Composition as the chemical structure and molecular weight affect the surface energy and phase separation, 2. Processing techniques like electrospinning, phase separation, and nanopatterning which dictate the nanoscale morphology, and 3. Environmental Conditions such as temperature, humidity, and solvent interactions that can lead to surface restructuring. To study and characterize polymer surface morphology at the nanoscale, various advanced analytical techniques are employed.

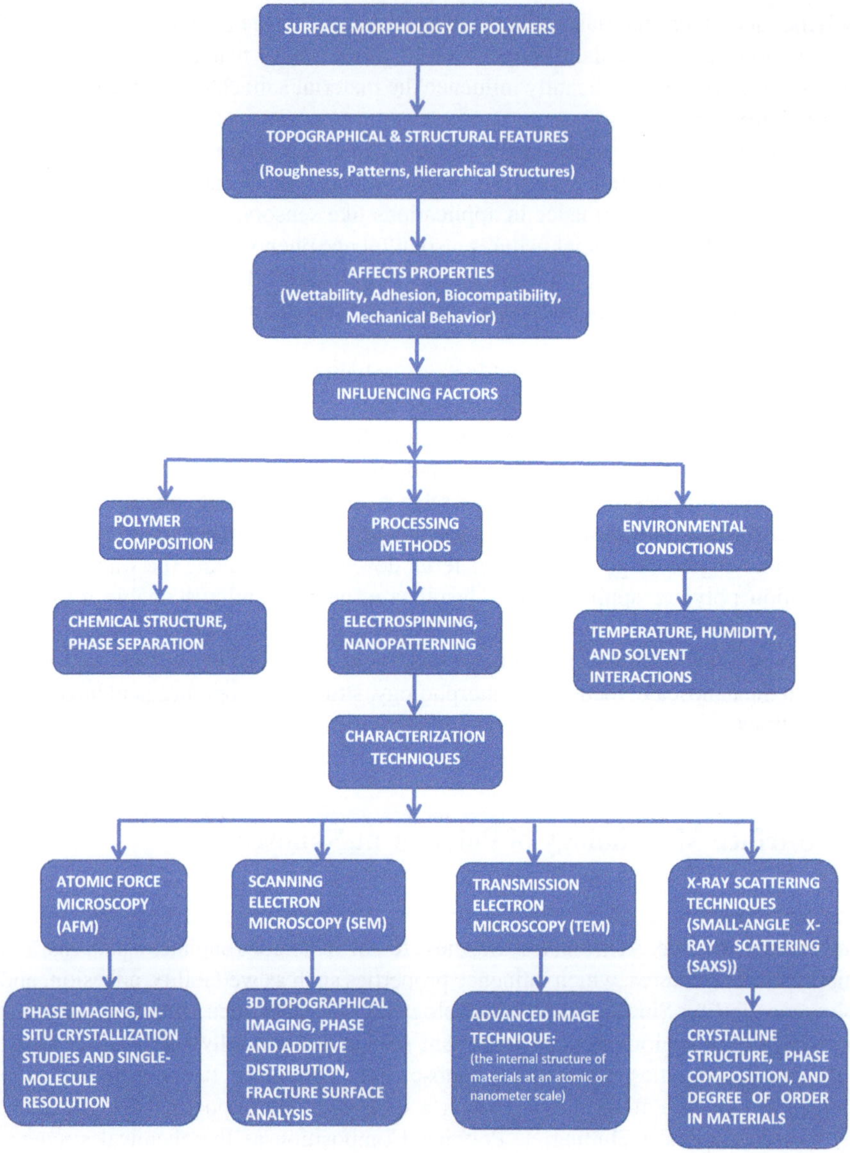

3 Atomic Force Microscopy (AFM)

Atomic Force Microscopy (AFM) is a powerful technique that provides high-resolution, three-dimensional topographic images and roughness measurements, enabling researchers to examine surfaces with nanometer precision. By generating

3D topographical maps with sub-nanometer resolution, AFM allows visualization of surface roughness, defects, and crystallinity in polymer materials. One of its advanced modes, phase imaging, detects variations in mechanical properties such as stiffness and adhesion across polymer surfaces by analyzing tip-sample interactions. AFM also supports in-situ crystallization studies, where researchers can observe real-time processes like polymer crystal growth, melting, and reorganization under elevated temperatures. Notably, AFM achieves single-molecule resolution, capable of imaging individual polymer strands (e.g., polyethylene) with resolutions down to 0.37 nm using specialized carbon-whisker tips. Additionally, nanoindentation via AFM enables probing of elastic and viscous properties at the nanoscale. For instance, polycarbonate films, which appear glassy in bulk, exhibit liquid-like behavior at the nanoscale due to confinement effects. Force-volume AFM further allows mapping of surface properties such as adhesion energy, as seen in studies of PEG-grafted silver nanocubes that reveal "octopus micelles" at cube corners [22].

Complementing experimental techniques, molecular dynamics simulations provide insights into surface morphology, particularly how chemically heterogeneous surfaces influence adhesion. Coarse-grained simulations show that surfaces with segregated weakly and strongly adsorbing sites form weaker adhesive layers compared to interdispersed configurations. At the nanoscale, cohesive failure transitions to a mixed cohesive-adhesive mechanism due to void nucleation at interfacial defects, highlighting the complex interplay between surface chemistry and mechanical behavior [23].

AFM requires minimal sample preparation, avoiding artifacts common in electron microscopy, and operates non-destructively on delicate polymers. AFM provides critical insights into how nanoscale morphological features—such as crystalline domains, amorphous regions, and surface roughness—affect the functional properties of polymers. By correlating topographical data with mechanical and compositional information, researchers can tailor polymer formulations, processing conditions, and surface modifications to optimize performance for applications ranging from flexible electronics, coatings and membranes to biomedical devices. AFM analysis conducted by Li et al. demonstrated well-ordered cylindrical nanodomains in PS-PEO block copolymer films, with lateral spacing around 29 nm. Such morphology is published in Biomaterials science [24] (Fig. 1).

4 Scanning Electron Microscopy (SEM)

Scanning Electron Microscopy (SEM) is an essential imaging technique that enables high-resolution analysis of surface morphology and composition in a wide range of materials. By directing a focused beam of electrons onto a sample, SEM detects signals such as secondary electrons, backscattered electrons, and characteristic X-rays, which are then used to generate detailed, three-dimensional-like images. These images reveal intricate surface features including texture, shape, and structural

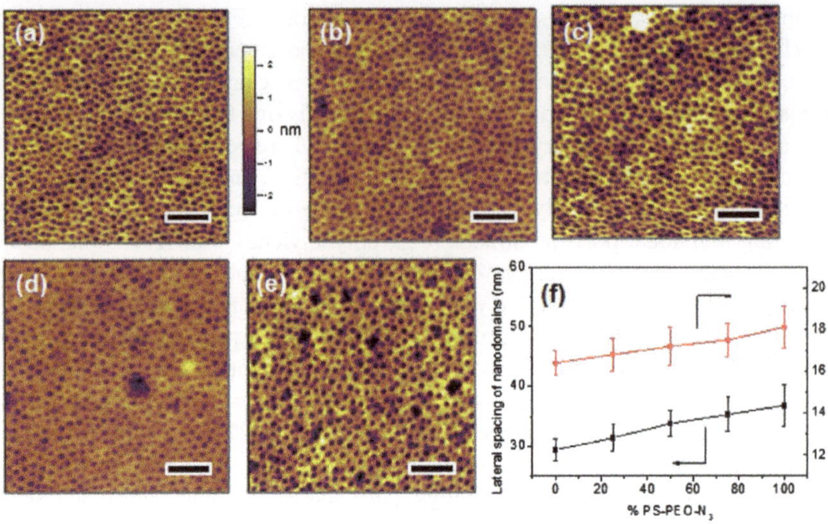

Fig. 1 AFM height images of polymer substrates obtained from self-assembly of **a** (100%) PS-PEO, and co-self-assembly of PS-PEO with **b** 25%, **c** 50%, **d** 75% of PS-PEO-N3, and **e** (100%) PS-PEO-N3; **f** lateral spacing and size of nanodomains in the resultant polymer films. The scale bar is 200 nm

composition. SEM surpasses optical microscopy in both depth of field and resolution—typically ranging from 1 to 10 nm—making it especially valuable in materials science, biology, nanotechnology, and forensics. However, non-conductive samples often require preparation, such as sputter-coating with conductive materials (e.g., gold or platinum), to prevent surface charging and enhance image clarity.

In polymer science, SEM is a cornerstone technique for investigating nanoscale surface morphology. It provides critical insights into structural features, phase distribution, and degradation mechanisms with resolution capabilities down to sub-nanometer levels. SEM's 3D topographical imaging, achieved through secondary electron (SE) detection, highlights surface roughness, crystallinity, and defects, while backscattered electron (BSE) imaging distinguishes between different phases based on atomic number contrast. It is particularly effective in identifying phases and additive distribution in polymer blends—such as differentiating continuous and dispersed phases in polypropylene composites—and mapping inorganic fillers through energy-dispersive X-ray spectroscopy (EDX).

Fracture surface analysis using SEM reveals underlying failure mechanisms (e.g., brittle versus ductile fracture) by examining cryogenically fractured samples, preserving nanoscale features. To accommodate sensitive or hydrated polymers, SEM can operate in low-vacuum modes, minimizing beam damage and eliminating the need for conductive coatings.

SEM also plays a vital role in nanocomposite characterization by visualizing nanoparticle dispersion—such as silica within poly(ethylene oxide)—and evaluating interphase regions critical for mechanical reinforcement. It can quantify filler homogeneity, which is often correlated with significant improvements in mechanical properties, such as tensile strength increases of up to 300%. Additionally, SEM facilitates degradation studies by tracking UV-induced surface cracks and erosion in polymers like poly(vinyl chloride) and polystyrene, and measuring surface roughness to assess photodegradation severity. It is also indispensable for examining porous polymer films, resolving nanopore structures ranging from 200 nm to 1 μm—features that are critical in applications like organic electronics [25]. Wei Li et al. report a scanning electron microscopy (SEM) image of polymer nanospheres synthesized at a template-to-resol mass ratio of 0.9, demonstrating uniform spherical morphology with an average diameter of ~200 nm (Fig. 2a). Following Pluronic F127 template removal and carbonization of the phenolic-formaldehyde resin at 700 °C under Ar, the spherical structure remains intact (Fig. 2b). This retention highlights the robust structural stability of the template/resin composite during high-temperature processing [26].

SEM's capability to image bulk samples and perform rapid, large-area nanoscale analysis makes it a preferred tool for routine polymer characterization. While Transmission Electron Microscopy (TEM) offers superior resolution at the atomic level and is ideal for investigating internal structures, SEM is often favored for its speed, ease of use, and ability to analyze surface morphology over larger areas.

In practical applications, SEM plays a critical role in assessing the performance and integrity of polymer materials. For instance, in UV-stabilized polymers, SEM imaging reveals significantly reduced surface cracking in poly(vinyl chloride) (PVC) samples blended with melamine-Schiff base UV absorbers, compared to untreated counterparts—demonstrating the effectiveness of stabilization additives. In the context of recycled polymers, SEM helps identify contaminants such as microplastics and evaluate morphological changes induced by reprocessing, which is essential for quality control and material reuse.

When integrated with complementary techniques like Atomic Force Microscopy (AFM) or Energy-Dispersive X-ray Spectroscopy (EDX), SEM enables a more comprehensive analysis of polymer systems. This multi-modal approach provides insights into surface topography, mechanical properties, and chemical composition, ultimately advancing innovations in materials engineering and supporting efforts toward sustainability in polymer development.

5 Transmission Electron Microscopy (TEM)

Transmission Electron Microscopy (TEM) is a sophisticated imaging technique designed to investigate the internal structure of materials at the atomic or nanometer scale. Unlike Scanning Electron Microscopy (SEM), which captures surface details, TEM operates by transmitting a high-energy electron beam through an ultra-thin sample—typically less than 100 nm thick. As the electrons interact with the internal features of the specimen, they generate highly magnified images that are projected

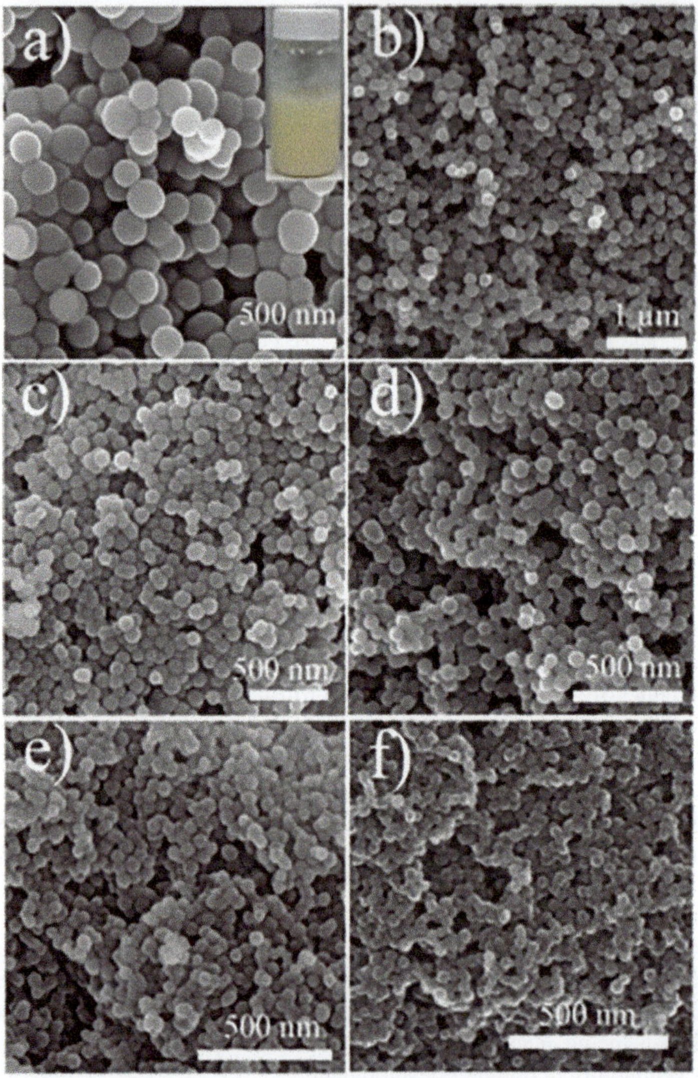

Fig. 2 Scanning electron microscopy (SEM) images of the polymer nanospheres before carbonization (**a, c, e**) and the corresponding mesoporous carbon nanospheres synthesized with different mass ratios of F127 to resol: **b** MCN-0.9; **d** MCN-1.0; **f** MCN-1.25. The inset in (**a**) is an optical image of polymer spheres in an aqueous solution

onto a detector or screen. This method delivers exceptional resolution, often below 1 nm, making it invaluable for analyzing crystal lattices, cellular structures, nanoparticles, and molecular arrangements. TEM is widely utilized in materials science, nanotechnology, and biological research to study structural defects and ultra-fine internal features. However, one of the main challenges of TEM lies in its complex and labor-intensive sample preparation, which requires specimens to be meticulously thinned and mounted on specialized grids for effective imaging. In the context of polymer analysis, TEM offers unparalleled insights into the structure of crystalline polymers. Modern advancements in TEM, such as aberration correction and low-dose focal series reconstruction, have enabled submolecular resolution. For example, in polycaprolactone (PCL), these techniques have resolved lattice deformations at a resolution of 2 Å, allowing direct imaging of oxygen atom positions within the polymer chains. Additionally, four-dimensional scanning TEM (4D-STEM), which utilizes nano-diffraction imaging, is capable of mapping nanoscale crystalline domains in semicrystalline polymers such as polyethylene (PE). This approach reveals lamellar orientations without the need for contrast agents like ruthenium tetroxide (RuO_4), thereby preserving the native morphology of the polymer. These capabilities make TEM an indispensable tool for advancing our understanding of polymer structure at the molecular level [27].

Transmission Electron Microscopy Tomography (TEMT) is a powerful technique that enables the reconstruction of three-dimensional nanostructures in complex, multiphase systems. It has been effectively used to visualize the spatial distribution of dual fillers, such as carbon black and silica, within rubber nanocomposites, providing insights into how these arrangements influence mechanical properties. Additionally, TEMT has captured detailed nanoscale features, including gyroid phases in block copolymers and phase separation in polymer blends, with a resolution of approximately 10 nm [28].

TEM is also instrumental in studying interfacial chemistry and mechanics in nanocomposites. By integrating Scanning Transmission Electron Microscopy (STEM) with Electron Energy Loss Spectroscopy (EELS), researchers have identified chemical bonding between silica nanoparticles and rubber matrices—critical to understanding reinforcement mechanisms [29]. Cryogenic TEM (cryo-TEM) further aids this analysis by preserving native structures, revealing complex interfacial networks in protein-polymer condensates. Moreover, in situ stretching experiments using TEM have enabled the real-time observation of cavity formation and strain distribution during nanocomposite fracture, offering valuable insight into mechanical behavior at the nanoscale.

One of the major challenges in TEM imaging of polymers is their sensitivity to electron beam damage. Recent innovations have addressed this limitation. These advancements have greatly expanded the utility of TEM in the non-destructive, high-resolution analysis of soft and beam-sensitive materials [30]. Wei Li et al. synthesized mesoporous carbon nanospheres (MCNs) via an acid-assisted organic–organic self-assembly method using F127 and resol, and characterized their morphology, pore structure, and surface area using TEM as shown in Fig. 3 and results revealed that the mesoporous carbon nanospheres exhibit uniform spherical morphology with tunable diameters (~120 nm for MCN-0.9), distinct mesoporous and microporous frameworks, and pore structure consistent with SEM analsysis results [26].

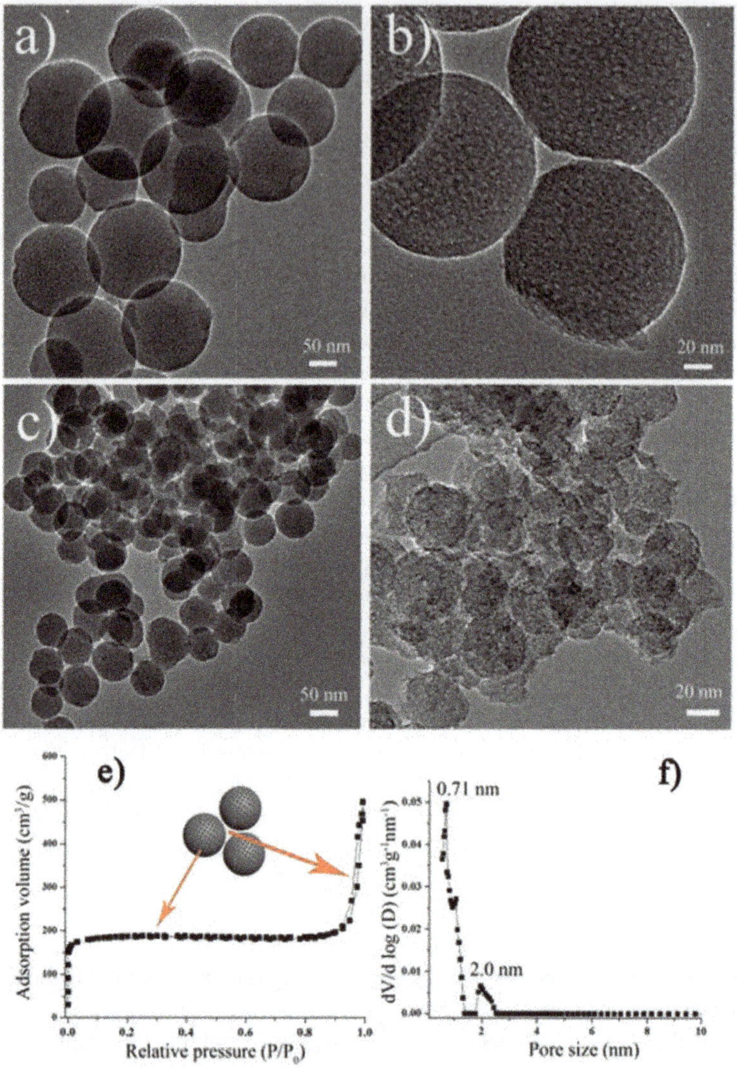

Fig. 3 Transmission electron microscopy (TEM) images of mesoporous carbon nanospheres (MCNs) synthesized with different mass ratios of F127 to resol: **a**, **b** MCN-0.9, **c** MCN-1.0, and **d** MCN-1.25. N2 sorption isotherms (**e**) and nonlocal density functional theory (NLDFT) pore size distributions (**f**) of typical mesoporous carbon nanospheres (MCN-1.0)

6 X-Ray Scattering Techniques

6.1 Small-Angle X-Ray Scattering (SAXS)

X-ray scattering techniques, such as Small-Angle X-ray Scattering (SAXS), are powerful tools for studying the structural properties of materials at the nanoscale. SAXS is particularly useful for investigating the size, shape, and organization of macromolecules, nanoparticles, and porous materials in solution or solid form. By measuring the scattering intensity of X-rays at small angles, SAXS provides insights into structural features ranging from a few nanometers to several hundred nanometers. Unlike traditional crystallographic methods, SAXS does not require crystalline samples, making it ideal for analyzing biological macromolecules, polymers, and colloidal systems in their native states. These techniques play a crucial role in materials science, biophysics, and nanotechnology, aiding in the development of novel materials and understanding biological processes at the molecular level.

6.2 Wide-Angle X-Ray Scattering (WAXS)

Wide-Angle X-ray Scattering (WAXS) is an advanced technique used to study the atomic and molecular structure of materials by analyzing X-ray scattering at large angles. Unlike Small-Angle X-ray Scattering (SAXS), which provides information about nanoscale structures, WAXS focuses on shorter length scales, typically in the range of interatomic distances. This makes it particularly useful for investigating the crystalline structure, phase composition, and degree of order in materials such as polymers, biomolecules, and nanomaterials. WAXS is widely used in fields like materials science, chemistry, and structural biology to examine molecular arrangements, polymorphism, and conformational changes in complex systems. By providing high-resolution structural information, WAXS plays a crucial role in the characterization and development of new materials with tailored properties.

Small-Angle X-ray Scattering (SAXS) and Wide-Angle X-ray Scattering (WAXS) are powerful and complementary techniques widely used for the structural analysis of polymers at the nanoscale. SAXS is particularly effective for probing features in the range of 1–100 nm, making it invaluable for characterizing nanoscale morphology such as lamellar structures, nanoparticle dispersion, and phase separation in polymer blends and nanocomposites. For instance, SAXS has been used to study the periodicity and thickness of crystalline and amorphous lamellae in semicrystalline polymers, providing insights into their hierarchical structure and its evolution under various conditions [31]. On the other hand, WAXS is sensitive to shorter length scales, typically below 1 nm, which allows for detailed examination of polymer crystallinity, chain packing, and molecular conformation. WAXS has been extensively employed to identify crystalline phases, determine interplanar spacings (d-spacings), and quantify the degree of crystallinity in various polymer systems [32]. When used together, SAXS and WAXS offer a comprehensive view of polymer structure, from the arrangement of crystalline domains to the organization of nanostructures, enabling researchers to correlate nanoscale morphology with macroscopic properties. The combined application of SAXS and WAXS, especially using synchrotron radiation, has proven particularly effective in elucidating the structural evolution of polymers during processing, deformation, and crystallization [31, 33]. Thus, SAXS and WAXS are indispensable tools for advancing the understanding of polymer structure–property relationships at the nanoscale.

6.3 X-Ray Diffraction

X-ray diffraction (XRD) is a fundamental technique for the structural analysis of polymers, providing detailed information about their crystalline and amorphous phases. When X-rays are directed onto a polymer sample, the resulting diffraction pattern reflects the arrangement of polymer chains and the degree of order within the material. Highly crystalline polymers produce sharp diffraction peaks, while amorphous or partially crystalline polymers exhibit broader halos or diffuse scattering, enabling clear differentiation between these states. By analyzing the positions and intensities of the diffraction peaks, researchers can determine key structural parameters such as d-spacing (the distance between crystallographic planes), crystallite size, and the percentage of crystallinity within the polymer. For example, studies on materials like polyethylene, polypropylene, and polyethylene terephthalate (PET) have shown that XRD patterns are sensitive to processing conditions and thermal history, which can significantly influence crystallinity and, consequently, material properties. The crystallite size can be calculated using the Scherrer equation, and the degree of crystallinity can be quantified by comparing the integrated intensities of crystalline peaks to the total scattering intensity [34]. Furthermore, XRD is widely used for phase identification by comparing measured patterns to reference databases such as the ICDD PDF4 + Organic database, facilitating the identification of different polymer types and polymorphs [35]. Overall, XRD serves as a rapid, non-destructive, and reliable method for elucidating the structural organization of polymers, making it indispensable in both research and industrial quality control. The various analysis carried out by XRD are explained below:

Crystallinity Analysis

XRD distinguishes between crystalline, semi-crystalline, and amorphous regions in polymers by analyzing diffraction patterns. For example, semi-crystalline polymers (e.g., polyethylene, nylon) show sharp peaks superimposed on broad amorphous halos. The relative intensity of these features quantifies the degree of crystallinity, which affects mechanical properties like tensile strength and creep resistance. Amorphous polymers (e.g., polystyrene) exhibit broad, diffuse peaks, as seen in pure PVC's XRD pattern. Scherrer's formula calculates crystallite size from peak broadening, revealing how processing parameters (temperature, draw ratio) influence crystallite dimensions [36].

Molecular Orientation

XRD detects alignment in polymer chains, particularly in fibers or films. The Hermans orientation function quantifies crystalline alignment, critical for understanding anisotropic properties in extruded or molded polymers. Stretching or annealing processes induce orientation, visible as changes in peak symmetry or intensity distribution [37].

Processing and Additive Effects

XRD reveals how processing methods or additives alter morphology. Plasticizers (e.g., ethylene carbonate) and ionic liquids reduce crystallinity in PVC-based electrolytes, enhancing ionic conductivity by increasing amorphous regions. Plasma treatment of PET films modifies surface crystallinity, shifting peak intensities and orientations. Composite materials (e.g., polymer-ceramic blends) show distinct diffraction patterns for crystalline fillers and amorphous matrices, aiding in thermal stability assessments [38].

Polymer Blends and Copolymers

XRD identifies structural interactions in blends and copolymers. Block copolymers vs. random copolymers exhibit distinct diffraction fingerprints due to differences in molecular packing. Blending polymers (e.g., PET with polystyrene) generates new diffraction features, indicating phase separation or intermolecular interactions.

XRD is often paired with Thermal analysis i.e. differential scanning calorimetry (DSC) to correlate crystallinity with melting behavior, FTIR spectroscopy for chemical structure validation and SEM for surface morphology comparisons. For instance, in PVC-$NH_4CF_3SO_3$ electrolytes, XRD confirmed amorphous dominance, while FESEM visualized structural homogeneity. XRD's ability to probe both crystalline and amorphous phases makes it indispensable for optimizing polymer performance in applications ranging from flexible electronics to high-strength composites [36–38].

7 Surface Roughness and Wettability

Surface roughness and wettability are closely interconnected properties that significantly impact the interaction between liquids and solid surfaces. Surface roughness refers to the microscopic texture of a material, characterized by variations in height and structure, which can influence adhesion, friction, and fluid behavior. Wettability, on the other hand, describes how a liquid spreads or beads on a surface, determined by the contact angle formed between the liquid and the solid. The relationship between these properties is governed by models such as Wenzel and Cassie-Baxter, which explain how roughness can enhance or reduce wettability depending on the intrinsic nature of the surface. A rough hydrophilic surface increases liquid spreading, while a rough hydrophobic surface enhances water repellency by trapping air pockets. This interplay is crucial in various applications, including self-cleaning materials, anti-fouling coatings, and biomedical devices, where surface modifications are designed to optimize performance.

8 Structural Features of Polymers at Nanoscale

8.1 Crystallinity and Molecular Arrangement

Polymers can exhibit amorphous or semi-crystalline structures depending on their molecular arrangement. The degree of crystallinity affects their mechanical strength, thermal stability, and optical properties. At the nanoscale, the crystallinity and molecular arrangement of polymers play a critical role in defining their mechanical, thermal, and functional properties. Polymers exhibit a semi-crystalline structure, consisting of both crystalline regions—where polymer chains are tightly packed in an ordered fashion—and amorphous regions, where chains are randomly arranged. At the nanoscale, the degree of crystallinity is influenced by factors such as molecular weight, processing conditions, and intermolecular interactions. Kongkhlang et al. demonstrated that how change in processing parameter can change morphology and molecular orientation at molecular level (Fig. 4) [39]. Nanoscale confinement can alter polymer crystallization, leading to size-dependent crystal growth, reduced crystallite size, and modified chain folding behavior. Additionally, molecular arrangement at this scale determines properties like tensile strength, barrier performance, and conductivity, which are crucial in applications such as nanocomposites, flexible electronics, and biomedical materials. Understanding and controlling nanoscale crystallinity in polymers enables the design of advanced materials with tailored performance for high-tech and industrial applications.

8.1.1 Chain Orientation and Molecular Packing

The orientation of polymer chains at the nanoscale affects their mechanical and electronic properties. Techniques such as X-ray diffraction (XRD) and polarized optical microscopy help in studying molecular alignment in thin polymer films. At the nanoscale, chain orientation and molecular packing of polymers significantly influence their mechanical strength, optical properties, and functional performance. Chain orientation refers to the alignment of polymer chains in a specific direction, which can occur during processing methods such as stretching, electrospinning, or nanoconfinement. Highly oriented polymer chains enhance tensile strength, thermal stability, and anisotropic conductivity, making them essential for high-performance fibers and nanocomposites. Molecular packing describes how polymer chains organize within crystalline or amorphous regions, affecting density, permeability, and

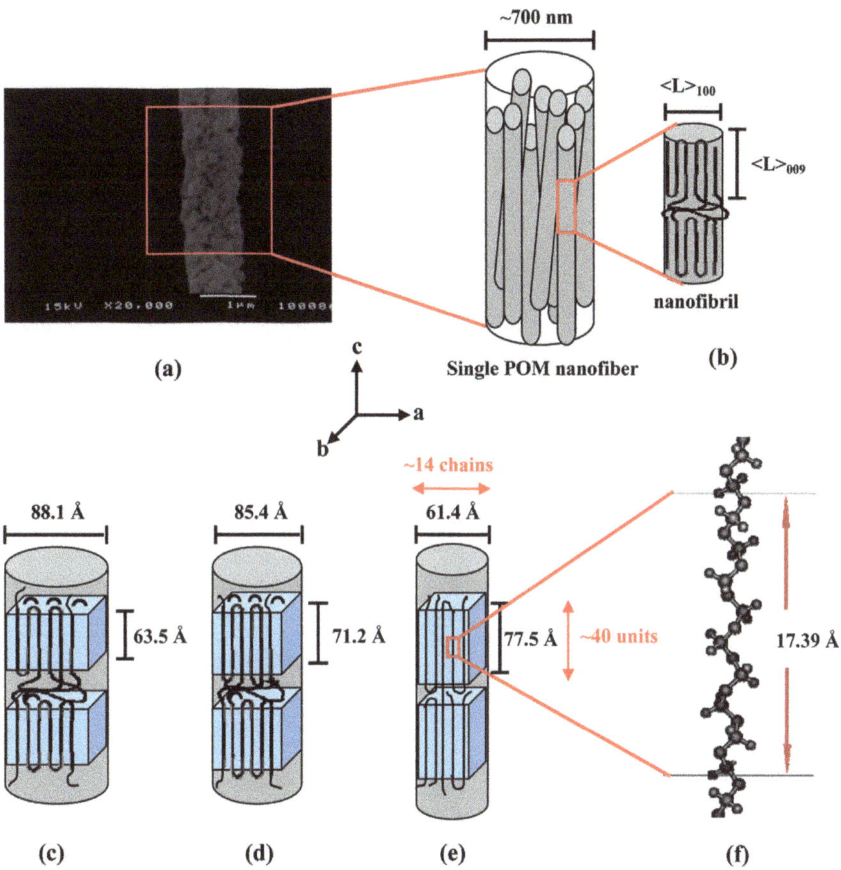

Fig. 4 **a** SEM micrograph of a single POM nanofiber collected at 630 m/min; **b** schematic representations of nanofibril in a single POM nanofiber; **c**–**e** schematic representations of the crystal orientation of **c** POM at 0 m/min, **d** POM at 630 m/min, and **e** POM at 1890 m/min; and **f** crystal conformation of the (9/5) helical structure of the POM chain

interaction with external stimuli. At nanoscale dimensions, restricted chain mobility and interface effects can lead to unique packing arrangements, altering crystallization behavior and phase transitions. Understanding nanoscale chain orientation and molecular packing is crucial for developing advanced materials for applications in flexible electronics, high-strength films, and biomedical nanodevices. The study carried out by Cheng et. al. shows that polymer crystallization affects ion movement in two ways: by making the path more twisted (structural effect) and by restricting chain motion (dynamic effect). Using a specially designed polymer crystal system, they found that ions mainly move along the folds in the crystal, and in-plane conductivity can be up to 2000 times higher than through-plane. Figure 5 depicts images by various instruments of PEO single crystals [40].

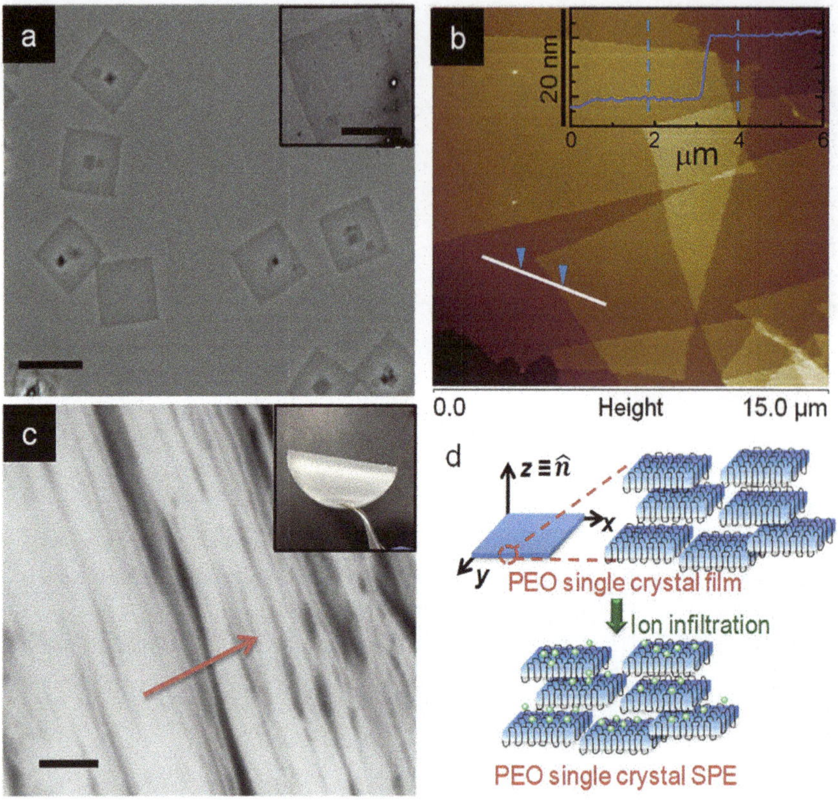

Fig. 5 **a** Phase contrast optical microscopy image of PEO single crystals grown from dilute solution before casting into films (scale bar, 20 µm). Inset shows a transmission electron micrograph of a typical single crystal; scale bar is 10 µm. **b** Atomic force microscopy height image of a 15 × 15 µm area scan of PEO single crystals. Inset shows the corresponding height profile along the white line. **c** Scanning electron micrograph of the cross section of a PEO single crystal film; red arrow shows the film normal; scale bar is 200 nm. Inset shows the optical image of the dry film. **d** Schematic of PEO single crystal SPE preparation

DSC, XPS, and DLS are complementary techniques for probing nanoscale interactions in polymer systems, offering insights into thermal behavior, surface chemistry, and colloidal stability. Here's how each method contributes.

9 Differential Scanning Calorimetry (DSC)

DSC is a widely used technique for analyzing the thermal transitions and energy changes in polymer-nanoparticle systems. DSC enables precise detection of key phase transitions such as the glass transition temperature (T_g), crystallization temperature (T_c), and melting temperature (T_m), all of which can shift significantly due to interactions between polymers and embedded nanoparticles. For instance, the presence of nanofillers in cyclic butylene terephthalate (CBT) has been shown to delay the onset of polymerization by 20–35 °C, highlighting the impact of nanoparticle additives on thermal behavior [41]. Additionally, DSC is instrumental in quantifying changes in polymer crystallinity induced by nanofillers. Studies on wood-polymer nanocomposites have demonstrated that the introduction of nanoparticles can alter decomposition enthalpies during pyrolysis, indicating modifications in the crystalline structure of the polymer matrix [42]. Beyond research applications, DSC is also valuable in industrial settings for rapidly assessing the homogeneity of nanoparticle dispersion within polymer batches, thus ensuring batch-to-batch consistency and quality control. Overall, DSC provides critical insights into the thermal and structural properties of polymer-nanoparticle composites, supporting both fundamental research and industrial development.

10 X-Ray Photoelectron Spectroscopy (XPS)

X-ray Photoelectron Spectroscopy (XPS) is a powerful tool for elucidating the surface composition and electronic interactions in nanomaterial systems. For example, XPS has been used to reveal charge transfer phenomena, such as platinum (Pt) nanoparticles smaller than 7 nm acting as electron acceptors when stabilized with polyvinylpyrrolidone (PVP), in contrast to bulk Pt, which tends to donate electrons [43]. This technique is also essential for identifying ligand binding and determining oxidation states; for instance, XPS studies have shown that ceria nanoparticles possess a higher degree of surface reduction compared to their microcrystalline counterparts [44]. Additionally, XPS can detect core-level binding energy shifts, with Pt nanoparticles exhibiting lower binding energies than bulk Pt due to pronounced size effects.

11 Dynamic Light Scattering (DLS)

Dynamic Light Scattering (DLS) complements XPS by measuring the hydrodynamic size and aggregation state of nanoparticles in suspension. DLS can resolve particle size distributions ranging from 0.5 nm to 2.5 μm, which is particularly important for characterizing nanomedicines such as lipid nanoparticles (LNPs) and viral vectors.

Furthermore, DLS is sensitive to the presence of aggregates in protein formulations, such as detecting adeno-associated virus aggregates larger than 500 nm, and is frequently used to monitor colloidal stability under various thermal or chemical stresses without the need for sample filtration.

The synergistic application of XPS and DLS, often in combination with other techniques such as Differential Scanning Calorimetry (DSC), provides a comprehensive understanding of nanomaterial systems. For example, correlating thermal stability data from DSC with surface oxidation states identified by XPS can elucidate the effects of nanoparticle surface chemistry in polymer-ceramic composites. Similarly, linking colloidal stability information from DLS with surface ligand interactions from XPS is invaluable in the development of drug-delivery systems. Moreover, combining DLS and DSC allows researchers to assess how nanoparticle size influences polymer crystallization kinetics. In Pt-PVP systems, XPS has revealed charge-transfer mechanisms, DSC has validated changes in thermal stability due to these interactions, and DLS has clarified the influence of nanoparticle size on polymer solution behaviour, demonstrating the value of an integrated analytical approach.

12 Interfacial Features of Polymers

Interfacial phenomena at the nanoscale determine the compatibility and stability of polymer blends and composites. This includes Interfacial tension and adhesion which affects phase separation and morphology in polymer blends and Interfacial bonding mechanisms such as van der Waals forces, hydrogen bonding, and covalent interactions. The interfacial interaction phenomenon is described here for polymer-substrate, polymer–polymer and polymer-nanoparticles.

12.1 Polymer-Substrate Interactions

Polymer-substrate interactions play a critical role in determining the structural and functional properties of polymer thin films, particularly in confined systems. These interactions are driven by forces between polymer chains and substrate surfaces, which influence phenomena such as adsorption, glass transition behavior, and surface morphology. For instance, annealing polymer films often leads to the formation of irreversibly adsorbed nanolayers, where monomer-substrate interactions stabilize polymer chains against desorption. This adsorbed layer significantly impacts film properties, including viscosity and glass transition temperature, as demonstrated through studies on homopolymers and random copolymers of styrene and methyl methacrylate [45]. Additionally, substrate adhesion coupled with intermolecular forces can alter the glass transition behavior of ultrathin films, where strong substrate interactions enhance thermal stability even for polymers with weaker interchain forces [46]. Surface energy and hydrophilicity of substrates also affect the morphology of polymer blends, influencing phase separation and dewetting behavior [47]. Understanding these interactions is crucial for optimizing polymer processing techniques in nanotechnology applications [46]. Surface energy and hydrophobicity of substrate also affect the morphology of polymer blends, influencing phase separation and dewetting behavior [47]. Understanding these interactions is crucial for optimizing polymer processing techniques in nanotechnology applications.

The adhesion and compatibility of polymers with different substrates depend on interfacial energy, chemical bonding, and van der Waals forces. Modifications such as plasma treatment and surface functionalization improve polymer-substrate interactions. The adhesion and compatibility of polymers with different substrates depend on interfacial energy, chemical bonding, and van der Waals forces. Modifications such as plasma treatment and surface functionalization improve polymer-substrate interactions. At the nanoscale, polymer-substrate interactions play a crucial role in determining the adhesion, mechanical properties, and functionality of polymer-based nanostructures. These interactions arise from van der Waals forces, hydrogen bonding, electrostatic attractions, and chemical bonding between polymer chains and the substrate surface. The degree of adhesion and interfacial compatibility depends on factors such as surface energy, roughness, and molecular affinity. At nanoscale dimensions, confinement effects and interfacial interactions can significantly alter polymer chain mobility, crystallinity, and molecular packing, leading to unique mechanical and thermal properties. Strong polymer-substrate interactions enhance stability and durability, which is essential in applications like thin-film coatings, nanoelectronics, and biomedical implants. Controlling these interactions through surface modifications or tailored polymer architectures enables the development of high-performance nanomaterials with optimized adhesion, flexibility, and functional properties.

Characterizing polymer-substrate interactions at the nanoscale requires advanced techniques to probe interfacial chemistry, morphology, and dynamics. Solid-State Nuclear Magnetic Resonance (NMR) distinguishes polymer chain dynamics at the substrate interface from bulk regions by analyzing spin–spin relaxation times (T_2) and reveals molecular mobility changes, such as slowed segmental motions in polymers adsorbed onto clay or silica substrates [48]. Infrared (IR) and Raman Spectroscopy identifies chemical bonding and conformational changes at polymer-substrate interfaces (e.g., hydrogen bonding, adsorption-induced shifts) [48]. Surface-Enhanced Raman Scattering (SERS) enhances sensitivity for detecting molecular interactions on plasmonic substrates [49]. Fluorescence Spectroscopy utilizes Förster Resonance Energy Transfer (FRET) to quantify filler dispersion and interfacial proximity in composites (e.g., cellulose-polyethylene systems) [48]. Small-Angle X-ray/Neutron Scattering (SAXS/SANS) analyze nanostructural changes in surfactant/polymer assemblies (e.g., micelle elongation under salt effects) and polymer-substrate packing density. SAXS detects electron density variations, while SANS distinguishes atomic nuclei, enabling complementary insights [49]. Molecular Dynamics Simulations predict orientation transitions and adsorption behavior of polymers on substrates, validated experimentally via spectroscopy. Nanolithography (Dip-Pen, LIPSS) creates controlled nanoscale patterns to study how substrate topography influences polymer adhesion and self-assembly [49]. Phase-Separated Block

Copolymer Templates technique used to investigate differential wetting and interfacial interactions of self-assembled diblock copolymers (e.g., PS-b-PMMA) forming periodic nanodomains [50]. These techniques collectively address critical aspects of polymer-substrate interactions, from molecular binding and chain dynamics to nanoscale morphology, enabling tailored material design for electronics, biomaterials, and catalysis. Characterizing polymer-substrate interactions at the nanoscale requires advanced techniques to probe interfacial chemistry, morphology, and dynamics. Solid-State Nuclear Magnetic Resonance (NMR) distinguishes polymer chain dynamics at the substrate interface from bulk regions by analyzing spin–spin relaxation times (T_2) and reveals molecular mobility changes, such as slowed segmental motions in polymers adsorbed onto clay or silica substrates [48]. Infrared (IR) and Raman Spectroscopy identifies chemical bonding and conformational changes at polymer-substrate interfaces (e.g., hydrogen bonding, adsorption-induced shifts) [48]. Surface-Enhanced Raman Scattering (SERS) enhances sensitivity for detecting molecular interactions on plasmonic substrates [49]. Fluorescence Spectroscopy utilizes Förster Resonance Energy Transfer (FRET) to quantify filler dispersion and interfacial proximity in composites (e.g., cellulose-polyethylene systems) [48]. Small-Angle X-ray/Neutron Scattering (SAXS/SANS) analyze nanostructural changes in surfactant/polymer assemblies (e.g., micelle elongation under salt effects) and polymer-substrate packing density. SAXS detects electron density variations, while SANS distinguishes atomic nuclei, enabling complementary insights [49]. Molecular Dynamics Simulations predict orientation transitions and adsorption behavior of polymers on substrates, validated experimentally via spectroscopy. Nanolithography (Dip-Pen, LIPSS) creates controlled nanoscale patterns to study how substrate topography influences polymer adhesion and self-assembly [49]. Phase-Separated Block Copolymer Templates technique used to investigate differential wetting and interfacial interactions of self-assembled diblock copolymers (e.g., PS-b-PMMA) forming periodic nanodomains [50]. These techniques collectively address critical aspects of polymer-substrate interactions, from molecular binding and chain dynamics to nanoscale morphology, enabling tailored material design for electronics, biomaterials, and catalysis.

12.2 Polymer–Polymer Interfaces

Polymer blends and multilayered polymer structures often exhibit distinct interfacial features that affect phase compatibility, mechanical integrity, and thermal stability. At the nanoscale, polymer–polymer interfaces play a critical role in determining the mechanical, thermal, and transport properties of polymer blends, composites, and multilayered structures. These interfaces are governed by intermolecular forces such as van der Waals interactions, hydrogen bonding, and entanglement between polymer chains. The degree of miscibility and interfacial adhesion depends on factors like molecular weight, chain compatibility, and interfacial tension. At nanoscale dimensions, confinement effects can lead to altered chain dynamics, restricted mobility, and phase separation, which influence properties such as toughness, permeability, and conductivity. Engineering polymer–polymer interfaces through compatibilizers, block copolymers, or nanostructuring techniques can enhance interfacial adhesion, improving material performance in applications such as high-performance films, nanocomposites, and bioinspired materials. Understanding and controlling these nanoscale interfaces is crucial for designing advanced multifunctional polymers with tailored properties for various technological applications.

Polymer–polymer interactions at the nanoscale exhibit unique characteristics distinct from bulk behavior, driven by interfacial effects, molecular mobility, and confinement. These interactions play a critical role in material properties and applications, from nanostructured polymers to advanced composites.

12.3 Nanoscale Phase Separation in Block Copolymers

Block copolymers (BCPs) like polystyrene-b-poly(methyl methacrylate) (PS-b-PMMA) self-assemble into periodic nanodomains (e.g., alternating PS and PMMA strips with ~45 nm spacing). This nanoscale phase separation creates sharp chemical interfaces that influence interfacial phenomena. For example, immunoglobulin G (IgG) proteins selectively adsorb onto PS domains, preferentially near PS-PMMA interfacial regions [51]. This highlights how nanoscale polymer-polymer interfaces can direct molecular interactions, even enabling spatially resolved biomolecule partitioning. For example, immunoglobulin G (IgG) proteins selectively adsorb onto PS domains, preferentially near PS-PMMA interfacial regions [51]. This highlights how nanoscale polymer-polymer interfaces can direct molecular interactions, even enabling spatially resolved biomolecule partitioning.

12.4 Interfiber Interactions in Polymeric Nanofibers

Molecular dynamics (MD) simulations reveal that aligned nanofibers (4.6 nm diameter) exhibit attraction when separated by $\leq$9 nm, leading to rapid coalescence below the glass transition temperature. In contrast, static energy calculations predict a shift from attraction to repulsion at 6 nm separation. This discrepancy underscores the role of molecular mobility in nanoscale polymer-polymer interactions, where dynamic rearrangements dominate over equilibrium predictions [52].

12.5 Surfactant and Polymer-Mediated Nanostructure Assembly

In plasmonic nanostar synthesis, polymers and surfactants act as shape-directing agents by modulating interfacial interactions. For instance, polymer-surface adsorption dynamics govern nanostar morphology, impacting applications like surface-enhanced Raman spectroscopy (SERS) [49]. The branched geometry of nanostars further complicates ligand-polymer interactions, necessitating tailored characterization methods to resolve interfacial behaviors. In plasmonic nanostar synthesis, polymers and surfactants act as shape-directing agents by modulating interfacial interactions. For instance, polymer-surface adsorption dynamics govern nanostar morphology, impacting applications like surface-enhanced Raman spectroscopy (SERS) [49]. The branched geometry of nanostars further complicates ligand-polymer interactions, necessitating tailored characterization methods to resolve interfacial behaviors.

12.6 Macroscopic Implications of Nanoscale Interactions

Nanoscale polymer-polymer interactions influence bulk material properties. For example, water diffusion in poly(ethylene oxide) (PEO) solutions deviates from classical models (e.g., Stokes-Einstein) due to localized water structuring near polymer chains [53]. Similarly, interfiber coalescence in nanofiber mats alters mechanical and functional properties, demonstrating the cascading effects of nanoscale dynamics on macroscopic performance [52]. Nanoscale polymer-polymer interactions influence bulk material properties. For example, water diffusion in poly(ethylene oxide) (PEO) solutions deviates from classical models (e.g., Stokes-Einstein) due to localized water structuring near polymer chains [53]. Similarly, interfiber coalescence in nanofiber mats alters mechanical and functional properties, demonstrating the cascading effects of nanoscale dynamics on macroscopic performance [52].

These findings emphasize the need for single-molecule and nanoscale-resolution techniques to unravel polymer-polymer interaction mechanisms, which are fundamental to designing next-generation nanomaterials.

13 Polymer-Nanoparticle Interactions

In nanocomposites, the interaction between polymer matrices and nanoparticles plays a crucial role in enhancing material properties. Surface functionalization of nanoparticles helps in improving dispersion and interfacial adhesion within the polymer matrix. At the nanoscale, polymer-nanoparticle interactions play a crucial role in defining the structural, mechanical, and functional properties of nanocomposites.

These interactions are influenced by factors such as surface chemistry, particle size, polymer chain mobility, and dispersion state which collectively dictate macroscopic material performance. Strong interactions, driven by van der Waals forces, hydrogen bonding, electrostatic attractions, or covalent bonding, enhance nanoparticle dispersion and improve composite properties such as mechanical strength, thermal stability, and conductivity.

The extreme surface-to-volume ratio of nanofillers (e.g., $10^6\times$ increase for spherical NPs vs microfillers) creates vast interfacial regions where polymer chains exhibit altered mobility, crystallinity, and entanglement [54]. This interphase zone (2–50 nm thick) forms through different methods, such as 1. Physical interactions include electrostatic forces, surface wetting, and mechanical interlocking [55], 2. Chemical bonding such as Covalent grafting in systems like PMMA-functionalized NPs [56] 3. Entanglement effects due to increased radius of gyration (Rg) near NPs due to chain stretching [57]. The extreme surface-to-volume ratio of nanofillers (e.g., $10^6\times$ increase for spherical NPs vs microfillers) creates vast interfacial regions where polymer chains exhibit altered mobility, crystallinity, and entanglement [54].

Nanofillers enhance stiffness by 40–60% through void reduction and restricted chain mobility, while excessive loading (>7 wt%) causes agglomeration-induced weakening. Poor compatibility, however, can lead to nanoparticle aggregation, reducing performance [56, 57]. The interfacial region between polymers and nanoparticles significantly affects crystallinity, chain relaxation, and phase behavior, altering overall material properties. High surface energy promotes NP clustering and creates dispersion issues that can be handled by surface modification (silane coupling agent) [56] and solvent assisted sonication [55]. The Hydrophilic NPs in hydrophobic matrices mitigated by plazma treatment and compatibilizer additives. These nanoscale interactions enable smart composites with stimuli-responsive behavior, leveraging the dual functionality of polymers and NPs. Optimal performance demands precise control over interfacial chemistry, NP spacing, and dispersion state, this enables researchers to develop advanced nanocomposites with enhanced durability, flexibility, and multifunctionality for applications in coatings, electronics, and biomedical fields.

Ganesan and Jayaraman reviews theoretical and computational approaches to understanding polymer nanocomposites (PNCs), focusing on how interactions between nanoparticles (NPs) and polymers dictate morphology and phase behavior, which in turn control macroscopic material properties [58]. The PRISM theory plays a central role, employing Ornstein–Zernike-like integral equations to analyze intermolecular and intramolecular correlations, such as radial distribution functions. This framework combines closures like the Percus–Yevick approximation for polymer-polymer interactions and hypernetted chain closures for NP-NP interactions, while the PRISM-MC hybrid method integrates Monte Carlo simulations to account for non-ideal chain conformations in grafted or matrix polymers. Complementing this, density functional theory (DFT) and self-consistent field theory (SCFT) model free energy landscapes to predict phase separation and NP dispersion.

Morphological outcomes depend critically on NP surface chemistry. Bare NPs exhibit dispersion behaviors dictated by competing depletion attraction and steric repulsion, with small NPs ((R∼RgR∼Rg) inducing chain stretching near interfaces and altering local dynamics. For grafted NPs, grafting density and chain length determine compatibility: low grafting promotes aggregation, while high grafting enhances dispersion by forming "corona" layers that mediate matrix interactions and enable microphase separation at specific loadings.

Simulations provide mechanistic insights: coarse-grained molecular dynamics reveal NP aggregation thresholds (e.g., 10 vol% for nanorods) and polymer alignment near NPs, while atomistic simulations quantify interfacial binding energies and reduced chain mobility in systems like PMMA/silica. These findings inform design principles, where optimal mechanical properties arise at 1–5 wt% NP loading due to enhanced stress transfer and minimized voids. Balancing matrix-induced macrophase separation with tether-mediated microphase separation requires careful tuning of grafted and matrix polymer lengths. The review highlights the necessity of integrating theory, simulations, and experiments to resolve nanoscale interactions and engineer PNCs with tailored functionalities.

14 Applications and Future Perspectives

Nanoscale surface morphology, structure, and interface engineering of polymers have profound implications in various fields. Polymers at the nanoscale have revolutionized various industries, offering enhanced mechanical, thermal, and functional properties for advanced applications. In nanomedicine, polymer-based nanoparticles enable targeted drug delivery, bioimaging, and tissue engineering. In electronics, nanoscale polymer structures are essential for flexible circuits, organic photovoltaics, and high-performance coatings. Nanocomposites with polymer matrices and nanofillers improve strength, conductivity, and thermal resistance in aerospace, automotive, and packaging applications. Looking ahead, future advancements in polymer nanotechnology will focus on self-assembling materials, bioinspired smart polymers, and sustainable nanomaterials for eco-friendly solutions. The integration of artificial intelligence and nano manufacturing techniques will further enable precise control over polymer properties, unlocking new possibilities in healthcare, energy storage, and soft robotics. As research progresses, the development of next-generation nanoscale polymers will drive innovation across multiple scientific and industrial domains.

Future research in this field will focus on developing novel characterization methods, improving interfacial engineering techniques, and designing new polymer nanocomposites with superior properties.

15 Conclusion

The surface morphology, structural characteristics, and interfacial features of polymers at the nanoscale play a pivotal role in defining their physical, chemical, and functional properties. Nanoscale surface morphology influences wettability, adhesion, and mechanical behavior, while the structural arrangement, including crystallinity and molecular packing, determines mechanical strength, thermal stability, and conductivity. Additionally, interfacial interactions, whether at polymer-substrate, polymer–polymer, or polymer-nanoparticle interfaces, significantly impact material performance by dictating adhesion, dispersion, and phase behavior. Advancements in nanotechnology and material engineering have enabled precise control over these features, leading to innovative applications in biomedicine, electronics, and nanocomposites. Future research will continue to explore novel strategies for tuning nanoscale polymer properties, paving the way for high-performance, multifunctional materials with enhanced durability, efficiency, and sustainability.

References

1. Crosby, A.J., Lee, J.: Polymer nanocomposites: the 'nano' effect on mechanical properties. Polym. Rev. **47**(2), 217–229 (2007). https://doi.org/10.1080/15583720701271278
2. Chen, X.C., Green, P.F.: Control of morphology and its effects on the optical properties of polymer nanocomposites. Langmuir **26**(5), 3659–3665 (2010). https://doi.org/10.1021/la903108u
3. Ma, W., Yang, C., Gong, X., Lee, K., Heeger, A.J.: Thermally stable, efficient polymer solar cells with nanoscale control of the interpenetrating network morphology. Adv Funct Materials **15**(10), 1617–1622 (2005). https://doi.org/10.1002/adfm.200500211
4. Keledi, G., Hári, J., Pukánszky, B.: Polymer nanocomposites: structure, interaction, and functionality. Nanoscale **4**(6), 1919–1938 (2012). https://doi.org/10.1039/C2NR11442A
5. Morphology, P.: Principles, characterization, and processing. Wiley, United Kingdom (2016)
6. Adhikari, R., Michler, G.H.: Influence of molecular architecture on morphology and micromechanical behavior of styrene/butadiene block copolymer systems. Prog. Polym. Sci. **29**(9), 949–986 (2004). https://doi.org/10.1016/j.progpolymsci.2004.06.002
7. Gentile, A., Ruffino, F., Grimaldi, M.: Complex-morphology metal-based nanostructures: fabrication, characterization, and applications. Nanomaterials **6**(6), 110 (2016). https://doi.org/10.3390/nano6060110
8. Bigg, D.M.: Mechanical property enhancement of semicrystalline polymers—A review. Polymer Engineering & Sci **28**(13), 830–841 (1988). https://doi.org/10.1002/pen.760281303
9. Hardick, O., Stevens, B., Bracewell, D.G.: Nanofibre fabrication in a temperature and humidity controlled environment for improved fibre consistency. J. Mater. Sci. **46**(11), 3890–3898 (2011). https://doi.org/10.1007/s10853-011-5310-5
10. Rasitha, T.P., Sofia, S., Anandkumar, B., Philip, J.: Long term antifouling performance of superhydrophobic surfaces in seawater environment: Effect of substrate material, hierarchical surface feature and surface chemistry. Colloids Surf. A: Phys.Chemical Eng. Asp., **647**, 129194 (2022). https://doi.org/10.1016/j.colsurfa.2022.129194
11. Howell, D., Behrends, B.: A review of surface roughness in antifouling coatings illustrating the importance of cutoff length. Biofouling **22**(6), 401–410 (2006). https://doi.org/10.1080/08927010601035738

12. Scardino, A.J., Zhang, H., Cookson, D.J., Lamb, R.N., Nys, R.D.: The role of nano-roughness in antifouling. Biofouling **25**(8), 757–767 (2009). https://doi.org/10.1080/08927010903165936
13. "österberg-et-al-2023-biobased-nanomaterials-the-role-of-interfacial-interactions-for-advanced-materials.pdf"
14. Meijer, H.E.H., Govaert, L.E.: Mechanical performance of polymer systems: The relation between structure and properties. Prog. Polym. Sci. **30**(8–9), 915–938 (2005). https://doi.org/10.1016/j.progpolymsci.2005.06.009
15. Sachan, R., Purwar, R.: Rheological study on blend solutions of non-mulberry silk fibroin and gelatin biopolymers. JPM **40**(3–4), 205–214 (2023). https://doi.org/10.32381/JPM.2023.40.3-4.6
16. Xue, J., Xie, J., Liu, W., Xia, Y.: Electrospun nanofibers: new concepts, materials, and applications. Acc. Chem. Res. **50**(8), 1976–1987 (2017). https://doi.org/10.1021/acs.accounts.7b00218
17. Wypych, G.: Chain mobility and progress in medicine, pharmaceuticals, and polymer science and technology. ChemTec Publishing, (2020). [Online]. https://books.google.co.in/books?id=ayjJDwAAQBAJ
18. Bao, L., Fan, H., Chen, Y., Yan, J., Yang, T., Guo, Y.: Effect of surface free energy and wettability on the adhesion property of waterborne polyurethane adhesive. RSC Adv. **6**(101), 99346–99352 (2016)
19. Mazumdar, P., Chockalingam, S., Rattan, S., Gupta, B.K.: Tunable mechanical, electrical, and thermal properties of polymer nanocomposites through GMA bridging at interface. ACS Omega **3**(4), 3675–3687 (2018)
20. Fang, Q., Lafdi, K.: Effect of nanofiller morphology on the electrical conductivity of polymer nanocomposites. Nano Express **2**(1), 010019 (2021)
21. Xuan, H., et al.: Tailoring nano-porous surface of aligned electrospun poly (L-Lactic acid) fibers for nerve tissue engineering. Int. J. Mol. Sci. **22**(7), 3536 (2021)
22. Shipley, W., Wang, Y., Chien, J., Wang, B., Tao, A.R.: Characterization of surface patterning on polymer-grafted nanocubes using atomic force microscopy and force volume mapping. Langmuir **40**(39), 20464–20473 (2024). https://doi.org/10.1021/acs.langmuir.4c01902
23. Baggioli, A., Casalegno, M., David, A., Pasquini, M., Raos, G.: Polymer-mediated adhesion: nanoscale surface morphology and failure mechanisms. Macromolecules **54**(1), 195–202 (2021). https://doi.org/10.1021/acs.macromol.0c02343
24. Li, H., Cooper-White, J.J.: Changing ligand number and type within nanocylindrical domains through kinetically constrained self-assembly—impacts of ligand 'redundancy' on human mesenchymal stem cell adhesion and morphology. Biomater. Sci. **2**(11), 1693–1705 (2014). https://doi.org/10.1039/C4BM00109E
25. Rydz, J., Šišková, A., Andicsová Eckstein, A.: Scanning electron microscopy and atomic force microscopy: topographic and dynamical surface studies of blends, composites, and hybrid functional materials for sustainable future. Adv. Mater. Sci. Eng., **2019**, 1–16 (2019). https://doi.org/10.1155/2019/6871785
26. Wei, J., et al.: Controllable synthesis of mesoporous carbon nanospheres and Fe–N/carbon nanospheres as efficient oxygen reduction electrocatalysts. Nanoscale **7**(14), 6247–6254 (2015). https://doi.org/10.1039/C5NR00331H
27. Biran, I., Houben, L., Kossoy, A., Rybtchinski, B.: Transmission electron microscopy methodology to analyze polymer structure with submolecular resolution. J. Phys. Chem. C **128**(14), 5988–5995 (2024). https://doi.org/10.1021/acs.jpcc.3c06977
28. Tencé-Girault, S., Woehling, V., Oikonomou, E.K., Karpati, S., Norvez, S.: About the art and science of visualizing polymer morphology using transmission electron microscopy. Macro Chem. & Phys. **219**(3), 1700483 (2018). https://doi.org/10.1002/macp.201700483
29. Cheng, Z., et al.: Experimental observation of localized interfacial phonon modes. Nat. Commun. **12**(1), 6901 (2021). https://doi.org/10.1038/s41467-021-27250-3
30. Rizvi, A., et al.: Revealing nanoscale structure and interfaces of protein and polymer condensates *via* cryo-electron microscopy. Nanoscale **16**(35), 16706–16717 (2024). https://doi.org/10.1039/D4NR01877J

31. Huang, Y.-J., et al.: In-situ synchrotron SAXS and WAXS investigation on the deformation of single and coaxial electrospun P(VDF-TrFE)-based nanofibers. IJMS **22**(23), 12669 (2021). https://doi.org/10.3390/ijms222312669
32. Lovell, R., Mitchell, G.R., Windle, A.H.: Wide-angle X-ray scattering study of structural parameters in non-crystalline polymers. Faraday Discuss. Chem. Soc. **68**, 46 (1979). https://doi.org/10.1039/dc9796800046
33. Chen, X., Schröder, J., Hauschild, S., Rosenfeldt, S., Dulle, M., Förster, S.: Simultaneous SAXS/WAXS/UV–Vis study of the nucleation and growth of nanoparticles: A test of classical nucleation theory. Langmuir **31**(42), 11678–11691 (2015). https://doi.org/10.1021/acs.langmuir.5b02759
34. Bhat, N.V., Deshmukh, R.R.: X-ray crystallographic studies of polymeric materials
35. Roe, R.: X-ray diffraction by polymers. In: Mark, H.F. (ed.) Encyclopedia of polymer science and technology, 3rd ed., pp. 1–47. Wiley, (2015). https://doi.org/10.1002/0471440264.pst635
36. Deraman, S.K., Tajuddin, N.A., Hanibah, H.: X-Ray Diffraction (XRD) and surface morphology studies of PVC complexes modified with doping salt and additives, vol. 18, no. 2, (2022)
37. Murthy, N.S.: X-ray diffraction from polymers. In: Guo, Q. (ed.) Polymer morphology, 1st ed., pp. 14–36. Wiley, (2016). https://doi.org/10.1002/9781118892756.ch2
38. Mustatea, A., Trusca, R., Trica, B., Dimonie, D.: XRD studies of the morphological changes generated by interface agents for obtaining new scalable sustainable blends based on starch and PCL. In: NeXT-Chem 2023, MDPI, p. 4 (2023). https://doi.org/10.3390/chemproc2023013004
39. Kongkhlang, T., Tashiro, K., Kotaki, M., Chirachanchai, S.: Electrospinning as a new technique to control the crystal morphology and molecular orientation of polyoxymethylene nanofibers. J. Am. Chem. Soc. **130**(46), 15460–15466 (2008). https://doi.org/10.1021/ja804185s
40. Cheng, S., Smith, D.M., Li, C.Y.: How does nanoscale crystalline structure affect ion transport in solid polymer electrolytes? Macromolecules **47**(12), 3978–3986 (2014). https://doi.org/10.1021/ma500734q
41. Lee, C.S., Dadmun, M.D.: Important thermodynamic characteristics of poly(3-hexyl thiophene). Polymer **55**(1), 4–7 (2014). https://doi.org/10.1016/j.polymer.2013.11.033
42. Rahman, M.R., Hamdan, S., Chang Hui, J.L.: Differential Scanning Calorimetry (DSC) and Thermogravimetric Analysis (TGA) of Wood polymer nanocomposites. MATEC Web Conf., vol. 87, p. 03013 (2017). https://doi.org/10.1051/matecconf/20178703013
43. Qiu, L., Liu, F., Zhao, L., Yang, W., Yao, J.: Evidence of a unique electron donor−acceptor property for platinum nanoparticles as studied by XPS. Langmuir **22**(10), 4480–4482 (2006). https://doi.org/10.1021/la053071q
44. Zhang, F., Wang, P., Koberstein, J., Khalid, S., Chan, S.-W.: Cerium oxidation state in ceria nanoparticles studied with X-ray photoelectron spectroscopy and absorption near edge spectroscopy. Surf. Sci. **563**(1–3), 74–82 (2004). https://doi.org/10.1016/j.susc.2004.05.138
45. Davis, M.J.B., Zuo, B., Priestley, R.D.: Competing polymer–substrate interactions mitigate random copolymer adsorption. Soft Matter **14**(35), 7204–7213 (2018). https://doi.org/10.1039/C8SM01433G
46. Xia, W., Keten, S.: Coupled effects of substrate adhesion and intermolecular forces on polymer thin film glass-transition behavior. Langmuir **29**(41), 12730–12736 (2013). https://doi.org/10.1021/la402800j
47. Han, X., Luo, C., Dai, Y., Liu, H.: Effect of polymer-substrate interactions on the surface morphology of polymer blend thin films. J Macromol Sci, Part B **47**(6), 1050–1061 (2008). https://doi.org/10.1080/00222340802266322
48. Bokobza, L.: Spectroscopic techniques for the characterization of polymer nanocomposites: A review. Polymers **10**(1), 7 (2017). https://doi.org/10.3390/polym10010007
49. Ferrari, D., Deriu, C., Fabris, L.: Surfactants and polymers on nanoscale surfaces: the interface landscape of plasmonic nanostars. Front. Nanotechnol. **6**, 1505304 (2024). https://doi.org/10.3389/fnano.2024.1505304

50. Hahm, J.: Fundamentals of nanoscale polymer-protein interactions and potential contributions to solid-state nanobioarrays. Langmuir **30**(33), 9891–9904 (2014). https://doi.org/10.1021/la404481t
51. Cho, D.H., Hahm, J.: Protein–polymer interaction characteristics unique to nanoscale interfaces: A perspective on recent insights. J. Phys. Chem. B **125**(23), 6040–6057 (2021). https://doi.org/10.1021/acs.jpcb.1c00684
52. Buell, S., Rutledge, G.C., Vliet, K.J.V.: Predicting polymer nanofiber interactions via molecular simulations. ACS Appl. Mater. Interfaces **2**(4), 1164–1172 (2010). https://doi.org/10.1021/am1000135
53. Moon, J.D., et al.: Nanoscale water–polymer interactions tune macroscopic diffusivity of water in aqueous poly(ethylene oxide) solutions. Chem. Sci. **15**(7), 2495–2508 (2024). https://doi.org/10.1039/D3SC05377F
54. Huang, J., Zhou, J., Liu, M.: Interphase in polymer nanocomposites. JACS Au **2**(2), 280–291 (2022). https://doi.org/10.1021/jacsau.1c00430
55. Bustamante-Torres, M., Romero-Fierro, D., Arcentales-Vera, B., Pardo, S., Bucio, E.: Interaction between filler and polymeric matrix in nanocomposites: magnetic approach and applications. Polymers **13**(17), 2998 (2021). https://doi.org/10.3390/polym13172998
56. Vera, M., Mella, C., Urbano, B.F.: Smart polymer nanocomposites: recent advances and perspectives. J. Chil. Chem. Soc. **65**(4), 4973–4981 (2020). https://doi.org/10.4067/S0717-97072020000404973
57. Ramesh, M., Rajeshkumar, L.N., Srinivasan, N., Kumar, D.V., Balaji, D.: Influence of filler material on properties of fiber-reinforced polymer composites: A review. e-Polymers, **22**(1), 898–916 (2022). https://doi.org/10.1515/epoly-2022-0080
58. Ganesan, V., Jayaraman, A.: Theory and simulation studies of effective interactions, phase behavior and morphology in polymer nanocomposites. Soft Matter **10**(1), 13–38 (2014). https://doi.org/10.1039/C3SM51864G

Tunable Shape and Size of Nano Polymers and Their Effect on Properties

Monika Siddharthan, Dharani Rajasekaran, and Jolly Bhadra

Abstract The integration of nanotechnology with polymer science has led to the development of nanopolymers, materials that exhibit novel properties distinct from their bulk counterparts. Among the key parameters influencing their behavior, the nanoscale size and shape play a pivotal role in determining their physicochemical and functional performance. Therefore, this chapter provides a comprehensive understanding on quantum confinement effects, increased surface-to-volume ratio, and shape-induced anisotropy and how these phenomena can be harnessed to tailor nanopolymer performance for specific applications. Further, the effect of size and shape tuning and their impact on mechanical, optical, thermal, electrical, surface and functional properties of nanopolymers are emphasized. Additionally, the chapter outlines the synthesis strategies to precisely control these parameters, and highlights their relevance with particular focus on biomedical, sensing, coating and energy storage applications.

1 Introduction

Polymers are macromolecules composed of repeating structural units called monomers that are covalently bonded to form long chains or complex network architectures. These polymers can be derived from diverse sources, including organic, inorganic, or mineral-organic, and can exist as natural materials (e.g., cellulose, proteins) or synthetic materials (e.g., polyethylene, polystyrene, poly vinylidene fluoride etc…) [1, 2]. The rapid evolution of polymer science and nanotechnology has witnessed the emergence of nanopolymers, materials that integrate the advantages of both the fields and leading to an entirely new class of functional materials with unique properties [3]. Nanopolymers are polymeric materials with at

M. Siddharthan · D. Rajasekaran · J. Bhadra (✉)
Qatar University Young Scientists Centre, Qatar University, Doha, Qatar
e-mail: jollybhadra@qu.edu.qa

B. Sharma et al. (eds.), *Nanostructured Polymers*, Engineering Materials,
https://doi.org/10.1007/978-981-95-3581-1_7

least one dimension structured at the nanometer scale (1–100 nm), where size-dependent effects such as high surface-area-to-volume ratios and quantum confinement significantly modify their physical, chemical, optical, and mechanical properties compared to their bulk counterparts [2]. As a result, nanopolymers possess multi-functional behavior that makes them highly suitable for a broad spectrum of applications, including drug delivery, sensors, optoelectronics and advanced coatings. Nanopolymers are typically prepared using a range of techniques including molecular self-assembly, incorporation of nanofillers, or controlled polymerization methods. During the preparation of nanopolymers, the formation of polymer chain confinement, crystal domains, interfaces, defects and internal phase separated regions influenced by nanoscale dimensions, can significantly alter the material's properties [4, 5]. Further, nanopolymers comprise of nanostructured polymers, polymer nanocomposites and polymeric nanoparticles [2, 6]. Nanostructured polymers are polymeric materials in the size of nanoscales, possess intrinsic nanoscale features such as nanodomains and self-assembled morphologies [7]. The size coupled properties of the nanostructured polymers demonstrate unique physico-chemical properties that meet the requirements of various applications including drug delivery, tissue engineering and biosensors. Further, polymer nanocomposites are a combination of polymer matrix combined with nanometer-sized fillers. These nanoscale fillers introduce a large interfacial area between polymer matrices and the fillers, which is the key distinguishing factor compared to traditional polymer composites such as carbon fiber-reinforced polymers. Their unique properties originate from the synergistic interaction between the host polymer matrix (e.g., polyaniline and polypyrrole) and the inorganic nanomaterial-based fillers (e.g., carbon nanotubes, and graphene). This leads to the enhancement of existing properties, as well as gives rise to novel functionalities that are absent in either of the components, exhibiting higher performance characteristics beyond those of traditional polymer composites [8]. The polymer nanocomposites have attracted attention across a broad spectrum of applications, including optoelectronics, sensors, energy storage systems, flexible electronics, and advanced structural materials. In the case of polymer nanocomposites, the polymer remains in a bulk state, whereas in polymeric nanoparticles, the polymer itself must be converted into nano-size material. Polymeric nanoparticles are usually in 10–200 nm range, and encompass various morphologies including nanoparticles, nanofibers, nanowires, nanospheres and other morphologies [9]. They are extensively used in biomedical applications for drug delivery, gene therapy, and diagnostics due to their biocompatibility, advantages of increasing the stability of drugs, controlled release profiles, and ability to target specific cells or tissues [10, 11]. Each of these categories offers distinct structural and functional attributes suitable for specific applications. Across all of these nanopolymers, optimizing the performance and functionality requires precise control over the size and shape, as these morphological parameters critically govern their physico-chemical properties and overall performance characteristics. Therefore, in this chapter, we examine the fundamental importance of size and shape, how they influence the properties of nanopolymers, outline synthesis strategies to precisely control these parameters, and highlight their impact across various advanced applications.

2 Fundamental Characteristics of Nanopolymers: Influence of Size and Shape Tuning

The essential characteristics that distinguish nanopolymers from conventional polymers is their size and morphology which highly influence the physical, chemical, mechanical, and functional properties of the materials. At the nanoscale, materials exhibit unique behaviors primarily due to their large surface area-to-volume ratio and quantum confinement effects, making size and morphology essential parameters for tuning nanopolymer performance. Notably, when particle sizes are reduced to nanometers, surface area per unit volume increases, leading to enhanced catalytic activity, diffusion, mechanical strength, optical behavior, and thermal stability [1, 12, 13]. Likewise, various shapes including spherical, rod or wire determines anisotropy, aspect ratio of the material which influence their interaction with surrounding environments, directional charge transport, and mechanical reinforcement [14, 15]. Consequently, polymers with the same chemical composition can provide different properties by reducing the dimension or by engineering their nanostructure into various shapes. Thus, tuning of size and shape of nanopolymers opens up vast possibilities in diverse fields such as nanomedicine, electronics, energy storage, environmental remediation, and advanced manufacturing.

2.1 Effect of Size Tuning in Nanopolymers

Unlike bulk polymers, where the properties are predominantly determined by factors such as crystallinity, chain entanglement, and macroscopic morphology, nanopolymers exhibit strong size-dependent effects due to quantum confinement of polymer chains and high surface-to-volume ratio [16]. As a result, precise control over particle size has become a crucial strategy for tuning functional performance of polymers across a wide range of applications.

2.2 Surface Area-to-Volume Ratio

Surface to volume ratio is the main reason for this size sensitivity in nanomaterials, that differ significantly from other materials. As particle size decreases into the nanoscale, their surface area-to-volume ratio increases, as a greater proportion of atoms or polymer chains reside on or near the surface rather than in the interior of the particle. For example, a particle of size 30 nm has 5% of its atoms on its surface, while particle size of 3 nm has 50% of its atoms on its surface [17, 18]. This increased surface dominance at the nanoscale enhances the role of surface energy, interfacial interactions, and surface-active sites, which in turn strongly influence the mechanical strength, chemical reactivity and interactions with the environment, heat resistance and optical properties of nanopolymers [19–21].

2.3 *Quantum Effects*

At nanoscale, quantum effects arise in the materials that the bulk form of the same material does not possess. One of the most significant phenomena is quantum confinement, which occurs when the dimensions of a material approach the exciton Bohr radius typically between 1 and 100 nm. This leads to the confinement of electron–hole pairs (excitons) within the physical dimensions of the particles with discrete energy levels [22]. These unique properties of nanoparticles arise from their intermediate position between bulk matter and single atoms, leading to distinct optical, electronic, and magnetic behaviors. In nanopolymers, the intrinsic, as well as functional properties of polymers can be amplified by the nanoscale dimensions of filler particles or when polymers itself are reduced to nanoscale dimensions. Nanopolymers exhibit enhanced mechanical strength, high surface energy, and surface reactivity when polymer chains or their composites are structured at nanoscale dimensions. For example, the electric conductivity of electrospun nanofibers conduct more current as the size are reduced below certain critical thickness within 100 nm [19, 23]. This is because of quantum confinement effect where, the nanopolymer's electronic energy levels become discrete, specifically properties such as bandgap, fluorescence emission, and carrier mobility become tunable based on size. This principle enables nanopolymer-based systems such as polymer-nanophosphor composites, dendrimers, and core–shell quantum nanogels to exhibit optical and electronic behaviors [2].

Therefore, precise tuning of nanopolymer size is a critical strategy in nanopolymer engineering due to the synergistic effects of quantum confinement and increased surface area. The next section explores the different nanopolymer dimensions and the significance of shape tuning in nanopolymers' performance.

2.4 *Effect of Shape Tuning in Nanopolymers*

As like particle size which plays a pivotal role in determining nanopolymer behavior at molecular level, shape also determines crucial parameters such as anisotropy, aspect ratio and interfacial interactions, thereby influencing intrinsic physical effects and the functional properties of nanopolymers. Anisotropy introduced by specific shapes directly affects diffusion kinetics, mechanical reinforcement and physicochemical interactions [15].

3 Shape Induced Anisotropy and Aspect Ratio

The shape of nanostructured materials inherently controls their anisotropy and aspect ratio, which in turn affect their overall performance. Anisotropy is a phenomenon where the material exhibits different physical properties along different directions, which is predominantly dictated by its shape. In isotropic particles like polymer nanospheres, properties such as charge distribution, light scattering, or diffusion remain uniform in all directions. However, anisotropic nanopolymers shapes such as rods, wires, or fibres can affect the surface polarization which in turn influences optical absorption, directional charge transport, or mechanical reinforcement of the composites due to their tendency to align preferentially [24]. Thus, shape-induced anisotropy enables preferential alignment of nanopolymers along different axes, which contributes to enhanced percolation networks, improved conductivity, and efficient stress-transfer pathways providing advantageous in applications such as sensors, photonic materials, and electronic devices.

The aspect ratio refers to the ratio of the length to the diameter (or width) of a nano-sized structure. For rod-like nanostructures such as nanotubes and nanowires, the aspect ratio is simply the ratio of the length to the diameter (L/D), whereas for planar structures such as nanosheets, it is the ratio of the lateral dimension to sheet thickness. Nanostructures such as nanowire, nanofiber and rods have a large length relative to their width, exhibit unique properties including superior mechanical reinforcement, electrical conductivity, and barrier properties. This is attributed to their extended interfacial area and ability to form percolating networks within the polymer matrix because of their high aspect ratio [25]. Accordingly, tuning the shape of nanopolymers affects their anisotropy and aspect ratio, enabling precise control over their overall performance. Further, based on their shape and dimension nanopolymers can be classified into following categories, each exhibiting distinct anisotropic behaviour and application potential.

3.1 Three-Dimensional (3D) Nanopolymers

In three-dimensional nanopolymers, charge carriers are confined in all directions. 3D- nanopolymers exhibit isotropic or extensively branched structures with internal network connectivity, conferring unique properties such as enhanced stability and functional versatility. The various dimensional nanopolymers are shown in Fig. 1 [26–30].

Dendrimers: Dendrimers are synthetic three-dimensional macromolecules, with highly branched architecture composed of a central core and numerous surface functional groups. The monodisperse nanoscale structure (typically 1–10 nm in diameter) and defined molecular weight of dendrimers impart unique physical and chemical properties [31]. These features make dendrimers versatile for applications in drug

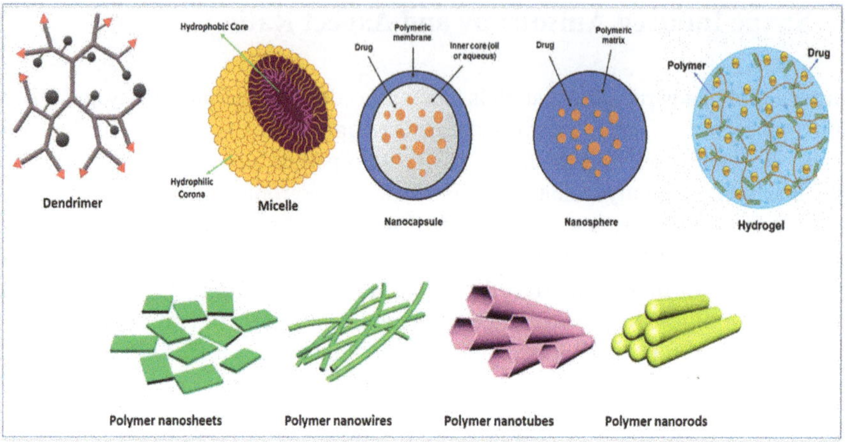

Fig. 1 Classification of nanopolymers based on dimensional structure

delivery (e.g., NSAIDs, anticancer agents, antimicrobials), advanced coatings, and water purification [10].

Micelles: Micelles are nanoscale, self-assembled colloidal structures formed by amphiphilic molecular compounds with both hydrophilic (water-attracting) and hydrophobic (water-repelling) regions in aqueous or nonaqueous solvents. These molecules are organized in a way that their hydrophobic segments face inward, away from the solvent, while their hydrophilic segments face outward, resulting in a core–shell architecture capable of encapsulating hydrophobic substances [32].

Hydrogel: Hydrogel nanoparticles also known as polymeric nanogels, made of a polymeric network with three dimensional configurations formed by crosslinked, swellable polymers that absorb large quantities of water or biological fluids. The presence of hydrophilic groups such as ether, amine, hydroxyl, sulfate and carboxyl in the polymer chains allows them to encapsulate and protect a wide range of therapeutic agents, making them highly versatile for biomedical applications. Nanogels can be composed of natural, synthetic, or hybrid polymers, and their properties such as size, charge, porosity, and degradability can be precisely tuned during synthesis [33].

Nanospheres and Nanocapsules: These type of nanopolymers are widely used in drug delivery. Nanospheres are ranging from 10 to 500 nm in diameter, comprising a continuous polymeric network with a regular sphere structure in which drug molecules are uniformly dispersed. Nanocapsules comprises of core–shell structured particles, where drug is confined to a cavity surrounded by a unique polymer shell membrane as a protection strategy for core materials. Their high surface area and isotropic shape make them well-suited for applications such as drug delivery, medical imaging, and biosensing. Additionally, nanospheres can be easily surface functionalized with ligands, antibodies, or fluorescent markers, enabling targeted delivery and diagnostic applications [34].

3.2 Two Dimensional (2D) Nanopolymers

Two dimensional nanopolymers consist of structures with two dimensions significantly larger than the nanoscale thickness. Such configurations include polymer-layered nanocomposites and polymer nanosheets and films.

Polymeric Nanosheets and nanofilms: The distinguishing feature of these nanopolymers is their free-standing structures, with distinctive advantages compared to other morphologies including a large specific surface area, abundant active sites and molecular permeability. These layered structures exhibit a high aspect ratio between their lateral dimensions and thickness, which imparts excellent selectivity, permeability, and chemical stability. They demonstrate exceptional barrier properties, making them ideal for packaging materials to prevent diffusion of pollutants, they are aslo used extensively in filtration membranes for water purification, wastewater treatment catalysis and corrosion resistance [35].

3.3 One-Dimensional (1D) Nanopolymers

One-dimensional nanopolymers have a distinct elongated structure, characterized by one dimension significantly greater outside the nanoscale (one dimension are larger than 100 nm). These include rod-shaped nanopolymers, polymeric nanotubes, nanofibres, and nanowires, often synthesized through techniques such as electrospinning, template synthesis, or self-assembly.

Nanotubes and Nanowires: The elongated morphology of polymer nanotubes and nanowires offer advantages due to their directionally extended shape, defined surface characteristics and anisotropic properties. These features contribute to their enhanced electrical, mechanical, and electrochemical performances. Nanowires are elongated, wire like nanostructures with diameters in the nanometer range demonstrating superior charge transport, high surface area, and tunable electronic properties that are vital for high-performance supercapacitors, chemical sensors, and flexible nanoelectronics [36, 37]. Polymer nanotubes are hollow, cylindrical nanostructures with diameters on the nanometer scale and lengths often reaching micrometers. Their tubular morphology provides a unique internal cavity paving way for additional internal surface area and enhanced ion diffusion kinetics suitable for encapsulating drugs, imaging agents, or catalytic substances [37].

Additionally, advanced synthesis strategies such as electrospinning, template-assisted growth, self-assembly, and electrochemical polymerization have enabled precise control over the dimensions, alignment, and surface functionalities of nanowires and nanotubes, further enhancing their applicability in energy storage, optoelectronic devices, stretchable transistors, and biomedical systems.

Nanorods and Nanofibers: Nanorods, another subclass of one-dimensional nanopolymers, are elongated nanostructures with rod-like morphology with lengths greater than their widths. Their elongated shape provides a larger contact area with cells or surfaces compared to spherical nanoparticles, enhancing adhesion and cellular interactions in biomedical contexts.

Further, polymer nanofibers have at least one dimension in the size of 100 nm or less. The highly elongated structure provides an enormous surface area-to-volume ratio, significant mechanical strength, high anisotropic electrical conductivity, enhanced light photoluminescence and thermodynamic properties. Electrospinning remains the most widely used technique for nanofiber fabrication, allowing precise control over fiber diameter, alignment, and porosity [9]. One of the most prominent applications of polymer nanofibers is in biomedical engineering, particularly in tissue engineering, drug delivery systems, and regenerative medicine where their morphology supports cellular adhesion, proliferation and nutrient diffusion [38].

Therefore, precise control over the shape of nanopolymers whether 1D (rods, tubes, fibres, wires), 2D (nanosheets), or 3D (nanospheres, capsules, dendrimers, micelles, hydrogels) and their size such as diameter, length, thickness significantly enhance their functional properties. This integrated tuning of size and shape, provides a powerful platform for developing innovative, high-performance nanopolymers tailored for advanced applications including biomedical, environmental, electronic, and catalytic applications. The following section provides insight on how changes in the size and shape of nanopolymers can improve their overall properties and performance.

4 Effects of Size and Shape on Various Nanopolymer Properties

The structural and functional behaviour of nanopolymers is primarily governed by their size and morphology. Thus, tuning them introduces nanoscale phenomena such as quantum confinement, surface area dominance and shape induced anisotropy, which directly influence their mechanical, optical, thermal, adsorption, electrical and functional properties. The following sections explore how size and shape specifically influence these properties.

5 Mechanical Properties

Nanopolymers offer the advantages of controllable structure, versatile design and superior compatibility with both organic and inorganic materials compared with classic nanomaterials [7]. Recent advances in molecular engineering and nanoscale architecture have enabled the development of nanopolymers with enhanced mechanical properties suitable for various applications like tissue engineering, corrosion resistant coatings [39, 40]. These nanostructured polymers often exhibit enhanced mechanical performance compared to their bulk counterparts due to size-induced effects, as size reduction reduces structural defects, increases surface area, and promotes more efficient polymer chain interactions leading to improved strength, elasticity, and toughness [41]. For instance, polymer nanocomposites containing silica nanoparticles exhibited increased in tensile strength, underscoring the reinforcement capability due to nano-level dispersion. Similarly, in 3D-printed polylactic acid specimens, tensile strength increased significantly for smaller geometries, confirming a strong inverse relationship between size and mechanical performance. This is because the smaller nanoparticles, especially those comparable in size to polymer monomers, enhance the toughness of polymer nanocomposites more effectively than larger particles [42]. Further, the nanoscale fillers provide higher surface-to-volume ratios, increasing interfacial interactions, contributing significant improvements in tensile, compressive, flexural strengths of the polymer material [15].

In addition to size, the shape of nanostructures particularly their aspect ratio and anisotropic nature play critical role in determining their mechanical performance. The smaller filler sizes combined with high-aspect-ratio nanostructures, significantly enhance stress transfer and crack resistance within the polymer matrix. The anisotropic properties of nanopolymers such as nanofibers, facilitate more efficient stress distribution and stronger interfacial bonding, leading to substantial improvements in Young's modulus and tensile strength, more effective than spherical or low-aspect-ratio fillers [43]. These enhancements are largely attributed to the increased interfacial area between nanopolymers and their surroundings which promotes more uniform and homogeneous stress distribution throughout the material. For example, electrospun nanofibers composed of electrospun polyurethane (PU), polystyrene (PS), polyacrylonitrile (PAN), and polyvinyl chloride (PVC) showed enhanced tensile strength and hardness compared to bulk counterparts reflecting size-dependent mechanical behaviors similar to other nano-objects [9]. Similarly, spherical nanogels with uniform morphology, significantly enhanced their water absorption, swelling kinetics, and flexibility in oil reservoir conformance control, under harsh temperature and salinity conditions than the bulk gels. This is attributed to the presence of nanoscale domains, which improved elasticity compared to conventional gels [33]. Notably, nanopolymer shape also influences surface roughness and adhesion properties, which are critical in mechanical joints and composite durability [44]. Recent studies on nanopolymer composites for energy-efficient automobile tires have demonstrated that reducing the nanoparticle size and optimizing their distribution within polymer chains substantially improves both the filler network stability and mechanical reinforcement [15].

The synergy between polymer chains and nanoscale fillers leads to remarkable improvements in strain-to-failure and load-bearing capacity [45]. The nature of this improvement is dictated by polymer matrix type, nanofiller size, nanofiller shape. Therefore, through careful tuning of size and shape, nanopolymers can be tailored to achieve high strength, and durability, enabling their use in load bearing, flexible, and impact-resistant structural applications.

6 Optical Properties

The optical properties of nanopolymers are highly influenced by both size and shape due to quantum confinement effects. As the size of the polymeric nanostructure approaches the exciton Bohr radius, the energy bandgap widens, resulting in size-dependent color changes, shifts in fluorescence and absorption spectra [19]. These effects are instrumental in designing optical chemosensors, imaging agents, and photonic materials. For example, in gold PMMA nanocomposites, variation in nanoparticle size leads to distinct shifts in optical absorption behavior due to surface plasmon resonance, a collective oscillation of free electrons at the interface between nanoparticles and polymers, induced by incident light. As the size of gold nanoparticles decreases, quantum confinement effects become significant, causing blue shifts, while larger nanoparticles exhibit red shifts [46]. The nanopolymers utilizing self-assembled lamellar block copolymers with sequestered gold nanocrystals offer tunable optical properties through nanoparticle size control, as the surface traps induced by high surface to volume ratio plays an important role in tuning optical properties [47]. For instance, the π-functional nanopolymers demonstrate enhanced quantum yields of deep-blue emissions due to quantum confinement and well-ordered structural organization on the nanoscale. These quantum effects enable the development of organic random lasers with exceptionally high-quality factors, which are unattainable in their bulk counterparts. Such phenomena underscore the fundamental importance of quantum effects arising from nanoscale dimensions and structural organization in determining the unique optical performance of nanopolymers [48]

The plasmon resonance bands can be modulated to the longer wavelength region by changing the size, shape and dielectric properties of the nanopolymer. Beyond size, the shape of nanostructures such as rods, discs, or core–shell morphologies affects anisotropy, polarization-dependent absorption, and scattering behavior. This can be seen in, non-spherical geometries such as nanorods, nanoprisms polymer composites (Ag/PVP/PAN) which exhibited distinct absorption features due to multipolar plasmon modes that differ drastically from spherical counterparts, emphasizing the significance of optical tuning through shape control. Shape-induced optical anisotropy stems from the interaction of light with the sharp edges, and aspect ratios of nanostructures. For instance, triangular nanoprisms concentrate electric fields more effectively at tips, enabling stronger surface-enhanced Raman scattering (SERS) signals compared to spherical isotropic shapes [24]. Therefore, optical properties of nanopolymers, result from an interplay of size-dependent quantum confinement, shape induced plasmonic resonances, and interfacial interactions with surrounding media, thus engineered nanopolymers hold vast promise in next-generation optoelectronic devices.

7 Thermal and Electrical Properties

The thermal properties of nanopolymers, such as glass transition temperature (Tg), melting point, thermal energy storage and thermal conductivity are significantly affected by nanoscale size and morphological confinement. These changes are often due to the restricted molecular mobility at the nanoscale and strong interfacial interactions between polymer chains and embedded nanoparticles. The glass transition temperature (Tg) is a thermal threshold at which a polymer transitions from a glassy, hard state to a more rubber-like, flexible state. Specifically, smaller nanoparticles exhibit stronger confinement effects on polymer chain mobility, resulting in elevated glass transition temperatures [3, 42]. However, studies on various polymer nanocomposites have reported that glass transition temperature (Tg) is also influenced by interfacial interactions and aggregation of nanofillers not just by size [49].

In case of morphology, nanofibers and aligned nanotubes facilitate directional heat transfer, whereas 3D structures such as polymer nanospheres due to internal phonon scattering between chains and at interfaces, may act as thermal insulators, as this increased phonon scattering reduces the mean free path for heat transfer. Additionally, at nanoscale the polymer chains become spatially confined and this reduced dimensionality suppresses crystallinity, resulting in lowering the Tg or broadening the melting point range [50]. The application of polymers in thermal energy storage systems, to store heat from renewable thermal energy sources, shows a significant advancement as various nanopolymers with controlled nanoscale size and tailored morphologies have demonstrated exceptional thermal stability and enhanced performance in low-temperature latent heat thermal energy storage applications [9]. Thus, tailoring not just nanoparticle size and shape but also surface functionality enables the control of Tg, offering a powerful strategy for tuning thermal performance of nanopolymeric systems.

Further, electrically conductive polymer nanocomposites are of great interest for a wide range of applications, as tuning their shape and size influences the charge carrier mobility, dielectric constant, and electrical conductivity. The aspect ratio of the filler has a very strong effect on the electrical percolation properties of nanopolymer, which denotes the efficient formation of interconnected network structure, which allows a continuous path across the material. High-aspect-ratio nanostructures like nanowires and nanofibers enhance percolation properties and provide continuous conductive pathways, enhancing current flow in polymer matrices than spherical or low-aspect-ratio fillers. Higher-aspect ratio fillers percolate at lower loadings due to more efficient network formation with fewer contacts required to form the network [25]. Similarly, sheet-like morphologies in 2D systems facilitate planar charge transport, which is critical in flexible electronics and sensors. The inclusion of conductive nanofillers such as carbon nanotubes, graphene, or metal nanoparticles into polymer matrices introduces size-dependent electrical properties improving electron transport pathways and achieving superior conductivity [51].

8 Surface Chemistry and Adsorption

In nanostructured polymers, the high surface to volume ratio due to size reduction, contributes to a higher density of reactive sites on the surface, this improves chemical reactivity, functionalization potential, and biomolecular interactions. This property is particularly advantageous for adsorption applications, where the adsorption efficacy of nanopolymers is strongly dictated by surface reactivity, morphology, and internal porosity, all of which are deeply influenced by nanoscale dimensions. These nanoscale features provide nanopolymer-based adsorbents a superior performance compared to conventional materials, offering faster adsorption kinetics, higher capacity, and enhanced selectivity toward target pollutants. For example, a nitrogen-enriched nanosized polymer 4Czpyz-poly exhibited an enhanced CO_2 adsorption capacity, attributed to synergistic effect of a controlled particle size distribution between 100 and 400 nm and a highly defined porous structure, nitrogen-rich chemistry, mechanical stability and good conductibility of the nanopolymer owing to its highly conjugated structure. Such integrated nanoscale features enable superior gas uptake properties defined as the material's ability to retain the gas molecules, and adsorption selectivity [52]. Similarly, chitosan-based nanopolymeric adsorbents exhibit high capacity for heavy metal removal due to the large number of functional groups exposed on the nanoparticle surface, while dendritic nanopolymers with hyperbranched, tree-like architectures, show exceptional performance in capturing organic dyes and pharmaceuticals from aqueous solutions due to their high external surface area and abundant functional active sites [53, 54].

The effect of size reduction on adsorption performance is also evident in polymer nanofibers. For example, PAN nanofibers with diameters in the range of 150–250 nm coated on window screens provided ~90.6% PM 2.5 filtration efficiency. Ionic liquid-modified PAN nanofibers improved surface roughness and hydrophilicity, enhancing adsorption of fine particulates. Further, polymeric nanofibers are also used for organic dye adsorption from water, with smaller fiber diameters enhancing surface adsorption sites [9]. Furthermore, the interaction between polymers and nanoparticles exhibits a synergistic effect that significantly enhances the rheological properties of the fluid, resulting in improved viscosity and superior performance in oil recovery applications than conventional polymers [55]. Thus, tuning of size and shape enhances the surface adsorption properties of nanopolymers, making them more effective for applications such as environmental remediation and air filtration applications [56].

9 Other Functional Properties

The size and shape of nanopolymers not only influence their physicochemical behaviors as discussed but also synergistic multifunctional behaviors which refer to the specific performances of nanopolymers that is desired in specific real-world application or devices, ranging from drug delivery to antimicrobial activity, diagnostics and imaging [57, 58]. Nanopolymers, namely liposomes, micelles, nanogels, and dendrimers allow dynamic shape transitions in response to external stimuli such as pH, temperature, or enzymatic activity. This in turn holds a potential advantage in stimuli-responsive drug delivery system that works based on the response of specific physiological or external stimuli to release therapeutic agents in a controlled manner [59]. The size of nanoparticles determines bio distribution, circulation time, and tissue penetration. Further, polymeric nanocarriers with diameters below 200 nm can avoid quick removal by the kidneys and those smaller than 100 nm can go deeper into tumors, while larger particles are limited to peripheral tumor regions. This enhanced penetration of smaller particles in tumors is due to their enhanced permeability and retention effects is useful for highly selective targeting applications [30]. Further, polymeric micelles with an average diameter between 5–100 nm, comprising of core shell structure, exhibit improved permeability and retention properties for the augmented uptake of anti-cancer drugs in malignant tissues [59]. Specifically, micelles, dendrimers, and nanogels owing to their spherical or globular configurations allow higher drug loading, better tissue permeability, and controlled release characteristics. Additionally, core–shell polymeric nanocapsules, demonstrate that controlled shell thickness and nanoscale uniformity allow fine-tuning of drug release rates while providing superior stability against premature degradation. Also, the shell-core architecture ensures high drug loading and maintains pharmacological activity with reduced systemic toxicity [60].

Polymeric nanoparticles employed for antimicrobial therapies benefit significantly from reduced sizes and modified surface charges. Because small-sized polymeric nanoparticles offer improved interaction with bacterial membranes due to higher surface-to-volume ratios, enabling more efficient cell membrane penetration and intracellular delivery. Especially, the nanoscale dimensions facilitate penetration of microbial membranes and biofilms while reducing antimicrobial resistance development [34].

Other than biological applications, size and shape play a crucial role in corrosion, adhesion, packaging and barrier properties of nanopolymers. In case of, PVC nanofibers (428–620 nm) in aluminum coatings, provided significant reduction in corrosion rate due to improved barrier effects [9]. Nanostructured polyaniline with particle sizes ranging from 2–60 nm was highly effective in enhancing adhesion, barrier performance, and improved the anticorrosion efficiency [39]. The drawback of inherent permeability in polymers can be mitigated by incorporation of nanofillers such as graphene, which leads to reduced permeability to oxygen, carbon dioxide, and water vapour which is crucial for prolonging shelf life [61]. Specifically, nanopolymers have significant advantage in catalysis, as smaller nanoparticles offer higher surface-to-volume ratios, thereby increasing the density of active catalytic sites [21].

Further, the size and shape of nanopolymers is influenced by a combination of synthesis methods and processing parameters. The following section will explore these synthesis strategies for tuning size and shape in nanopolymers.

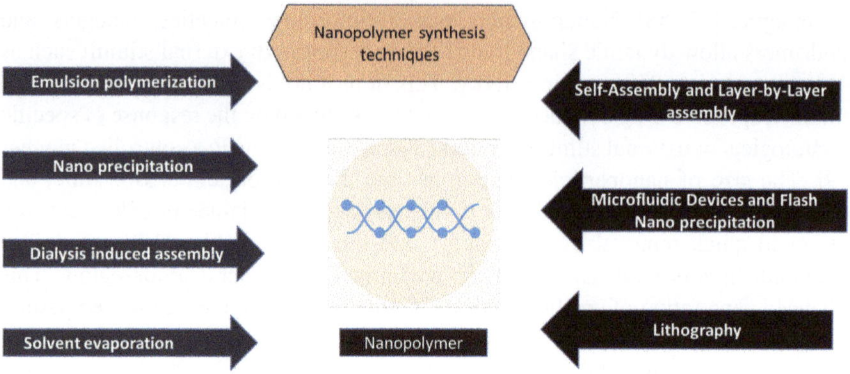

Fig. 2 Schematic representation of various nanopolymer synthesis techniques

9.1 Synthesis and Other Techniques to Tune Size and Shape

Recent literature reports numerous advanced methods for fabricating polymer nanostructures using diverse approaches, including polymerization, self-assembly routes, and soft lithography as listed in Fig. 2. These methods are divided as "bottom-up" and "top-down" synthesis according to the processes involved in creating nanoscale structures. Bottom-up approaches rely on the spontaneous organization or self-assembly of molecular or monomeric precursors into stable nanostructures. Top-down methods start from bulk polymers or polymer matrices and progressively reduce their dimensions to nanoscale structures [62]. These synthesis strategies are used to produce nanopolymers with controlled nanoscale dimensions and morphologies. The physical, chemical and functional properties of nanopolymers are strongly influenced by their size. Hence, size-controlled preparation techniques have emerged as a key strategy for achieving nanopolymers with application specific performance in areas like drug delivery, sensing, and energy conversion.

10 Bottom-Up Approaches

10.1 Emulsion Polymerization

Emulsions are systems with liquid droplets dispersed in a continuous liquid phase. Nanopolymers can be prepared either from performed polymers or by direct polymerization of monomers using various polymerization techniques such as microemulsion, mini-emulsion, surfactant-free emulsion and interfacial polymerization [13, 63]. The emulsion polymerization method is commonly used to synthesize polymer nanoparticles ranging from 80 to 900 nm [64], mainly for applications such as drug and gene delivery, where the polymer and drug are first dissolved in an organic solvent, then emulsified into water containing a surfactant using mixing or sonication, followed by evaporation of the solvent to form nanoparticles, which are then collected and purified. The size and shape of the nanoparticles can be tuned by adjusting the polymer concentration, solvent-to-water ratio, surfactant level, and mixing speed [65].

10.2 Nanoprecipitation or Solvent Displacement

Nanoprecipitation involves dissolving the polymer and drug in a water-miscible organic solvent, which is then rapidly added to an aqueous phase under stirring, causing the polymer to precipitate as nanoparticles. This method commonly yields nanoparticles with mean diameters around 170 nm [64], but the size can be tuned from below 50 nm up to 300 nm or more by adjusting formulation and process conditions such as polymer concentration, solvent type, mixing parameters including rate of addition. Nanoprecipitation is widely applied for drug delivery, enabling encapsulation of various therapeutic agents in nanoparticles with controlled size for improved targeting and release profiles [66].

10.3 Solvent Evaporation

The solvent evaporation technique is a widely used bottom-up approach for synthesizing polymeric nanoparticles, especially for drug encapsulation. In this method, the polymer is first dissolved in a volatile organic solvent such as dichloromethane, or chloroform, forming the dispersed phase then emulsified into an aqueous continuous phase containing a stabilizer like polyvinyl alcohol (PVA). Once the emulsion is formed, the organic solvent gradually evaporated. There are various critical Parameters that influence the size such as polymer concentration in the organic phase, solvent type and volatility, and stirring rate [13].

10.4 Dialysis Induced Assembly

The dialysis induced assembly is a simple and gentle technique for synthesizing nanopolymers, which involves placing a polymer in a water miscible organic solvent (e.g., acetone, DMSO, DMF) into a semi-permeable membrane and then immersing it in water. As the solvent gradually diffuses out and water diffuses in, the change in solvent environment induces the self-assembly or precipitation leading to the formation of nanostructures such as nanoparticles, micelles or nanogels and dendrimers [67]. This method typically produces spherical nanoparticles in the size range of 100–300 nm, depending on factors such as polymer concentration, solvent type, dialysis time, and temperature. Dialysis-based nanoparticles are widely used in drug delivery (for controlled and targeted release), gene therapy, vaccine delivery, diagnostic imaging, and tissue engineering, owing to their biocompatibility, controlled size, and ability to encapsulate both hydrophobic and hydrophilic molecules [68].

10.5 Self-Assembly and Layer-by-Layer Assembly

Self-assembly is a bottom-up approach in which amphiphilic polymers or block copolymers spontaneously organize into nanostructures (such as micelles, vesicles, rods, or worms) when exposed to selective solvents or changes in environmental conditions like pH, temperature, or ionic strength. During synthesis, block copolymers are typically dissolved in an organic solvent and added to water (or vice versa), leading to spontaneous aggregation due to hydrophobic-hydrophilic interactions [69]. This process does not require harsh conditions or surfactants. The resulting nanostructures generally have sizes ranging from 10 to 200 nm [70], depending on polymer composition, concentration, solvent polarity, and block ratio. Self-assembled nanoparticles are widely applied in drug delivery, particularly for hydrophobic drug encapsulation, gene delivery, and stimuli-responsive systems, due to their tunable morphology, biodegradability, and ability to disassemble at the target site [71].

Layer-by-Layer (LbL) is a stepwise deposition method for creating multilayered thin films or capsules. This method enables the fabrication of nanoparticles with precisely controlled sizes, typically in the range of 50–200 nm, depending on the core template and number of deposited layers [72].

10.6 Microfluidic Devices and Flash Nanoprecipitation

Microfluidic devices and Flash Nanoprecipitation (FNP) are advanced bottom-up techniques that enable the production of nanoparticles with highly precise control over size and uniformity. In microfluidic systems, fluids containing polymers or drugs and solvents are pumped through microscale channels, allowing for rapid and controlled mixing, which leads to reproducible nanoparticle formation via nanoprecipitation or solvent displacement. These platforms facilitate continuous, scalable production and allow for fine-tuning of particle size, morphology, and polydispersity, often yielding nanoparticles in the 20–200 nm range [73]. FNP, in particular, uses specialized mixers to achieve instantaneous mixing of solvent and anti-solvent streams, resulting in rapid nucleation and highly uniform nanoparticles. The ability to precisely control mixing and reaction parameters in these systems makes them especially valuable for applications such as drug delivery, mRNA/lipid nanoparticle formulations, gene therapy, where speed, reproducibility, and scalability are critical for successful translation from laboratory to clinical settings [74].

11 Top-Down Approaches

11.1 Lithography

Top-down approaches utilize nanofabrication techniques to create nanostructures with the desired shapes and characteristics from bulk materials. Soft lithography is emerging as a low-cost alternative technique suitable for the fabrication of polymeric arrays to conventional lithography techniques such as electron-beam (e-beam) X-ray or extreme-UV lithography [62]. PRINT (Particle Replication in Non-Wetting Templates) method employs soft lithography to generate monodisperse particles with precisely controlled size, shape, and composition. The process uses a mold made from fluorinated perfluoropolyether elastomers, which contains nanoscale cavities tailored to the desired particle geometry. These molds are filled with a liquid polymer or biomaterial and then solidified often by UV curing or thermal crosslinking. The unique non-wetting properties of the mold material allow for easy harvesting of the particles without adhesion or deformation. PRINT technology enables the fabrication of particles ranging from 50 nm to several micrometers, in diverse shapes such as spheres, rods, cubes, and disks. This high degree of control over particle attributes is particularly valuable for applications in targeted drug delivery, vaccine development, imaging, and diagnostics, where particle shape and size can significantly influence biological interactions and therapeutic outcomes [75].

11.2 Post-Synthesis Techniques to Tune Size and Shape

After the initial synthesis of nanopolymers, several post-synthesis techniques are employed to further refine and control the size and shape of nanoparticles. These methods are essential for ensuring product uniformity, optimizing performance, and meeting the specific requirements of biomedical and industrial applications.

11.3 Density Gradient Centrifugation

This is an advanced centrifugation technique widely used to separate nanoparticles based on their size and shape. By applying centrifugal force, particles of different sizes or densities sediment at different rates, allowing for the purification of uniform nanoparticles. This technique is particularly valuable in biomedical applications where consistency and reproducibility are critical [76, 77].

11.4 Filtration

Filtration utilizes membranes with defined pore sizes to remove nanoparticles that are either too large or too small. This process ensures that only particles within a desired size range are retained, which is especially important for drug delivery and diagnostic applications where uniformity impacts efficacy and safety [78].

11.5 Size-Exclusion Chromatography

Size-exclusion chromatography isolates nanoparticles according to their hydrodynamic size. By passing a nanoparticle suspension through a porous matrix, particles are separated and collected in fractions of similar size. This technique is commonly used in research and clinical settings to produce monodisperse nanoparticle samples [79]. Thus, the nanopolymers synthesized by various approaches with controlled morphology and functionality find widespread use in biomedical, sensing, energy storage, and coating applications.

11.6 Application

As demonstrated in earlier sections, the ability to tune the size and shape of nanopolymers enables a high degree of control over key material properties such as surface area, conductivity, and anisotropy etc., These enhancements in properties has opened up new frontiers across a wide range of advanced applications from biomedical applications, environmental remediation to advanced energy technologies, and smart electronic devices.

11.7 Biomedical Applications

In the last two decades unprecedented development of nanopolymers has been observed in various biomedical applications such as drug delivery, tissue engineering and antimicrobial coatings. Some of the widely used nanopolymers in targeted drug delivery are nanotubes, hydrogels, dendrimers and micelles. The size of the nanopolymers in drug delivery are typically engineered within the 10–200 nm range, as this interval offers the best balance between avoiding rapid clearance by the kidneys (which occurs for particles smaller than 10 nm) and evading filtration by the liver, spleen, and bone marrow (which is more common for particles larger than 200 nm) [80]. And also the size and shape of nanopolymers directly affect their circulation time in the bloodstream, cellular uptake, biodistribution, and ability to penetrate biological barriers. For example, rod-shaped or disc-shaped nanoparticles often exhibit different adhesion and uptake properties compared to spherical ones, with studies showing that rods and discs can have higher binding rates and longer circulation times, which can be advantageous for targeted drug delivery [81]. Mehandole et al. [60] demonstrated that nanocapsules integrating hydrophobic and hydrophilic domains enable controlled interaction with pathogens, ensuring selective antimicrobial activity while minimizing off-target effects. Therefore, by tuning the physical parameters, nanoparticles can be optimized for specific therapeutic goals, such as deeper tumor penetration or prolonged circulation, ultimately improving drug delivery efficiency and other biomedical applications.

11.8 Sensing and Biosensing

In sensing and biosensing, nanopolymers with sizes typically in the 1–100 nm range are favored because their high surface area-to-volume ratio enhances sensitivity and reactivity, enabling the detection of biomolecules at very low concentrations [82]. The shape of nanoparticles whether spherical, rod-like, plays a crucial role in determining their optical and electronic properties, which can be harnessed for highly sensitive and selective biosensors [83]. Thus, nanopolymers can be tailored for specific sensing applications, such as multiplexed detection, real-time diagnostics, or improved signal amplification [80].

11.9 Energy Storage and Conversion

For energy storage and conversion, nanopolymers are generally designed within the 1–100 nm range [82], with specific shapes such as rods, wires, or platelets being used to maximize surface area and optimize pathways for electron and ion transport. These morphological characteristics directly influence catalytic activity, charge storage capacity, and overall device efficiency [84]. By tuning the size and shape, nanomaterials can be engineered for higher energy density in batteries, more efficient catalysts in fuel cells, and improved performance in solar cells and supercapacitors.

11.10 Nanocoating and Other Applications

The size and shape of nanopolymers determine the uniformity, barrier properties, and functional characteristics of coatings, though larger aggregates may be employed for certain mechanical or optical effects, smaller particles yield finer, more homogeneous coatings, while different shapes such as platelets or spheres can impart specific optical, mechanical, or hydrophobic properties [85]. This tunability allows for the development of advanced coatings with tailored functionalities, including anti-corrosive, self-cleaning, surfaces.

11.11 Future Outlook and Conclusion

The integration of nanotechnology into polymeric materials has led to significant advancements in the development of nanopolymers with enhanced mechanical, optical, thermal, and functional properties. As explored throughout this chapter, precise control over size and shape at the nanoscale is a key strategy for tailoring polymer performance at the molecular level. This structural tunability enables nanopolymers to meet the requirements of advanced applications, including drug delivery, energy storage, sensing technologies, and environmental remediation.

Despite these advances, there are several challenges including achieving uniform particle size and narrow size distribution during synthesis, which continues to hinder scalable manufacturing. Further, the environmental and biological implications of nanopolymers require careful consideration, particularly the need to ensure their biocompatibility, biodegradability and the adoption of sustainable synthesis methods. Looking ahead, the development of stimuli-responsive nanopolymers and shape-transformable polymer systems that adapt to their environment or function autonomously with computational approach accelerating the rational design of nanopolymers with programmable properties. In summary, the ability to tune the size and shape at the nanoscale offers a strategic pathway to engineer polymers with wide functional properties in developing smart and sustainable materials for various applications.

Acknowledgements This book chapter was supported by grant CCEC 01-1108-230163.

References

1. Anbusagar, N.R.R., Palanikumar, K., Ponshanmugakumar, A.: Preparation and properties of nanopolymer advanced composites: a review. Polym. Based Nanocomposites Energy Environ. Appl. 27–73 (2018)
2. Stoyanov, O.V., Haghi, A.K., Zaikov, G.E. (eds.): Nanopolymers and modern materials: preparation, properties, and applications. CRC Press (2013)
3. Paul, D.R., Robeson, L.M.: Polymer nanotechnology: nanocomposites. Polymer **49**(15), 3187–3204 (2008)
4. Nakagawa, S., Marubayashi, H., Nojima, S.: Crystallization of polymer chains confined in nanodomains. Eur. Polymer J. **70**, 262–275 (2015)
5. Peponi, L., Puglia, D., Torre, L., Valentini, L., Kenny, J.M.: Processing of nanostructured polymers and advanced polymeric based nanocomposites. Mater. Sci. Eng. R. Rep. **85**, 1–46 (2014)
6. Kannan, B., Cha, H., Hosie, I.C., Fakirov, S.: Nano-size polymers: preparation, properties, applications (2016)

7. Cheng, X., Han, Q., Yue, B., Shi, M., Yang, C.: Self-assembled nano-polymers modified water-based sizing agent for enhancing the dual interfacial properties of carbon fibre/epoxy resin composites. Compos. B Eng. **262**, 110828 (2023)
8. Hossain, S.K.S., Hoque, M.E.: Polymer nanocomposite materials in energy storage: properties and applications. In: Polymer-based nanocomposites for energy and environmental applications, pp. 239–282. Woodhead Publishing (2018)
9. Lu, Y., Shah, K.W., Xu, J.: Synthesis, morphologies and building applications of nanostructured polymers. Polymers **9**(10), 506 (2017)
10. Mathesvaran, T., Senthilkumar, S.K., Aravind, E., Aruna, T., Dineshkumar, S., Durgadevi, M., Thilagavathy, P.: A review on nanoparticles and its application (2024)
11. Mortazavi, S., Sillanpaa, M., Bose, D.: Innovative nanopolymers for adsorption of pollutants from water. Int. J. Environ. Anal. Chem. 1–28 (2024)
12. Adaveesh, B., Mahesh, V., Rakesh, M., Nithin, H.R., Channabasavaradhya, S.M., Disha, I.G.: Unveiling the impact of particle size on the physio-mechanical properties of eco-friendly polymer composites. Iran. Polym. J. **34**(5), 689–701 (2025)
13. Rao, J.P., Geckeler, K.E.: Polymer nanoparticles: preparation techniques and size-control parameters. Prog. Polym. Sci. **36**(7), 887–913 (2011)
14. Nemati, N.H., Moghadam, K.R., Afra, N.: Optical, morphological, and thermal properties of polymer nanocomposite films and coatings. In: Polymer Nanocomposite Films and Coatings, pp. 71–101. Woodhead Publishing (2024)
15. Yue, T., Zhang, Z., Li, S., Zhao, H., Duan, P., Zhang, G., Zhang, L., Liu, J.: Designing and fabricating nanopolymer composites beyond traditional polymer nanocomposites toward fuel saving of automobile tires. Nano Energy **101**, 107584 (2022)
16. Chrissopoulou, K., Anastasiadis, S.H.: Effects of nanoscopic-confinement on polymer dynamics. Soft Matter **11**(19), 3746–3766 (2015)
17. Kamal, I., Al-Naimi, S.: Polymers and nanotechnology. In: Proceedings of the Knowledge Based Industries & Nanotechnology Conference, vol. 2 (2008)
18. Singh, S., Ahmed, R.: Vital role of nanopolymers in drilling and stimulations fluid applications. In: SPE Annual Technical Conference and Exhibition?, pp. SPE-130413. SPE (2010)
19. Godovski, D.Y.: Electron behavior and magnetic properties of polymer nanocomposites. Therm. Electr. Conduct. Polym. Mater. 79–122 (2005)
20. Marquis, D.M., Guillaume, E., Chivas-Joly, C.: Properties of nanofillers in polymer. Nanocomposites Polym. Anal. Methods **59**, 261 (2011)
21. Zhou, Z.Y., Tian, N., Li, J.T., Broadwell, I., Sun, S.G.: Nanomaterials of high surface energy with exceptional properties in catalysis and energy storage. Chem. Soc. Rev. **40**(7), 4167–4185 (2011)
22. Ramalingam, G., Kathirgamanathan, P., Ravi, G., Elangovan, T., Arjunkumar, B., Manivannan, N., Kasinathan, K.: Quantum confinement. IntechOpen (2020)
23. He, J.H., Wan, Y.Q., Xu, L.: Nano-effects, quantum-like properties in electrospun nanofibers. Chaos Solitons Fractals **33**(1), 26–37 (2007)
24. Kudryashov, M., Logunov, A., Gogova, D., Mashin, A., De Filpo, G.: Ag/PVP/PAN nanocomposites with triangular nanoprisms of silver synthesized by UV-induced polymerization: morphology manipulation and optical properties tuning. Opt. Mater. **101**, 109746 (2020)
25. Mutiso, R.M., Winey, K.I.: Electrical conductivity of polymer nanocomposites. Polym. Sci. Compr. Ref. 327–344 (2012)
26. Anwar, M.Z., Kathuria, H., Chiu, G.N.: Development of a Eudragit-based hydrogel for the controlled release of hydrophobic and hydrophilic drugs for the treatment of wound infections. J. Drug Deliv. Sci. Technol. **94**, 105484 (2024)
27. Dai, X., Hou, C., Xu, Z., Yang, Y., Zhu, G., Chen, P., Huang, Z., Yan, L.T.: Entropic effects in polymer nanocomposites. Entropy **21**(2), 186 (2019)
28. Pontes, A.P., Welting, T.J., Rip, J., Creemers, L.B.: Polymeric nanoparticles for drug delivery in osteoarthritis. Pharmaceutics **14**(12), 2639 (2022)
29. Serras, A., Faustino, C., Pinheiro, L.: Functionalized polymeric micelles for targeted cancer therapy: steps from conceptualization to clinical trials. Pharmaceutics **16**(8), 1047 (2024)

30. Subjakova, V., Oravczova, V., Hianik, T.: Polymer nanoparticles and nanomotors modified by DNA/RNA aptamers and antibodies in targeted therapy of cancer. Polymers **13**(3), 341 (2021)
31. Pérez-Ferreiro, M., Abelairas, A.M., Criado, A., Gómez, I.J., Mosquera, J.: Dendrimers: exploring their wide structural variety and applications. Polymers **15**(22), 4369 (2023)
32. Fateh, S.T., Moradi, L., Kohan, E., Hamblin, M.R., Dezfuli, A.S.: Comprehensive review on ultrasound-responsive theranostic nanomaterials: mechanisms, structures and medical applications. Beilstein J. Nanotechnol. **12**(1), 808–862 (2021)
33. Seidy Esfahlan, M., Khodapanah, E., Tabatabaei-Nezhad, S.A., Salami-Kalajahi, M.: Fabrication, optimization and characterization of preformed-particle-gel containing nanogel particles for conformance control in oil reservoirs. Polym. Bull. **79**(9), 7137–7159 (2022)
34. Spirescu, V.A., Chircov, C., Grumezescu, A.M., Andronescu, E.: Polymeric nanoparticles for antimicrobial therapies: an up-to-date overview. Polymers **13**(5), 724 (2021)
35. Wei, S., Wang, L., Yue, J., Wu, R., Fang, Z., Xu, Y.: Recent progress in polymer nanosheets for photocatalysis. J. Mater. Chem. A **11**(44), 23720–23741 (2023)
36. Tatum, W.K., Luscombe, C.K.: π-Conjugated polymer nanowires: advances and perspectives toward effective commercial implementation. Polym. J. **50**(8), 659–669 (2018)
37. Wang, K., Wu, H., Meng, Y., Wei, Z.: Conducting polymer nanowire arrays for high performance supercapacitors. Small **10**(1), 14–31 (2014)
38. Al-Abduljabbar, A., Farooq, I.: Electrospun polymer nanofibers: processing, properties, and applications. Polymers **15**(1), 65 (2022)
39. Xu, H., Zhang, Y.: A review on conducting polymers and nanopolymer composite coatings for steel corrosion protection. Coatings **9**(12), 807 (2019)
40. Mohammadi Nasr, S., Rabiee, N., Hajebi, S., Ahmadi, S., Fatahi, Y., Hosseini, M., Bagherzadeh, M., Ghadiri, A.M., Rabiee, M., Jajarmi, V., Webster, T.J.: Biodegradable nanopolymers in cardiac tissue engineering: from concept towards nanomedicine. Int. J. Nanomed. 4205–4224 (2020)
41. Wang, G., Najafi, F., Ho, K., Hamidinejad, M., Cui, T., Walker, G.C., Singh, C.V., Filleter, T.: Mechanical size effect of freestanding nanoconfined polymer films. Macromolecules **55**(4), 1248–1259 (2022)
42. Tjong, S.C.: Structural and mechanical properties of polymer nanocomposites. Mater. Sci. Eng. R. Rep. **53**(3–4), 73–197 (2006)
43. Balasundaram, G., Webster, T.J.: An overview of nano-polymers for orthopedic applications. Macromol. Biosci. **7**(5), 635–642 (2007)
44. Budhe, S., Ghumatkar, A., Birajdar, N., Banea, M.D.: Effect of surface roughness using different adherend materials on the adhesive bond strength. Appl. Adhes. Sci. **3**, 1–10 (2015)
45. Crosby, A.J., Lee, J.Y.: Polymer nanocomposites: the "nano" effect on mechanical properties. Polym. Rev. **47**(2), 217–229 (2007)
46. Srivastava, S., Haridas, M., Basu, J.K.: Optical properties of polymer nanocomposites. Bull. Mater. Sci. **31**, 213–217 (2008)
47. Bockstaller, M.R., Thomas, E.L.: Proximity effects in self-organized binary particle–block copolymer blends. Phys. Rev. Lett. **93**(16), 166106 (2004)
48. Lin, D., Li, Y., Zhang, H., Zhang, S., Gao, Y., Zhai, T., Hu, S., Sheng, C., Guo, H., Xu, C., Wei, Y.: In situ super-hindrance-triggered multilayer cracks for random lasing in π-functional nanopolymer films. Research **6**, 0027 (2023)
49. Serenko, O.A., Roldughin, V.I., Askadskii, A.A, Serkova, E.S., Strashnov, P.V., Shifrina, Z.B.: The effect of size and concentration of nanoparticles on the glass transition temperature of polymer nanocomposites. RSC Adv. **7**(79), 50113–50120 (2017)
50. Zhan, H., Nie, Y., Chen, Y., Bell, J.M., Gu, Y.: Thermal transport in 3D nanostructures. Adv. Func. Mater. **30**(8), 1903841 (2020)
51. Barpuzary, D., Kim, K., Park, M.J.: Two-dimensional conducting polymers: synthesis and charge transport. J. Polym. Sci. Part B: Polym. Phys. **57**(18), 1169–1176 (2019)
52. Al-Hetlani, E., Rajendran, N., BabuVelappan, A., Amin, M.O., Ghazal, B., Makhseed, S.: Design and synthesis of a nanopolymer for CO2 capture and wastewater treatment. Ind. Eng. Chem. Res. **60**(24), 8664–8676 (2021)

53. Li, B., Li, K., Li, X.: Fabrication of abundantly functionalized dendritic biochar composites as adsorbents for the high-efficiency removal of heavy metal ions and dyes. Sep. Purif. Technol. **337**, 126368 (2024)
54. Naeimi, A., Amini, M., Okati, N.: Removal of heavy metals from wastewaters using an effective and natural bionanopolymer based on Schiff base chitosan/graphene oxide. Int. J. Environ. Sci. Technol. **19**(3), 1301–1312 (2022)
55. Sun, Y., Zhang, W., Li, J., Han, R., Lu, C.: Mechanism and performance analysis of nanoparticle-polymer fluid for enhanced oil recovery: a review. Molecules **28**(11), 4331 (2023)
56. Zhang, R., Liu, C., Hsu, P.C., Zhang, C., Liu, N., Zhang, J., Lee, H.R., Lu, Y., Qiu, Y., Chu, S., Cui, Y.: Nanofiber air filters with high-temperature stability for efficient PM2. 5 removal from the pollution sources. Nano Lett. **16**(6), 3642–3649 (2016)
57. Alharbi, H.M.: Exploring the frontier of biopolymer-assisted drug delivery: advancements, clinical applications, and future perspectives in cancer nanomedicine. Drug Des. Dev. Ther. 2063–2087 (2024)
58. Paramanick, D., Modi, D., Farheen, Nagarani, K.: A comprehensive review on advancement in nano polymer system for drug targeting. Multifunct. Mater. Eng. Biol. Appl. 245–271 (2025)
59. Das, S.S., Bharadwaj, P., Bilal, M., Barani, M., Rahdar, A., Taboada, P., Bungau, S., Kyzas, G.Z.: Stimuli-responsive polymeric nanocarriers for drug delivery, imaging, and theragnosis. Polymers **12**(6), 1397 (2020)
60. Mehandole, A., Walke, N., Mahajan, S., Aalhate, M., Maji, I., Gupta, U., Mehra, N.K., Singh, P.K.: Core–shell type lipidic and polymeric nanocapsules: the transformative multifaceted delivery systems. AAPS PharmSciTech **24**(1), 50 (2023)
61. Bratovčić, A., Odobašić, A., Ćatić, S., Šestan, I.: Application of polymer nanocomposite materials in food packaging. Croat. J. Food Sci. Technol. **7**(2), 86–94 (2015)
62. Mijangos, C., Hernández, R., Martin, J.: A review on the progress of polymer nanostructures with modulated morphologies and properties, using nanoporous AAO templates. Prog. Polym. Sci. **54**, 148–182 (2016)
63. Antonietti, M., Landfester, K.: Polyreactions in miniemulsions. Prog. Polym. Sci. **27**(4), 689–757 (2002)
64. Zielińska, A., Carreiró, F., Oliveira, A.M., Neves, A., Pires, B., Venkatesh, D.N., Durazzo, A., Lucarini, M., Eder, P., Silva, A.M., Santini, A.: Polymeric nanoparticles: production, characterization, toxicology and ecotoxicology. Molecules **25**(16), 3731 (2020)
65. Kumari, A., Yadav, S.K., Yadav, S.C.: Biodegradable polymeric nanoparticles based drug delivery systems. Colloids Surf. B **75**(1), 1–8 (2010)
66. Zander, A.J., Ehrlich, M.S., ur Rehman, S., Schneider, M.: Evaporation-triggered nanoprecipitation for PLGA nanoparticle formation using a spinning-disc system. J. Drug Deliv. Sci. Technol. **108**, 106901 (2025)
67. Fruntke, A., Blümbott, B., Koschella, A., Heinze, T., Wilke, T.: From current research to chemistry education: preparation of polysaccharide-based nanoparticles by dialysis. In: Conference Proceedings. New Perspectives in Science Education 2023 (2023)
68. Soppimath, K.S., Aminabhavi, T.M., Kulkarni, A.R., Rudzinski, W.E.: Biodegradable polymeric nanoparticles as drug delivery devices. J. Control. Release **70**(1–2), 1–20 (2001)
69. Nsairat, H., Lafi, Z., Al-Najjar, B.O., Al-Samydai, A., Saqallah, F.G., El-Tanani, M., Oriquat, G.A., Sa'bi, B.M., Ibrahim, A.A., Dellinger, A.L., Alshaer, W.: How advanced are self-assembled nanomaterials for targeted drug delivery? A comprehensive review of the literature. Int. J. Nanomed. **31**, 2133–61 (2025)
70. Dev Ammini, G., Weerarathna, L.J., Junkers, T.: Continuous flow block copolymer nanoaggregate synthesis and their flow dialysis purification. Chem. Methods e202500025 (2025)
71. Mishra, P.K., Ekielski, A.: The self-assembly of lignin and its application in nanoparticle synthesis: A short review. Nanomaterials **9**(2), 243 (2019)
72. Correa, S., Choi, K.Y., Dreaden, E.C., Renggli, K., Shi, A., Gu, L., Shopsowitz, K.E., Quadir, M.A., Ben-Akiva, E., Hammond, P.T.: Highly scalable, closed-loop synthesis of drug-loaded, layer-by-layer nanoparticles. Adv. Func. Mater. **26**(7), 991–1003 (2016)

73. Nunziata, G., Borroni, A., Rossi, F.: Advanced microfluidic strategies for core-shell nanoparticles: the next-generation of polymeric and lipid-based drug nanocarriers. Chem. Eng. J. Adv. **21**, 100759 (2025)
74. Thomée E. Microfluidic nanoparticle synthesis: a short review. Elveflow. 2021 Feb 5.
75. Rowland, S., Aghakhani, A., Whalley, R.D., Ferreira, A.M., Kotov, N., Gentile, P.: Layer-by-layer nanoparticle assembly for biomedicine: mechanisms, technologies, and advancement via acoustofluidics. ACS Appl. Nano Mater. **7**(14), 15874–15902 (2024)
76. Kumari, S., Raturi, S., Kulshrestha, S., Chauhan, K., Dhingra, S., András, K., Thu, K., Khargotra, R., Singh, T.: A comprehensive review on various techniques used for synthesizing nanoparticles. J. Market. Res. **1**(27), 1739–1763 (2023)
77. Li, P., Kumar, A., Ma, J., Kuang, Y., Luo, L., Sun, X.: Density gradient ultracentrifugation for colloidal nanostructures separation and investigation. Sci. Bull. **63**(10), 645–662 (2018)
78. Khan, Y., Sadia, H., Ali Shah, S.Z., Khan, M.N., Shah, A.A., Ullah, N., Ullah, M.F., Bibi, H., Bafakeeh, O.T., Khedher, N.B., Eldin, S.M.: Classification, synthetic, and characterization approaches to nanoparticles, and their applications in various fields of nanotechnology: a review. Catalysts **12**(11), 1386 (2022)
79. Robertson, J.D., Rizzello, L., Avila-Olias, M., Gaitzsch, J., Contini, C., Magoń, M.S., Renshaw, S.A., Battaglia, G.: Purification of nanoparticles by size and shape. Sci. Rep. **6**(1), 27494 (2016)
80. Dolai, J., Mandal, K., Jana, N.R.: Nanoparticle size effects in biomedical applications. ACS Appl. Nano Mater. **4**(7), 6471–6496 (2021)
81. Öztürk, K., Kaplan, M., Çalış, S.: Effects of nanoparticle size, shape, and zeta potential on drug delivery. Int. J. Pharm. **5**, 124799 (2024)
82. Khan, I., Saeed, K., Khan, I.: Nanoparticles: properties, applications and toxicities. Arab. J. Chem. **12**(7), 908–931 (2019)
83. Altammar, K.A.: A review on nanoparticles: characteristics, synthesis, applications, and challenges. Front. Microbiol. **17**(14), 1155622 (2023)
84. Gupta, A., Allison, C.A., Kumar, A., Srivastava, R., Lin, W., Sultana, J., Mishra, S.R., Perez, F., Gupta, R.K., Dawsey, T.: Tuned morphology configuration to augment the electronic structure and durability of iron phosphide for efficient bifunctional electrocatalysis and charge storage. J. Energy Storage **1**(73), 108824 (2023)
85. Tiwari, J.N., Tiwari, R.N., Kim, K.S.: Zero-dimensional, one-dimensional, two-dimensional and three-dimensional nanostructured materials for advanced electrochemical energy devices. Prog. Mater. Sci. **57**(4), 724–803 (2012)

Polysaccharide Nanoparticles-Based Functional Constructs

Surbhi Kharwar, Sheetal Jaiswal, Vinod Kumar, and Sunil Kumar

Abstract Nanotechnology, especially the development of nanoparticles (NPs), has revolutionized diverse fields including medicine, agriculture, and environmental science. Polysaccharide-based nanoparticles (PNPs) have gained significant interest due to their natural origin, biocompatibility, biodegradability, and functional versatility. Derived from sources such as plants, algae, animals, and microbes, polysaccharides like chitosan, alginate, starch, and cellulose serve as green alternatives in nanoparticle synthesis. These biopolymers offer abundant reactive groups such as hydroxyl, carboxyl, and amino that enable efficient binding and functionalization, eliminating the need for toxic reagents. Various fabrication techniques, including ionic gelation, nanoprecipitation, and electrospinning, allow for the synthesis of tailored PNPs with specific size, morphology, and surface characteristics. Functionalization approaches, such as ligand conjugation, PEGylation, and covalent crosslinking, further enhance the stability, targeting ability, and therapeutic efficacy of these nanomaterials. Surface modification plays a vital role in optimizing drug delivery, diagnostics, and tissue regeneration applications. Extraction and purification of polysaccharides from natural sources are critical for ensuring the quality and performance of the resulting nanoparticles. The unique physicochemical and biological properties of PNPs, including controlled release behavior, mechanical strength, and biocompatibility, position them as vital tools in modern nanomedicine and environmental sustainability. As research advances, the development of smart, stimuli-responsive, and eco-friendly PNPs is expected to drive innovations in targeted therapeutics and sustainable material science.

S. Kharwar (✉)
Department of Botany, University of Lucknow, Lucknow, Uttar Pradesh, India
e-mail: surbhi10bhu@gmail.com

S. Jaiswal
Department of Chemistry, Institute of Science, Banaras Hindu University, Varanasi, India

V. Kumar
Department of Botany, Pt. Prithi Nath (PG) College, Kanpur, India

S. Kumar (✉)
Central Ayurveda Research Institute, Bhubaneswar, Odisha, India
e-mail: sunilkumarbhu08@gmail.com

B. Sharma et al. (eds.), *Nanostructured Polymers*, Engineering Materials,
https://doi.org/10.1007/978-981-95-3581-1_8

Keywords Nanoparticles · Polysaccharide-based nanoparticles · Fabrication techniques · Functionalization · Properties

1 Introduction

Nanotechnology has emerged as a transformative field across various scientific domains, with nanoparticles (NPs) at the forefront of innovation in medicine, agriculture, environmental remediation, and materials science. Nanoparticles are defined as materials with dimensions typically ranging from 1 to 100 nm, possessing unique physical, chemical, and biological properties that differ significantly from their bulk counterparts due to their high surface area to volume ratio and quantum effects [1]. These properties include enhanced reactivity, mechanical strength, optical behavior, and electrical conductivity, which make nanoparticles highly desirable for applications in drug delivery, imaging, catalysis, and biosensing [2].

In recent years, the integration of biopolymers, especially polysaccharides, in nanoparticle synthesis has garnered significant attention due to their natural abundance, biocompatibility, biodegradability, and functional versatility. Polysaccharides are naturally occurring long-chain carbohydrates composed of monosaccharide units linked by glycosidic bonds. Polysaccharides such as chitosan, alginate, cellulose, starch, and guar gum serve as excellent platforms for the development of green and sustainable nanomaterials [3]. Their molecular structures, rich in hydroxyl, carboxyl, and amino groups, enable efficient interaction with metal ions and other active agents, thereby facilitating the formation and stabilization of nanoparticles without the use of toxic chemicals [4]. Additionally, polysaccharide-based nanoparticles have been widely explored in biomedical applications including controlled drug release, tissue engineering, and wound healing, due to their low toxicity and ability to modulate cellular interactions [5].

The synergistic combination of nanoparticles and polysaccharides has enabled the development of hybrid materials with improved biomedical, pharmaceutical, and environmental applications. For instance, chitosan-coated silver nanoparticles exhibit improved antibacterial activity and biocompatibility compared to uncoated silver nanoparticles [6]. Thus, understanding the fundamental properties of nanoparticles and the role of polysaccharides in their synthesis and functionalization is crucial for advancing green nanotechnology and sustainable material design.

As the demand for eco-friendly and biologically safe nanomaterials grows, polysaccharide-based nanoparticles represent a promising avenue for sustainable nanotechnology. Their synthesis aligns with green chemistry principles and offers an alternative to conventional synthetic approaches, ensuring minimal environmental impact and improved biocompatibility.

2 Synthesis and Functionalization of Polysaccharide Nanoparticles (PNPs)

The synthesis and functionalization of PNPs have emerged as a pivotal area of research due to their biocompatibility, biodegradability, and potential in targeted delivery systems. Various synthesis methods are employed depending on the type of polysaccharide and intended application, including ionic gelation, nanoprecipitation, solvent evaporation, emulsification, and electrospinning. Among these, ionic gelation is frequently used for chitosan, where crosslinking occurs between positively charged amino groups and negatively charged polyanions like tripolyphosphate (TPP), forming stable nanoscale networks [7]. Polysaccharides are typically derived from marine (e.g., alginate, chitosan), plant (e.g., pectin, starch), or microbial sources (e.g., dextran, pullulan). These natural polymers contain reactive functional groups (hydroxyl, carboxyl, amino) that allow for diverse chemical modifications, enhancing stability, solubility, and biological targeting. Functionalization involves grafting functional molecules such as peptides, drugs, or ligands onto the nanoparticle surface to facilitate specific interactions with cells or tissues [8]. For instance, folic acid or RGD peptides can be conjugated to chitosan nanoparticles to improve targeting of cancer cells expressing specific receptors. Further chemical modification includes surface coating with polyethylene glycol (PEGylation) to enhance circulation time in the bloodstream and reduce immunogenicity. Covalent crosslinking using agents like glutaraldehyde or genipin can also enhance mechanical strength and control degradation rate [9]. Thus, the combination of tailored synthesis techniques and surface functionalization strategies enables the development of smart PNPs for controlled drug release, gene delivery, imaging, and regenerative medicine, positioning them as vital tools in modern nanomedicine.

3 Extraction and Purification

The first step in synthesizing PNPs is the isolation of polysaccharides from natural sources. This involves extraction using hot water, acid, or alkali solutions followed by centrifugation, filtration, and concentration [10]. The chosen method depends on the nature of the raw material and the type of polysaccharide targeted. For example, hot water extraction is common for starch and gums, while chitosan requires chemical modification of chitin extracted from crustacean shells.

Once extracted, crude polysaccharides contain proteins, pigments, minerals, and other impurities that must be removed to enhance nanoparticle functionality. This is achieved through a series of purification steps, namely:

- **Deproteinization**: Typically performed using Sevag reagent (chloroform:butanol) or enzymatic hydrolysis to eliminate proteins.
- **Demineralization**: Conducted using dilute hydrochloric acid (HCl) to remove calcium, magnesium, and other inorganic salts.

- **Bleaching**: Carried out using hydrogen peroxide (H_2O_2) or ethanol to eliminate pigments and improve the color and purity of the polysaccharide [11, 12].

4 Sources of Polysaccharides

Polysaccharides are high-molecular-weight biopolymers composed of monosaccharide units linked by glycosidic bonds. They are widely distributed in nature and serve structural, storage, and functional roles in various biological systems. Based on their origin, polysaccharides can be categorized into four major sources: plant-based, animal-based, algal, and microbial. Each source provides unique structural features and physicochemical properties that influence their biological and industrial applications.

4.1 Plant-Based Polysaccharides

Plants are the most abundant source of polysaccharides, producing both structural and storage types. Common plant-derived polysaccharides include:

- **Cellulose**: The most abundant organic polymer, found in plant cell walls, composed of β-(1 $\rightarrow$ 4)-linked D-glucose units [13].
- **Starch**: A storage polysaccharide comprising amylose (linear) and amylopectin (branched), present in cereals, tubers, and legumes [14].
- **Pectin**: Found in the middle lamella of plant cell walls, primarily extracted from citrus peels and apple pomace, rich in galacturonic acid [15].
- **Plant gums**: Such as guar gum and gum arabic, are secreted from trees (e.g., *Acacia senegal*) and used as emulsifiers and stabilizers [16].

4.2 Animal-Based Polysaccharides

Animals primarily contribute structural polysaccharides, notably:

- **Chitin**: A major component of the exoskeletons of crustaceans, insects, and fungal cell walls. It is composed of N-acetyl-D-glucosamine units [12].
- **Chitosan**: Obtained through deacetylation of chitin and widely studied for biomedical applications due to its biocompatibility and antimicrobial properties [17].

4.3 Algal Polysaccharides

Marine algae, especially brown algae, produce various anionic polysaccharides used in food, pharmaceutical, and biomedical fields:

- **Alginate**: Extracted from brown seaweeds (*Macrocystis pyrifera, Laminaria* spp.), it consists of mannuronic and guluronic acid residues [18].
- **Carrageenan and agar**: Extracted from red algae and used as gelling and thickening agents in food and drug formulations [19].

4.4 Microbial Polysaccharides

Certain bacteria and fungi synthesize extracellular polysaccharides with unique physicochemical properties:

- **Dextran**: Produced by *Leuconostoc mesenteroides* during sucrose fermentation; used in biomedical and pharmaceutical applications [20].
- **Pullulan**: Produced by *Aureobasidium pullulans*, known for its film-forming and oxygen-barrier properties [21].
- **Xanthan gum**: Secreted by *Xanthomonas campestris* and widely used in food, cosmetics, and pharmaceutical industries [22].

Each source offers distinct advantages depending on the desired application, including biodegradability, film-forming ability, water solubility, and interaction with biological tissues.

5 Types of Polysaccharides

Polysaccharides are classified based on their origin, structure, and function. Among the numerous polysaccharides, chitin, chitosan, alginate, and cellulose are of particular interest due to their biocompatibility, biodegradability, and wide-ranging applications in biomedical, pharmaceutical, and environmental fields.

5.1 Chitin and Chitosan

Chitin is the second most abundant natural polysaccharide after cellulose and is primarily found in the exoskeletons of crustaceans, insects, and fungal cell walls. It consists of β-(1 $\rightarrow$ 4)-linked N-acetyl-D-glucosamine units (Fig. 1). Due to its insolubility in water and most solvents, its practical applications are limited in its native form [12].

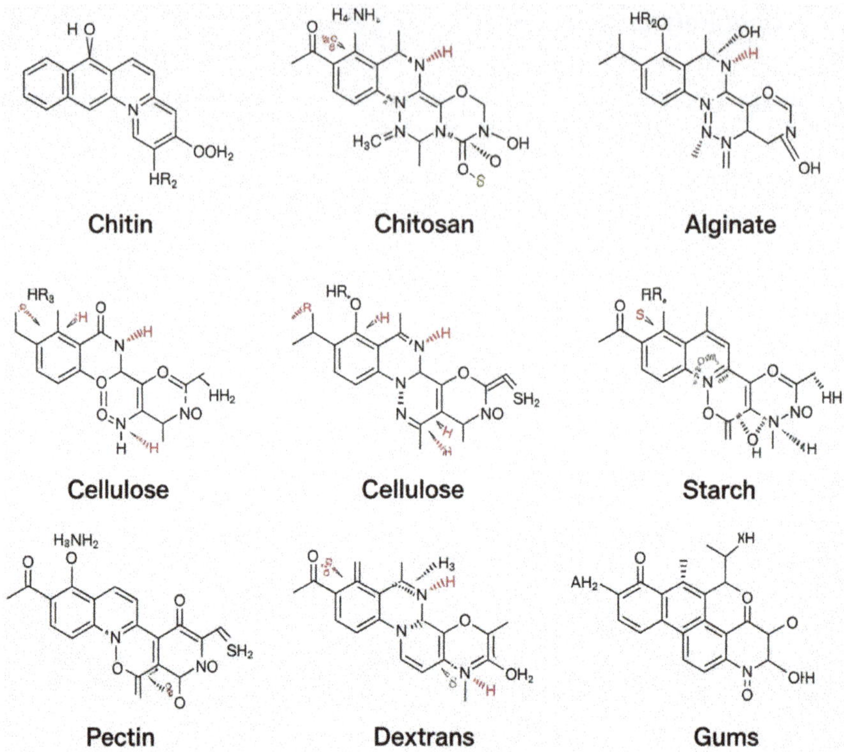

Fig. 1 Image showing the chemical structures of polysaccharides such as chitin and chitosan, alginate, cellulose, starch, pectin, dextrans, and gums

Chitosan is derived from chitin via a deacetylation process that removes acetyl groups, converting it into a more soluble polymer consisting primarily of D-glucosamine units. Chitosan is soluble in acidic conditions, exhibits antimicrobial properties, and is extensively used in drug delivery, wound healing, and tissue engineering [17].

5.2 *Alginate*

Alginate is a naturally occurring anionic polysaccharide predominantly extracted from brown seaweeds (e.g., *Laminaria, Macrocystis, Ascophyllum* spp.). It comprises linear copolymers of β-D-mannuronic acid (M) and α-L-guluronic acid (G) residues (Fig. 1). The ratio and sequence of M and G blocks determine the physicochemical properties of alginate, such as gel strength and viscosity [23].

Owing to its mild gelation properties with divalent cations (especially calcium), biocompatibility, and low toxicity, alginate is widely used in controlled drug release systems, wound dressings, and tissue scaffolds [24].

5.3 Cellulose

Cellulose is the most abundant biopolymer on Earth and is the primary structural component of plant cell walls. It is composed of linear chains of β-(1 → 4)-linked D-glucose units forming strong intermolecular and intramolecular hydrogen bonds (Fig. 1), resulting in high crystallinity and insolubility in water [13].

Microcrystalline cellulose and nanocellulose (including cellulose nanocrystals and nanofibrils) are widely used in pharmaceutical formulations as excipients, binders, and drug carriers. Moreover, its biodegradability and mechanical strength make cellulose ideal for use in packaging and biomedical materials [25].

5.4 Starch

Starch is a storage polysaccharide in plants, composed of two polysaccharides: amylose (linear) and amylopectin (branched) (Fig. 1). It is commonly derived from sources like maize, potato, and rice, and used in food, pharmaceutical tablets, and as a biodegradable polymer [14].

5.5 Pectin

Pectin is a complex polysaccharide rich in galacturonic acid units (Fig. 1) and is primarily extracted from citrus peels and apple pomace. Due to its gelling properties, it is widely used in food and pharmaceutical formulations for controlled drug release [15].

5.6 Dextrans

Produced by bacteria such as *Leuconostoc mesenteroides*, dextrans are branched glucans with α-(1 → 6) and some α-(1 → 3) linkages (Fig. 1). They are used in plasma expanders, drug carriers, and nanoparticle synthesis due to their biocompatibility and solubility [26].

5.7 Gums

Natural gums like guar gum, gum arabic, and xanthan gum are complex polysaccharides derived from plants or microbial fermentation (Fig. 1). These are used as emulsifiers, thickeners, and stabilizers in food, pharmaceuticals, and cosmetics [16, 22].

6 Nanofabrication Techniques of Polysaccharide Nanoparticles

Polysaccharide-based nanoparticles have emerged as promising carriers in biomedical, pharmaceutical, and food applications due to their biocompatibility, biodegradability, and functional versatility. The method of nanoparticle fabrication significantly influences their size, morphology, surface characteristics, and drug release profiles. Several nanofabrication techniques are employed to prepare PNPs, including nanoprecipitation, electrospinning, solvent evaporation, and template-assisted methods.

6.1 Nanoprecipitation

Nanoprecipitation, also referred to as solvent displacement, is a widely employed technique for fabricating polymeric nanoparticles, including those based on polysaccharides. This method is particularly suitable for hydrophobically modified polysaccharides and is prized for its simplicity, reproducibility, and scalability.

The nanofabrication by nanoprecipitation is based on the interfacial deposition of a polymer after the spontaneous diffusion of an organic solvent (in which the polymer is dissolved) into a non-solvent (typically water). As the organic solvent diffuses into the aqueous phase, the polymer precipitates, forming nanoparticles.

Steps

1. A polysaccharide or its hydrophobic derivative is dissolved in a water-miscible organic solvent (e.g., acetone, ethanol, or acetonitrile).
2. The organic solution is added dropwise into an aqueous phase under moderate stirring.
3. Immediate diffusion of the solvent occurs, followed by the formation of nanoparticles.
4. The organic solvent is removed by evaporation, and nanoparticles are collected by centrifugation or filtration (Fig. 2).

Example: Hydrophobically modified chitosan (e.g., N-palmitoyl chitosan) or dextran derivatives have been effectively used to prepare nanoparticles by this technique [7, 27].

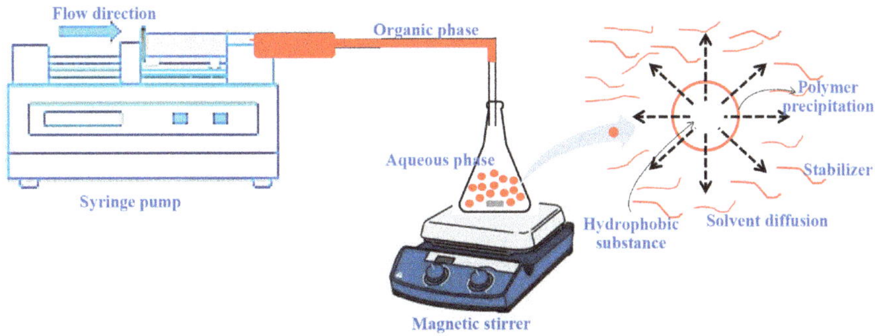

Fig. 2 The schematic diagram represents the synthesis of nanostructure using nanoprecipitation

6.2 Electrospinning

Electrospinning is an electrohydrodynamic technique primarily used for fabricating nanofibers, but under specific conditions, it can be adapted for producing PNPs or nano-aggregates. The technique leverages electrostatic forces to draw a charged polymer solution or melt into nanostructures.

In electrospinning, a high-voltage electric field is applied between a polymer-containing syringe needle and a grounded collector. As voltage increases, the polymer solution at the tip of the syringe forms a *Taylor cone*. Once the electrostatic repulsion overcomes the surface tension, a fine jet is ejected [28]. As the jet travels through the air, solvent evaporates, resulting in the formation of nanofibers or particles, depending on the solution properties and process parameters.

The morphology (fibers vs. particles) is dictated by factors such as polymer concentration, molecular weight, viscosity, surface tension, and electrical conductivity [29].

Steps

1. Prepare a homogeneous polymer solution.
2. Load the solution into a syringe with a metal needle.
3. Apply a high-voltage power supply (typically 10–30 kV).
4. Adjust the flow rate (e.g., 0.5–2 mL/h) and distance to collector (10–20 cm).
5. Collect nanostructures on an aluminum foil or rotating drum (Fig. 3).

For nanoparticle formation rather than nanofibers, low-viscosity solutions or emulsion electrospinning can be employed [30].

6.3 Solvent Evaporation

Solvent evaporation is a widely used method for fabricating nanoparticles, particularly suited for encapsulating hydrophobic drugs within biopolymeric carriers.

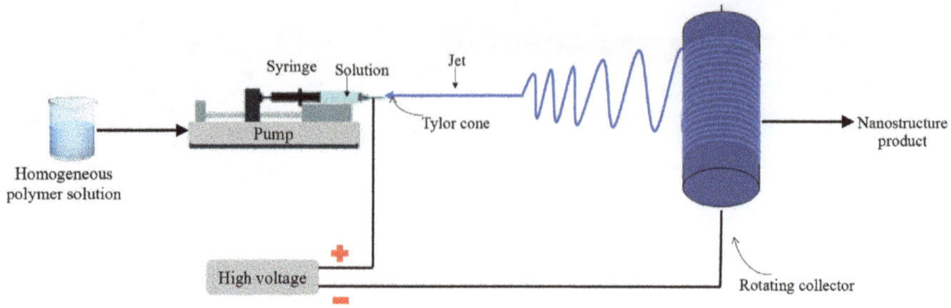

Fig. 3 The diagram represents the synthesis of nanostructure using the method of electrospinning

Although originally developed for synthetic polymers, this method has been adapted for modified polysaccharides (e.g., hydrophobic derivatives of chitosan, starch, and cellulose) to form stable nanoscale particles [31].

The solvent evaporation technique involves:

- Dissolving a polysaccharide or a hydrophobic polysaccharide derivative in an organic solvent.
- Emulsifying this organic phase in an aqueous phase containing surfactants or stabilizers.
- Removing the organic solvent through evaporation, leading to nanoparticle solidification and formation.

This process results in solid nanoparticles with controlled size and morphology. Since many native polysaccharides are water-soluble, hydrophobic modification or copolymer blending is usually required [32].

Steps

1. **Polymer/Drug Dissolution**: The polysaccharide and drug are dissolved in a volatile organic solvent (e.g., dichloromethane, chloroform, ethyl acetate).
2. **Emulsification**: This solution is emulsified into an aqueous phase using mechanical stirring or sonication, forming oil-in-water (O/W) emulsion.
3. **Solvent Evaporation**: The organic solvent is evaporated under reduced pressure or stirring, leading to nanoparticle hardening.
4. **Purification**: Nanoparticles are separated by centrifugation, washed, and redispersed in an appropriate medium (Fig. 4).

For example, chitosan and hydrophobically modified dextran have been processed via this method for drug delivery applications [33].

6.4 Template-Assisted Method

The template-assisted method is a versatile nanofabrication approach that uses preformed nanostructured templates to guide the shape, size, and morphology of

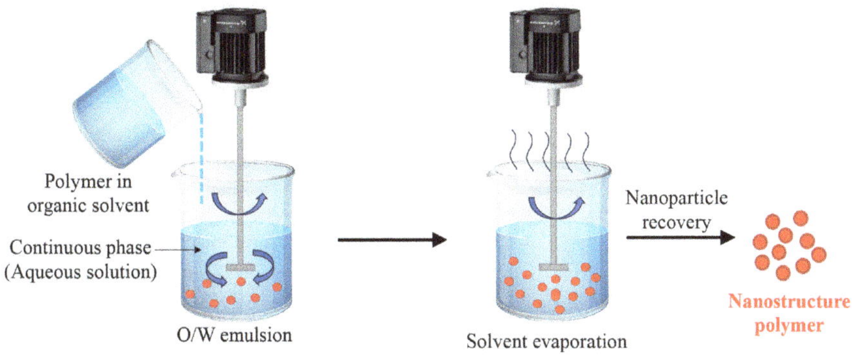

Fig. 4 Schematic of nanostructure formed during solvent evaporation process

nanoparticles. In the context of PNPs, this technique offers precise control over dimensions and uniformity, which is essential for drug delivery, biosensing, and tissue engineering applications. In this method, a physical or chemical template (such as nanoporous membranes, micelles, or sacrificial scaffolds) is used to direct the deposition or assembly of polysaccharide molecules. The template determines the size and shape of the resulting nanoparticles. After formation, the template is removed (e.g., by dissolution, etching, or degradation), leaving behind the desired nanostructures [34]. This technique is particularly useful for polysaccharides that do not readily self-assemble into uniform nanostructures [35].

6.4.1 Types of Templates Used

- **Hard Templates**: Anodic aluminum oxide (AAO) membranes, mesoporous silica, polycarbonate track-etched membranes.
- **Soft Templates**: Surfactant micelles, emulsions, liposomes.
- **Soft Templates**: Surfactant micelles, emulsions, liposomes.

Steps

1. **Template Preparation**: Select a template with nanoscale pores or cavities.
2. **Polysaccharide Deposition**: Introduce the polysaccharide solution into the template, often by infiltration, spin coating, or vacuum-assisted methods.
3. **Gelation or Crosslinking**: Solidify the material via physical or chemical crosslinking.
4. **Template Removal**: Dissolve or degrade the template using solvents or enzymatic treatment.
5. **Nanoparticle Collection**: Recover the free-standing PNPs (Fig. 5).

Example: Chitosan has been fabricated into uniform nanoparticles using nanoporous AAO templates [36].

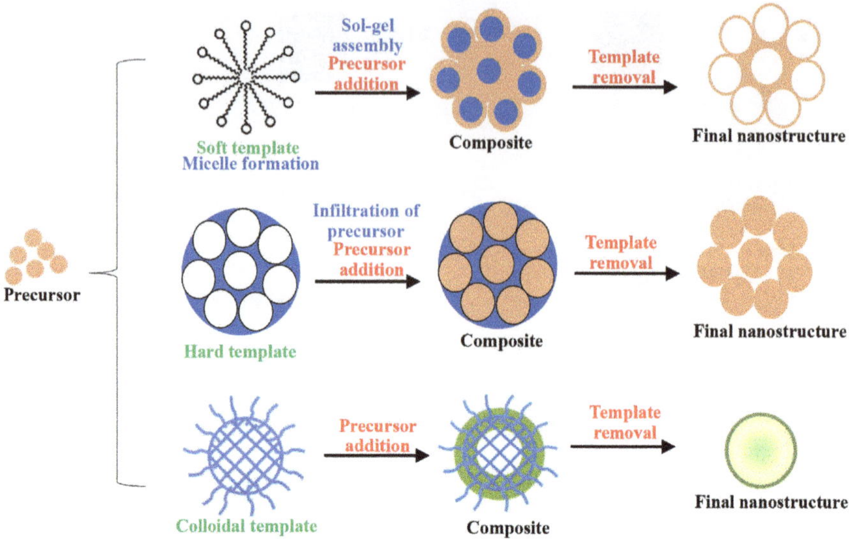

Fig. 5 Pictorial representation of different template assisted methods for nanostructure synthesis

7 Surface Modification and Functionalization

Surface modification of PNPs is crucial for enhancing their physicochemical properties, biocompatibility, and application potential in drug delivery, diagnostics, and tissue engineering. Various techniques such as chemical crosslinking, ionic gelation, and functional group grafting are employed to alter surface characteristics and improve performance. Chemical crosslinking using agents like glutaraldehyde or genipin strengthens nanoparticle stability by forming covalent bonds between polymer chains [37]. Ionic gelation, a milder method, utilizes electrostatic interactions, such as between chitosan and tripolyphosphate, to create stable nanoparticle systems suitable for encapsulating bioactives [7]. Furthermore, surface functionalization through PEGylation or ligand conjugation enhances solubility, prolongs circulation time, and enables targeted delivery [38]. Polysaccharide coatings such as dextran can also be applied to metallic or magnetic cores to improve biocompatibility and reduce toxicity [39]. Additionally, modern techniques like click chemistry allow for precise and responsive modifications tailored to environmental stimuli or specific therapeutic needs [40]. These surface engineering strategies significantly expand the biomedical applications of PNPs by improving their interaction with biological environments.

7.1 Chemical Crosslinking

Chemical crosslinking is a widely used strategy for the surface modification of PNPs to enhance their stability, mechanical strength, and functional performance. This method involves the formation of covalent bonds between polysaccharide chains or between the polymer and a crosslinking agent. Such modifications are crucial in biomedical applications like drug delivery, tissue engineering, and biosensing, where stability under physiological conditions is essential.

Chitosan, a cationic polysaccharide derived from chitin, is frequently crosslinked using agents like glutaraldehyde, genipin, or sodium tripolyphosphate (TPP). Glutaraldehyde facilitates Schiff base formation between its aldehyde groups and the amino groups of chitosan, resulting in a stable nanoparticle matrix. However, glutaraldehyde's toxicity limits its biomedical applicability, prompting the use of natural crosslinkers such as genipin, which forms biocompatible and less cytotoxic networks through amine-mediated reactions [41]. Another approach involves ionic crosslinking, where TPP interacts electrostatically with protonated amine groups in chitosan, enabling nanoparticle formation under mild conditions without harmful solvents [7].

Alginate, a naturally occurring anionic polysaccharide, can be crosslinked with divalent cations like calcium ions to form stable gel-like nanoparticles. This technique enhances encapsulation efficiency and controlled drug release. Similarly, starch and dextran-based nanoparticles are chemically crosslinked with epichlorohydrin or carbodiimide agents to introduce structural rigidity and modulate degradation kinetics [42].

Overall, chemical crosslinking not only stabilizes PNPs but also offers avenues for further functionalization by introducing reactive groups or ligands for targeted delivery. The choice of crosslinker and method significantly affects the nanoparticle's biocompatibility and performance in biomedical settings.

7.2 Ionic Gelation

Ionic gelation is a widely employed technique for the fabrication and surface modification of PNPs, particularly those derived from biopolymers like chitosan and alginate. This method relies on the electrostatic interaction between oppositely charged molecules, enabling the formation of stable nanostructures under mild and aqueous conditions—ideal for encapsulating sensitive bioactive compounds such as proteins, peptides, and nucleic acids.

Chitosan, a cationic polysaccharide, forms nanoparticles when it interacts with polyanions such as tripolyphosphate (TPP). The ionic interaction between the protonated amino groups of chitosan and the negatively charged phosphate groups of TPP leads to the formation of a crosslinked network, stabilizing the nanoparticle surface [7]. This process does not require the use of organic solvents or high temperatures, making it particularly suitable for biomedical applications including drug delivery, gene therapy, and vaccine delivery.

In addition to nanoparticle formation, ionic gelation can be exploited for surface modification by incorporating functional molecules such as targeting ligands (e.g., folic acid), polymers (e.g., polyethylene glycol), or antimicrobial agents during the gelation process. This allows for the enhancement of particle stability, mucoadhesiveness, controlled release properties, and cell-specific targeting [43].

Alginate, an anionic polysaccharide, similarly undergoes gelation upon contact with divalent cations like calcium ions. This results in the formation of hydrogel networks, which can also be modified by incorporating other functional agents to alter the surface characteristics for improved therapeutic performance.

Ionic gelation thus presents a versatile, green, and efficient method for both nanoparticle synthesis and surface engineering, with broad applicability in the design of polysaccharide-based nanocarriers for biomedical use.

7.3 Attachment/Grafting of Functional Groups

Attachment or grafting of functional groups onto the surface of PNPs is a key strategy to enhance their physicochemical and biological performance, especially for targeted drug delivery, improved biocompatibility, and prolonged systemic circulation. This approach involves covalently linking functional molecules—such as ligands, polymers, or bioactive compounds—onto reactive groups present on the polysaccharide backbone (e.g., hydroxyl, amino, or carboxyl groups).

Chitosan, a widely used cationic polysaccharide, contains primary amine groups that offer versatile modification sites. One common modification is PEGylation, where polyethylene glycol (PEG) chains are grafted onto the nanoparticle surface. This imparts "stealth" characteristics to the nanoparticles by minimizing protein adsorption and recognition by the mononuclear phagocyte system, thereby extending circulation time [38]. Additionally, targeting ligands such as folic acid, peptides, or antibodies can be grafted to enable receptor-mediated cellular uptake, improving therapeutic efficacy [44].

Similarly, alginate and dextran nanoparticles can be modified via carbodiimide chemistry, enabling the conjugation of carboxyl or hydroxyl-containing ligands. These grafting strategies improve mucoadhesion, site-specific delivery, and stimuli-responsiveness [45].

Advanced grafting techniques like "click chemistry" (e.g., azide-alkyne cycloaddition) offer highly specific and efficient surface functionalization under mild conditions, preserving the bioactivity of sensitive molecules. Such methods are especially useful for tailoring nanoparticles for theranostic applications—combining therapeutic and diagnostic functions.

Overall, the grafting of functional groups is an effective and versatile method for customizing PNPs, broadening their applications across biomedical, pharmaceutical, and environmental fields.

7.4 Nanoparticle Integration

Surface modification of PNPs through nanoparticle integration involves the incorporation or coating of functional nanoparticles—such as metallic, magnetic, or polymeric nanoparticles—onto or within the polysaccharide matrix. This hybridization significantly enhances the physicochemical properties of PNPs, such as mechanical strength, magnetic responsiveness, imaging capability, and controlled drug release behavior, thus expanding their applications in theranostics, targeted drug delivery, and biosensing.

A notable example is the integration of magnetic nanoparticles like iron oxide (Fe_3O_4) into chitosan or dextran-based PNPs. This modification enables magnetic field-guided drug delivery and real-time imaging through magnetic resonance imaging (MRI). The biocompatible polysaccharide shell reduces the toxicity of the magnetic core and provides functional groups for further surface conjugation [39]. Similarly, gold nanoparticles (AuNPs) integrated into alginate or chitosan matrices have been explored for photothermal therapy and biosensing due to their unique optical properties [46].

PNPs can also serve as carriers for other functional nanoparticles, forming core–shell or matrix-dispersed systems. For instance, silica or polymeric nanoparticles embedded into polysaccharide matrices enhance drug loading and enable pH-sensitive release, making them ideal for oral or tumor-targeted drug delivery [47].

These integrated systems can be further surface-functionalized with targeting ligands, antibodies, or responsive polymers, creating multifunctional nano platforms. The presence of the polysaccharide matrix ensures biocompatibility and biodegradability, while the embedded nanoparticles confer enhanced functionality. Thus, nanoparticle integration is a powerful surface modification strategy that combines the biological advantages of polysaccharides with the functional versatility of synthetic or inorganic nanoparticles for advanced biomedical applications.

8 Characteristics of Polysaccharide Nanoparticles

PNPs are nanostructures derived from natural polysaccharides such as chitosan, alginate, dextran, starch, and gums. They exhibit a variety of physicochemical and biological characteristics that make them attractive for applications in drug delivery, biomedical imaging, and tissue engineering.

One of the primary characteristics of PNPs is their biodegradability and biocompatibility, owing to their natural origin. This allows them to be safely used in biological systems without eliciting significant toxicity or immune responses [48]. The hydrophilicity of polysaccharides further improves their compatibility with biological environments, facilitating cellular uptake and mucosal adhesion. PNPs typically possess a nano-size range (10–500 nm), which is crucial for enhanced permeability and retention (EPR) effect in tumor tissues and for cellular internalization. Their surface charge, usually modifiable, plays a significant role in interactions with cells and biological barriers. For instance, chitosan-based nanoparticles, having a positive surface charge, can interact with negatively charged cell membranes, enhancing cellular uptake [49]. Another significant feature is the chemical versatility of polysaccharides. Functional groups such as hydroxyl, carboxyl, and amino groups allow easy chemical modification and fictionalization. This enables the conjugation of targeting ligands, drugs, imaging agents, or other polymers to tailor the nanoparticles for specific biomedical applications [50]. PNPs also show controlled drug release properties, which can be influenced by environmental triggers such as pH, temperature, or enzymes. This is particularly useful in achieving site-specific delivery, minimizing systemic side effects.

In summary, PNPs offer an advantageous combination of safety, modifiability, and functional performance, making them suitable candidates for a broad range of therapeutic and diagnostic applications.

8.1 Biodegradability

PNPs exhibit excellent biodegradability due to their natural polymeric origin. Their backbone consists of glycosidic linkages that are readily cleaved by enzymes such as amylases, cellulases, and lysozymes present in biological systems. This makes PNPs ideal for biomedical applications where temporal control over degradation is essential, such as in drug delivery or tissue engineering. Upon degradation, they yield non-toxic monosaccharides or oligosaccharides that are either metabolized or excreted without eliciting inflammatory responses. The degradation rate can be modulated by modifying the molecular weight, crystallinity, and degree of substitution of the polysaccharides. For instance, chitosan and dextran nanoparticles degrade slowly in vivo, enabling sustained therapeutic effects [28]. This property also minimizes the need for surgical removal, improving patient compliance.

8.2 Biocompatibility

Biocompatibility is a critical requirement for nanocarriers used in biomedical applications. PNPs are inherently biocompatible due to their natural origin and structural similarity to components of the extracellular matrix. These particles do not provoke adverse immune responses, making them safe for applications such as injectable drug delivery systems or wound healing scaffolds. Polysaccharides like alginate, chitosan, and dextran have been shown to exhibit excellent cell adhesion, proliferation, and differentiation support. Moreover, their hydrophilic surfaces reduce protein adsorption and opsonization, thereby enhancing circulation time [51]. Surface modifications, such as PEGylation or ligand conjugation, further improve their compatibility and functional versatility in vivo.

8.3 Mechanical Strength

Although natural polysaccharides are generally soft, their nanoparticles can exhibit sufficient mechanical strength for biomedical applications when properly crosslinked or composited. Chitosan and alginate nanoparticles, for example, demonstrate considerable mechanical integrity due to ionic or covalent crosslinking, which stabilizes the structure under physiological conditions. This mechanical robustness ensures that the nanoparticles maintain their architecture during drug delivery or under biological stress. Moreover, mechanical properties can be enhanced by blending polysaccharides with synthetic polymers or reinforcing with inorganic nanoparticles like silica or calcium phosphate [52]. Such reinforcements allow for tunable mechanical performance that is essential for load-bearing tissue scaffolds or durable coatings.

8.4 Responsive Behavior

PNPs can be engineered to respond to various physiological stimuli, such as pH, temperature, enzymes, or redox potential. This responsive behavior is crucial for targeted and controlled drug delivery. For instance, chitosan nanoparticles swell and release their payload under acidic conditions, which is beneficial for tumor targeting or intracellular delivery [53]. Similarly, dextran can be modified to respond to oxidative stress, allowing site-specific release in inflamed or cancerous tissues. Such stimuli-responsive systems improve therapeutic efficacy and reduce side effects by minimizing drug release in non-target areas.

8.5 Renewability

The raw materials for PNPs are derived from renewable plant and microbial sources, such as algae (alginate), fungi (chitin/chitosan), and plants (starch, cellulose, gums). This sustainable sourcing minimizes environmental impact and aligns with the principles of green chemistry. Moreover, their renewability supports a circular economy by offering biodegradable alternatives to petroleum-derived polymers in packaging, pharmaceuticals, and cosmetics [54]. Cultivation and extraction technologies are also improving, making large-scale and eco-friendly production more feasible. This property is particularly valuable in industries aiming for sustainable materials with reduced carbon footprints.

8.6 *Non-Toxicity*

PNPs are widely recognized for their non-toxic nature, which makes them suitable for oral, injectable, and transdermal applications. Their degradation products are typically natural monosaccharides or oligosaccharides, which are either absorbed by cells or eliminated harmlessly. Toxicity studies on chitosan, alginate, and dextran nanoparticles have shown minimal cytotoxicity and hemolytic activity even at relatively high concentrations [55]. Furthermore, their chemical inertness reduces the likelihood of reactive oxygen species (ROS) generation or DNA damage, making them safe for long-term biomedical use. Regulatory agencies often classify them as GRAS (Generally Recognized As Safe), further supporting their safe use.

9 Functional Constructs of Polysaccharides

Polysaccharides, naturally occurring biopolymers composed of long chains of monosaccharide units, have gained significant attention as versatile functional constructs in biomedical, pharmaceutical, and environmental applications. Their unique physicochemical properties such as biodegradability, biocompatibility, non-toxicity, and the ability to be chemically modified enable the fabrication of diverse functional materials including nanoparticles, nanofibers, hydrogels, films, and scaffolds [56].

One of the key advantages of polysaccharides lies in their abundance and renewability from natural sources such as plants (cellulose, starch, and pectin), animals (chitin, chitosan), algae (alginate), and microbes (dextrans, xanthan gum). This sustainable origin makes polysaccharide-based constructs eco-friendly alternatives to synthetic polymers. For instance, chitosan nanoparticles have been extensively studied for drug delivery due to their mucoadhesive properties, positive surface charge, and antimicrobial effects, facilitating enhanced drug uptake and sustained release [57].

Polysaccharide nanofibers produced via electrospinning mimic the extracellular matrix structure, providing ideal scaffolds for tissue engineering and wound healing by promoting cell attachment and proliferation [58]. Meanwhile, hydrogels made from polysaccharides form three-dimensional, hydrophilic networks capable of retaining large amounts of water, offering excellent platforms for controlled drug delivery, regenerative medicine, and biosensing [58].

Furthermore, the rich functional groups present in polysaccharides (hydroxyl, carboxyl, amino groups) facilitate chemical modification such as grafting, crosslinking, and conjugation with bioactive molecules, allowing customization of mechanical strength, degradation rate, and biological activity. This tunability enables the design of stimuli-responsive and targeted delivery systems.

In summary, polysaccharides serve as highly adaptable functional constructs with broad applications in nanomedicine, tissue engineering, and sustainable materials science, supported by ongoing research exploring their full potential.

9.1 Nanoparticles

Nanoparticles (NPs) are nanoscale materials, typically ranging from 1 to 100 nm in size, widely utilized as functional constructs in various fields including medicine, environmental science, and industry. Due to their small size and high surface-area-to-volume ratio, nanoparticles exhibit unique physicochemical properties such as enhanced reactivity, improved solubility, and the ability to cross biological barriers. These features make them excellent vehicles for drug delivery, diagnostics, imaging, and therapy [59].

In biomedical applications, polysaccharide-based nanoparticles—such as those fabricated from chitosan, alginate, dextran, or starch—offer biocompatibility, biodegradability, and reduced toxicity, making them preferable over synthetic polymeric nanoparticles. Functionalization of these nanoparticles with targeting ligands (e.g., antibodies, peptides) allows for site-specific delivery of therapeutic agents, enhancing efficacy and minimizing side effects. Additionally, surface modifications enable stimuli-responsive behavior, such as pH or temperature sensitivity, which triggers drug release at desired locations [3].

Nanoparticles can also integrate multiple functionalities; for instance, magnetic or gold nanoparticles are used for theranostics, combining therapy and diagnostics by enabling imaging-guided drug delivery or photothermal therapy. Furthermore, nanoparticles can be engineered for sustained and controlled drug release, improving patient compliance [60]. Their modular design permits customization to a wide range of therapeutic challenges, including cancer, infectious diseases, and gene therapy.

Overall, nanoparticles represent a versatile and powerful platform in nanomedicine and other advanced technologies, continually evolving with advances in material science and biotechnology.

9.2 Nanofibers

Nanofibers are ultrafine fibers with diameters typically below 500 nm, generated primarily through electrospinning and other fabrication techniques. They provide a unique combination of high surface area, porosity, and mechanical flexibility, which has made them highly valuable in biomedical engineering, filtration, sensing, and catalysis [29].

Polysaccharide-based nanofibers, such as those derived from chitosan, cellulose, and alginate, are particularly attractive due to their inherent biocompatibility, biodegradability, and bioactivity. In tissue engineering, nanofibers mimic the native extracellular matrix (ECM) by providing a fibrous scaffold that promotes cellular adhesion, proliferation, and differentiation [28]. This biomimicry supports tissue regeneration and wound healing, as nanofiber mats can also be loaded with drugs, growth factors, or antimicrobial agents to facilitate controlled release at the injury site.

Additionally, the surface chemistry of nanofibers can be tailored by chemical modification or co-electrospinning with other polymers or nanoparticles, imparting multifunctionality such as enhanced mechanical strength, electrical conductivity, or stimuli responsiveness. For example, nanofibers containing silver nanoparticles exhibit antimicrobial properties useful in wound dressings [61]. The porosity of nanofiber networks also makes them suitable for filtration applications, capturing particulates or biological agents effectively. The versatility and tunability of nanofibers continue to drive innovation in biomedical applications, environmental technologies, and advanced material development.

9.3 Hydrogels

Hydrogels are three-dimensional, hydrophilic polymer networks capable of absorbing and retaining large quantities of water or biological fluids. They closely resemble natural tissue environments; making them ideal functional constructs in drug delivery, wound healing, and tissue engineering [62].

Polysaccharide-based hydrogels, such as those formed from alginate, chitosan, pectin, and dextran—are particularly favored due to their biocompatibility, biodegradability, and tunable physical properties. These hydrogels can be engineered to respond to various stimuli, including pH, temperature, or enzymatic activity, enabling controlled and site-specific drug release [63]. For example, chitosan-based hydrogels show pH-sensitive swelling, allowing them to release drugs preferentially in acidic tumor environments or inflamed tissues.

In tissue engineering, hydrogels serve as scaffolds that provide a moist, protective environment facilitating cell infiltration and tissue regeneration. They can be reinforced with nanoparticles, nanofibers, or growth factors to enhance mechanical strength and biological activity. Injectable and self-healing hydrogels represent advanced classes that allow minimally invasive delivery and dynamic repair of tissue defects [64]. Moreover, hydrogel dressings maintain an optimal moist environment conducive to wound healing while providing antimicrobial and anti-inflammatory effects when loaded with appropriate agents. Their high water content also supports nutrient and oxygen transport critical for tissue viability.

Table 1 Applications of PNPs

SI. no	Application	Description	Polysaccharides used	Citation
1.	Drug delivery systems	Enhance drug solubility, bioavailability, and controlled release	Chitosan, alginate, dextran	[73, 74]
2.	Gene delivery	Safe, biodegradable vectors for gene therapy	Chitosan, hyaluronic acid	[74]
3.	Cancer therapy	Targeted delivery of anticancer agents, reduced side effects	Chitosan, pectin, starch	[75, 74]
4.	Vaccine delivery	Stabilization and controlled release of antigens	Chitosan, dextran	[73]
5.	Antimicrobial delivery	Controlled release of antimicrobial compounds for wound healing	Chitosan, alginate	[75]
6.	Tissue engineering	Scaffolds for cell growth and tissue regeneration	Chitosan, hyaluronic acid	[74]
7.	Food packaging	Edible coatings, antimicrobial films to enhance food shelf-life	Chitosan, starch	[73]
8.	Biosensors	Carrier matrices for enzymes or biomolecules in biosensing devices	Cellulose, alginate, chitosan	[74]
9.	Cosmetics and personal care	Controlled delivery of active compounds in creams and lotions	Chitosan, hyaluronic acid	[75]
10.	Environmental remediation	Removal of heavy metals and dyes from wastewater	Chitosan, cellulose	[74]

9.4 *Applications of Polysaccharide Nanoparticles*

PNPs have gained significant attention in recent years due to their natural abundance, biocompatibility, biodegradability, and functional versatility. Derived from natural polymers such as chitosan, alginate, cellulose, dextran, pectin, and starch, these nanoparticles are suitable for a wide range of applications across various sectors, including biomedicine, environmental science, food and packaging (Table 1 and Fig. 6).

9.5 *Biomedicine*

PNPs are extensively applied in biomedicine due to their biocompatibility, biodegradability, and ability to encapsulate therapeutic molecules (Table 1). Chitosan, alginate, and hyaluronic acid-based nanoparticles are used for drug delivery, tissue engineering, and vaccine delivery systems. Chitosan nanoparticles enhance mucosal drug delivery due to their cationic nature, promoting cellular uptake and controlled release [50]. Alginate nanoparticles protect sensitive drugs and enable sustained delivery in targeted therapy, such as in cancer treatment [56]. Moreover, PNPs serve as scaffolds

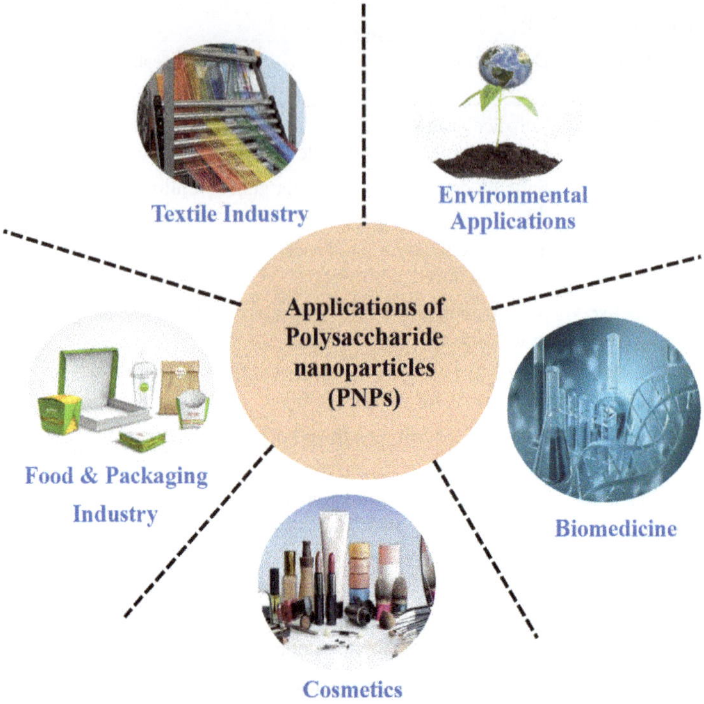

Fig. 6 The schematic diagram shows various applications of PNPs

in tissue engineering due to their ECM-mimicking structure and bioactivity, aiding in cell proliferation and differentiation [8]. These characteristics make PNPs promising candidates for advanced therapeutic applications.

9.6 Environmental Applications

PNPs offer eco-friendly solutions in environmental remediation. Their functional groups (–OH, –COOH, $–NH_2$) enable binding of heavy metals and dyes from wastewater (Table 1). Chitosan and cellulose-based nanoparticles, for instance, effectively remove contaminants like lead, cadmium, and methylene blue due to their high surface area and chelating ability [65]. Moreover, starch and alginate nanoparticles are biodegradable and do not introduce toxic residues, making them ideal for sustainable treatment processes. Nanoparticle-based biosorbents are also being developed for the removal of pesticides and pharmaceutical residues from water systems [66]. These materials contribute to low-cost, scalable, and green remediation technologies.

9.7 Food and Packaging Industry

In the food industry, PNPs are employed for the encapsulation of nutraceuticals, stabilizing bioactives such as polyphenols, vitamins, and probiotics (Table 1). Chitosan and pectin nanoparticles protect sensitive compounds from degradation and improve their bioavailability in functional foods [67]. In food packaging, polysaccharide-based nanocomposites exhibit excellent film-forming capabilities, antimicrobial activity, and barrier properties. For example, chitosan nanoparticles embedded in edible coatings extend the shelf life of perishable products by preventing microbial growth and moisture loss [68]. Such innovations support sustainable, biodegradable alternatives to synthetic plastics in packaging.

9.8 Textile Industry

Polysaccharide nanoparticles have emerged as multifunctional agents in the textile industry. They are used to impart antimicrobial, UV-protection, and anti-wrinkle properties to fabrics. Chitosan nanoparticles (Table 1), for instance, are applied to cotton textiles for antibacterial activity without affecting fabric quality, making them suitable for medical and hygiene-related textiles [69]. Additionally, these nanoparticles enhance dye uptake of natural dyes and eliminate the need for toxic mordants, promoting greener processing techniques [70]. Their film-forming and adhesive properties also improve the durability and wash-fastness of coatings applied on fibers.

9.9 Cosmetics

PNPs are widely used in cosmetic formulations for the controlled delivery of active ingredients such as antioxidants, vitamins, and anti-aging agents (Table 1). Chitosan and hyaluronic acid nanoparticles improve skin permeability and protect bioactives from oxidation and degradation [71]. Their moisturizing properties and biodegradability make them ideal for sensitive skin products. These nanoparticles can also be engineered for sustained release, enhancing the efficacy of skin-lightening, anti-wrinkle, and UV-protection formulations [72]. Additionally, their natural origin and non-toxicity align with the rising consumer demand for green and safe cosmetics.

9.10 Conclusions and Future Prospectives

PNPs represent a promising class of biomaterials owing to their biocompatibility, biodegradability, non-toxicity, and eco-friendliness. These nanoparticles, derived from renewable natural sources such as chitosan, alginate, cellulose, starch, and pectin, have shown great potential across a range of applications, including biomedicine, environmental remediation, food packaging, textiles, and cosmetics. Their versatile chemical structure allows for surface modifications, enabling targeted delivery, controlled release, and responsive behavior under physiological conditions. Moreover, their intrinsic biological activity and the ability to encapsulate diverse therapeutic and functional agents make PNPs valuable tools in nanomedicine and sustainable materials science. Future research should prioritize green, solvent-free, and energy-efficient synthesis techniques for PNPs. The use of water-based systems, enzymatic methods, and plant extracts for nanoparticle synthesis can minimize environmental impact and enhance scalability for industrial applications [73]. Emphasizing eco-friendly routes aligns with global sustainability goals and regulatory trends demanding safer materials. The integration of PNPs with inorganic or synthetic materials (e.g., metallic nanoparticles, carbon-based nanomaterials, synthetic polymers) can create hybrid nanostructures with synergistic properties. Such hybrids can offer improved mechanical strength, enhanced antimicrobial activity, and multifunctionality for use in sensors, smart drug delivery systems, and regenerative medicine [74]. Advancements in stimulus-responsive PNPs-capable of reacting to pH, temperature, light, or enzymes open the door for intelligent systems in diagnostics, precision medicine, and smart packaging. These "smart" PNPs can provide real-time feedback or controlled release of payloads in response to specific physiological or environmental triggers, greatly enhancing their functionality and efficacy [75]. In conclusion, PNPs are at the forefront of sustainable nanotechnology. With continued innovation in synthesis, functionalization, and application development, they are poised to revolutionize multiple industries while promoting environmental and human health.

Acknowledgements Sunil Kumar is thankful to Central Ayurveda Research Institute, Bhubaneswar Odisha, CCRAS, New Delhi for providing necessary facilities and grammarly tools during this study. Vinod Kumar and Surbhi Kharwar express their sincere gratitude to Pt. Prithi Nath (PG) College, Kanpur (affiliated to CSJM University) in Kanpur and the University of Lucknow for their invaluable support and encouragement.

Author Contribution The study was conceptualized by SK, SK, and VK. Together with SJ and SK found relevant literature and contributed to the manuscript's writing. SK and VK reviewed and made edits to the manuscript final draft.

Conflict of Interest Statement Every author state that there were no financial or commercial ties that would raise the possibility of a conflict of interest during the research.

References

1. Iravani, S.: Green synthesis of metal nanoparticles using plants. Green Chem. **13**(10), 2638–2650 (2011). https://doi.org/10.1039/C1GC15386B
2. Ibrahim Khan, K.S., Khan, I.: Nanoparticles: properties, applications and toxicities. Arab. J. Chem. **12**(7), 908–931 (2019)

3. Kumari, A., Yadav, S.K., Yadav, S.C.: Biodegradable polymeric nanoparticles based drug delivery systems. Colloids Surf. B **75**(1), 1–18 (2010). https://doi.org/10.1016/j.colsurfb.2009.09.001
4. Raveendran, P., Fu, J., Wallen, S.L.: Completely "green" synthesis and stabilization of metal nanoparticles. J. Am. Chem. Soc. **125**(46), 13940–13941 (2006). https://doi.org/10.1021/ja029267j
5. Chaturvedi, V.K., Singh, A., Singh, V.K., Singh, M.P.: Cancer nanotechnology: a new revolution for cancer diagnosis and therapy. Curr. Drug Metab. **20**(6), 416–429 (2019)
6. He, X., Liu, X., Yang, J., Du, H., Chai, N., Sha, Z., He, C.: Tannic acid-reinforced methacrylated chitosan/methacrylated silk fibroin hydrogels with multifunctionality for accelerating wound healing. Carbohyd. Polym. **247**, 116689 (2020)
7. Calvo, P., Remuñán-López, C., Vila-Jato, J.L., Alonso, M.J.: Chitosan and chitosan/ethylene oxide–propylene oxide block copolymer nanoparticles as novel carriers for proteins and vaccines. Pharm. Res. **14**(10), 1431–1436 (1997). https://doi.org/10.1023/A:1012155028404
8. Kumar, M.N.V.R., Muzzarelli, R., Muzzarelli, C., Sashiwa, H., Domb, A.J.: Chitosan chemistry and pharmaceutical perspectives. Chem. Rev. **104**(12), 6017–6084 (2004)
9. Isik, S., Yeman Kiyak, B., Akbayir, R., Seyhali, R., Arpaci, T.: Microglia mediated neuroinflammation in Parkinson's disease. Cells **12**(7), 1012 (2023)
10. Ji, X., Han, L., Liu, F., Yin, S., Peng, Q., Wang, M.: A mini-review of isolation, chemical properties and bioactivities of polysaccharides from buckwheat (Fagopyrum Mill). Int. J. Biol. Macromol. **127**, 204–209 (2019)
11. Hackenberg, S., Leitner, T., Jekle, M., Becker, T.: Maltose formation in wheat dough depending on mechanical starch modification and dough hydration. Carbohyd. Polym. **185**, 153–158 (2018)
12. Kurita, K.: Chitin and chitosan: functional biopolymers from marine crustaceans. Mar. Biotechnol. **8**(3), 203–226 (2006). https://doi.org/10.1007/s10126-005-0097-5
13. Klemm, D., Heublein, B., Fink, H.P., Bohn, A.: Cellulose: fascinating biopolymer and sustainable raw material. Angew. Chem. Int. Ed. **44**(22), 3358–3393 (2005). https://doi.org/10.1002/anie.200460587
14. Singh, N., Singh, J., Kaur, L., Sodhi, N.S., Gill, B.S.: Morphological, thermal and rheological properties of starches from different botanical sources. Food Chem. **81**(2), 219–231 (2003). https://doi.org/10.1016/S0308-8146(02)00416-8
15. Sriamornsak, P.: Chemistry of pectin and its pharmaceutical uses: a review. Silpakorn Univ. Int. J. **3**(1–2), 206–228 (2003)
16. Glicksman, M.: Gum technology in the food industry. Academic Press (1969)
17. Rinaudo, M.: Chitin and chitosan: properties and applications. Prog. Polym. Sci. **31**(7), 603–632 (2006). https://doi.org/10.1016/j.progpolymsci.2006.06.001
18. Bokkhim, H., Bansal, N., Grøndahl, L., Bhandari, B.: Characterization of alginate–lactoferrin beads prepared by extrusion gelation method. Food Hydrocolloids **53**, 270–276 (2016)
19. Campo, V.L., Kawano, D.F., da Silva, D.B., Carvalho, I.: Carrageenans: biological properties, chemical modifications and structural analysis—a review. Carbohyd. Polym. **77**(2), 167–180 (2009). https://doi.org/10.1016/j.carbpol.2009.01.020
20. Le Barbier, C.: Mixing medicines with foods and drinks-latest regulatory considerations. Int. J. Pharm. **457**(1), 341–342 (2013)
21. Leathers, T.D.: Biotechnological production and applications of pullulan. Appl. Microbiol. Biotechnol. **62**, 468–473 (2003). https://doi.org/10.1007/s00253-003-1323-2
22. Sworn, G.: Xanthan gum. In: Phillips, G.O., Williams, P.A. (eds.), Handbook of Hydrocolloids, 2nd ed., pp. 186–203. Woodhead Publishing (2009)
23. Draget, K.I., Skjåk-Bræk, G., Stokke, B.T.: Similarities and differences between alginic acid gels and ionically crosslinked alginate gels. Food Hydrocoll. **20**(2–3), 170–175 (2006)
24. Lee, K.Y., Mooney, D.J.: Alginate: properties and biomedical applications. Prog. Polym. Sci. **37**(1), 106–126 (2012). https://doi.org/10.1016/j.progpolymsci.2011.06.003
25. Moon, R.J., Martini, A., Nairn, J., Simonsen, J., Youngblood, J.: Cellulose nanomaterials review: structure, properties and nanocomposites. Chem. Soc. Rev. **40**(7), 3941–3994 (2011). https://doi.org/10.1039/c0cs00108b

26. Dahiya, D., Nigam, P.S.: Dextran of diverse molecular-configurations used as a blood-plasma substitute, drug-delivery vehicle and food additive biosynthesized by Leuconostoc, Lactobacillus and Weissella. Appl. Sci. **13**(22), 12526 (2023)
27. Vila, A., Sánchez, A., Janes, K., Behrens, I., Kissel, T., Jato, J.L.V., Alonso, M.J.: Low molecular weight chitosan nanoparticles as new carriers for nasal vaccine delivery in mice. Eur. J. Pharm. Biopharm. **57**(1), 123–131 (2004)
28. Bhattarai, N., Edmondson, D., Veiseh, O., Matsen, F.A., Zhang, M.: Electrospun chitosan-based nanofibers and their cellular compatibility. Biomaterials **26**(31), 6176–6184 (2005)
29. Greiner, A., Wendorff, J.H.: Electrospinning: a fascinating method for the preparation of ultra-thin fibers. Angew. Chem. Int. Ed. **46**(30), 5670–5703 (2007). https://doi.org/10.1002/anie.200604646
30. Lu, T., Li, Y., Chen, T.: Techniques for fabrication and construction of three-dimensional scaffolds for tissue engineering. Int. J. Nanomed. 337–350 (2013)
31. Khan, Y., Sadia, H., Ali Shah, S.Z., Khan, M.N., Shah, A.A., Ullah, N., Khan, M.I.: Classification, synthetic, and characterization approaches to nanoparticles, and their applications in various fields of nanotechnology: a review. Catalysts **12**(11), 1386 (2022)
32. Gillespie, R.J., Adams, D.R., Bebbington, D., Benwell, K., Cliffe, I.A., Dawson, C.E., Williamson, D.S.: Antagonists of the human adenosine A2A receptor. Part 1: discovery and synthesis of thieno [3, 2-d] pyrimidine-4-methanone derivatives. Bioorganic Med. Chem. Lett. **18**(9), 2916–2919 (2008)
33. Calderón, M., Welker, P., Licha, K., Fichtner, I., Graeser, R., Haag, R., Kratz, F.: Development of efficient acid cleavable multifunctional prodrugs derived from dendritic polyglycerol with a poly (ethylene glycol) shell. J. Control. Release **151**(3), 295–301 (2011)
34. Li, Y., Maciel, D., Rodrigues, J., Shi, X., Tomás, H.: Biodegradable polymer nanogels for drug/nucleic acid delivery. Chem. Rev. **115**(16), 8564–8608 (2015)
35. Sill, T.J., Von Recum, H.A.: Electrospinning: applications in drug delivery and tissue engineering. Biomaterials **29**(13), 1989–2006 (2008)
36. Dan, X., Chunhua, L., Dunpu, Z., Jianbin, S., Yaru, N., Zhongzi, X.: Fabrication and optical properties of self-assembled films with micro needlelike arrays under magnetic field. Mater. Lett. **71**, 94–97 (2012)
37. Piras, C.C., Smith, D.K.: Multicomponent polysaccharide alginate-based bioinks. J. Mater. Chem. B **8**(36), 8171–8188 (2020)
38. Bamberger, D., Hobernik, D., Konhäuser, M., Bros, M., Wich, P.R.: Surface modification of polysaccharide-based nanoparticles with PEG and dextran and the effects on immune cell binding and stimulatory characteristics. Mol. Pharm. **14**(12), 4403–4416 (2017)
39. Uthaman, S., Cherukula, K., Cho, K., Jeong, Y.Y., Park, I.K.: Polysaccharide-coated magnetic nanoparticles for imaging and gene therapy. Biomed. Res. Int. **2015**, 959175 (2015). https://doi.org/10.1155/2015/959175
40. Yang, W., Zhuang, Y., Wang, X., Tan, Y., Li, Z., Liu, Q., Xia, H.: Nanoparticle surface engineering with heparosan polysaccharide reduces serum protein adsorption and enhances cellular uptake. Nano Lett. **22**(5), 2103–2111 (2022). https://doi.org/10.1021/acs.nanolett.2c00349
41. Zhang, M., Liu, W., Gao, X., Cui, P., Zou, T., Hu, G., Zhai, L.: Preparation and characterization of semi-alicyclic polyimides containing trifluoromethyl groups for optoelectronic application. Polymers **12**(7), 1532 (2020)
42. Li, J., Xiang, H., Zhang, Q., Miao, X.: Polysaccharide-based transdermal drug delivery. Pharmaceuticals **15**(5), 602 (2022)
43. Pan, Y., Li, Y.J., Zhao, H.Y., Zheng, J.M., Xu, H., Wei, G., Hao, J.S.: Bioadhesive polysaccharide in protein delivery system: chitosan nanoparticles improve the intestinal absorption of insulin in vivo. Int. J. Pharm. **249**(1–2), 139–147 (2002). https://doi.org/10.1016/S0378-5173(02)00480-2
44. Pattanayak, R., Basak, P., Sen, S., Bhattacharyya, M.: Interaction of KRAS G-quadruplex with natural polyphenols: a spectroscopic analysis with molecular modeling. Int. J. Biol. Macromol. **89**, 228–237 (2016)

45. Sirisha, V.L., D'Souza, J.S.: Polysaccharide-based nanoparticles as drug delivery systems. In Marine OMICS, pp. 663–702. CRC Press (2016)
46. Karayianni, M., Sentoukas, T., Skandalis, A., Pippa, N., Pispas, S.: Chitosan-based nanoparticles for nucleic acid delivery: technological aspects, applications, and future perspectives. Pharmaceutics **15**(7), 1849 (2023)
47. De Anda-Flores, Y., Carvajal-Millan, E., Campa-Mada, A., Lizardi-Mendoza, J., Rascon-Chu, A., Tanori-Cordova, J., Martínez-López, A.L.: Polysaccharide-based nanoparticles for colon-targeted drug delivery systems. Polysaccharides **2**(3), 626–647 (2021)
48. Kang, B.K., Lee, J.S., Chon, S.K., Jeong, S.Y., Yuk, S.H., Khang, G., Cho, S.H.: Development of self-microemulsifying drug delivery systems (SMEDDS) for oral bioavailability enhancement of simvastatin in beagle dogs. Int. J. Pharm. **274**(1–2), 65–73 (2004)
49. Guo, H., Su, Y., Guo, C., Chen, Q., Liu, Z., Geng, H., Chen, D.: Polysaccharide based drug delivery systems for Chinese medicines. Biocatal. Agric. Biotechnol. **44**, 102441 (2022)
50. Sogias, I.A., Williams, A.C., Khutoryanskiy, V.V.: Why is chitosan mucoadhesive? Biomacromol **9**(7), 1837–1842 (2008)
51. Masood, F.: Polymeric nanoparticles for targeted drug delivery system for cancer therapy. Mater. Sci. Eng. C **60**(Suppl C), 569–578 (2016)
52. Alvarez Echazú, M.I., Renou, S.J., Alvarez, G.S., Desimone, M.F., Olmedo, D.G.: Synthesis and evaluation of a chitosan–silica-based bone substitute for tissue engineering. Int. J. Mol. Sci. **23**(21), 13379 (2022)
53. Zhao, L., Seth, A., Wibowo, N., Zhao, C.X., Mitter, N., Yu, C., Middelberg, A.P.J.: Nanoparticle vaccines. Vaccine **32**(3), 327–337 (2014). https://doi.org/10.1016/j.vaccine.2013.11.069
54. Bhatia, S.K., Gurav, R., Choi, T.R., Kim, H.J., Yang, S.Y., Song, H.S., Yang, Y.H.: Conversion of waste cooking oil into biodiesel using heterogenous catalyst derived from cork biochar. Biores. Technol. **302**, 122872 (2020)
55. George, M., Abraham, T.E.: Polyionic hydrocolloids for the intestinal delivery of protein drugs: alginate and chitosan—a review. J. Control. Release **114**(1), 1–14 (2006). https://doi.org/10.1016/j.jconrel.2006.04.017
56. Šebenik, U., Lapasin, R., Krajnc, M.: Rheology of aqueous dispersions of Laponite and TEMPO-oxidized nanofibrillated cellulose. Carbohyd. Polym. **240**, 116330 (2020)
57. Sashiwa, H., Aiba, S.: Chemically modified chitin and chitosan as biomaterials. Prog. Polym. Sci. **29**(9), 887–908 (2004). https://doi.org/10.1016/j.progpolymsci.2004.04.002
58. Shao, P., Shao, J., Han, L., Lv, R., Sun, P.: Separation, preliminary characterization, and moisture-preserving activity of polysaccharides from Ulva fasciata. Int. J. Biol. Macromol. **72**, 924–930 (2015)
59. Farokhzad, O.C., Langer, R.: Impact of nanotechnology on drug delivery. ACS Nano **3**(1), 16–20 (2009). https://doi.org/10.1021/nn900002m
60. Peer, D., Karp, J.M., Hong, S., Farokhzad, O.C., Margalit, R., Langer, R.: Nanocarriers as an emerging platform for cancer therapy. Nano-Enabled Med. Appl. 61–91 (2020)
61. Goh, Y.F., Shakir, I., Hussain, R.: Electrospun fibers for tissue engineering, drug delivery, and wound dressing. J. Mater. Sci. **48**, 3027–3054 (2013)
62. Peppas, N.A., Bures, P., Leobandung, W., Ichikawa, H.: Hydrogels in pharmaceutical formulations. Eur. J. Pharm. Biopharm. **50**(1), 27–46 (2000). https://doi.org/10.1016/S0939-6411(00)00094-2
63. Hoffman, A.S.: Hydrogels for biomedical applications. Adv. Drug Deliv. Rev. **64**, 18–23 (2012). https://doi.org/10.1016/j.addr.2012.09.010
64. Li, J., Mooney, D.J.: Designing hydrogels for controlled drug delivery. Nat. Rev. Mater. **1**(12), 1–17 (2016)
65. Varma, A.J., Deshpande, S.V., Kennedy, J.F.: Metal complexation by chitosan and its derivatives: a review. Carbohyd. Polym. **55**(1), 77–93 (2004)
66. Bolisetty, S., Peydayesh, M., Mezzenga, R.: Sustainable technologies for water purification from heavy metals: review and analysis. Chem. Soc. Rev. **48**(2), 463–487 (2019)
67. Wang, C., Gao, X., Chen, Z., Chen, Y., Chen, H.: Preparation, characterization and application of polysaccharide-based metallic nanoparticles: a review. Polymers **9**(12), 689 (2017)

68. Sowmya, S., Bumgardener, J.D., Chennazhi, K.P., Nair, S.V., Jayakumar, R.: Role of nanostructured biopolymers and bioceramics in enamel, dentin and periodontal tissue regeneration. Prog. Polym. Sci. **38**(10–11), 1748–1772 (2013)
69. Perelshtein, I., Applerot, G., Perkas, N., Wehrschetz-Sigl, E., Hasmann, A., Guebitz, G.M., Gedanken, A.: Antibacterial properties of an in situ generated and simultaneously deposited nanocrystalline ZnO on fabrics. ACS Appl. Mater. Interfaces **1**(2), 361–366 (2009)
70. Tang, H., Han, Z., Zhao, C., Jiang, Q., Tang, Y., Li, Y., Cheng, Z.: Preparation and characterization of Aloe vera polysaccharide-based packaging film and its application in blueberry preservation. Prog. Org. Coat. **177**, 107445 (2023)
71. Hasnain, M.S., Nayak, A.K. (eds.): Natural polysaccharides in drug delivery and biomedical applications. Academic Press (2019)
72. Plucinski, A., Lyu, Z., Schmidt, B.V.: Polysaccharide nanoparticles: from fabrication to applications. J. Mater. Chem. B **9**(35), 7030–7062 (2021)
73. Mody, V.V., Siwale, R., Singh, A., Mody, H.R.: Introduction to metallic nanoparticles. J. Pharm. Bioallied Sci. **2**(4), 282–289 (2010). https://doi.org/10.4103/0975-7406.72127
74. Shokrani, H., Shokrani, A., Sajadi, S.M., Yazdi, M.K., Seidi, F., Jouyandeh, M., Ramakrishna, S.: Polysaccharide-based nanocomposites for biomedical applications: a critical review. Nanoscale Horiz. **7**(10), 1136–1160 (2022)
75. Ahmed, E.M.: Hydrogel: preparation, characterization, and applications: a review. J. Adv. Res. **6**(2), 105–121 (2015). https://doi.org/10.1016/j.jare.2013.07.006

Nanoengineering of Polymer Particles for Traversing Bio, Nano Interactions

Saahil Hembrom, Soumya Pandey, and Neeta Kumari

Abstract Polymer nanoparticles have emerged as transformative tools at the interface of nanotechnology and biomedicine, driven by advances in nanoengineering that enable precise control over their size, shape, and surface functionality. This chapter critically explores how these engineered particles traverse biological barriers and respond to complex physiological cues, transforming passive carriers into dynamic agents for targeted therapy, diagnostics, and real, time monitoring. Key design strategies such as emulsion polymerization, nanoprecipitation, and microfluidics enable tunable control over particle architecture, while surface modifications, through ligand conjugation, PEGylation, and zwitterionic coatings, extend circulation times and enhance targeting precision. Responsive and smart polymers, capable of reacting to pH, redox potential, and enzymatic activity, and innovations like "sheddable" stealth coatings further improve site, specific delivery. Core–shell and multilayer architectures, assembled via techniques like layer, by, layer deposition and hybrid inorganic–polymer design, integrate therapeutic and imaging functions within single nano systems. Case studies highlight their practical translation: crossing the blood–brain barrier with ligand, targeted carriers, leveraging the EPR effect in tumours, penetrating mucosal barriers, and overcoming endosomal entrapment to deliver payloads directly into the cytosol. Applications extend from targeted drug and gene delivery, including siRNA, mRNA, and CRISPR systems, to theranostics and biosensing platforms capable of real, time disease monitoring. Despite these advances, challenges in reproducibility, large, scale manufacturing, and in vitro versus in vivo discrepancies remain significant hurdles. Emerging trends such as AI, guided nanoparticle design, biomimetic and programmable particles, and a stronger integration of data science promise to bridge these gaps, propelling the field toward truly personalised and adaptive nanomedicine. Ultimately, the future success of polymer nanoparticles relies on interdisciplinary collaboration bringing together chemists, engineers, clinicians, and regulatory scientists to translate laboratory breakthroughs into safe, scalable, and

S. Hembrom · N. Kumari (✉)
Department of Civil and Environmental Engineering, Birla Institute of Technology, Ranchi, Jharkhand, India
e-mail: neetak@bitmesra.ac.in

S. Pandey
Department of Civil Engineering, Kalinga University, Raipur, Chhattisgarh, India

B. Sharma et al. (eds.), *Nanostructured Polymers*, Engineering Materials,
https://doi.org/10.1007/978-981-95-3581-1_9

clinically effective therapies, reshaping the paradigm of diagnosis and treatment in modern medicine.

Keywords Bio · Nano interactions · Biomimetic particles · Drug delivery · Gene delivery · Polymer nanoparticles · Responsive polymers · Theranostics

1 Introduction

In recent decades, polymer particles have emerged as a crucial class of materials in nanotechnology and biomedicine, driven by their versatility in chemical design, tunable physical properties, and biocompatibility [1]. These particles, often ranging from tens to hundreds of nanometers, are synthesised using diverse techniques such as emulsion polymerisation, self, assembly, and microfluidic methods, allowing precise control over size, shape, and surface chemistry. Their applications span targeted drug delivery, imaging, vaccine platforms, and tissue engineering scaffolds, where the nanoscale dimension offers advantages like prolonged circulation, enhanced permeability and retention (EPR) effects, and cellular internalisation [2]. Unlike inorganic nanoparticles, polymer particles can be engineered to degrade under physiological conditions, releasing therapeutic payloads in a controlled fashion while minimising toxicity. Polymers such as poly (lactic, co, glycolic acid) (PLGA), polyethene glycol (PEG), and chitosan have become cornerstones of this field due to their FDA approval history and favourable safety profiles [3]. Yet, despite substantial advancements, translating polymer, based nanosystems from laboratory prototypes to clinical use remains challenging, highlighting a persistent gap between design and real, world functionality. This tension highlights the increasing need for nanoengineering—a multidisciplinary approach that strategically integrates polymer chemistry, molecular biology, and computational modelling to tailor particles at the nanoscale for specific biological environments.

The need for advanced nanoengineering stems from the complex and dynamic nature of biological systems that nanoparticles must navigate. Once administered, polymer particles encounter a series of biological barriers: protein adsorption forming the "protein corona," opsonisation leading to rapid clearance by the mononuclear phagocyte system, cellular uptake regulated by receptor, mediated pathways, and intracellular trafficking that can redirect particles to degradation pathways rather than therapeutic targets. Each of these steps presents unique challenges [1]. For instance, the protein corona can drastically alter the intended targeting capability of functionalized surfaces, as the biological identity perceived by cells often differs from the synthetic design. While particle size and shape influence biodistribution, smaller particles (<100 nm) often penetrate tumours more effectively but may also face faster renal clearance, whereas larger particles enjoy prolonged circulation yet risk accumulation in the liver and spleen [3]. The surface charge of polymer particles also modulates cell interaction; cationic particles typically achieve higher internalisation but can induce toxicity and inflammatory responses. Nanoengineering addresses these challenges by designing "stealth" coatings, pH, responsive or redox, responsive release systems, and surface ligands that exploit disease, specific biomarkers, all grounded in molecular, level understanding of bio, nano interactions [4]. Innovations such as controlled polymer architectures (e.g., star polymers, dendrimers) and hybrid particles combining organic polymers with inorganic cores further illustrate the sophistication required to navigate biological complexity [5].

At the heart of these engineering efforts lies the concept of "bio, nano interactions," an umbrella term encompassing the multitude of physical, chemical, and biological processes occurring when nanostructures interface with living systems. This includes interactions at molecular (protein–particle binding), cellular (endocytosis, exocytosis), tissue (extravasation through vascular walls), and systemic (immune response modulation) levels [4]. Critically, these interactions are reciprocal: while polymer particles are engineered to achieve specific biological functions, the biological environment actively shapes particle fate through dynamic processes such as enzymatic degradation, immune surveillance, and mechanical transport. The scope of bio, nano interactions is therefore not limited to in vitro assessments of cytotoxicity or cellular uptake, but extends to in vivo pharmacokinetics, immunogenicity, and long, term biodistribution [6]. For example, the immune system's recognition of "foreign" nanoscale objects can trigger complement activation or production of anti, PEG antibodies, compromising therapeutic efficacy and safety. Similarly, differences in cellular receptor expression across healthy and diseased tissues can lead to highly variable targeting performance, complicating clinical translation [7]. Understanding these interactions demands advanced characterisation tools from high, resolution microscopy and single, particle tracking to proteomic analyses of the protein corona and computational models that can predict biological behaviour from physicochemical properties [8].

A critical analysis of current trends reveals both achievements and limitations in the field. On one hand, polymer particle design has progressed remarkably, yielding systems capable of co, delivering multiple therapeutics, stimuli, responsive release, and real, time imaging. On the other hand, the biological complexity remains underestimated in many preclinical models [9]. Conventional in vitro assays often fail to capture the influence of blood flow, immune cell heterogeneity, and disease microenvironmental factors, leading to discrepancies between preclinical promise and clinical outcomes. Furthermore, while surface functionalization strategies have become increasingly elaborate—using antibodies, peptides, and small molecules for active targeting—the real impact on therapeutic index has sometimes been modest due to protein corona masking or off, target effects [10]. Emerging strategies are thus shifting toward "immune, instructive" materials that modulate host responses rather than simply evading them, and "smart" particles that adapt their behaviour dynamically in response to local cues. Such approaches exemplify the maturation of nanoengineering from passive carriers to active participants in biological systems [11].

The ethical and regulatory considerations are increasingly shaping the future of polymer, based nanomedicine. The translation of nanoengineered particles necessitates rigorous characterisation of reproducibility, scalability, and long, term safety, areas where academia and industry must collaborate closely [12]. Recent advances in artificial intelligence and machine learning hold promise to accelerate this process, enabling data, driven optimisation of particle design by learning from large experimental datasets on bio, nano interactions [13]. Ultimately, the intersection of polymer science, biology, and engineering is transforming our approach to complex diseases, from cancer to neurodegenerative disorders, where precise modulation of biological pathways at the nanoscale offers unprecedented therapeutic possibilities [14].

The polymer particles occupy a pivotal role in modern nanotechnology and biomedicine due to their customizability and functional versatility. Yet, the intricate challenges posed by the biological milieu highlight the critical need for nanoengineering an interdisciplinary effort to design particles capable of predictable and controllable interactions with complex living systems [15]. The concept of bio, nano interactions, encompassing every level from molecular binding events to systemic immune responses, provides a unifying framework to understand and innovate in this field. Continued progress depends not only on novel materials and sophisticated design strategies but equally on deep, quantitative insights into how nanoscale objects and biological systems mutually influence each other, a dialogue at the heart of nanomedicine's promise and its challenges [16].

2 Fundamentals of Bio, Nano Interactions

In recent decades, polymer particles have emerged as an essential class of materials in nanotechnology and biomedicine, driven by their versatility in chemical design, tunable physical properties, and biocompatibility. These particles, often ranging from tens to hundreds of nanometers, are synthesised using diverse techniques such as emulsion polymerisation, self, assembly, and microfluidic methods, allowing precise control over size, shape, and surface chemistry [17]. Their applications span targeted drug delivery, imaging, vaccine platforms, and tissue engineering scaffolds, where the nanoscale dimension offers advantages like prolonged circulation, enhanced permeability and retention (EPR) effects, and cellular internalisation. Unlike inorganic nanoparticles, polymer particles can be engineered to degrade under physiological conditions, releasing therapeutic payloads in a controlled fashion while minimising toxicity [18]. Polymers such as poly (lactic-co-glycolic acid) (PLGA), polyethene glycol (PEG), and chitosan have become cornerstones of this field due to their FDA-approved history and favourable safety profiles [19]. Yet, despite substantial advancements, translating polymer, based nanosystems from laboratory prototypes to clinical use remains challenging, highlighting a persistent gap between design and real, world functionality [20]. This tension highlights the increasing need for nanoengineering—a multidisciplinary approach that strategically integrates polymer chemistry, molecular biology, and computational modelling to tailor particles at the nanoscale for specific biological environments [21].

The need for advanced nanoengineering stems from the complex and dynamic nature of biological systems that nanoparticles must navigate. Once administered, polymer particles encounter a series of biological barriers: protein adsorption forming the "protein corona," opsonisation leading to rapid clearance by the mononuclear phagocyte system, cellular uptake regulated by receptor, mediated pathways, and intracellular trafficking that can redirect particles to degradation pathways rather than therapeutic targets [22]. Each of these steps presents unique challenges. For instance, the protein corona can drastically alter the intended targeting capability of functionalized surfaces, as the biological identity perceived by cells often differs from the synthetic design [23]. Furthermore, particle size and shape influence biodistribution: smaller particles (<100 nm) often penetrate tumours more effectively but may also face faster renal clearance, whereas larger particles enjoy prolonged circulation yet risk accumulation in the liver and spleen. The surface charge of polymer particles also modulates cell interaction; cationic particles typically achieve higher internalisation but can induce toxicity and inflammatory responses [24]. Nanoengineering addresses these challenges by designing "stealth" coatings, pH, responsive or redox, responsive release systems, and surface ligands that exploit disease, specific biomarkers, all grounded in molecular, level understanding of bio, nano interactions [25]. Innovations such as controlled polymer architectures (e.g., star polymers, dendrimers) and hybrid particles combining organic polymers with inorganic cores further illustrate the sophistication required to navigate biological complexity [26].

At the heart of these engineering efforts lies the concept of "bio, nano interactions," an umbrella term encompassing the multitude of physical, chemical, and biological processes occurring when nanostructures interface with living systems. This includes interactions at molecular (protein–particle binding), cellular (endocytosis, exocytosis), tissue (extravasation through vascular walls), and systemic (immune response modulation) levels [27]. Critically, these interactions are reciprocal: while polymer particles are engineered to achieve specific biological functions, the biological environment actively shapes particle fate through dynamic processes such as enzymatic degradation, immune surveillance, and mechanical transport. The scope of bio, nano interactions is therefore not limited to in vitro assessments of cytotoxicity or cellular uptake, but extends to in vivo pharmacokinetics, immunogenicity, and long, term biodistribution [28]. For example, the immune system's recognition of "foreign" nanoscale objects can trigger complement activation or production of anti, PEG antibodies, compromising therapeutic efficacy and safety. Similarly, differences in cellular receptor expression across healthy and diseased tissues can lead to highly variable targeting performance, complicating clinical translation [7]. Understanding these interactions demands advanced characterisation tools from high, resolution microscopy and single, particle tracking to proteomic analyses of the protein corona and computational models that can predict biological behaviour from physicochemical properties [29].

A critical analysis of current trends reveals both achievements and limitations in the field. On one hand, polymer particle design has progressed remarkably, yielding systems capable of co, delivering multiple therapeutics, stimuli, responsive release, and real, time imaging. On the other hand, the biological complexity remains underestimated in many preclinical models [9]. Conventional in vitro assays often fail to capture the influence of blood flow, immune cell heterogeneity, and disease microenvironmental factors, leading to discrepancies between preclinical promise and clinical outcomes. Furthermore, while surface functionalization strategies have become increasingly elaborate using antibodies, peptides, and small molecules for active targeting, the real impact on therapeutic index has sometimes been modest due to protein corona masking or off, target effects [30]. Emerging strategies are thus shifting toward "immune, instructive" materials that modulate host responses rather than simply evading them, and "smart" particles that adapt their behaviour dynamically in response to local cues. Such approaches exemplify the maturation of nanoengineering from passive carriers to active participants in biological systems [31].

Moreover, ethical and regulatory considerations are increasingly shaping the future of polymer, based nanomedicine. The translation of nanoengineered particles necessitates rigorous characterisation of reproducibility, scalability, and long, term safety, areas where academia and industry must collaborate closely [32]. Recent advances in artificial intelligence and machine learning hold promise to accelerate this process, enabling data, driven optimisation of particle design by learning from large experimental datasets on bio, nano interactions. Ultimately, the intersection of polymer science, biology, and engineering is transforming our approach to complex diseases, from cancer to neurodegenerative disorders, where precise modulation of biological pathways at the nanoscale offers unprecedented therapeutic possibilities [33].

3 Strategies for Nanoengineering of Polymer Particles

3.1 Size and Shape Control

A central strategy in nanoengineering polymer particles for biomedical applications revolves around the precise control of their size and shape, since these parameters profoundly dictate in vivo fate, circulation time, and biodistribution. Modern synthesis techniques have evolved to meet this need, each offering unique capabilities and limitations [34]. Emulsion polymerisation, for instance, is a widely used technique where monomers polymerise within nanoscale droplets stabilised by surfactants in an aqueous phase. By adjusting surfactant concentration, monomer feed, and reaction kinetics, researchers can finely tune particle sizes, often within the 50–500 nm range [35]. This method has been especially valuable for producing spherical particles with relatively narrow polydispersity indices, supporting consistent performance in drug delivery applications [36]. Yet, traditional emulsion approaches are limited in shaping particles beyond simple spheres, prompting interest in alternative techniques.

Nanoprecipitation offers another versatile route, relying on the rapid mixing of a polymer solution in a solvent with a non, solvent, leading to spontaneous nanoparticle formation via solvent displacement. This technique allows quick fabrication of nanoparticles typically between 50 and 300 nm, suitable for encapsulating hydrophobic drugs, and offers relatively straightforward scalability. Importantly, the process parameters, such as polymer concentration, solvent/non, solvent ratio, and stirring speed enable further refinement of particle size and even some influence on morphology. Despite these advantages, nanoprecipitation is often restricted to creating spherical particles, and the dynamic nature of precipitation can sometimes produce broader size distributions if mixing isn't optimised [37].

To overcome the limitations of shape control inherent in bulk processes, microfluidics has emerged as a transformative nanoengineering tool. Microfluidic devices channel fluids through micron, scale geometries, enabling highly controlled mixing and particle assembly. By manipulating flow rates, shear stress, and channel design, researchers can not only produce particles with highly uniform sizes but also generate non, spherical morphologies such as rods, disks, or even more complex anisotropic shapes. This level of precision is unmatched by traditional batch methods, and recent advances allow the production of particles down to tens of nanometers with tunable aspect ratios. Microfluidics also supports continuous production, which holds promise for clinical translation and industrial scalability [38].

The significance of controlling size and shape through these methods becomes clear when examining their biological implications. Smaller nanoparticles (typically<100 nm) often exhibit prolonged circulation times because they evade rapid clearance by the spleen and liver, while their small size facilitates deeper penetration into tumours via the enhanced permeability and retention (EPR) effect. Conversely, larger particles may show greater accumulation in organs like the liver and spleen, where resident macrophages filter out foreign materials [39]. Beyond size, shape has emerged as a critical, though historically underappreciated, design parameter. Rod, shaped or filamentous nanoparticles often demonstrate superior margination towards vascular walls and reduced uptake by macrophages compared to spherical counterparts, resulting in longer circulation and potentially improved targeting efficiency. Disc, like particles can also exhibit unique flow dynamics, sometimes favouring accumulation at disease sites [40].

Importantly, these effects are not merely academic observations—they translate into measurable therapeutic benefits. For instance, nanoparticles engineered via microfluidics to have elongated shapes have shown enhanced tumour targeting in animal models, attributed to both prolonged systemic circulation and preferential alignment with blood flow. Meanwhile, controlling particle size through emulsion polymerisation or nanoprecipitation can directly affect pharmacokinetics, reducing off, target toxicity by minimising rapid clearance and improving accumulation in disease tissues [41].

Thus, the convergence of size and shape control through advanced nanoengineering techniques such as emulsion polymerisation, nanoprecipitation, and microfluidics represents a cornerstone of modern nanomedicine. It illustrates how subtle manipulation at the nanoscale can yield substantial gains in therapeutic efficacy and safety, bridging the gap between synthetic design and complex biological realities. By continually refining these techniques and deepening our understanding of bio, nano interactions, researchers are moving closer to realising truly personalised, efficient, and clinically translatable polymer nanoparticle systems [42].

3.2 Surface Functionalization

Surface functionalization represents one of the most powerful strategies in nanoengineering, transforming passive polymer particles into sophisticated bioresponsive tools capable of navigating the complex biological milieu. At the forefront of these modifications is ligand conjugation, wherein bioactive molecules such as antibodies, peptides, or sugars are covalently attached to the nanoparticle surface to endow them with targeting capabilities [43]. Antibodies, with their high specificity and affinity for cell, surface antigens, enable nanoparticles to recognize and bind selectively to diseased cells, such as cancer cells overexpressing HER2 or EGFR. Peptides, being smaller and synthetically versatile, offer similar targeting precision but often with reduced immunogenicity and better tissue penetration, making them attractive for diseases where deep or diffuse tissue targeting is required [21]. Sugar moieties, or glycan structures, tap into lectin, mediated pathways, allowing nanoparticles to exploit natural carbohydrate recognition mechanisms; for instance, mannose, functionalized nanoparticles can target macrophages, while sialic acid modifications can modulate interactions with the immune system. While ligand conjugation has shown substantial promise, its effectiveness hinges critically on controlling ligand density, orientation, and stability: overly dense ligand coverage can trigger rapid clearance or off, target binding, whereas sparse or poorly oriented ligands may fail to achieve sufficient affinity [44, 45].

Parallel to active targeting, stealth strategies such as PEGylation have been widely adopted to prolong circulation times and reduce immune clearance. Polyethene glycol (PEG), a hydrophilic and flexible polymer, forms a hydrated shell around nanoparticles, sterically hindering adsorption of plasma proteins (opsonins) that otherwise mark particles for phagocytosis by the mononuclear phagocyte system [46]. This stealth effect significantly enhances pharmacokinetic profiles, allowing nanoparticles to remain in circulation longer and increasing their chances of reaching target tissues through the enhanced permeability and retention (EPR) effect. PEGylation has underpinned the clinical translation of several nanoparticle, based therapeutics; however, it is not without limitations [47]. Recent studies have highlighted the emergence of anti, PEG antibodies in some patients, which can lead to accelerated blood clearance upon repeated dosing and potential hypersensitivity reactions. These findings have motivated research into alternative stealth strategies that retain the benefits of PEG while mitigating immune responses [10].

One such alternative is the use of zwitterionic coatings—materials bearing both positive and negative charges nearby on the same molecule, resulting in a net neutral surface charge. Zwitterionic polymers, such as phosphorylcholine, sulfobetaine, and carboxybetaine, mimic the outer surface of naturally zwitterionic cell membranes, thereby promoting biocompatibility and resistance to non, specific protein adsorption [48]. The unique hydration properties of these materials, arising from strong ionic solvation, create an energetically unfavourable environment for protein binding, effectively minimising protein corona formation and subsequent immune recognition. Compared to PEG, zwitterionic coatings have demonstrated even lower levels of non, specific adsorption in some studies and may offer greater stability in complex biological fluids. Additionally, because zwitterionic materials are structurally similar to natural cell surfaces, they are generally less immunogenic, reducing the risk of anti, polymer antibody production [49].

Critically, these surface functionalization strategies are often combined to harness synergistic effects. For example, a nanoparticle may be simultaneously PEGylated for stealth, functionalized with targeting peptides for receptor, mediated uptake, and carry zwitterionic segments to suppress residual protein binding. Yet, integrating multiple functionalities is not trivial: each modification can alter particle size, surface charge, and hydrophobicity, potentially affecting biodistribution and cellular uptake in unpredictable ways. Moreover, the dynamic nature of the protein corona can still mask targeting ligands, highlighting the ongoing challenge of ensuring that engineered functionalities remain biologically active in vivo [50, 51].

Ultimately, surface functionalization transforms polymer nanoparticles from passive carriers into actively navigating and communicating agents within biological systems. Whether through the precision of ligand conjugation, the steric shielding of PEGylation, or the biomimetic neutrality of zwitterionic coatings, these strategies aim to reconcile synthetic design with the realities of the biological environment, pushing nanomedicine closer to the goal of safe, effective, and personalised therapy [52].

3.3 Responsive and Smart Polymers

The quest to design polymer nanoparticles that can intelligently adapt to complex and dynamic biological environments has led to the emergence of responsive and smart polymers—materials engineered to sense and react to specific physiological cues, thereby improving therapeutic precision and safety [53]. Among the most explored strategies are pH, sensitive polymers, which exploit the acidic microenvironment of tumors or endosomes (pH ~ 5–6) compared to normal tissues (pH~7.4). These polymers remain stable at physiological pH but undergo conformational changes, swelling, or degradation in acidic conditions, triggering controlled release of encapsulated drugs precisely where they are needed [54]. For instance, polymers containing acid, labile linkers such as hydrazones, orthoesters, or acetal groups disassemble in acidic conditions, facilitating endosomal escape and intracellular delivery of sensitive cargoes like siRNA or proteins. Complementing pH, responsiveness, redox, sensitive polymers are designed to exploit intracellular environments rich in reducing agents like glutathione (present at millimolar concentrations inside cells but micromolar) [55]. Incorporating disulfide bonds into polymer backbones or crosslinkers enables these particles to remain stable during circulation but degrade rapidly once internalised, releasing therapeutic payloads in the cytosol. Similarly, enzyme, sensitive polymers integrate peptide sequences or other moieties cleavable by disease, associated enzymes (e.g., matrix metalloproteinases or cathepsins), adding an extra layer of spatial specificity, particularly valuable in cancer or inflammatory diseases where overexpressed enzymes act as local triggers [9].

A fascinating extension of this responsiveness is the development of "sheddable" coatings: protective layers that detach under certain biological conditions to reveal functional surfaces or targeting ligands. For instance, nanoparticles may be coated with PEG to evade immune clearance during circulation, but are designed to shed this layer in acidic tumour environments or upon enzymatic activation, thereby exposing positively charged or ligand, functionalized cores that enhance cellular uptake [56]. This dynamic modulation reconciles the conflicting demands of stealth (avoiding clearance) and targeting (promoting internalisation), a longstanding challenge in nanoparticle design. Sheddable coatings exemplify the shift from static to dynamic nanomaterials, mirroring the adaptive nature of biological systems themselves [57].

Beyond surface responsiveness, core–shell and multilayer architectures further enhance nanoparticle functionality by spatially segregating roles within the same particle. Core–shell particles consist of an inner core—often a drug reservoir or imaging agent—encapsulated by an outer shell engineered for stability, biocompatibility, or responsiveness [53]. This architecture not only protects sensitive cargoes from premature degradation but also allows independent tuning of surface chemistry and release kinetics. Multilayer assemblies push this concept further, building nanoparticles layer by layer through electrostatic attraction, hydrogen bonding, or covalent interactions [58]. The layer, by, layer (LbL) technique enables precise deposition of alternating polyelectrolytes, peptides, or nucleic acids, offering nanoscale control over composition and thickness. Such modularity supports the co, delivery of multiple therapeutics with staggered release profiles, facilitates surface functionalization with targeting ligands, and even allows particles to respond to sequential biological cues [59].

Adding inorganic elements to these architectures creates hybrid inorganic–polymer nanoparticles that combine the versatility of polymers with the unique properties of inorganic cores such as gold, iron oxide, or silica. For example, iron oxide cores confer magnetic responsiveness, enabling external field, guided delivery or magnetic resonance imaging (MRI) contrast, while gold cores support photothermal therapy or imaging through localized surface plasmon resonance. Encapsulating these cores within polymer shells preserves biocompatibility and supports surface functionalization, effectively merging diagnostic and therapeutic functions into a single "theranostic" platform [60].

Collectively, these strategies illustrate a broader shift in nanomedicine—from static carriers to dynamic, multifunctional systems capable of sensing, adapting, and responding to the complex biological environments they encounter. Whether through pH, redox, or enzyme sensitivity; detachable stealth layers; or sophisticated core–shell and multilayer structures, these innovations highlight the power of nanoengineering to reconcile competing design requirements—stealth versus targeting, stability versus responsiveness, and single, agent delivery versus combination therapy. Ultimately, this evolving toolkit moves polymer nanoparticles closer to realising their full clinical potential: as smart, adaptive agents able not only to carry drugs but also to orchestrate their delivery with spatial and temporal precision within living systems [61, 62].

4 Traversing Biological Barriers: Case Studies

Successfully delivering polymer nanoparticles to precise biological targets requires navigating a labyrinth of formidable biological barriers—each shaped by evolution to protect the body from pathogens, but equally adept at blocking therapeutic nanoparticles. Among the most challenging is the blood–brain barrier (BBB), a tightly regulated interface formed by endothelial cells linked with tight junctions, pericytes, and astrocytic end, feet [63]. Traditional small molecules often fail to cross the BBB, and nanoparticles, by their size, face even greater obstacles. Yet, innovative nanoengineering strategies have demonstrated promising breakthroughs. For instance, polymeric nanoparticles functionalized with transferrin or apolipoprotein E peptides exploit receptor, mediated transcytosis, ferrying particles across endothelial cells into the central nervous system. Another example includes PEGylated PLGA nanoparticles decorated with angiopep, 2, which targets low, density lipoprotein receptor, related protein, 1 (LRP1) overexpressed on the BBB, successfully delivering chemotherapeutic agents to glioblastoma models in vivo [64, 65]. More recently, researchers have combined magnetic cores with polymeric shells to guide nanoparticles across the BBB under an external magnetic field, a technique that, while still experimental, illustrates how physical forces can complement biochemical targeting [66].

The tumour microenvironment presents a different but equally complex challenge. Characterised by leaky vasculature and impaired lymphatic drainage, tumours allow nanoparticles—typically between 50 and 200 nm—to accumulate through the enhanced permeability and retention (EPR) effect. Yet, the EPR effect is notoriously heterogeneous; some tumours exhibit dense stroma and high interstitial pressure that hinder penetration. To address this, active targeting strategies have been deployed alongside size optimisation [67]. For example, nanoparticles surface, functionalized with folic acid, RGD peptides, or antibodies against overexpressed receptors like HER2 or EGFR can selectively bind tumour cells, enhancing local accumulation beyond passive EPR alone. Recent case studies show hybrid core–shell nanoparticles carrying both imaging agents and chemotherapeutics actively targeting integrins, enabling real, time imaging of delivery and higher intratumoral drug concentrations. Additionally, enzyme, responsive coatings that shed in the protease, rich tumour milieu have improved deep tissue penetration by exposing positively charged surfaces, enhancing electrostatic interactions with cancer cells [68, 69].

Mucosal barriers, including those in the pulmonary and gastrointestinal tracts, add another dimension of complexity. Thick mucus layers, rapid turnover, and chemical gradients collectively impede nanoparticle diffusion and absorption. Nanoengineered strategies have responded with mucus, penetrating particles, often employing dense, low, molecular, weight PEG coatings to minimise adhesive interactions with mucin glycoproteins [70]. For pulmonary delivery, dry powder formulations of polymer nanoparticles have been developed to optimise aerosolisation and deposition in distal airways, followed by mucus penetration and cellular uptake. In gastrointestinal applications, pH, sensitive polymer coatings that resist stomach acid but dissolve in the neutral, to, basic pH of the intestine have improved oral bioavailability of sensitive therapeutics. For instance, chitosan, based nanoparticles functionalized with trimethyl groups enhance mucoadhesion and transiently open tight junctions, facilitating paracellular transport of macromolecules like insulin, demonstrating the promise of non, invasive delivery of biologics [26, 71].

Finally, even after successful navigation through systemic and tissue, level barriers, nanoparticles must overcome the cellular hurdle of endosomal entrapment to deliver payloads into the cytoplasm. Endocytosed particles typically follow endosomal maturation into lysosomes, where acidic pH and hydrolytic enzymes risk degrading therapeutic cargo. To circumvent this, nanoengineers have devised pH, responsive polymers that exploit the "proton sponge effect," buffering endosomal acidification and inducing osmotic swelling that ruptures membranes [72]. Polymers incorporating amines, such as polyethyleneimine (PEI), or histidine, rich sequences have shown such effects, though balancing efficacy with cytotoxicity remains critical. Alternatively, membrane, disruptive peptides can be grafted onto nanoparticles, triggered by acidic pH to transiently destabilise endosomal membranes. Hybrid core–shell nanoparticles with redox, sensitive disulfide linkages can release their payload in the cytosol, where glutathione concentrations are high, ensuring that even after endosomal escape the cargo remains bioactive [73].

Taken together, these case studies illustrate how sophisticated nanoengineering transforms polymer nanoparticles into dynamic agents capable of sensing, responding to, and traversing the body's most sophisticated defences. By integrating ligand targeting, stealth coatings, stimulus, responsive elements, and architectural innovations, researchers have moved beyond generic delivery vehicles to design particles tailored for precise biological challenges—from crossing the BBB to penetrating mucus, from homing in on tumours to escaping endosomes. Each barrier demands a unique strategy, but collectively, these advances highlight the transformative potential of polymer nanoparticles when engineered not simply to survive in the body, but to actively interact with and adapt to its complex biological landscape [74, 75].

5 Applications in Therapeutics and Diagnostics

The convergence of polymer science, molecular biology, and nanotechnology has propelled polymer nanoparticles into a central role in modern therapeutics and diagnostics, enabling interventions that were previously inconceivable. Perhaps the most widely pursued application remains targeted drug delivery, where nanoengineered polymer particles are designed to ferry chemotherapeutics directly to diseased tissues, sparing healthy cells and minimising systemic toxicity [76]. In cancer, this concept materialises through both passive targeting—leveraging the enhanced permeability and retention (EPR) effect of leaky tumour vasculature—and active targeting via ligands that recognise overexpressed receptors like HER2, EGFR, or folate receptors on tumour cells. Sophisticated systems combine these approaches: PEGylated polymeric micelles decorated with tumour, homing peptides have shown prolonged circulation and enhanced tumour accumulation in preclinical models, while redox, sensitive cores ensure that cytotoxic agents like doxorubicin are released preferentially in the reductive intracellular milieu [77]. Beyond oncology, polymer nanoparticles are being adapted for neurodegenerative diseases such as Alzheimer's and Parkinson's, where crossing the blood–brain barrier remains a formidable challenge. Targeting ligands like transferrin, angiopep, 2, or even cell, penetrating peptides guide nanoparticles through receptor, mediated transcytosis, delivering neuroprotective agents or anti, aggregative peptides directly into the brain parenchyma. Similarly, in inflammatory disorders such as rheumatoid arthritis or inflammatory bowel disease, nanoparticles functionalized with ligands targeting adhesion molecules or inflammatory cytokines localise to inflamed tissues, releasing corticosteroids or biologics in situ to reduce systemic exposure [78].

Parallel to drug delivery, polymer nanoparticles have opened new horizons in gene delivery, overcoming many limitations of viral vectors such as immunogenicity and size constraints. Synthetic polymers like poly (β, amino esters), chitosan derivatives, and PEGylated polyethylenimine have been engineered to complex and protect nucleic acids—siRNA, mRNA, or CRISPR–Cas9 systems—against degradation in circulation and facilitate cellular uptake [79]. Responsive designs, such as pH, or redox, sensitive linkages, enable cargo release once particles enter the cytoplasm, bypassing lysosomal degradation. Recent breakthroughs include lipid–polymer hybrid nanoparticles delivering mRNA vaccines that elicit strong and durable immune responses, a concept validated by the success of mRNA, based COVID, 19 vaccines. Likewise, CRISPR–Cas9 ribonucleoprotein complexes packaged in biodegradable polymer nanoparticles have demonstrated precise gene editing in vivo, correcting genetic defects in models of liver and muscle disease without triggering significant immune responses—an advance that moves gene therapy closer to routine clinical application [80].

Complementing these therapeutic functions, polymer nanoparticles also serve as powerful diagnostic and theranostic tools, uniting therapy and imaging within a single platform. Fluorescently labelled polymer particles enable high, resolution visualisation of biodistribution and cellular trafficking, while nanoparticles incorporating MRI, active components like gadolinium chelates or superparamagnetic iron oxide cores provide non, invasive whole, body imaging capabilities [81]. For instance, core–shell nanoparticles with iron oxide cores and polymeric shells functionalized for tumour targeting have achieved simultaneous drug delivery and MRI, based tracking of accumulation in tumours, facilitating treatment monitoring and real, time dose optimisation. This theranostic capability not only enhances treatment precision but also provides critical feedback on therapeutic response, enabling adaptive treatment strategies [82].

Beyond systemic applications, nanoengineered polymer particles are increasingly being harnessed for biosensing and real, time monitoring, capitalising on their capacity to respond to specific biological cues. Smart polymers that undergo fluorescence changes upon binding analytes or experiencing pH, temperature, or redox shifts have been integrated into nanosensors capable of detecting biomarkers associated with cancer, infection, or metabolic disorders at remarkably low concentrations [27]. For instance, nanoparticles functionalized with aptamers or antibodies detect circulating proteins or nucleic acids, generating optical or electrochemical signals in real time. In some designs, these nanoparticles are embedded within wearable devices or implantable systems, continuously monitoring physiological parameters and transmitting data wirelessly, transforming disease management from episodic to continuous care [83].

The integration of polymer nanoparticles into therapeutics and diagnostics marks one of the most dynamic frontiers of nanomedicine, shaped by a convergence of advances in polymer chemistry, molecular targeting, and bio, nano interface engineering. Among therapeutic applications, targeted drug delivery remains the most mature and widely studied area [84]. Polymer, based nanocarriers such as PEGylated micelles and PLGA nanoparticles have shown consistent benefits in extending drug circulation times and enhancing accumulation at disease sites through the enhanced permeability and retention (EPR) effect. Yet, critical studies reveal that passive targeting alone often yields limited clinical efficacy due to tumor heterogeneity and variable vascular permeability [85, 86]. To address these limitations, researchers have turned to active targeting by conjugating ligands—antibodies, peptides, or small molecules—that bind to overexpressed receptors on tumor or inflamed tissues. For example, folate, or RGD, modified polymer particles have demonstrated increased intratumoral uptake and reduced systemic toxicity in preclinical cancer models [87]. Despite promising results, challenges remain: the dynamic formation of the protein corona can mask targeting ligands, and differences between animal models and human tumors complicate translation, often leading to underwhelming clinical outcomes.

Beyond oncology, polymer nanoparticles have shown promise in treating neurodegenerative disorders, where crossing the blood–brain barrier (BBB) poses a formidable challenge. Studies using transferrin, conjugated or angiopep, 2, functionalized nanoparticles have reported successful delivery of neuroprotective agents and genes into the central nervous system [88]. However, these approaches face hurdles of scale, safety, and patient, specific BBB variability. In inflammatory disorders like rheumatoid arthritis and inflammatory bowel disease, nanoparticles engineered to home to inflamed endothelium or immune cells have shown potential to localise drugs at disease sites, offering a path to reduced systemic immunosuppression [89].

Gene delivery, once dominated by viral vectors, has been reshaped by advances in polymeric carriers. Cationic polymers such as polyethyleneimine and poly (β, amino esters) can compact nucleic acids into stable nanoparticles, protecting them from degradation while promoting cellular uptake. Recent research has focused on reducing toxicity through biodegradable polymers and incorporating stimuli, responsive linkages for controlled release [90]. Notably, polymer nanoparticles have enabled the delivery of siRNA and mRNA, as well as CRISPR–Cas9 systems, showing efficacy in correcting genetic disorders in animal models [78]. Despite these advances, balancing transfection efficiency with biocompatibility remains a persistent challenge, and long, term safety data are still limited.

In diagnostics and theranostics, polymer nanoparticles have moved beyond passive imaging agents to multifunctional systems that can simultaneously diagnose and treat disease. For example, nanoparticles incorporating magnetic cores within polymer shells enable real, time MRI tracking of drug delivery while releasing therapeutic payloads [91]. Fluorescently labelled polymer particles have been used for intraoperative imaging to delineate tumour margins, improving surgical precision [92]. Recent trends also include smart particles that change optical or magnetic signals in response to local pH, enzymes, or redox conditions, offering real, time readouts of disease activity or treatment response.

Finally, polymer nanoparticles have found roles in biosensing and real, time monitoring, often by integrating aptamers or antibodies that detect specific biomarkers, triggering measurable fluorescence or electrochemical changes [93]. These systems promise earlier diagnosis and more personalised treatment adjustments but face technical challenges in stability, sensitivity, and integration into wearable or implantable devices.

6 Challenges and Emerging Trends

Despite the striking progress in nanoengineering polymer particles for medicine, the path from laboratory innovation to clinical practice remains fraught with challenges that demand both technological refinement and systemic change. Foremost among these is the issue of reproducibility and scalability in manufacturing. While laboratory protocols can yield well, defined nanoparticles under tightly controlled conditions, scaling production to industrial levels introduces variability in particle size, shape, surface chemistry, and drug loading—all of which critically influence biological performance [94]. Techniques like emulsion polymerisation, nanoprecipitation, and even advanced methods like microfluidics, though powerful, must be adapted to large, scale continuous processes with stringent quality controls to meet Good Manufacturing Practice (GMP) standards. Variations in raw materials, reactor design, and process parameters can lead to subtle but significant differences in product batches, complicating regulatory approval and potentially undermining therapeutic consistency [95]. This challenge is amplified by the complexity of multifunctional nanoparticles, where each additional feature—targeting ligands, stealth coatings, or responsive elements—introduces another layer of variability.

Against this backdrop, emerging trends in the field seek to reconcile complexity with reproducibility and move beyond empirical design toward data, driven, rational approaches. One of the most promising directions is the integration of artificial intelligence (AI) and machine learning into nanoparticle design and manufacturing [96]. By analyzing large datasets of physicochemical properties, biological outcomes, and synthesis parameters, AI models can identify patterns and optimise design variables to predict nanoparticle behaviour in vivo, drastically accelerating discovery cycles and reducing reliance on trial, and, error experimentation. This computational approach also aids in anticipating patient, to, patient variability by simulating how nanoparticles might interact with different protein coronas or immune profiles, guiding the customisation of particles for personalised medicine [97].

In parallel, biomimetic strategies are redefining how polymer nanoparticles engage with biological systems. Instead of merely evading immune detection through synthetic coatings, researchers now coat nanoparticles with natural cell membranes derived from red blood cells, leukocytes, or even cancer cells themselves. These biomimetic particles inherit complex surface proteins and glycans, enabling them to camouflage as "self," exploit native trafficking pathways, and evade immune clearance more effectively than synthetic stealth polymers alone [98]. Early studies have demonstrated remarkable improvements in circulation time and targeting precision, suggesting that nature's designs may offer solutions where synthetic strategies fall short. Programmable particles, another frontier, push this concept further: these systems are engineered to respond to multiple sequential biological cues, effectively executing a predefined "logic" as they travel through the body [99]. For example, a particle might circulate stealthily under physiological pH but reveal targeting ligands in the acidic tumour microenvironment, then release its drug payload upon sensing high intracellular glutathione. This capacity to sense and react to complex, dynamic biological environments heralds a new era of precision nanomedicine [100].

Yet, even the most elegant engineering must ultimately navigate the complex landscape of regulatory and translational considerations to reach patients. Nanoparticles, by their very nature, blur traditional regulatory categories: they can function simultaneously as drugs, devices, and diagnostics, challenging established approval pathways. Regulators must evaluate not only the safety of individual components but also their emergent behaviour as multifunctional systems, a task complicated by the limited predictive power of standard in vitro assays for complex in vivo interactions like protein corona formation or immune activation [101]. Establishing validated, standardised characterisation techniques and defining critical quality attributes is essential to ensure consistent product performance. Furthermore, translating from bench to bedside requires scalable, cost, effective manufacturing methods that preserve the intricate nanoarchitectures and functionalities developed in the laboratory, without compromising safety or efficacy [32].

These challenges underscore that the next phase of nanomedicine will require more than technological innovation alone; it demands an integrated effort bridging materials science, biology, engineering, data science, and regulatory policy. The convergence of AI, guided design, biomimetic strategies, and programmable functionalities holds immense promise, but only through robust manufacturing processes and transparent regulatory frameworks can these breakthroughs be reliably and safely translated to clinical practice. As the field matures, the vision of polymer nanoparticles as truly personalised, adaptive agents capable of precise diagnosis, targeted therapy, and real, time monitoring edges closer to reality, not as isolated laboratory curiosities, but as reproducible, scalable, and approved tools reshaping the future of medicine [102, 103] (Table 1).

7 Conclusion and Future Directions

In tracing the evolution of polymer nanoparticles from simple carriers to intelligent, adaptive nanosystems, several key insights emerge that underscore both the promise and complexity of this field. At its foundation, the success of polymer, based nanomedicine hinges on a nuanced understanding of bio, nano interactions: the dynamic and often unpredictable interplay between synthetic design and the living biological environment. Advances in nanoengineering—from precise size and shape control using emulsion polymerisation, nanoprecipitation, and microfluidics, to sophisticated surface functionalization with ligands, stealth coatings, and zwitterionic materials—have transformed how these particles navigate biological barriers. Responsive and smart polymers add a further layer of functionality, enabling particles to sense and react to pH, redox states, or enzymatic activity, while core–shell and multilayer architectures permit the compartmentalisation of diagnostic and therapeutic functions. Collectively, these strategies have led to tangible applications across targeted drug delivery for cancer and neurodegenerative diseases, gene therapy platforms delivering siRNA and CRISPR systems, and even real, time biosensing and theranostics, demonstrating the field's remarkable breadth.

Looking forward, the next generation of nanoengineered polymer particles will likely embrace even greater complexity and precision. AI, guided design promises to shift nanoparticle development from largely empirical experimentation to data, driven optimisation, uncovering subtle patterns linking physicochemical properties to biological performance and accelerating discovery timelines. Biomimetic particles, cloaked with natural cell membranes or engineered to mimic biological surfaces, may finally reconcile the need for stealth with efficient targeting, achieving immune evasion without sacrificing specificity. Meanwhile, programmable nanoparticles capable of executing "logic gates"—sequentially responding to different biological cues—could unlock unprecedented spatial and temporal control over therapeutic delivery. These advances hint at a future where polymer nanoparticles do not merely carry drugs but actively participate in disease sensing, decision, making, and intervention.

Yet, realising this vision demands a fundamentally interdisciplinary approach. Polymer chemists, materials scientists, and biomedical engineers must work alongside immunologists, pharmacologists, and clinicians to anticipate and navigate biological complexity—from immune responses and protein corona effects to patient, to, patient variability. Computational scientists and data analysts will be indispensable in harnessing machine learning tools to guide design and predict in vivo

Table 1 Key discrepancies between in vitro and in vivo results for polymer nanoparticles

Aspect	In vitro findings	In vivo observations	Reason for discrepancy
Protein corona formation	Often under, characterised; may appear minimal in simple media	Rapid, dynamic adsorption of diverse plasma proteins alters particle identity	In vitro media lack the complexity of blood plasma; they don't reproduce real protein diversity and dynamics
Cellular uptake	High apparent uptake, especially for cationic particles	Reduced or variable uptake due to opsonisation, clearance, or altered biodistribution	Absence of immune components and flow conditions in vitro; neglects protein corona masking effects
Targeting ligand activity	Consistent receptor binding and specificity	Targeting efficiency reduced or lost due to protein corona masking	Corona can shield ligands; in vitro tests don't fully capture competitive binding with plasma proteins
Circulation time	Cannot be measured	Rapid clearance by the liver, spleen, or immune cells	In vitro systems lack the reticuloendothelial system and organ, level clearance mechanisms
Toxicity profiles	May appear low in static assays	Immunogenicity or inflammatory responses are possible in vivo	In vitro lacks immune complexity; doesn't measure complement activation, cytokine release, or hypersensitivity
Endosomal escape	Often efficient in controlled conditions	Variable and sometimes poor in living organisms	In vivo complexity adds barriers: Tissue penetration, heterogeneous cell types, and dynamic environments
Biodistribution	Uniform cell exposure is assumed	Highly heterogeneous distribution among tissues and within tumours	Limited by vascularisation, tissue density, and biological barriers not present in vitro
Stability	Appears stable in buffered solutions	Aggregation, degradation, or transformation under physiological conditions	Blood shear stress, enzymes, and ionic strength alter particle behaviour beyond what static tests predict

Sources Literature Review

behaviour. Moreover, regulatory scientists and industry partners are needed to translate promising laboratory prototypes into reproducible, scalable, and clinically approved therapies, ensuring that technological sophistication does not come at the expense of safety, cost, effectiveness, or patient access.

In essence, the future of nanoengineered polymer particles lies at the intersection of creativity and constraint: designing ever more intelligent and multifunctional systems, while grounding innovation in robust science and scalable manufacturing. As these fields converge, polymer nanoparticles are poised not only to extend the frontiers of targeted therapy and diagnostics but to reshape medicine itself, shifting from reactive treatment to proactive, precision, guided care tailored to the biology of each disease and each patient.

References

1. Cortez-Jugo, C., Czuba-Wojnilowicz, E., Tan, A., Caruso, F.: A focus on "bio" in bio–nanoscience: the impact of biological factors on nanomaterial interactions. Adv. Healthc. Mater. **10**(16), 2100574 (2021)
2. Jin, X., Wan, H., Jiao, Z., Li, J., Yan, L.T.: Physical principles at bio-nano interfaces with active matter. Natl. Sci. Open **3**(4), 20230079 (2024)
3. Li, Z., Xiao, C., Yang, X., Li, Z.: Progress in the mechanical properties of nanoparticles for tumor-targeting delivery. Chem. Soc. Rev. (2025)
4. Mamidi, N., De Silva, F.F., Salehi, A.O.M.: Advanced disease therapeutics using engineered living drug delivery systems. Nanoscale (2025)
5. Nienhaus, K., Xue, Y., Shang, L., Nienhaus, G.U.: Protein adsorption onto nanomaterials engineered for theranostic applications. Nanotechnology **33**(26), 262001 (2022)
6. Parvin, N., Joo, S.W., Mandal, T.K.: Biodegradable and stimuli-responsive nanomaterials for targeted drug delivery in autoimmune diseases. J. Funct. Biomater. **16**(1), 24 (2025)
7. Moghimi, S.M., Haroon, H.B., Yaghmur, A., Hunter, A.C., Papini, E., Farhangrazi, Z.S., Simberg, D., Trohopoulos, P.N.: Perspectives on complement and phagocytic cell responses to nanoparticles: from fundamentals to adverse reactions. J. Control. Release **356**, 115–129 (2023)
8. Sun, Y., Zhou, Y., Rehman, M., Wang, Y.F., Guo, S.: Protein corona of nanoparticles: isolation and analysis. Chem. Bio. Eng. **1**(9), 757–772 (2024)
9. Santhamoorthy, M., Asaithambi, P., Ramkumar, V., Elangovan, N., Perumal, I., Kim, S.C.: A review on the recent advancements of polymer-modified mesoporous silica nanoparticles for drug delivery under stimuli-trigger. Polymers **17**(12), 1640 (2025)
10. Souri, M., Soltani, M., Kashkooli, F.M., Shahvandi, M.K., Chiani, M., Shariati, F.S., Mehrabi, M.R., Munn, L.L.: Towards principled design of cancer nanomedicine to accelerate clinical translation. Mater. Today Bio. **13**, 100208 (2022)
11. Raheem, A., Mandal, K., Biswas, S., Ahari, A., Hassani Najafabadi, A., Farhadi, N., Zehtabi, F., Gangrade, A., Mecwan, M., Maity, S., Sharma, S.: Smart biomaterials in healthcare: breakthroughs in tissue engineering, immunomodulation, patient-specific therapies, and biosensor applications. Appl. Phys. Rev. **12**(1) (2025)
12. Souto, E.B., Blanco-Llamero, C., Krambeck, K., Kiran, N.S., Yashaswini, C., Postwala, H., Severino, P., Priefer, R., Prajapati, B.G., Maheshwari, R.: Regulatory insights into nanomedicine and gene vaccine innovation: safety assessment, challenges, and regulatory perspectives. Acta Biomater. (2024)
13. Kopac, T.: Leveraging artificial intelligence and machine learning for characterizing protein corona, nanobiological interactions, and advancing drug discovery. Bioengineering **12**(3), 312 (2025)

14. Mukherjee, S., Madamsetty, V.S., Bhattacharya, D., Roy Chowdhury, S., Paul, M.K., Mukherjee, A.: Recent advancements of nanomedicine in neurodegenerative disorders theranostics. Adv. Funct. Mater. **30**(35), 2003054 (2020)
15. Heidari, M., Al Hasani, S., Thangavel, S., Kumar, A.: Challenges and perspectives of polymeric nanocomposites for biomedical applications. In: Nanomanufacturing Techniques in Sustainable Healthcare Applications, pp. 317–340 (2024)
16. Li, S., Cortez-Jugo, C., Ju, Y., Caruso, F.: Approaching two decades: biomolecular coronas and bio–nano interactions. ACS Nano **18**(49), 33257–33263 (2024)
17. Bendre, A., Bhat, M.P., Lee, K.H., Altalhi, T., Alruqi, M.A., Kurkuri, M.: Recent developments in microfluidic technology for synthesis and toxicity-efficiency studies of biomedical nanomaterials. Mater. Today Adv. **13**, 100205
18. Malek-Khatabi, A., Tabandeh, Z., Nouri, A., Mozayan, E., Sartorius, R., Rahimi, S., Jamaledin, R.: Long-term vaccine delivery and immunological responses using biodegradable polymer-based carriers. ACS Appl. Bio Mater. **5**(11), 5015–5040 (2022)
19. Jahani, A., Nassira, H.: Polyethylene glycol-based materials: transformative applications in biomedicine and the food industry. Mater. Chem. **3**, 1074 (2024)
20. Kumar, A., Shahvej, S.K., Yadav, P., Modi, U., Yadav, A.K., Solanki, R., Bhatia, D.: Clinical applications of targeted nanomaterials. Pharmaceutics **17**(3), 379 (2025)
21. Sitia, L., Sevieri, M., Signati, L., Bonizzi, A., Chesi, A., Mainini, F., Corsi, F., Mazzucchelli, S.: HER-2-targeted nanoparticles for breast cancer diagnosis and treatment. Cancer **14**(10), 2424 (2022)
22. Panico, S., Capolla, S., Bozzer, S., Toffoli, G., Dal Bo, M., Macor, P.: Biological features of nanoparticles: protein corona formation and interaction with the immune system. Pharmaceutics **14**(12), 2605 (2022)
23. Jiang, Z., Chu, Y., Zhan, C.: Protein corona: challenges and opportunities for targeted delivery of nanomedicines. Expert Opin. Drug Deliv. **19**(7), 833–846 (2022)
24. Yuan, L., Chen, Q., Riviere, J.E., Lin, Z.: Pharmacokinetics and tumor delivery of nanoparticles. J. Drug Deliv. Sci. Technol. **83**, 104404 (2023)
25. Chen, X., Xu, L., Shi, H.: Emerging biomaterials and bio-nano interfaces in pulmonary hypertension therapy: transformative strategies for personalized treatment. Front. Bioeng. Biotechnol. **13**, 1567783 (2025)
26. Yang, C., Lin, Z.I., Chen, J.A., Xu, Z., Gu, J., Law, W.C., Yang, J.H.C., Chen, C.K.: Organic/inorganic self-assembled hybrid nano-architectures for cancer therapy applications. Macromol. Biosci. **22**(2), 2100349 (2022)
27. Son, M.H., Park, S.W., Sagong, H.Y., Jung, Y.K.: Recent advances in electrochemical and optical biosensors for cancer biomarker detection. Bio Chip J. **17**(1), 44–67 (2023)
28. Giannopoulos-Dimitriou, A., Saiti, A., Petrou, A., Vizirianakis, I.S.: Fatouros, D.G.: liposome stability and integrity. In: Liposomes in Drug Delivery, pp. 89–121. Academic (2024)
29. De Luca, E., Coricciati, C., Trombetta, M., Gigli, G., Mozetic, P., Rainer, A.: Optical nanoscopy for elucidating nano–bio interactions: a one health perspective. Adv. Opt. Mater. **13**(13), 2401397 (2025)
30. Mitchell, M.J., Billingsley, M.M., Haley, R.M., Wechsler, M.E., Peppas, N.A., Langer, R.: Engineering precision nanoparticles for drug delivery. Nat. Rev. Drug Discov. **20**(2), 101–124 (2021)
31. Gokcekuyu, Y., Ekinci, F., Guzel, M.S., Acici, K., Aydin, S., Asuroglu, T.: Artificial intelligence in biomaterials: a comprehensive review. Appl. Sci. **14**(15), 6590 (2024)
32. Desai, N., Rana, D., Patel, M., Bajwa, N., Prasad, R., Vora, L.K.: Nanoparticle therapeutics in clinical perspective: classification, marketed products, and regulatory landscape. Small 2502315 (2025)
33. Rodríguez, C.F., Guzmán-Sastoque, P., Santacruz-Belalcazar, A., Rodriguez, C., Villamarin, P., Reyes, L.H., Cruz, J.C.: Magnetoliposomes for nanomedicine: synthesis, characterization, and applications in drug, gene, and peptide delivery. Expert Opin. Drug Deliv. 1–30 (2025)
34. Kim, T., Han, H.S., Yang, K., Kim, Y.M., Nam, K., Park, K.H., Choi, S.Y., Park, H.W., Choi, K.Y., Roh, Y.H.: Nanoengineered polymeric rna nanoparticles for controlled biodistribution and efficient targeted cancer therapy. ACS Nano **18**(11), 7972–7988 (2024)

35. Lohmann, V., Rolland, M., Truong, N.P., Anastasaki, A.: Controlling size, shape, and charge of nanoparticles via low-energy miniemulsion and heterogeneous RAFT polymerization. Eur. Polym. J. **176**, 111417 (2022)
36. Bhardwaj, A.K., Kant, A., Rehalia, A., Singh, V., Sharma, R.: A review on nanomaterials for drug delivery systems and application of carbon based nanomaterials. ES Mater. Manuf. **21**(2), 824 (2023)
37. Fatehbasharzad, P., Fatehbasharzad, P., Sillanpää, M., Shamsi, Z.: Investigation of bioimpacts of metallic and metallic oxide nanostructured materials: size, shape, chemical composition, and surface functionality: a review. Part. Part. Syst. Charact. **38**(10), 2100112 (2021)
38. Dash, S., Das, T., Patel, P., Panda, P.K., Suar, M., Verma, S.K.: Emerging trends in the nanomedicine applications of functionalized magnetic nanoparticles as novel therapies for acute and chronic diseases. J. Nanobiotechnology **20**(1), 393 (2022)
39. Ejigah, V., Owoseni, O., Bataille-Backer, P., Ogundipe, O.D., Fisusi, F.A., Adesina, S.K.: Approaches to improve macromolecule and nanoparticle accumulation in the tumor microenvironment by the enhanced permeability and retention effect. Polymers **14**(13), 2601 (2022)
40. Kapate, N., Clegg, J.R., Mitragotri, S.: Non-spherical micro-and nanoparticles for drug delivery: progress over 15 years. Adv. Drug Deliv. Rev. **177**, 113807 (2021)
41. Naser, S.S., Gupta, A., Choudhury, A., Yadav, A., Sinha, A., Kirti, A., Singh, D., Kujawska, M., Kaushik, N.K., Ghosh, A., De, S.: Biophysical translational paradigm of polymeric nanoparticle: embarked advancement to brain tumor therapy. Biomed. Pharmacother. **179**, 117372 (2024)
42. Sadeghi, A., PourEskandar, S., Askari, E., Akbari, M.: Polymeric nanoparticles and nanogels: how do they interact with proteins? Gels **9**(8), 632 (2023)
43. Díez-Pascual, A.M.: Surface engineering of nanomaterials with polymers, biomolecules, and small ligands for nanomedicine. Materials **15**(9), 3251 (2022)
44. Gadekar, A., Bhowmick, S., Pandit, A.: A glycotherapeutic approach to functionalize biomaterials-based systems. Adv. Funct. Mater. **30**(44), 1910031 (2020)
45. Koide, R., Hirane, N., Kambe, D., Yokoi, Y., Otaki, M., Nishimura, S.I.: Antiadhesive nanosome elicits role of glycocalyx of tumor cell-derived exosomes in the organotropic cancer metastasis. Biomaterials **280**, 121314 (2022)
46. Guo, C., Yuan, H., Wang, Y., Feng, Y., Zhang, Y., Yin, T., He, H., Gou, J., Tang, X.: The interplay between PEGylated nanoparticles and blood immune system. Adv. Drug Deliv. Rev. **200**, 115044 (2023)
47. Gawali, P., Saraswat, A., Bhide, S., Gupta, S., Patel, K.: Human solid tumors and clinical relevance of the enhanced permeation and retention effect: a 'golden gate'for nanomedicine in preclinical studies? Nanomedicine **18**(2), 169–190 (2023)
48. Lv, W., Wang, Y., Fu, H., Liang, Z., Huang, B., Jiang, R., Wu, J., Zhao, Y.: Recent advances of multifunctional zwitterionic polymers for biomedical application. Acta Biomater. (2024)
49. Luozhong, S., Liu, P., Li, R., Yuan, Z., Debley, E., Chen, Y., Hu, Y., Cao, Z., Cui, M., McIlhenny, K., McCurdy, C.: Poly (carboxybetaine) lipids enhance mRNA therapeutics efficacy and reduce their immunogenicity. Nat. Mater. **1–10** (2025)
50. Overby, C., Park, S., Summers, A., Benoit, D.S.: Zwitterionic peptides: tunable next-generation stealth nanoparticle modifications. Bioact. Mater. **27**, 113–124 (2023)
51. Ren, J., Andrikopoulos, N., Velonia, K., Tang, H., Cai, R., Ding, F., Ke, P.C., Chen, C.: Chemical and biophysical signatures of the protein corona in nanomedicine. J. Am. Chem. Soc. **144**(21), 9184–9205 (2022)
52. Ly, P.D., Ly, K.N., Phan, H.L., Nguyen, H.H., Duong, V.A., Nguyen, H.V.: Recent advances in surface decoration of nanoparticles in drug delivery. Front. Nanotechnol. **6**, 1456939 (2024)
53. Shi, Y., Zhang, Y., Zhu, L., Miao, Y., Zhu, Y., Yue, B.: Tailored drug delivery platforms: stimulus-responsive core–shell structured nanocarriers. Adv. Healthc. Mater. **13**(1), 2301726 (2024)
54. Wu, W., Luo, L., Wang, Y., Wu, Q., Dai, H.B., Li, J.S., Durkan, C., Wang, N., Wang, G.X.: Endogenous pH-responsive nanoparticles with programmable size changes for targeted tumor therapy and imaging applications. Theranostics **8**(11), 3038 (2018)

55. Zhou, H., Chen, D.S., Hu, C.J., Hong, X., Shi, J., Xiao, Y.: Stimuli-responsive nanotechnology for RNA delivery. Adv. Sci. **10**(36), 2303597 (2023)
56. Zhou, W., Jia, Y., Liu, Y., Chen, Y., Zhao, P.: Tumor microenvironment-based stimuli-responsive nanoparticles for controlled release of drugs in cancer therapy. Pharmaceutics **14**(11), 2346 (2022)
57. Nikezić, A.V., Novaković, J.G.: Nano/microcarriers in drug delivery: moving the timeline to contemporary. Curr. Med. Chem. **30**(26), 2996–3023 (2023)
58. Zhang, X., Liang, T., Ma, Q.: Layer-by-layer assembled nano-drug delivery systems for cancer treatment. Drug Deliv. **28**(1), 655–669 (2021)
59. Sarode, A., Annapragada, A., Guo, J., Mitragotri, S.: Layered self-assemblies for controlled drug delivery: a translational overview. Biomaterials **242**, 119929 (2020)
60. Karagözlü, M., Aşır, S., Türkmen, D., Zidan, A., Hacıoğulları, F., Denizli, A.: Functionalized magnetic nanoparticles based hybrid systems for theranostics. In: Functionalized Magnetic Nanoparticles for Theranostic Applications, pp. 173–212 (2024)
61. Sun, Y., Davis, E.: Nanoplatforms for targeted stimuli-responsive drug delivery: a review of platform materials and stimuli-responsive release and targeting mechanisms. Nano **11**(3), 746 (2021)
62. Yang, Y., Zeng, W., Huang, P., Zeng, X., Mei, L.: Smart materials for drug delivery and cancer therapy. View **2**(2), 20200042 (2021)
63. Lu, H., Cai, Z., Hu, P.: Recent advances in polymeric delivery vehicles for controlled and sustained drug release. Pharmaceutics **16**(9), 1184 (2024)
64. Gao, X., Xu, J., Yao, T., Liu, X., Zhang, H., Zhan, C.: Peptide-decorated nanocarriers penetrating the blood-brain barrier for imaging and therapy of brain diseases. Adv. Drug Deliv. Rev. **187**, 114362 (2022)
65. Thomsen, M.S., Johnsen, K.B., Kucharz, K., Lauritzen, M., Moos, T.: Blood–brain barrier transport of transferrin receptor-targeted nanoparticles. Pharmaceutics **14**(10), 2237 (2022)
66. Grzegorzewski, J., Michalak, M., Wołoszczuk, M., Bulicz, M., Majchrzak-Celińska, A.: Nanotherapy of glioblastoma—where hope grows. Int. J. Mol. Sci. **26**(5), 1814 (2025)
67. Sindhwani, S., Syed, A.M., Ngai, J., Kingston, B.R., Maiorino, L., Rothschild, J., MacMillan, P., Zhang, Y., Rajesh, N.U., Hoang, T., Wu, J.L.: The entry of nanoparticles into solid tumours. Nat. Mat. **19**(5), 566–575 (2020)
68. Jahan, S., Karim, M.E., Chowdhury, E.H.: Nanoparticles targeting receptors on breast cancer for efficient delivery of chemotherapeutics. Biomedicine **9**(2), 114 (2021)
69. Ma, K., Yin, M., Chen, K.Z., Qiao, S.L.: Targeted nanoprobes enabled precision theranostics in triple-negative breast cancer. Chem. Rec. **25**(2), e202400171 (2025)
70. Watchorn, J., Clasky, A.J., Prakash, G., Johnston, I.A., Chen, P.Z., Gu, F.X.: Untangling mucosal drug delivery: engineering, designing, and testing nanoparticles to overcome the mucus barrier. ACS Biomater Sci. Eng. **8**(4), 1396–1426 (2022)
71. Feczko, T.: Polymeric nanotherapeutics acting at special regions of body. J. Drug Deliv. Sci. Technol. **64**, 102597 (2021)
72. Mehta, M.J., Kim, H.J., Lim, S.B., Naito, M., Miyata, K.: Recent progress in the endosomal escape mechanism and chemical structures of polycations for nucleic acid delivery. Macromol. Biosci. **24**(4), 2300366 (2024)
73. Xu, L., Cao, Y., Xu, Y., Li, R., Xu, X.: Redox-responsive polymeric nanoparticle for nucleic acid delivery and cancer therapy: progress, opportunities, and challenges. Macromol. Biosci. **24**(3), 2300238 (2024)
74. Beach, M.A., Nayanathara, U., Gao, Y., Zhang, C., Xiong, Y., Wang, Y., Such, G.K.: Polymeric nanoparticles for drug delivery. Chem. Rev. **124**(9), 5505–5616 (2024)
75. Cui, J., Xu, Y., Tu, H., Zhao, H., Wang, H., Di, L., Wang, R.: Gather wisdom to overcome barriers: well-designed nano-drug delivery systems for treating gliomas. Acta Pharm. Sin. B. **12**(3), 1100–1125 (2022)
76. Zeng, Z., Tian, J., Xu, W., Liu, H., Xiang, D., Liao, D., Wu, J., Chen, C.: Light-responsive conjugated polymer nanoparticles with spatial-controlled camptothecin release via π–π stacking for improved combinatorial therapy of breast cancer. Mater. Des. **245**, 113270 (2024)

77. Gralewska, P., Gajek, A., Marczak, A., Rogalska, A.: Targeted nanocarrier-based drug delivery strategies for improving the therapeutic efficacy of PARP inhibitors against ovarian cancer. Int. J. Mol. Sci. **25**(15), 8304 (2024)
78. Xu, C.F., Chen, G.J., Luo, Y.L., Zhang, Y., Zhao, Y., Cao, Z.T., Yang, X.Z., Xia, H., Wang, J.: Rational designs of in vivo CRISPR-Cas delivery systems. Adv. Drug Deliv. Rev. **168**, 3–29 (2021). https://doi.org/10.1016/j.addr.2020.11.009
79. Kaupbayeva, B., Tsoy, A., Safarova, Y., Nurmagambetova, A., Murata, H., Matyjaszewski, K., Askarova, S.: Unlocking genome editing: advances and obstacles in CRISPR/Cas delivery technologies. J. Funct. Biomater. **15**(11), 324 (2024)
80. Lee, H., Rho, W.Y., Kim, Y.H., Chang, H., Jun, B.H.: RISPR-Cas9 gene therapy: non-viral delivery and stimuli-responsive nanoformulations. Molecules **30**(3), 542 (2025)
81. Łopuszyńska, N., Węglarz, W.P.: Contrasting properties of polymeric nanocarriers for mri-guided drug delivery. Nano **13**(15), 2163 (2023)
82. Qiao, R., Fu, C., Forgham, H., Javed, I., Huang, X., Zhu, J., Whittaker, A.K., Davis, T.P.: Magnetic iron oxide nanoparticles for brain imaging and drug delivery. Adv. Drug Deliv. Rev. **197**, 114822 (2023)
83. Janghorban, M., Aradanas, I., Kazemi, S., Ngaju, P., Pandey, R.: Recent advances, opportunities, and challenges in developing nucleic acid integrated wearable biosensors for expanding the capabilities of wearable technologies in health monitoring. Biosensors **12**(11), 986 (2022)
84. Essa, M.L., El-Kemary, M.A., Ebrahem Saied, E.M., Leporatti, S., Nemany Hanafy, N.A.: Nano targeted therapies made of lipids and polymers have promising strategy for the treatment of lung cancer. Materials **13**(23), 5397 (2020)
85. Danhier, F.: To exploit the tumor microenvironment: since the EPR effect fails in the clinic, what is the future of nanomedicine? J. Control. Release **244**, 108–121 (2016). https://doi.org/10.1016/j.jconrel.2016.11.015
86. Wilhelm, S., Tavares, A.J., Dai, Q., Ohta, S., Audet, J., Dvorak, H.F., Chan, W.C.W.: Analysis of nanoparticle delivery to tumours. Nat. Rev. Mater. **1**(5), 16014 (2016). https://doi.org/10.1038/natrevmats.2016.14
87. Peer, D., Karp, J.M., Hong, S., Farokhzad, O.C., Margalit, R., Langer, R.: Nanocarriers as an emerging platform for cancer therapy. Nat. Nanotechnol. **2**(12), 751–760 (2007). https://doi.org/10.1038/nnano.2007.387
88. Saraiva, C., Praça, C., Ferreira, R., Santos, T., Ferreira, L., Bernardino, L.: Nanoparticle-mediated brain drug delivery: overcoming blood–brain barrier to treat neurodegenerative diseases. J. Control. Release **235**, 34–47 (2016). https://doi.org/10.1016/j.jconrel.2016.05.044
89. Mura, S., Nicolas, J., Couvreur, P.: Stimuli-responsive nanocarriers for drug delivery. Nat. Mater. **12**(11), 991–1003 (2013). https://doi.org/10.1038/nmat3776
90. Pack, D.W., Hoffman, A.S., Pun, S., Stayton, P.S.: Design and development of polymers for gene delivery. Nat. Rev. Drug Discov. **4**(7), 581–593 (2005). https://doi.org/10.1038/nrd1775
91. Sun, C., Lee, J.S.H., Zhang, M.: Magnetic nanoparticles in MR imaging and drug delivery. Adv. Drug Deliv. Rev. **60**(11), 1252–1265 (2008). https://doi.org/10.1016/j.addr.2008.03.018
92. Kobayashi, H., Ogawa, M., Alford, R., Choyke, P.L., Urano, Y.: New strategies for fluorescent probe design in medical diagnostic imaging. Chem. Rev. **110**(5), 2620–2640 (2010). https://doi.org/10.1021/cr900263j
93. Zhu, G., Ye, X., Dong, Z., Lu, Q., Zhang, H.: Nanoscale engineering of biosensors for in vivo real-time detection. ACS Sens. **4**(5), 1152–1169 (2019). https://doi.org/10.1021/acssensors.9b00177
94. Gupta, J., Sharma, G.: Nanogel: a versatile drug delivery system for the treatment of various diseases and their future perspective. Drug Deliv. Transl. Res. **15**(2), 455–482 (2025)
95. Mahmud, M.M., Pandey, N., Winkles, J.A., Woodworth, G.F., Kim, A.J.: Toward the scale-up production of polymeric nanotherapeutics for cancer clinical trials. Nano Today **56**, 102314 (2024)
96. Tao, H., Wu, T., Aldeghi, M., Wu, T.C., Aspuru-Guzik, A., Kumacheva, E.: Nanoparticle synthesis assisted by machine learning. Nat. Rev. Mater. **6**(8), 701–716 (2021)

97. Akhtar, M., Nehal, N., Gull, A., Parveen, R., Khan, S., Khan, S., Ali, J.: Explicating the transformative role of artificial intelligence in designing targeted nanomedicine. Expert Opin. Drug Deliv. 1–21 (2025)
98. Zhang, S., Zhang, X., Gao, H., Zhang, X., Sun, L., Huang, Y., Zhang, J., Ding, B.: Cell membrane-coated biomimetic nanoparticles in cancer treatment. Pharmaceutics **16**(4), 531 (2024)
99. Li, X., Li, M., Tang, L., Shi, D., Lam, E., Bae, J.: 3D shape morphing of stimuli-responsive composite hydrogels. Mater. Chem. Front. **7**(23), 5989–6034 (2023)
100. Ferrero, C., Casas, M., Caraballo, I.: Redox-responsive polymersomes as smart doxorubicin delivery systems. Pharmaceutics **14**(8), 1724 (2022)
101. Youden, B., Jiang, R., Carrier, A.J., Servos, M.R., Zhang, X.: A nanomedicine structure–activity framework for research, development, and regulation of future cancer therapies. ACS Nano **16**(11), 17497–17551 (2022)
102. El-Tanani, M., Satyam, S.M., Rabbani, S.A., El-Tanani, Y., Aljabali, A.A., Al Faouri, I., Rehman, A.: Revolutionizing drug delivery: the impact of advanced materials science and technology on precision medicine. Pharmaceutics **17**(3), 375 (2025)
103. Sarma, L., Singh, S., Dutta, D.: Challenges and Future Directions for Next-Generation Drug Delivery Systems, pp. 571–596. In Next-Generation Drug Delivery Systems. Humana, New York (2025)

Protein and Lipid Nanomaterial Based Functional Construct

Yamini Thakur, Anuradha Choudhary, Mamta Tripathi, Rama Sarojinee, Amit Kumar Sharma, and Reetu Sharma

Abstract Protein and lipid nanomaterial-based functional constructs represent a cutting-edge platform for targeted drug delivery, diagnostics, and therapeutic modulation. These hybrid nanomaterials provide improved stability, controlled release, and precise targeting capabilities by fusing the specificity and functional variety of proteins with the biocompatibility and customisable structure of lipids. Protein components, such as antibodies, enzymes, or ligands, allow receptor-mediated targeting and immunological regulation, while lipid nanoparticles make it easier to transport hydrophobic medications, nucleic acids, and peptides. These constructions are being used more and more in the creation of vaccines, infectious diseases, cancer treatment, and neurological disorders. Stimulus-sensitive release and enhanced pharmacokinetics are further made possible by the combination of responsive design and sophisticated surface engineering. Protein and lipid hybrid nanoplatforms are anticipated to help with the existing issues of therapeutic resistance, off-target toxicity, and bioavailability as research progresses, opening the door for next-generation

Y. Thakur · A. Choudhary
Department of Chemistry, Govt. J Yoganandam Chhattisgarh College, Raipur, Chhattisgarh, India
e-mail: yamini.thakur2010@gmail.com

A. Choudhary
e-mail: anuradha.cgcollege@gmail.com

M. Tripathi
School of Studies in Chemistry, Pt. Ravishankar Shukla University, Raipur, Chhattisgarh, India
e-mail: mamtat320@gmail.com

R. Sarojinee
Department of Chemistry, Govt. D.B. Girls College, Raipur, Chhattisgarh, India
e-mail: rama.sarojinee@gmail.com

A. K. Sharma
Department of Chemistry, Atma Ram Sanatan Dharma College, University of Delhi, New Delhi, India

R. Sharma (✉)
Department of Chemistry, Netaji Subhas University of Technology, New Delhi, India
e-mail: ritus1066@gmail.com

B. Sharma et al. (eds.), *Nanostructured Polymers*, Engineering Materials,
https://doi.org/10.1007/978-981-95-3581-1_10

precision medicine. Recent developments and possible uses of these multipurpose nanostructures in several biomedical fields are highlighted in this abstract.

Keywords Lipid nanoparticles (LNPs) · Protein-based nanoparticles (PBNPs) · Functional nanoconstructs · Drug delivery · Nanomedicine

1 Introduction

Nanotechnology has permeated virtually every field and its potential continues to expand. Its integration with biomolecules like proteins and lipids amplifies its utility, leading to transformative advancements in medicine, agriculture, and environmental science [1]. Different biomolecules like proteins, polysaccharides, and synthetic polymers can be used for nanoparticle synthesis [2]. Protein nanotechnology is an emerging field that merges the unique physicochemical features of proteins with nanoscale technology. This integration leads to protein-based nanoparticles (PBNPs) of a colloidal nature, size ranging from 10–100 nm. Shape and size depend upon the method used for synthesis [3]. Their unique physicochemical properties make them highly suitable for application in drug delivery, diagnosis, and tissue engineering. Their amphiphilic nature facilitates interactions with both hydrophilic and hydrophobic solvents and drugs/molecules showing efficacy as versatile carriers in nanoparticles-based delivery system. PBNPs are of biological origin, biodegradable, metabolizable, and surface modifiable with conjugation of ligands and therapeutic agents [4].

Advancements in nanotechnology have revolutionized medical science, particularly in the development of novel drug delivery systems. Among the various nanomaterials explored, lipid-based nanostructures have gained prominence for their ability to form multifunctional constructs. These constructs not only facilitate the efficient delivery of therapeutic agents but also enhance bioavailability, reduce systemic toxicity, and enable targeted therapy [5]. The structural resemblance of lipid nanomaterials to biological membranes facilitates cellular uptake and intracellular trafficking, making them ideal candidates for drug and gene delivery applications.

Lipid nanomaterial-based functional constructs are engineered platforms where lipid-based nanoscale carriers are modified or integrated with functional components such as targeting ligands, imaging agents, and stimuli-responsive elements. These constructs enable site-specific drug delivery, real-time imaging, and controlled therapeutic release. Their multifunctionality makes them essential tools in modern medicine, especially in precision oncology, infectious disease management, and personalized therapeutics.

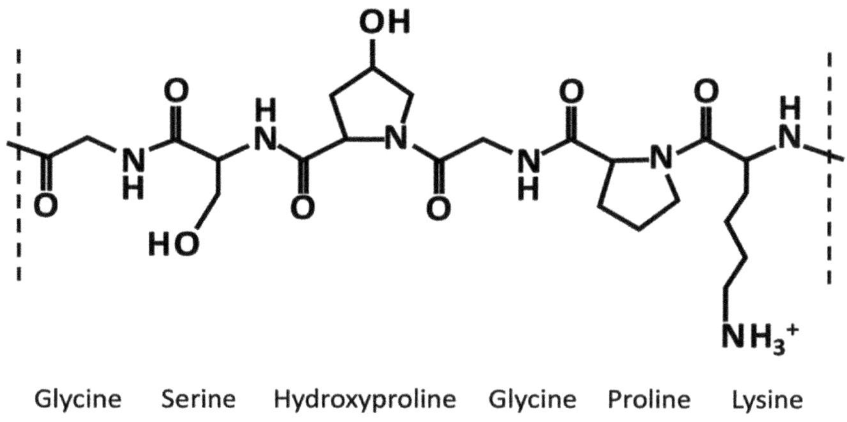

Fig. 1 Schematic chemical structure of a fragment of α-chain of gelatin showing the repetition of the triplet (Gly-X-Y)n, where X and Y are usually proline and hydroxyproline, respectively [9]

2 Protein Nanomaterial Based Functional Construct

2.1 *Types of Protein Nanoparticles*

2.1.1 Gelatin Nanoparticles (GNPs)

Gelatin is derived from the hydrolysis of collagen. It is a low molecular weight polypeptide consisting of 18 amino acid residues except cystine and cysteine. It is polyampholyte in nature as it contains both cationic and anionic groups along with hydrophobic groups. Around 13% of the polypeptide chain of gelatin has positively charged amino acids mainly lysine and arginine residues, 12% has negatively charged amino acid residues mainly glutamic and aspartic acid, and 11% hydrophobic residues i.e. valine, methionine, leucine and isoleucine (Fig. 1) [6]. Acidic and basic hydrolysis of collagen gives gelatin type A and type B respectively. The IEP of type A is 7–9 whereas type B has IEP around 4–5 [7]. Studies show that type B has more drug-releasing and delivery potential than type A [8].

2.1.2 Albumin

It is the most abundant and one of the most important plasma proteins. It is derived mainly from human serum (HSA), bovine serum (BSA), or egg white (ovalbumin) [10]. Albumin is a globular protein composed of a single polypeptide chain of approximately 585 amino acids, with a molecular weight of 66.5 kDa [11]. HSA has a 3D heart-shaped structure (Fig. 2). It has three main parts called domains: Domain I (first section, amino acids 1–195), Domain II (middle section, amino acids 196–383), Domain III (last section, amino acids 384–585). Each domain is further divided into

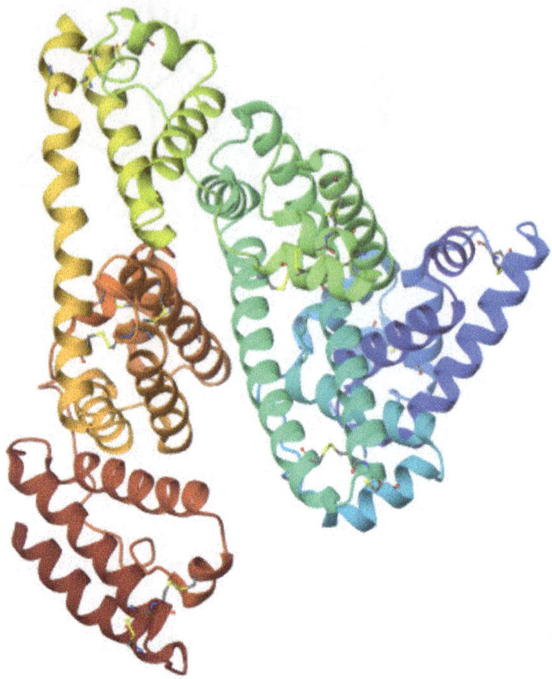

Fig. 2 Structure of human serum albumin (PDB ID: 1AO6)

subdomains A and B. [12]. The whole protein is a single chain and is about 80 × 30 Å. Stability of this structure is attributed to 17 disulfide bonds formed by 34 cysteine amino acids. It contains mainly charged amino acids (like lysine, arginine, glutamic acid, and aspartic acid) which help it interact with drugs. Because it has positive, negative, and hydrophobic groups it can carry and bind various types of drugs easily [13, 14].

2.1.3 Casein

Casein (Latin word '*caseus*' meaning *cheese*) is the primary protein found in milk. Approximately 80% of all milk protein is casein. The protein fraction which is approx. 93% of the dry mass is composed of different types- α_{s1-}, α_{s2-}, $\beta-$, and k-casein, which differ in primary structure and the remaining 7% of the dry mass is colloidal calcium phosphate [15]. Different models have been proposed to explain casein micelle structure. Submicelle Model proposes that casein micelles are made of small, spherical submicelles (Figs. 3 and 4). The core, rich in α and β casein is hydrophobic, while the outer surface is hydrophilic due to the presence of k-casein. Colloidal calcium phosphate is present between the submicelles crosslinking them [16]. The Nanocluster Model suggests that the homogenous protein matrix is embedded with calcium phosphate like cherrystone. It indicates that micelle

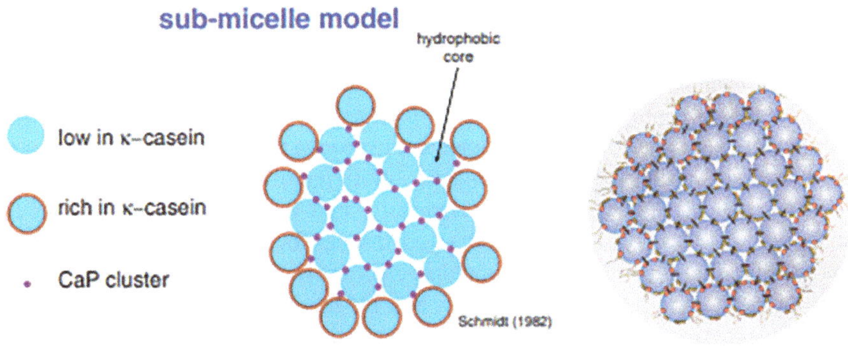

Fig. 3 Structure of casein micelles (sub-micelle model) [16]

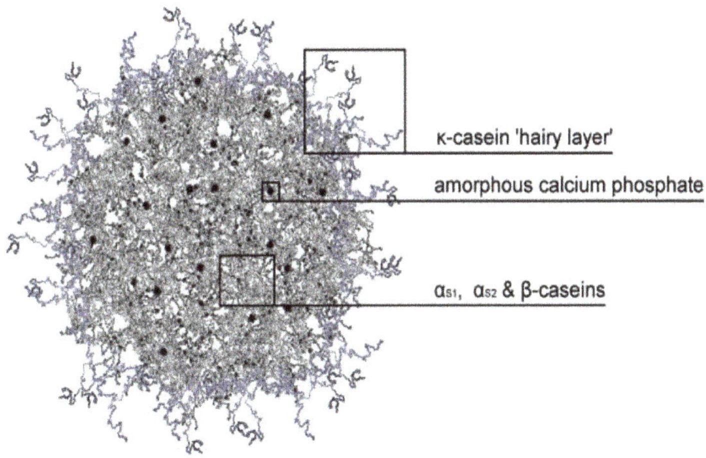

Fig. 4 A schematic representation of casein micelle model proposed by Holt [17]

assembly is formed due to weak interactions such as hydrophobic, hydrogen bonding, electrostatic interactions, and others [17]. The hydrophobic core and hydrophilic surface enable it to encapsulate hydrophobic drugs/molecules and protect them from oxidation, temperature, or UV light [18–20].

2.1.4 Zein

Zein, a class of prolamin protein, is a plant protein mainly found in maize. John Gorman was the first to isolate and characterize it in 1821. Zein consists of four peptides: α, β, γ, and δ. α-zein (19-22 kDa) is the most abundant form, with nearly 80% of the total zein, and consists of two prolamin groups. About 10-15 % of the total protein is β-zein of molecular weight 17 kDa with methionine as the main constituent. γ (27-28 kDa) and δ zein (9-10 kDa) are present in small fractions of 5-10 % and 3% respectively. The presence of non-polar amino acids such as proline, leucine, and alanine contributes to its insolubility in water, facilitating self-assembly into nanoparticles. The helical rod-like conformation of zein protein is stabilized by hydrophobic interactions and intra-molecular hydrogen bonding. These interactions also offer reactive sites for drug binding and surface modification, enabling the fabrication of stable zein NPs capable of encapsulating both hydrophobic and hydrophilic bioactives. ZNPs are generally synthesized through bottom-up methods such as nanoprecipitation, anti-solvent precipitation, electrospraying, solvent emulsification-evaporation, heat-induced self-assembly, and pH-driven methods. A corn-derived protein, ideal for oral drug delivery due to its resistance to gastric degradation.

2.1.5 Silk Fibroin

Offers mechanical strength and biocompatibility, useful in tissue engineering applications.

3 Synthesis of Protein NPs

3.1 GNPs

The following processes are used to synthesize GNPs:

(a) Physicochemical Process: It involves precipitation/flocculation of colloids. It includes desolvation, precipitation, and coacervation.

(b) Mechanical Process: Specific equipment is used in the mechanical process. It includes electrospray, spray drying, and emulsion [21].

Desolvation Method: In this method, a desolvating agent (e.g. alcohol or acetone) is added to the aqueous solution of gelatin to dehydrate it and bring conformational changes (from stretched to coil). The next step is crosslinking it with an agent to stabilize the nanoparticles formed [22]. In the two-step desolvation process, the high molecular weight (HMW) gelatin is precipitated and the low molecular weight (LMW) gelatin is discarded in the first step followed by redissolution and solvation of HMW gelatin. This gives smaller and uniform NPs [23].

Coacervation Method: Coacervation involves the phase separation of a polymer-rich colloidal solution into two distinct phases, a dense phase at the bottom and a dilute liquid phase supernatant. In simple coacervation ethanol or salt is added gradually to the gelatin solution thereby reducing the solubility of gelatin leading to phase separation and formation of coacervates (NPs) [24]. In complex coacervation, electrostatic interaction between two oppositely charged polymers (polyelectrolytes) causes coacervation [25]. A crosslinking agent like GA is then added to stabilize the GNPs formed [26].

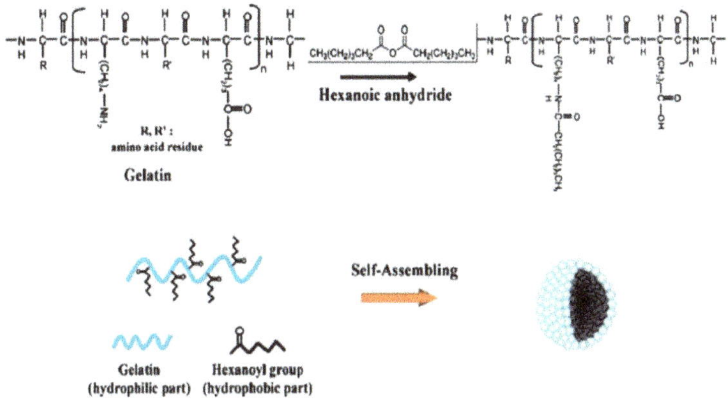

Fig. 5 Showing the formation of GNPs through self-assembly [30]

Emulsion: In this method aqueous solution of gelatin is emulsified into an oil phase e.g. paraffin oil or an organic solvent polymethylmethacrylate [27, 28]. To stabilize the gelatin droplets, crosslinking agents such as glutaraldehyde or genipin are added [15].

Reverse-phase microemulsion: In this method, an aqueous solution of gelatin is added to the surfactant solution, sod bis(2-ethylhexyl) sulfosuccinate in n-hexane followed by crosslinking with GA. The solvent (n-hexane) is then evaporated to get the GNPs [29].

Self- Assembly: Gelatin NPs can be formed through a self-assembly process by means of chemical modification in which the hydrophilic part of gelatin is modified to attach with hydrophobic group (Fig. 5). One common approach involves grafting hydrophobic hexanoyl groups to the amino groups of gelatin. In aqueous solution, this helps gelatin to self-assemble into micelles or NPs [30].

Nanoprecipitation: It involves the precipitation of gelatin from an organic solvent into an aqueous phase in which gelatin is not soluble. Aqueous solution of gelatin and drug (solvent phase) is added slowly to ethanol (non-solvent phase) containing poloxamer as a stabilizer. NPs formed due to interfacial turbulence during solvent displacement are then crosslinked with GA [31].

3.2 Synthesis of Albumin NPs

Following are the common methods of preparation of albumin NPs:

Desolvation Method: In this method ethanol (desolvating agent) is added to the aqueous solution of albumin with constant stirring until the whole solution becomes turbid (Fig. 6). As the solubility of albumin decreases due to the addition of ethanol, it separates as the albumin phase and aggregates [32]. Then it is crosslinked with glutaraldehyde to get the albumin nanoparticles (Fig. 7). Studies show that the pH of the medium greatly affects the size of nanoparticles, the higher the pH, the smaller the nanoparticle size [33].

Emulsification Method: Scheffel and his coworkers discovered this method in 1972 to prepare albumin spheres. In this method, aqueous solution of albumin (aqueous phase) is added to the organic phase (oil). Homogenization at high speed or ultrasonication of the resulting mixture consequently forms nanoemulsion which can be stabilized through crosslinking with glutaraldehyde or thermal treatment at temp above 120 °C [35, 36, 37]. Organic solvent is evaporated by solvent diffusion or low-pressure solvent evaporation with the help of rotary evaporator [38]. Thus NPs are obtained.

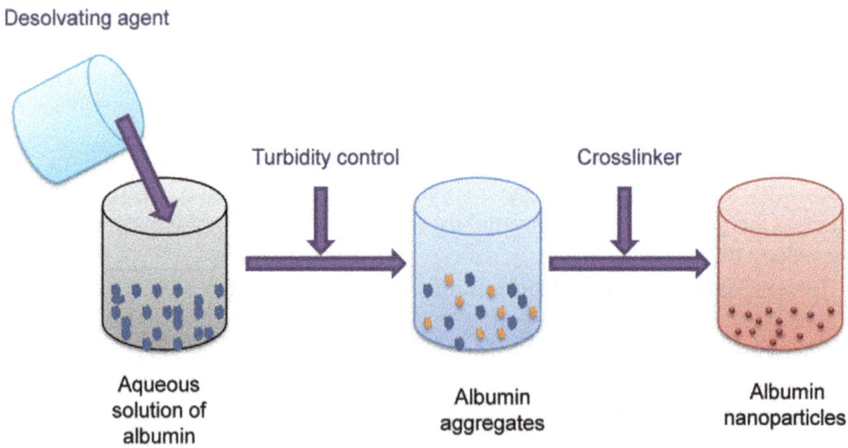

Fig. 6 Desolvation process for the formation of albumin NPs [34]

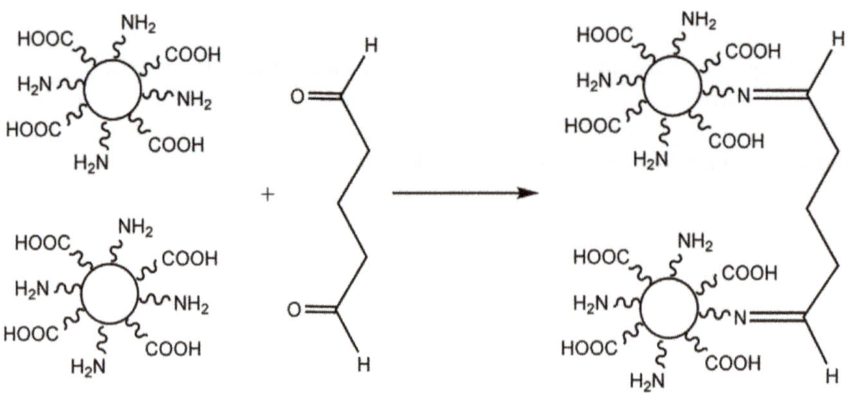

Fig. 7 Crosslinking of albumin nanoparticles with glutaraldehyde [34]

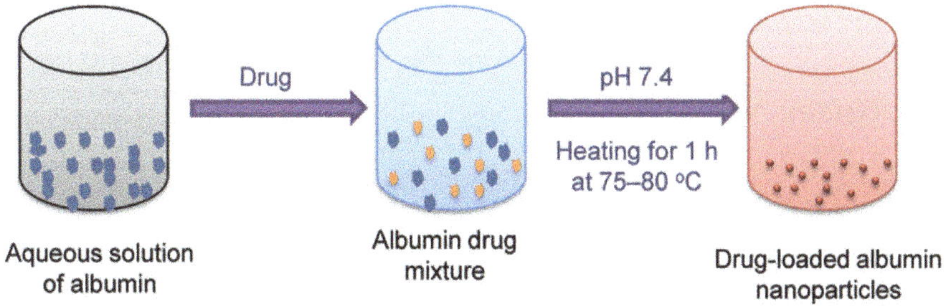

Fig. 8 Thermal gelation method for albumin NPs formation [34]

Thermal Gelation: In this method, heat is used to prepare albumin nanoparticles from albumin (Fig. 8). Heating causes irreversible changes in the structure of albumin from alpha helix to beta sheets. These beta-sheets aggregate due to hydrophobic and electrostatic interactions, hydrogen bonds, and disulfide bonds creating stable nanoparticles [35, 39].

Nab (Nanoparticle albumin-bound) Technique: It is a special method to prepare albumin nanoparticles that encapsulate and carry lipophilic drugs [40]. Firstly, aqueous solution of albumin acting as aqueous phase is mixed with organic phase containing drug. Then the water in oil (w/o) phase is homogenized under high pressure to get nanoemulsion. Organic solvent is then evaporated using rotary evaporator under vacuum. Albumin nanopowder can be collected after freeze-drying the supernatant of the centrifuge [41, 42]. Abraxane (nab-paclitaxel) is the first nab technology-based approved drug and is used in the treatment of breast cancer, lung cancer, pancreas, and advanced gastric cancer [40].

Nanospray Drying: This method involves the following steps: The first step is formation of droplets of albumin solution by atomization. A vibrating mesh containing large number of uniform laser-drilled holes is used for this. When the solution passes through these holes, tiny droplets are formed. In the second step, a flow of hot gases (nitrogen and a small amount of CO_2) is made to pass through the droplets drying and converting them into solid particles [43]. In the last step, particles get negatively charged after being given high voltage (15 kV) for collecting them efficiently [40].

3.3 Methods of Synthesis of Casein

There are several methods of synthesis of casein nanoparticles. Some important methods are Self-Assembly, Heat-induced Gelation, Ionic Gelation, Nanoprecipitation, and Emulsification-Solvent Evaporation [18].

4 Application of PBNPs

4.1 Gelatin NPs

Drug delivery: A drug or gene may be incorporated into the GNPs by being trapped within the nanoparticle matrix during the preparation or by adsorption onto the surface of the already formed NPs. Drug loading in GNPs can occur due to physical entrapment of the drug, electrostatic interaction, covalent conjugation, hydrophobic interaction, or hydrogen bonding between the drug and GNPs [44]. GNPs have been significantly used for the targeted delivery of anticancer drugs like cisplatin, curcumin, cytarabine, doxorubicin, methotrexate, 17- AAG, paclitaxel, resveratrol, cycloheximide, and noscapine [5].

Protein, vaccine, and gene delivery: Proteins and peptides can be delivered into the body using GNPs. Li et al. have studied a composite system inclusive of BSA-loaded GNPs encapsulated in poly(lactic-co-glycolic acid) microspheres and found that they help release the protein very slowly and safely [45, 46]. GNPs can also deliver vaccines by targeting special immune cells called dendritic cells (DCs). When used to deliver tetanus toxoid (TT) vaccine in mice, GNPs produced a stronger immune response than regular vaccine materials like alum TT [47]. Multiple plasmids can be carried by GNPs to the cells thus protecting DNA from being broken down [48].

Tissue Engineering and Regenerative Medicine: Electrospun gelatin nanofibers mimic the extracellular matrix, promoting cell adhesion and proliferation. These structures are employed in wound healing, bone regeneration, and as scaffolds for tissue engineering applications [49, 50].

Food Industry and Agricultural Applications: GNPs are incorporated into biodegradable films for food packaging and preservation of food for a long time inhibiting microbial growth [51]. GNPs are explored as carriers for bioactive compounds in functional foods, enabling controlled release and improved stability of nutrients and supplements. GNPs have been used in agriculture as their small size and positive surface charge facilitate better adhesion to plant surface leading to enhanced pest control and reduced environmental impact [52].

4.2 Albumin Nanoparticles

Cancer therapy: Albumin is a natural polymer found in blood plasma. Its properties, such as biocompatibility, non-immunogenicity, and non-toxicity, make it a suitable candidate in cancer theranostics such as gas therapy, chemodynamic therapy (CDT), sonodynamic therapy (SDT), phototherapy (PTT/PDT), radiotherapy, chemotherapy, immunotherapy and cancer imaging [53]. Albumin-based nanoparticles can target tumor tissues through enhanced permeability and retention (EPR) effect, facilitating

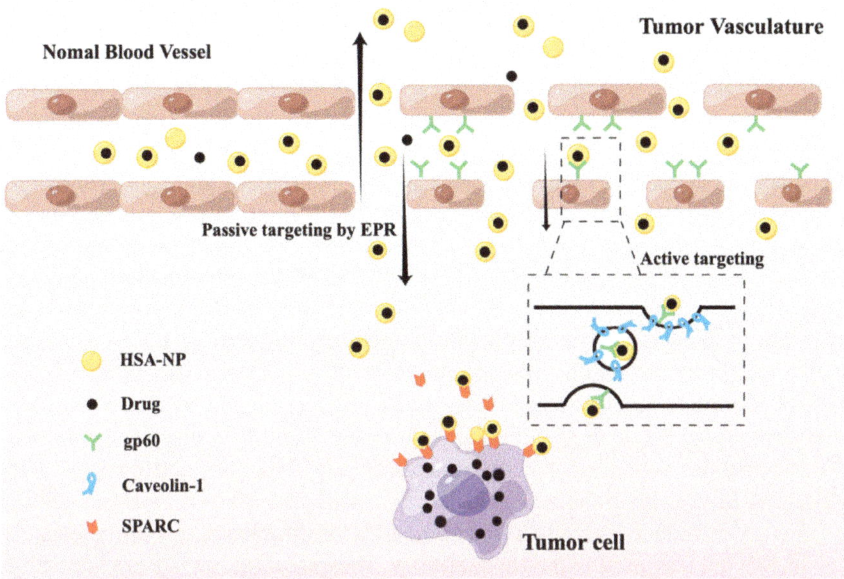

Fig. 9 A schematic representation of drug delivery by albumin NPs to tumor cells [54]

passive targeting. Another mode is active targeting. Receptors like gp60/albondin (a 60 kDa glycoprotein), found on the tumor endothelial cell surface, bind with albumin-bound drug, activate caveolin-1, and facilitate its transport across the cell through endocytosis or transcytosis. In the interstitial fluid space, SPARC (secreted protein acidic rich in cysteine) (Fig. 9) protein attracts albumin to bind with it, enhancing its accumulation inside the tumor [54].

Inflammatory Disease Treatment: Inflammatory diseases occur when neutrophils adhere to the lining of blood vessel walls and migrate into tissues. When using albumin nanocarriers for drug delivery, they are bound to Fcγ receptors present on the neutrophil surface. This is called hitchhiking. By attaching to these cells, the NPs can travel along with the neutrophils and get delivered to the target. Li et al. used modified BSA nanoparticles loaded with celastrol drug, which induced apoptosis in inflammatory neutrophils to cure rheumatoid arthritis. Researchers have prepared BSA-loaded NPs to treat rheumatoid arthritis and endometriosis successfully [55, 56].

Regeneration Process: Albumin-based drug delivery system has emerged as a great tool for wound healing [57], regeneration of bone tissues [19], and spinal cord injury repair [20].

Gene and Nucleic Acid Delivery: Albumin NPs serve as potential vectors for the delivery of siRNA, miRNA, and plasmid DNA [58, 59].

4.3 Applications of Casein NPs

Owing to unique physicochemical properties casein NPs have emerged as a potential drug delivery system for targeted delivery and controlled release of drugs to tumors or specific tissues. Zahariev et al. developed casein-based NPs for the effective delivery of Daunorubicin in acute Lymphocytic Leukemia and found significant cytotoxicity in the Reh lymphoblastic cell line. The effect of Curcumin encapsulated casein NPs was investigated on MCF-7 and A549 cell lines by Barick et al. and Jenifer et al. in separate studies, and it was found that casein loaded drugs have enhanced toxicity towards tumor cells than the pure drug. In research performed by Zhen et al., inhibition of hepatic H22 tumor growth in mice was found to be 1.5 times greater when cisplatin-loaded casein NPs were used than free cisplatin. Casein-coated iron oxide NPs or functionalized casein surface-modified iron oxide NPs have been used as biomarkers for enhancing MRI contrast imaging. These have also been employed as nano-capsular vehicles for nutraceuticals such as lycopene and omega-3 fatty acids. Casein micelle stabilized gold nanoparticles have shown catalytic activity and are used to reduce aromatic nitro compounds (4-nitrophenol to 4-aminophenol). Casein NPs are also used in the food industry. They are used in biodegradable films to improve the oxidative stability of food materials.

5 Lipid Nanomaterial Based Functional Construct

Lipid-based nanomaterials have emerged as a transformative platform in biomedical engineering, offering multifunctional constructs that enable targeted therapy, controlled release, and integrated diagnostics. Due to their inherent biocompatibility, amphiphilic nature, and structural similarity to biological membranes, lipid nanomaterials serve as versatile carriers for therapeutic and diagnostic agents. Herein is a comprehensive overview of lipid nanomaterial-based functional constructs, exploring their design strategies, classification, surface modifications, and biomedical applications with a focus on cancer therapy, gene delivery, and vaccine development.

5.1 Types of Lipid Nanomaterials

The classification of lipid nanomaterials is primarily based on their morphology, lipid composition, and physical state. The main types include liposomes, solid lipid nanoparticles (SLNs), nanostructured lipid carriers (NLCs), and lipid-polymer hybrid nanoparticles (Fig. 10).

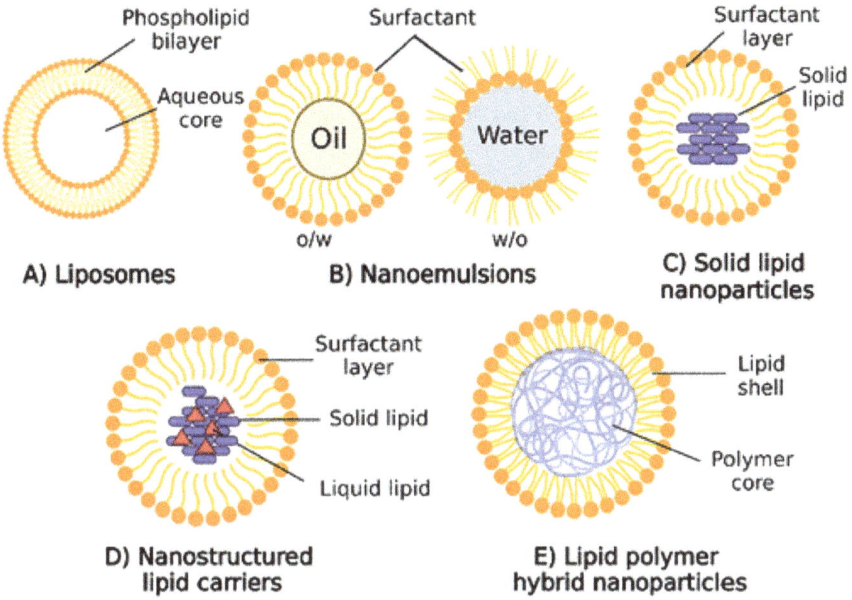

Fig. 10 The structure of several LNP formulations used in drug delivery is shown in the schematic picture. **A** Hydrophilic medications are encapsulated in the aqueous core of liposomes, which are spherical vesicles with phospholipid bilayers. Hydrophobic drugs are included into the lipid bilayers. **B** Oil droplets dispersed in an aqueous phase stabilized by surfactants make up nanoemulsions, which allow lipophilic medicines to be present in the oil phase while inhibiting droplet agglomeration. **C** Hydrophobic medicines are entrapped by solid lipid matrices to generate solid lipid nanoparticles (SLNs), which are nanoscale particles with a lipid core. **D** Nanostructured lipid carriers (NLCs) are comparable to SLNs, but they have a better drug loading capacity and a more stable matrix since they contain both liquid and solid lipids. **E** Lipid polymer hybrid nanoparticles offer the advantages of both systems by combining components based on polymers and lipids [60]

5.1.1 Liposomes

Liposomes are spherical vesicles comprising one or more phospholipid bilayers surrounding an aqueous core. Discovered in the 1960s, they have since become one of the most widely used nanocarriers in drug delivery [61]. Liposomes can encapsulate both hydrophilic and lipophilic drugs, protecting them from degradation and enhancing therapeutic efficacy. Their surface can be modified with polyethylene glycol (PEG), antibodies, or peptides to improve circulation time and target specificity.

5.1.2 Nanoemulsions

Nanoemulsions are a different kind of LNP made up of 50–500 nm-sized spherical biphasic liquid droplets. They are made up of an exterior continuous phase that covers an internally distributed oil phase. Oil-in-water (o/w) or water-in-oil (w/o) droplets can be used to create nano-emulsions that contain hydrophilic or hydrophobic active ingredients, respectively. The colloidal nature of nanoemulsions allows hydrophobic medications to be encapsulated and dissolved, minimising side effects.

5.1.3 Solid Lipid Nanoparticles (SLNs)

SLNs consist of a solid lipid core stabilized by surfactants. These nanoparticles are solid at both room and body temperatures, offering better physical stability and controlled drug release compared to traditional liposomes. SLNs are suitable for the delivery of lipophilic drugs and have been explored in the treatment of various diseases, including cancer and neurodegenerative disorders [62].

5.1.4 Nanostructured Lipid Carriers (NLCs)

NLCs represent the second generation of lipid nanoparticles. They are composed of a mixture of solid and liquid lipids, which improves drug loading capacity and prevents drug expulsion during storage. NLCs exhibit improved bioavailability and have been applied in dermal, oral, and parenteral drug delivery systems [63].

5.1.5 Lipid-Polymer Hybrid Nanoparticles

These are composite systems that combine the advantages of lipid-based and polymeric nanoparticles. The lipid shell offers biocompatibility and bioadhesiveness, while the polymeric core provides structural integrity and controlled release [64]. This hybrid approach has shown great promise in delivering chemotherapeutics and genetic material.

5.2 Synthesis of Lipid Nanoparticles

Nanoparticle synthesis generally follows one of two strategies: bottom-up or top-down approaches. The bottom-up approach involves assembling nanoparticles from atomic or molecular units, employing methods such as gas-phase synthesis, block copolymer templating, the Turkeyvich method, and microbial synthesis. In contrast, the top-down approach breaks down bulk materials into nanoscale particles using physical techniques like milling, spark ablation, or laser ablation [65–67].

For lipid nanoparticles (LNPs), synthesis is predominantly achieved through bottom-up wet chemistry methods, reflecting their colloidal and self-assembling nature. This section outlines key LNP synthesis techniques, including nanoprecipitation, single and double emulsification, nonsolvent emulsification, thin-film hydration, microfluidics, and impingement jet mixing [68].

The selected synthesis method critically affects the physicochemical properties of LNPs—such as size, surface charge, and encapsulation efficiency—which in turn dictate their stability, biodistribution, and therapeutic efficacy. Therefore, choosing the appropriate method is essential for optimizing LNP performance across various biomedical applications.

Nanoprecipitation

Nanoprecipitation, first introduced by Fessi et al. in 1989, is a widely used method for encapsulating hydrophobic drugs within lipid nanoparticles (LNPs) [69]. This technique relies on a spontaneous precipitation process, where two miscible phases—an organic phase and an aqueous phase—are continuously mixed under moderate magnetic stirring to induce LNP self-assembly. The organic phase typically contains film-forming polymers, drug molecules, lipophilic surfactants, and organic solvents, while the aqueous phase comprises water and a stabilizer. Polymers used in this method can be either non-biodegradable, such as Eudragit, or biodegradable, such as polylactide (PLA), poly (lactide-co-glycolide) (PLGA), and poly-ε-caprolactone (PCL) [70, 71].

Following nanoparticle formation, the organic solvent must be removed to obtain a purified LNP suspension. This can be achieved through methods like dialysis, ultracentrifugation, rotary evaporation, or freeze-drying. Dialysis, the most common method, uses a semipermeable membrane to gradually diffuse the solvent into a surrounding buffer. Ultracentrifugation separates LNPs by density, allowing for recovery of particles while discarding unincorporated drugs and free lipids. Rotary evaporation removes solvent under reduced pressure and elevated temperature, leaving behind a concentrated nanoparticle formulation. Alternatively, freeze-drying involves freezing the suspension and removing the solvent via sublimation, resulting in a dry powder suitable for long-term storage [72–75].

The physicochemical characteristics of LNPs formed via nanoprecipitation—particularly particle size and encapsulation efficiency—are highly sensitive to process parameters, including stirring speed, organic-to-aqueous phase ratio, and the concentrations of lipids, surfactants, and drugs. Inadequate mixing or non-uniform lipid saturation can hinder spontaneous nucleation, leading to irregular particle formation and size heterogeneity. Incomplete mixing of the two phases before precipitation further contributes to inconsistent particle size distributions. Therefore, rigorous characterization of particle size, morphology, and surface properties is essential prior to biological testing or therapeutic application.

Single and Double Emulsification

The single emulsification method, originally developed by Gasko, is widely employed for formulating lipid nanoparticles (LNPs) encapsulating hydrophobic drugs [73]. This approach uses low-melting lipids such as stearic acid and Compritol 888 ATO, combined with surfactants like Epikuron 200 and Tween 80, and emulsifiers including polysorbate 20 and polysorbate 600 [75]. In this process, the lipid-solvent mixture is preheated, then emulsified with water to form an oil-in-water (o/w) emulsion. The emulsion is subsequently added to cold water under continuous stirring, inducing lipid crystallization and LNP formation.

A variant of this technique is the solvent-based emulsification method, which includes strategies such as emulsion-solvent evaporation, solvent diffusion, solvent displacement, and solvent injection [75]. In these approaches, lipids and hydrophobic drugs are first dissolved in an organic solvent, then emulsified with an aqueous solution to form o/w emulsions. The solvent is then removed—typically through evaporation under agitation—followed by purification via centrifugation. While effective, these techniques can pose concerns regarding residual solvent toxicity, particularly when lipid solubility in the chosen solvent is suboptimal.

For hydrophilic drug delivery, the double emulsification technique is preferred. This method involves the formation of a water-in-oil (w/o) primary emulsion by dispersing an aqueous solution into a melted lipid-solvent phase. This w/o emulsion is then emulsified into another aqueous phase containing a hydrophilic emulsifier to form a water-in-oil-in-water (w/o/w) double emulsion. The process can be facilitated using sonication or membrane emulsification techniques [76]. Typically, two surfactants are used: a hydrophobic surfactant stabilizes the internal w/o interface, while a hydrophilic surfactant stabilizes the outer o/w interface. This method has been successfully used to encapsulate drugs such as Baclofen, Idarubicin, Ketoprofen, Nevirapine, and Tobramycin [77].

Despite its advantage of low mechanical energy input, emulsification methods also present limitations, including low dispersion efficiency, emulsion instability, lipid insolubility in certain organic solvents, and the need for additional solvent removal steps. Like other LNP synthesis methods such as nanoprecipitation and microfluidic mixing, solvent removal is essential to minimize toxicity and ensure the safety and functionality of the final LNP formulation.

Nonsolvent Emulsification

Nonsolvent emulsification offers a solvent-free strategy for producing oil-in-water (o/w) emulsions using melted lipids as the lipid phase [78]. This method avoids the use of organic solvents, making it more environmentally friendly and potentially safer for biomedical applications. Commonly employed techniques include high-pressure homogenization and ultrasonication. In the homogenization process, melted lipids are mixed with drug compounds and a preheated aqueous phase containing surfactants. The mixture is then subjected to several cycles of high-pressure homogenization to achieve nanoparticles of the desired size. Alternatively, ultrasonication involves high-speed stirring of melted lipids and surfactant-rich aqueous phase, followed by application of ultrasonic waves to reduce droplet size and form nanoscale emulsions.

Despite its solvent-free nature, this technique involves complex, multistep processes that can introduce batch-to-batch variability. Key factors—including homogenization/sonication time, lipid-to-surfactant ratio, drug concentration, and the chemical nature of lipids and surfactants—can significantly influence particle size, polydispersity index (PDI), zeta potential, and drug encapsulation efficiency. To ensure product uniformity and reproducibility, meticulous optimization and control of process parameters are essential.

Thin Film Hydration

Thin film hydration is a widely established method for lipid nanoparticle (LNP) synthesis in both research laboratories and industrial settings. In this process, lipids are first dissolved in an organic solvent to ensure uniformity and are then subjected to rotary evaporation, which removes the solvent and deposits a thin lipid film on the walls of a round-bottom flask. Upon hydration with an aqueous buffer, the film spontaneously forms multilamellar vesicles, which are then passed through polycarbonate membrane filters using extrusion to produce uniform-sized nanoparticles [74].

Several parameters—such as applied pressure, hydration temperature, membrane pore size, extrusion force, and number of extrusion cycles—directly impact the mean particle size and PDI of the resulting LNPs. The extrusion process is preferred due to its low energy requirements, minimal risk of contamination, and consistent production of nanoparticles with narrow size distribution. Industrial-scale extruders such as LIPEX and LiposoFast LF-50 have been successfully used to produce liposomes and LNPs encapsulating a wide range of payloads, including small molecules, peptides, proteins, and larger biological agents.

Microfluidic Process

The microfluidic process is widely recognized as one of the most effective and scalable methods for the industrial production of lipid nanoparticles (LNPs). It operates on the principle of nanoprecipitation, where controlled and rapid mixing of lipid and drug solutions occurs within microchannels. Unlike conventional bulk mixing methods, microfluidics offers superior control over mixing kinetics, enabling the formation of nanoparticles with greater uniformity and reproducibility. The microscale dimensions of the channels enhance mass and heat transfer, which further contributes to the generation of LNPs with well-defined size, morphology, and narrow polydispersity.

A key advantage of the microfluidic approach is the ability to precisely tune formulation parameters—including flow rate, solvent-to-aqueous phase ratio, and component concentrations—to optimize LNP characteristics for specific applications. This tunability allows researchers to customize particle size, drug loading capacity, encapsulation efficiency, and colloidal stability, making microfluidics a versatile tool in the development of tailored nanomedicines.

Microfluidic systems typically consist of microchannels with dimensions in the tens of microns, where fluids are manipulated with high precision. Two primary types of microfluidic platforms are commonly used: (a) Chip-based systems, where organic and aqueous phases are mixed within a microfabricated chip containing precisely designed channels, and (b) Capillary-based systems, in which the organic solvent is introduced through a central capillary and surrounded by outer tubes that deliver the aqueous solution for rapid dilution and nanoparticle assembly. These systems offer scalable, automated, and reproducible synthesis, positioning microfluidics as a key technology in the production of LNPs for both pharmaceutical research and commercial drug development [79, 80].

Impingement Jet Mixer (IJM)

Impingement jet mixing (IJM), often referred to as the "tea stirrer" approach, is a highly efficient microfluidic mixing technique increasingly utilized for the small-scale production of mRNA-loaded lipid nanoparticles (LNPs). This method operates by directing two high-velocity fluid streams to collide head-on, generating intense turbulent mixing and shear forces that rapidly homogenize the lipid and aqueous phases. During LNP synthesis, lipids dissolved in an organic solvent are mixed with an aqueous phase containing the drug or nucleic acid payload. The resulting high-shear environment promotes spontaneous self-assembly of lipid droplets, which are then stabilized by surfactants to form uniform LNPs. By precisely tuning key processing parameters such as flow rate, fluid pressure, and temperature, the size, polydispersity, and encapsulation efficiency of the nanoparticles can be effectively controlled.

One of the core advantages of IJM technology is its scalability. Despite its compact design, the system can be scaled up using parallelization strategies without significant redesign, making it highly suitable for pharmaceutical manufacturing—especially during high-demand scenarios such as the COVID-19 pandemic. Companies such as Knauer and Pfizer have successfully implemented IJM systems in commercial-scale vaccine production. At Knauer, the IJM system operates at pressures of up to 400 psi, enabling efficient mixing of lipid and mRNA solutions. Pfizer, in its large-scale production at the Kalamazoo facility, scaled IJM systems by parallelizing 100 static mixers, achieving a production capacity of over 100 million mRNA vaccine doses per month. This large-scale operation is fully automated, with computerized control of pressure and flow rates to ensure process consistency and product quality [81].

Overall, IJM represents a powerful, scalable, and high-throughput solution for producing uniform, stable, and therapeutically effective LNPs, and continues to play a pivotal role in the advancement of mRNA therapeutics and nanomedicine.

5.3 *Applications of LNPs*

The versatility of lipid nanomaterials allows them to be applied across a broad spectrum of medical disciplines.

Cancer Therapy

Lipid nanomaterial-based constructs are extensively studied in oncology for their ability to enhance the delivery of chemotherapeutic agents, overcome multidrug resistance, and reduce systemic toxicity. PEGylated liposomes such as Doxil® are approved for clinical use and have demonstrated significant improvement in patient outcomes [82]. Recent developments include lipid nanoparticles delivering siRNA or immunomodulators for combination therapies in solid tumors.

Gene Delivery

Cationic lipid nanoparticles can complex with negatively charged nucleic acids, facilitating gene transfer into target cells. Lipid-based vectors are used to deliver DNA, siRNA, and mRNA, with significant success in recent mRNA vaccine development [83]. These systems are being further developed for gene editing technologies like CRISPR-Cas9.

Vaccine Delivery

The COVID-19 pandemic highlighted the potential of lipid nanoparticles in delivering mRNA vaccines. LNPs used in Pfizer-BioNTech and Moderna vaccines provided stability, cellular uptake, and endosomal escape for efficient translation of mRNA into antigenic proteins [84]. This success has spurred research into lipid-based vaccines for influenza, Zika, and cancer.

Neurological Applications

Lipid nanoparticles can cross the blood–brain barrier, offering new treatment possibilities for neurological diseases. Functional constructs are being explored for the delivery of neuroprotective agents, anti-inflammatory drugs, and genetic material to treat disorders like Alzheimer's, Parkinson's, and glioblastoma [85–87].

Infectious Diseases and Immunomodulation

Beyond vaccines, lipid nanomaterials are used for delivering antimicrobial agents and immunotherapies [88, 89]. They enhance drug solubility, protect labile compounds, and modulate immune responses to achieve better control over infectious and inflammatory diseases [90, 91].

6 Future Perspectives

The future of protein and lipid nanomaterials lies in smart, multifunctional systems capable of performing complex biological tasks. Integrating artificial intelligence, personalized medicine, and biomimetic strategies can enhance targeting accuracy and therapeutic outcomes. Research is also focusing on biodegradable and naturally derived protein and lipid sources to reduce toxicity and environmental impact.

Emerging technologies such as microfluidics and 3D printing offer new opportunities for the fabrication of tailored lipid nano constructs. Combining these platforms with omics data and patient-specific profiles will enable the next generation of precision nanomedicine.

7 Conclusion

Protein- and lipid-based nanomaterials have emerged as transformative platforms for designing functional constructs in drug delivery, diagnostics, and regenerative medicine. Their biocompatibility, tunable surface properties, and ability to encapsulate both hydrophilic and hydrophobic agents make them ideal candidates for precision nanomedicine. Lipid nanoparticles (LNPs), in particular, have demonstrated tremendous clinical relevance—especially in mRNA vaccine delivery—owing to their capacity for safe and efficient gene and drug delivery. Similarly, protein-based nanocarriers offer unique advantages such as enzymatic responsiveness, structural diversity, and inherent targeting capabilities. Advancements in formulation strategies—ranging from nanoprecipitation and emulsification to microfluidics and impingement jet mixing—have significantly enhanced the control over particle size, stability, and payload release. The integration of synthetic and biological components has further expanded the functional landscape of these nanosystems, enabling stimuli-responsive, targeted, and personalized therapeutic solutions. As the field progresses, continued interdisciplinary collaboration will be essential to overcome current challenges related to large-scale production, long-term stability, and regulatory compliance. With robust research and translational momentum, protein- and lipid-based nanoconstructs are poised to redefine the future of nanomedicine.

References

1. Altammar, K.A.: A review on nanoparticles: characteristics, synthesis, applications and challenges. Front. Microbiol. **14**, 1155622 (2023)
2. Mohanraj, V.J., Chen, Y.: Nanoparticles—a review. Tropical J. Pharm. Res. **5**(1), 561–573 (2006)
3. Saif, A., Anjum, L., Faisal, Z., Akram, N., Shah, Y.A., Irfan, R., Saeed, F., Afzaal, M., Shah, M.A.: Recent advances in protein-based nanoparticles and their applications in the delivery of bioactive compounds. Int. J. Food Prop. **26**(2), 2866–2880 (2023)
4. Mohanty, A., Park, I.K.: Protein—caged nanoparticles: a promising nanomedicine against cancer. PubMed Central **59**(1), 1–12 (2023)
5. Torchilin, V.P.: Recent advances with liposomes as pharmaceutical carriers. Nat. Rev. Drug Discov. **4**(2), 145–160 (2005)
6. Elzoghby, A.O.: Gelatin-based nanoparticles as drug and gene delivery systems: reviewing three decades of research. J. Control. Release **172**(3), 1075–1091 (2013)
7. Yasmin, R., Shah, M., Khan, S.A., Ali, R.: Gelatin nanoparticles: a potential candidate for medical applications. Nanotechnol. Rev. **6**(2), 191–207 (2017)
8. Bajpai, A., Choubey, J.: In vitro release dynamics of an anticancer drug from swellable gelatin nanoparticles. J. Appl. Polymer Sci. **101**, 2320–2332 (2006)
9. Milano, F., Masi, A., Madaghiele, M., Sannino, A., Salvadore, L., Gallo, N.: Current trends in gelatin-based drug delivery systems. Pharmaceutics **15**(5), 1499 (2023)
10. Copolovici, D.M., Langel, K., Eriste, E., Langel, U.: Cell-penetrating peptides: design, synthesis and applications. ACS Nano **8**(3), 1972–1994 (2014)
11. Elzoughby, A.O., Samy, W.M., Elgindy, N.A.: Albumin based nanoparticles as potential controlled release drug delivery system. J. Control. Release **157**(2), 168–182 (2012)

12. He, X.M., Carter, D.C.: Atomic structure and chemistry of human serum albumin. Nature **358**, 209–215 (1992)
13. Peters, T.: Serum albumin. Adv. Protein Chem. **37**, 161–245 (1985)
14. Peters, T., Jr.: All about albumin. Academic Press, Cambridge, MA, USA (1995)
15. De Kruif, C.G., Huppertz, T., Urban, V.S., Petukhov, A.V.: Casein micelles and their internal structure. Adv. Colloid Interf. Sci. **171–172**, 36–52 (2012)
16. Schmidt, D.G.: Association of caseins and casein micelle structure. In: Fox, P.F. (ed.) Developments in Dairy Chemistry, pp. 61–86. Applied Science Publishers, Barking, UK (1982)
17. De Kruif, C.G., Holt, C.: Casein micelle structure, functions and interactions. In: Fox, P.F., McSweeney, P.L.H. (eds.) Advanced Dairy Chemistry. Proteins, 3rd edn. Kluwer Academic/ Plenum Publishers, New York (2003)
18. Spanu, C., Camorani, S., Tortorella, S., Agnello, L., Maturi, M., Franchini, M.C., Laura, C., Locatelli, E.: Synthesis and functionalization of casen nanoparticles with aptamers for triple-negative breast cancer targeting. New J. Chem. **46**, 21995 (2022)
19. Farjaminejad, S.; Farjaminejad, R.; Garcia-Godoy, F. Nanoparticles in Bone Regeneration: A Narrative Review of Current Advances and Future Directions in Tissue Engineering. J Funct Biomater. 15(9), 241 (2024)
20. Lin, Y., Wan, Y., Du, X., Li, J., Wei, J., Li, T., Li, C., Liu, Z., Zhou, M., Zhong, Z.: TAT-modified serum albumin nanoparticles for sustained-release of tetramethylpyrazine and improved targeting to spinal cord injury. J. Nanobiotechnol. **19**, 28 (2021)
21. Milano, F., Masi, A., Madaghiele, M., Sannino, A., Salvadore, L., Gallo, N.: Current trends in gelatin-based drug delivery systems. Pharmaceuticals **15**(5), 1499 (2023)
22. Farrugia, C.A., Groves, M.J.: Gelatin behaviour in dilute aqueous solution: designing a nanoparticle formulation. J. Pharm. Pharmacol. **51**, 643–649 (1999)
23. Coester, C., Langer, K., Briesen, H.V., Kreuter, J.: Gelatin nanoparticles by two-step disolvation- a new preparation method, surface modification and cell uptake. J. Microencapsul. **17**, 187–193 (2000)
24. Mohanty, B., Aswal, V.K., Kohlbrecher, J., Bohidar, H.B.: Synthesis of gelatin nanoparticles via simple coacervation. J. Surface. Sci. Technol. **21**(3–4), 149–160 (2005)
25. Marty, J., Oppenheim, R., Speiser, P.: Nanoparticles—a new colloidal drug delivery system. Pharm. Acta Helv. **53**, 17 (1978)
26. Leo, E., Vandelli, M.A., Cameroni, R., Forni, F.: Doxorubicin-loaded gelatin nanoparticles stabilized by glutaraldehyde: involvement of the drug in the crosslinking process. Int. J. Pharm. **155**, 75–82 (1997)
27. Choubey, J., Bajpai, A.: Investigation on magnetically controlled delivery of doxorubicin from superparamagnetic nanocarriers of gelatin crosslinking with genipin. J. Mater. Sci. Mater. Med. **21**, 1573–1586 (2010)
28. Avossa, J., Herwig, G., Toncelli, C., Itel, F., Rossi, R.M.: Electrospinning based on benign solvents: current definitions, implications and strategies. Green Chem. **24**, 2347–2375 (2022)
29. Gupta, A.K., Gupta, M., Yarwood, S.J., Curtis, A.S.G.: Effect of cellular uptake of gelatin nanoparticles on adhesion, morphology and cytoskeleton organization of human fibroblasts. J. Control. Release **95**, 197–207 (2004)
30. Li, W.M., Liu, D.M., Chen, S.Y.: Amphiphilically—modified gelatin nanoparticles: self-assembly behavior, controlled biodegradability, and rapid cellular uptake for intracellular drug delivery. J. Mater. Chem. **21**, 12381–12388 (2011)
31. Lee, E.J., Khan, S.A., Park, J.K., Lim, K.H.: Studies on the characteristics of drug-loaded gelatin nanoparticles formed by nanoprecipitation. Bioprocess Biocyst. Eng. **35**, 297–307 (2012)
32. Karimi, M., Avci, P., Mobasseri, R., Hamblin, M.R., Naderi-Manesh, H.: The novel albumin-chitosan core shell nanoparticles for gene delivery; preparation, optimization and cell uptake investigation. J. Nanoparticle Res. **15**, 1–14 (2013)
33. Langer, K., Balthasar, S., Vogel, V., Dinauer, H., Briesen, H., Schubert, D.: Optimization of the preparation process for human serum albumin (HAS) nanoparticles. Int. J. Pharm. **275**, 169–180 (2003)

34. Shrivastava, A., Prajapati, A.: Albumin and functionalized albumin nanoparticles: production strategies, characterization, and target indications. Asian Biomed. **14**(6), 217–242 (2020)
35. Elzoghby, A.O., Samy, W.M., Elgindt, N.A.: Albumin-based nanoparticles as potential controlled drug delivery systems. J. Control. Release **157**, 168–182 (2012)
36. Tarhini, M., Greige-Gerges, H., Elaissari, A.: Protein-based nanoparticles: from preparation to encapsulation of active molecules. Int. J. Pharm. **522**, 172–197 (2017)
37. Rai, A., Jenifer, J., Upputuri, R.T.P.: Nanoparticles in therapeutic applications and role of albumin and casein nanoparticles in cancer therapy. Asian Biomed. **11**, 3–20 (2017)
38. Szczech, M., Szczepanowicz, K.: Polymeric core-shell nanoparticles prepared by spontaneous emulsification solvent evaporation and functionalized by the layer-by-layer method. Nanomaterials (Basel) **10**, 496 (2020)
39. Loureiro, A., Azoiaa, N.G., Gomesb, A.C., Cavaco-Pauloa, A.: Albumin-based nanodevices as drug carriers. Curr. Pharm. Des. **22**, 1371–1390 (2016)
40. Desai, N.: Nanoparticle albumin-bound paclitaxel (Abraxane(R)). In: Albumin in Medicine, pp. 101–119. Springer (2016)
41. Thao, L.Q., Byeon, H.J., Lee, S., Lee, E.S., Choi, H.-G., et al.: Pharmaceutical potential of tacrolimus loaded albumin nanoparticles having targetability to rheumatoid arthritis tissues. Int. J. Pharm. **497**, 268–276 (2016)
42. Yu, X., Di, Y., Xie, C., Song, Y., He, H., Li, H., et al.: An in vitro and in vivo study of gemcitabine loaded albumin nanoparticles in a pancreatic cancer cell line. Int. J. Nanomed. **10**, 6825–6834 (2015)
43. Arpagaus, C.: PLA/PLGA nanoparticles prepared by nano spray drying. J. Pharm. Invest. **49**, 405–426 (2019)
44. Kommareddy, S., Shenoy, D.B., Amiji, M.M.: Gelatin nanoparticles and their biofunctionalization. In: Kumar C.S.S.R. (ed.) Nanotechnologies for the Life Sciences, vol. 1, pp. 330–352. Biofunctionalization of Nanomaterials, WILEY-VCH Verlag GmbH & Co. KGaA, Weinheim (2005)
45. Li, J.K., Wang, N., Wu, X.S.: A novel biodegradable system based on gelatin nanoparticles and poly (lactic-co-glycolic acid) microspheres for protein and peptide drug delivery. J. Pharm. Sci. **86**, 891–895 (1997)
46. Li, J.K., Wang, N., Wu, X.S.: Gelatin nanoparticles for encapsulation of protein/peptide drugs using an emulsifier-free emulsion method. J. Microencapsul. **15**, 163–172 (1998)
47. Sudheesh, M.S., Vyas, S.P., Kohli, D.V.: Nanoparticles based immunopotentiation via tetanus toxoid loaded gelatin and animated gelatin nanoparticles. Drug Deliv. **18**, 320–330 (2011)
48. Kaul, G., Amiji, M.: Protein nanoparticles for gene delivery. In: Amiji, M. (ed.) Polymeric Gene Delivery: Principles and Applications, pp. 429–447. CRC Press LLC, Boca Raton, FL (2005)
49. Mohite, P., Puri, A., Munde, S., Dave, R., Khan, S., Patil, R., Singh, A.K., Tipduangta, P., Singh, S., Chittasupho, C.: Potential of chitosan/gelatin-based nanofibers in delivering drugs for the management of varied complications: a review. Polymers **17**, 435 (2025)
50. El-Seedi, H.R., Said, N.S., Yosri, N., Hawash, H.B., El-Sherif, D.M., Yaseen, M., Omar, H., Shou, Q., Attia, N.F., Zou, X., Guo, Z., Khalifa, S.A.M.: Gelatin nanofibers: recent insights in synthesis, bio-medical applications and limitations. Heliyon **9**(5), e16228 (2023)
51. Wu, X., Liu, Z., He, S., Liu, J., Shao, W.: Development of an edible food packaging gelatin/zein based nanofiber film for the shelf-life extension of strawberries. Food Chem. **426**, 136652 (2023)
52. Park, S., et al.: Gelatin nanoparticles can improve pesticide delivery performance to plants. Small **20**(10), e2401234 (2024)
53. Shen, X., Liu, X., Li, T., Chen, Y., Chen, Y., Wang, P., Zheng, L., Yang, H., Wu, C., Deng, S., Liu, Y.: Nanovehicles towards potential cancer diagnosis and therapy. Front. Chem. **9**, 746646 (2021)
54. Qu, N., Song, K, Ji, Y., Liu, M., Chen, L., Lee, R.J., Teng, L.: Albumin nanoparticles-based drug delivery systems. Int. J. Nanomed. **19**, 6945–6980 (2024)

55. Liu, S., Liu, M., Xiu, J., et al.: Celastrol-loaded bovine serum albumin nanoparticles target inflamed neutrophils for improved rheumatoid arthritis therapy. Acta Biomater **174**, 345–357 (2024)
56. Zhu, S., Zhang, J., Xue, N., et al.: Highly specific neutrophil-mediated delivery of albumin nanoparticles to ectopic lesion for endometriosis therapy. J Nanobiotechnology. **21**(1), 81 (2023)
57. Huang, Z., Xie, J., Gao, J., Feng, H., Liu, C.: Bovine serum albumin modified Prussian blue nanoparticles enhance wound healing through the promotion of skin angiogenesis. J. Drug Deliv. Sci. Technol. **107**, 106833 (2025)
58. Li, J., Wang, Y., Zhu, Y., Oupicky, D.: Recent advances in delivery of drug-nucleic acid combinations for cancer treatment. J. Control. Release **172**, 589–600 (2013)
59. Izquierdo, M.: Short interfering RNAs as a tool for cancer gene therapy. Cancer Gene Ther. **12**, 217–227 (2005)
60. Mehta, M., Bui, T.A., Yang, X., Aksoy, Y., Goldys, E.M., Deng, W.: Lipid-based nanoparticles for drug/gene delivery: an overview of the production techniques and difficulties encountered in their industrial development. ACS Mater. Au **3**, 600–619 (2023)
61. Allen, T.M., Cullis, P.R.: Liposomal drug delivery systems: from concept to clinical applications. Adv. Drug Deliv. Rev. **65**(1), 36–48 (2013)
62. Mehnert, W., Mäder, K.: Solid lipid nanoparticles: production, characterization and applications. Adv. Drug Deliv. Rev. **47**(2–3), 165–196 (2001)
63. Müller, R.H., Radtke, M., Wissing, S.A.: Nanostructured lipid matrices for improved microencapsulation of drugs. Int. J. Pharm. **242**(1–2), 121–128 (2002)
64. Zhang, L., Chan, J.M., Gu, F.X., Rhee, J.W., Wang, A.Z., Radovic-Moreno, A.F., Farokhzad, O.C.: Self-assembled lipid–polymer hybrid nanoparticles: A robust drug delivery platform. ACS Nano **2**(8), 1696–1702 (2008)
65. Shafique, M., Ur Rehman, M., Kamal, Z., Alzhrani, R.M., Alshehri, S., Alamri, A.H., Bakkari, M.A., Sabei, F.Y., Safhi, A.Y., Mohammed, A.M., et al.: Formulation development of lipid polymer hybrid nanoparticles of doxorubicin and its in-vitro, in-vivo and computational evaluation. Front Pharmacol. **14**, 1025013 (2023)
66. Hadinoto, K., Sundaresan, A., Cheow, W.S.: Lipid-polymer hybrid nanoparticles as a new generation therapeutic delivery platform: a review. Eur. J. Pharm. Biopharm. **85**(3 Pt A), 427−443 (2013)
67. Venishetty, V.K., Komuravelli, R., Kuncha, M., Sistla, R., Diwan, P.V.: Increased brain uptake of docetaxel and ketoconazole loaded folate-grafted solid lipid nanoparticles. Nanomedicine **9**(1), 111–121 (2013)
68. Musielak, E., Feliczak-Guzik, A., Nowak, I.: Synthesis and potential applications of lipid nanoparticles in medicine. Materials (Basel) **15**(2), 682 (2022)
69. Martinez Rivas, C.J., Tarhini, M., Badri, W., Miladi, K., Greige-Gerges, H., Nazari, Q.A., Galindo Rodriguez, S.A., Roman, R., Fessi, H., Elaissari, A.: Nanoprecipitation process: from encapsulation to drug delivery. Int. J. Pharm. **532**(1), 66–81 (2017)
70. Zhang, L.I., Zhang, L.: Lipid-polymer hybrid nanoparticles: synthesis, characterization and applications. Nano LIFE **01**(01n02), 163−173 (2010)
71. Bazylińska, U., Lewińska, A., Lamch, L., Wilk, K.A.: Polymeric nanocapsules and nanospheres for encapsulation and long sustained release of hydrophobic cyanine-type photosensitizer. Colloids Surf. A **442**, 42−49 (2014)
72. Streck, S., Neumann, H., Nielsen, H.M., Rades, T., McDowell, A.: Comparison of bulk and microfluidics methods for the formulation of poly-lactic-co-glycolic acid (PLGA) nanoparticles modified with cell penetrating peptides of different architectures. Int. J. Pharm. X **1**, 100030 (2019)
73. Gupta, A., Eral, H.B., Hatton, T.A., Doyle, P.S.: Nanoemulsions: formation, properties and applications. Soft Matter **12**(11), 2826–2841 (2016)
74. Ganesan, P., Narayanasamy, D.: Lipid nanoparticles: different preparation techniques, characterization, hurdles, and strategies for the production of solid lipid nanoparticles and nanostructured lipid carriers for oral drug delivery. Sustain. Chem. Pharm. **6**, 37–56 (2017)

75. Battaglia, L., Trotta, M., Gallarate, M., Carlotti, M.E., Zara, G.P., Bargoni, A.: Solid lipid nanoparticles formed by solvent-in-water emulsion-diffusion technique: development and influence on insulin stability. J. Microencapsul. **24**(7), 672 (2007)
76. Vandergraaf, S., Schroen, C., Boom, R.: Preparation of double emulsions by membrane emulsification? a review. J. Membr. Sci. **251**(1–2), 7–15 (2005)
77. Gibaud, S., Attivi, D.: Microemulsions for oral administration and their therapeutic applications. Expert Opin. Drug Deliv. **9**(8), 937–951 (2012)
78. Dong, Y., Ng, W.K., Shen, S., Kim, S., Tan, R.B.: Solid lipid nanoparticles: continuous and potential large scale nanoprecipitation production in static mixers. Colloids Surf. B Biointerfaces **94**, 68–72 (2012)
79. Ding, S., Anton, N., Vandamme, T.F., Serra, C.A.: Microfluidic nanoprecipitation systems for preparing pure drug or polymeric drug loaded nanoparticles: an overview. Expert Opin. Drug Deliv. **13**(10), 1447–1460 (2016)
80. Liu, Y., Yang, G., Hui, Y., Ranaweera, S., Zhao, C.X.: Microfluidic nanoparticles for drug delivery. Small **18**(36), e2106580 (2022)
81. Hao, Y., Seo, J.H., Hu, Y., Mao, H.Q., Mittal, R.: Flow physics and mixing quality in a confined impinging jet mixer. AIP Adv. **10**(4), 045105 (2020)
82. Barenholz, Y.: Doxil®—the first FDA-approved nano-drug: lessons learned. J. Control. Release **160**(2), 117–134 (2012)
83. Hou, X., Zaks, T., Langer, R., Dong, Y.: Lipid nanoparticles for mRNA delivery. Nat. Rev. Mater. **6**(12), 1078–1094 (2021)
84. Pardi, N., Hogan, M.J., Porter, F.W., Weissman, D.: MRNA vaccines—a new era in vaccinology. Nat. Rev. Drug Discov. **17**(4), 261–279 (2018)
85. Dong, Y., Liu, Y., Su, X., Zhang, Y., Luo, Y., Pan, X., Sahay, G.: Lipid nanoparticles for mRNA delivery to the brain through the blood–brain barrier in vivo and in human brain organoids. Nat. Mater. (2025). https://doi.org/10.1038/s41563-024-01885-0
86. Tiwari, R., Yadav, D., Kumar, S., Sharma, A., Jadhav, J.: Lipid-based nanoformulations for the treatment of Alzheimer's disease: recent advances and perspectives. J. Nanobiotechnology **23**(59) (2025)
87. Praveen, G., Maheswari, R.U.: Surface-engineered lipid-based nanocarriers for targeting brain disorders: a comprehensive review. Front. Bioeng. Biotechnol. **10**, 963265 (2022)
88. Centre for Pharmaceutical Innovation: Lipid nanocarriers-enabled delivery of antibiotics and antimicrobial adjuvants to overcome bacterial biofilms. Pharmaceutics **16**(3), 396 (2024)
89. Meng, M., Li, Y., Wang, J., Zhang, H., Liu, X.: Innovative nebulization delivery of lipid nanoparticle–encapsulated siRNA: a therapeutic advance for Staphylococcus aureus–induced pneumonia. J. Transl. Med. **22**, 942 (2024)
90. Tang, C., Jing, W., Han, K., Yang, Z., Zhang, S.: MRNA-laden lipid nanoparticle-enabled in situ CAR macrophage engineering for eradication of multidrug-resistant bacteria in sepsis mouse models. ACS Nano **18**(3), 2261–2278 (2024)
91. Mohammadi, M., Arabestani, M.R., Kalhori, F., Soleimani Asl, S., Asgari, M., Hosseini, S.M.: Antibiotics encapsulated solid lipid nanoparticles as antimicrobial agents for wound infection treatment. Front. Immunol. **15**, 1435151 (2024)

Applications of Nanostructured Polymers

Nanopolymers for Energy Harvesting

Md. Arafat Hossain, Md. Wasikur Rahman, Victoria Padilla, Nazmul Islam, and Mohammed Jasim Uddin

Abstract Nanopolymers are a special class of nanomaterials that combine polymers with nanoscale features to provide improved mechanical, electrical, optical, chemical, and thermal properties. Nanopolymers have an essential role in our everyday life. They are used in clean energy harvesting due to their flexibility, biocompatibility, easy processability, lightweight, low cost, and friendliness. The innovative techniques have enabled scientists to synthesize nanopolymers with unique properties. Synthesizing nanopolymers and tailoring for desired properties broaden the application of nanopolymers. Nanopolymers are promising candidates for flexible and wearable energy harvesting devices. The flexible, self-powered, and wearable nanopolymer-based triboelectric and piezoelectric nanogenerators show outstanding performances in scavenging electrical energy from human body movement, wind energy, rain-drop energy, wave energy, and vibrational energy. It is crucial to tailor the properties of nanopolymers, such as the charge density, piezoelectricity, ferroelectricity, and dielectricity, to enhance the performance of nanopolymer-based energy harvesting devices.

Keywords Energy harvesting · Nanopolymers · TENG · PENG · Ecoflex 00-30

Md. A. Hossain · Md. W. Rahman · V. Padilla · M. J. Uddin (✉)
Photonics and Energy Research Laboratory, The University of Texas Rio Grande Valley, Edinburg, TX, USA
e-mail: uddinm3@sunypoly.edu

Department of Mechanical Engineering, The University of Texas Rio Grande Valley, Edinburg, TX, USA

N. Islam
Department of Electrical and Computer Engineer, The University of Texas Rio Grande Valley, Edinburg, TX, USA

Md. W. Rahman · M. J. Uddin
Department of Physics and Astronomy, The University of Texas Rio Grande Valley, Edinburg, TX, USA

M. J. Uddin
Department of Mechanical Engineering, SUNNY Polytechnic Institute, Utica, NY, USA

B. Sharma et al. (eds.), *Nanostructured Polymers*, Engineering Materials,
https://doi.org/10.1007/978-981-95-3581-1_11

1 Introduction

The term “nano” is the prefix used in conjugation with System International d’unites (SI units). It is the measuring unit for tiny objects which are one billionth of a meter (1 nm (nm) = 10^{-9} m). Nanostructure is the study of materials having at least one dimension within 1 to 100 nm. A nanoparticle can be considered as a zero-dimensional (0D), one-dimensional (1D), two-dimensional (2D), and three-dimensional (3D). 0D nanomaterials have all three-dimensions in the nanoscale range. Examples of OD nanomaterials are quantum dots, carbon nanodots, etc. 1D nano-materials have at least one of their three dimensions in the nanometer range. For example, nanowires, nanotubes, nanorods, and nanofibers. In 2D nanomaterials only one dimension (thickness) is on the nanoscale range. For example, graphene, hexagonal boron nitride (h-BN), and transition metal dichalcogenides (TMDs). On the other hand, three-dimensional materials have all three dimensions in nanometer range. Liposome, dendrimer, and polycrystalline are the examples of 3D nanomaterials. The classification of nanomaterials based on their dimensionality is shown in Fig. 1.

Nanopolymers specially combine the properties of repeating unit (monomer) and nanotechnology to enhance electro-chemical and mechanical properties. The dimension of nanopolymers could be in the range of 1 nm to 100 nm. In nano scale, nanopolymers have a large surface area to volume ratio providing them enhanced chemical reactivity. Example of nanopolymers are nanohydrogels, dendrimers, nanocells, micelles, and liposomes [1]. Nanopolymers have a crucial role in our everyday life due to their unique properties and diverse application. These materials are used in automotive, food packaging, adhesive, electrical insulation, medical equipment, pipe and hose, bulletproof vest, and many more. The manipulation of polymer structure at nanoscale significantly enhances their mechanical, electrical, thermal, and optical properties. For instance, polydimethylsiloxane is used as a sealant, adhesive, catheters, tubing, and flexible and biocompatible implants in their bulk form. Whereas, in the nano range, PDMS film behaves as a negative triboelectric materials owing to its large surface area and greater electron affinity [2]. One of the unique characteristics of nanopolymers is that their properties are tunable. The innovative processing technologies have provided excellent nanopolymers with various remarkable properties. Combining nanomaterial with nano-polymer enhances the mechanical and electrical properties. The nanopolymers exhibit piezoelectric [3], ferroelectric [4], pyroelectric properties [5, 6], and conductive properties [7, 8]. The classification of nanopolymers based on their properties and their application is shown in Table 1. Tuning the properties of nanopolymers made them suitable for a wide range of applications from energy harvesting [9–11] to in-vivo [8, 12] and in-vitro sensing [13, 14] for biomedical applications.

Recently, nanopolymer-based energy harvesting has become a popular technique for scavenging electrical energy from mechanical energy or ambient sources. The piezoelectric nanogenerator (PENG) [9] and triboelectric nanogenerator (TENG) [10] are promising nanopolymer-based energy harvesting candidates owing to their

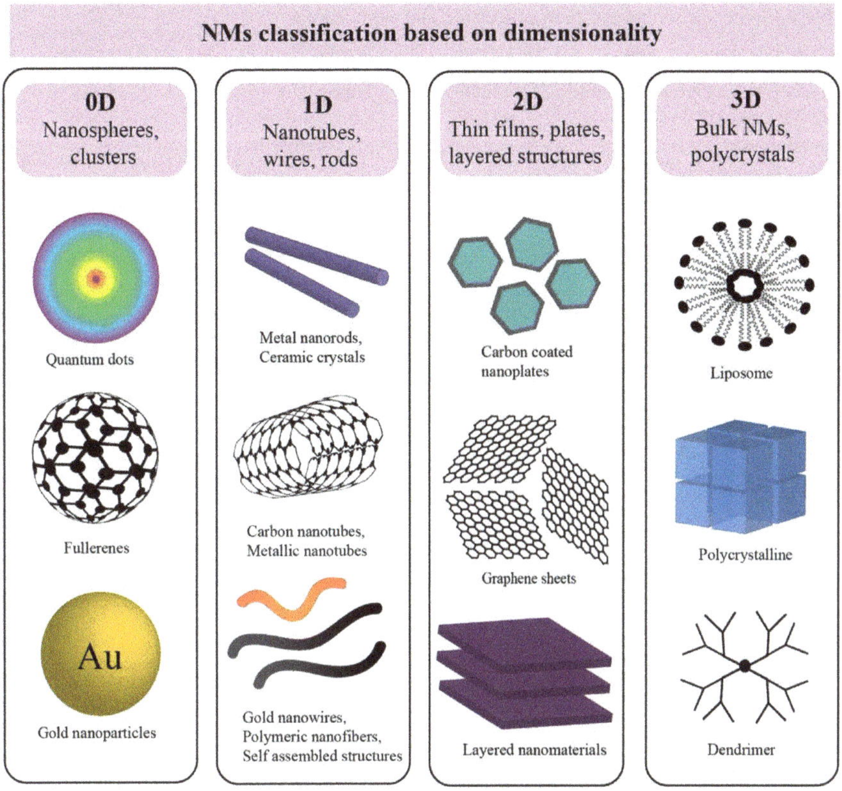

Fig. 1 Classification of nanomaterials based on their dimensionality: zero (OD), one (1D), two (2D) and three dimensional (3D) [22]

flexibility, biocompatibility, and wearability [11, 15]. These devices generate electricity from human walking, running, knee bending-stretching, and elbow bending-stretching [16–18], finger tapping [19, 20] are a few examples of them. After the invention of the Internet of Things (IoT), nanopolymers have been widely used in flexible, biocompatible, and wearable energy harvesting, portable electronic, and health monitoring devices [21]. Nanopolymer-based energy harvesting has advantages over ceramic-based owing to its flexibility, stretchability, and cyclic bending and loading. Nanopolymer-based energy harvesters have significantly advanced self-powered wearable technologies, enabling devices to convert ambient energy into electrical power efficiently. The continuous development of nanopolymers and their composites is expected to further revolutionize sustainable clean energy harvesting in the coming years.

This chapter provides an overview of nanopolymers, including their classification, their role in energy harvesting, and examples of some of the most widely used nanopolymers in energy harvesting devices.

Table 1 Some nanopolymers, their properties, and applications

Nanopolymers	Properties	Applications	References
Polyacetylene (PA), polyaniline (PANI), polypyrrole (PPy), polythiophene (PTH), poly(para-phenylene) (PPP), poly(phenylenevinylene) (PPV), and polyfuran (PF), poly (3,4-ethylenedioxythiophene)	Conductive	Supercapacitors, Corrosion inhibition, photocatalytic, Biomedical and antimicrobial, Electroch1emical gas sensors, Bioelectronics	[7, 8, 23]
(PEDOT) Polyvinylidene fluoride (PVDF), Poly-L-lactic acid (PLLA), Polyhydroxybutyrate (PHB), Polyamide, Collagen, Cellulose, Chitosan, Silk fibroin, Cellulose nanocrystals	Piezoelectric	Energy harvesters, Sensors, Transducer and actuators, and Physiological signal monitor	[3, 4, 24, 25]
PVDF, P(VDF-TrFE) Pyroelectric	Ferroelectric, Wearable bioelectric	Energy harvesters	[5, 6, 26–28]

2 Nanopolymers

Nanopolymers are an advanced class of materials that combine polymer with nanotechnology at nanoscale resulting unique properties. Incorporation of nanomaterials into polymer significantly enhance the chemical, mechanical, thermal, and electrical properties of the nanopolymers [29]. Nanopolymers have emerged as a promising material in energy harvesting, offering innovative solutions for alternative green energy. These materials, which combine polymers with nanoscale fillers, enhance the efficiency and functionality of energy-harvesting devices. The integration of nanopolymers into various energy systems, such as triboelectric [10], piezoelectric [9, 30], solar energy [31, 32] exhibits not only the output performance but also the required flexibility and processability required for self-powered wearable devices. The popular energy harvesting techniques are triboelectric nanogenerator (TENG) [10], piezoelectric nanogenerator (PENG) [9], and hybrid nanogenerator (HNG) [30]. When used as triboelectric materials, nanopolymers can convert mechanical energy such as human motion, vibration, and applying force/pressure into electrical energy [33]. Nanopolymers such as polytetrafluoroethylene (PTFE) flim [34], polydimethylsiloxane (PDMS) film [35], fluorinated ethylene propylene (FEP) film [36], nylon 6,6 fiber [37], and Ecoflex 00-30 film [38] have demonstrated excellent performance in TENGs. The addition of nanoparticles (carbon nanotube [39], multiwalled carbon nanotube [40], $BaTiO_3$ [41], $SrTiO_3$ [42]) enhance the triboelectric properties by increasing the surface charge density and improving the mechanical strength.

At nanoscale, polymers exhibit remarkable electrical conductivity [8], piezoelectricity [43, 44], ferroelectricity [26], and mechanical properties such as flexibility and stretchability [45, 46]. One of the advantages of nanopolymers is their tunable properties. Adding conductive, piezoelectric nanofiller, and doping increases the electrical conductivity of nanopolymers [47]. The properties of nanopolymers vary with their phases. For instance, PVDF is widely used as a fire and heat-resistance, optical wear resistance, biocompatible vascular sutures, pipes, and food packaging [48]. In nanoscale, PVDF film have several phases such as α-, β-, γ-, δ, and ε-phases [49]. The polar phases β-, γ-, and δ are the reason for the electroactive properties of nanopolymer, where all dipoles are parallel, resulting in a non-zero dipole moment [50]. In electroactive nanopolymers, β-phase has a crucial role in piezoelectric, pyroelectric, and ferroelectric properties [51]. The α-phase can be transformed into β-phase by cold spraying [52, 53]. The innovative fabrication techniques, electrospinning [54–56] and poling [57, 58], form the β-phase of nanopolymers. The piezo-polymers are widely used in PENGs due to their flexibility, piezoelectricity, and ferroelectricity [24]. Piezo-polymer PVDF [59] and its copolymers such as PVDF-HFP [60], P(VDF-TrFE) [61] are widely used in PENGs owing to their piezoelectric, ferroelectric, and dielectric properties. Piezo-polymers combined with piezoelectric nanomaterials significantly improve the performance of the PENG [62, 63]. For instance, Addition of piezoelectric ZnO to PVDF increases the piezoelectricity of ZnO@PVDF piezo-composite. The ZnO@PVDF piezo-composite is used for breathable and wearable nanogenerators due to the enhanced piezoelectricity as well as the output performance [64].

Nanopolymer-based PENGs have advantages over ceramic-based PENGs. Ceramic-based PENGs are stiff and brittle in nature, limiting their application in durable, reliable, and long term cyclic mechanical loading and bending applications. Meanwhile, nanopolymer-based PENGs are flexible, lightweight, stretchable, durable, and reliable for cyclic bending-stretching capabilities [65]. Nanopolymer-based HNGs integrate multiple energy harvesting techniques, such as triboelectric, piezoelectric, electromagnetic, and thermoelectric, into a single system to enhance energy harvesting efficiency [66]. These HNGs utilize the unique properties of nanopolymers, such as flexibility, high conductivity, piezoelectricity, and tunable electronic properties, to optimize energy scavenging from multiple sources like mechanical vibrations, external force, heat, and electromagnetic. Nanopolymer-based energy harvesters have emerged as a promising candidate for converting ambient mechanical energy into electrical energy, owing to the growing demand for flexible, stretchable, and self-powered wearable and portable electronic devices. They have many applications, such as energy harvesting, self-powered electronics, and biomedical implants. Despite these advancements, challenges remain in optimizing the energy conversion efficiency and long-term durability of nanopolymer-based energy harvesters. Recently, researchers have focused on improving the output performance of energy harvesters, developing novel nanopolymers, and exploring new device structures to overcome these limitations.

3 Nanopolymers in Nanogenerators

The nanopolymer-based energy harvesting devices are TENG, PENG, and HNG. TENG uses two nanopolymers: one is electron donating and another one is electron accepting. Nanopolymers in energy harvesting devices are used in the form of nanofilms, nanofibers, or nanocomposites. Usually, nanopolymers with higher electron affinity difference are selected in TENG for better performances. PENG uses a single nanopolymer having high piezoelectric constant or nanopolymer composite contaning piezoelectric material. HNGs are the combination of triboelectric, piezoelectric, electromagnetic, or thermoelectric effect. These energy nanogenerators are discussed below.

4 Energy Harvesting Nanogenerators

The most common nanopolymer-based energy nanogenerators are TENG, PENG, and HNG. The construction and working principle of these nanogenerators are discussed below.

4.1 Triboelectric Nanogenerator

Recently, TENG has emerged as a promising technology for harnessing various types of renewable energy from the ambient environment. TENG uses two different types of polymers, i.e., positive triboelectric and negative triboelectric polymers. When these two different layers make physical contact with each other, electrons transfer from the negative layer to the positive layer by triboelectrification. The positive charges accumulated on the positive layer and negative charges accumulate on the negative layer creating a potential difference between the layers [10]. The open circuit current generates through the external circuit due to the developed potential difference between the layers. TENG can be used to generate electricity from human motion, i.e., stepping, elbow bending-stretching, breathing, to name a few. TENG devices can operate in several modes-contact-separation, lateral sliding, freestanding, and single electrodes [67]; as shown in Fig. 2.

Contact-Separation Mode

In contact-separation mode, the tribolayers are placed on top of each other separated by spacers as shown in Fig. 2a. The positive and negative layers are separated by a certain distance and external circuit is connected electrodes attached on top of the layers. The commonly used electrodes are copper, aluminum, silver, etc. The spacers used in TENG are polyurethane foam [68] and compression springs [69]. The variable force is applied on top of either one of the layers. The positive and

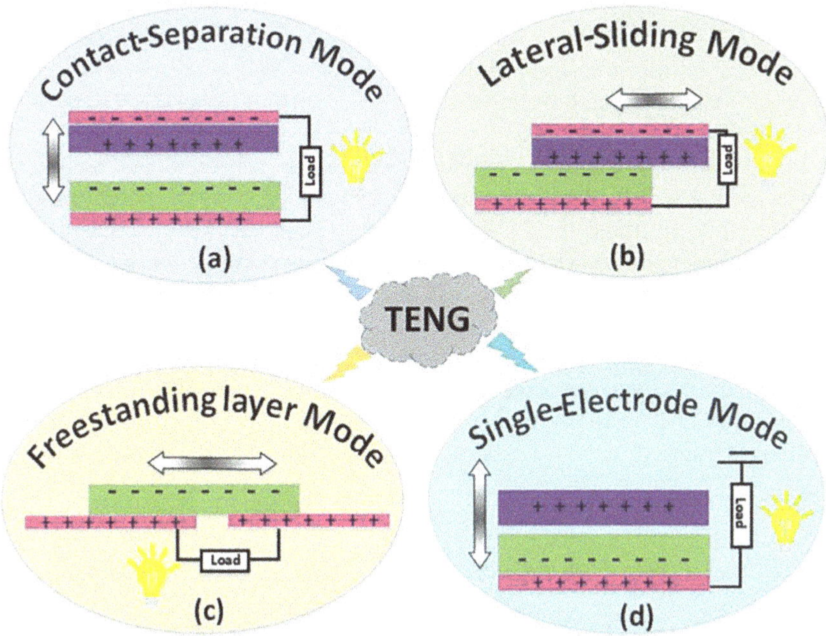

Fig. 2 Operating modes of triboelectric nanogenerators [67]

negative layers repeatedly contact each other due to the external applied force or pressure.

Later-Sliding Mode

In later-sliding mode, tribo-layers move relative to each other making friction between the contact interfaces, as shown in Fig. 2b [70]. In this mode, two layers rub each other in the tangential direction. The triboelectrification process occurred by sliding friction between the layers. A periodic change in the contact area between the two surfaces leads to lateral separation of the charge centers, which creates a voltage drop for driving the flow of electrons in the external load.

Freestanding Mode

The freestanding working mode is given in Fig. 2c. In this mode, triboelectric layer is forced to slide along/ or contact a similar plane with other stationary terminals [71]. Energy generation depends on the relative movement between dielectric and electrode, in which triboelectrification and electrostatic induction develop potential difference between electrodes. Electricity generation increases with increasing the relative movement of the tribolayer and electrodes.

Single-Electrode Mode

Single electrode mode consists of only one tribolayer [72], as shown in Fig. 2d. Triboelectrification depends on the periodic changes in the contact surface area of the tribolayer. The electrostatic induction creates free electrons and drives current to the external circuit. TENGs effectively converts mechanical energy into electrical energy. TENGs are most effective for converting low-frequency and low-amplitude mechanical energy into electrical energy, making them suitable for harvesting energy from the ambient environment. TENGs have been widely used in nano-energy harvesting [73–75], self-powered sensor [76], and high voltage sources [77, 78]. Recently, researchers have used binary oxides, such as ZnO nanoflakes [79–81], ZnO flowers [52], ZnO nanorods [82–84], ZnO nanowires [85–87], VO_2 [88], and TiO_2 [89, 90], to increase the performance of TENGs. Adding ternary oxides, $BaTiO_3$ [41, 91, 92] and $SrTiO_3$ [42] in TENGs promote the relative permittivity and charge density of the tribomaterials.

4.2 *Piezoelectric Nanogenerator*

A PENG is an energy harvesting device that converts mechanical energy into electrical energy under mechanical stress. PENG uses the piezoelectric phenomenon to generate electricity. When a piezoelectric material is subject to stress or external pressure, its crystalline structure deforms. The external stress creates a charge imbalance in the unit crystals, accumulating positive and negative charges on the opposite faces of the crystal, leading to generating a potential difference between the surface, as shown in Fig. 3. The working mechanism of film-based PENGs is shown in Fig. 4 as (a) the dipole makes placed alignment in a regular direction after poling process; (b) the piezopotential occurred after applied compressive forces and (c) reaches the maximum polarization density; (d) electronics flow back to equilibrium after the force is release [93]. PENGs readily convert mechanical energy from ambient energy into electrical energy, which provides a potential source of self-powered portable electronic devices. PENGs convert mechanical energy into electrical energy in the nanometer range, where the large surface area in the piezoelectric materials with microstructure exhibits better performance in PENGs. Compared to ceramics-based PENGs, nanopolymer-based PENGs have better flexibility and are able to absorb considerable stress without breaking the constituent piezoelectric materials. These properties made them ideal candidates for low power energy sources in the field of microelectronic systems. Theapplications of PENGs are [94–97], pressure sensor [98, 99], cardiovascular sensor [100, 101], etc. The application of PENG as energy harvester is shown in Fig. 5.

Recently, PENGs are emerging as promising technology for harvesting mechanical energy, making them a sustainable and eco-friendly alternative to the conventional energy sources. Often PENGs required high output performance for practical applications. The output performance of PENGs can be enhanced by tailoring the

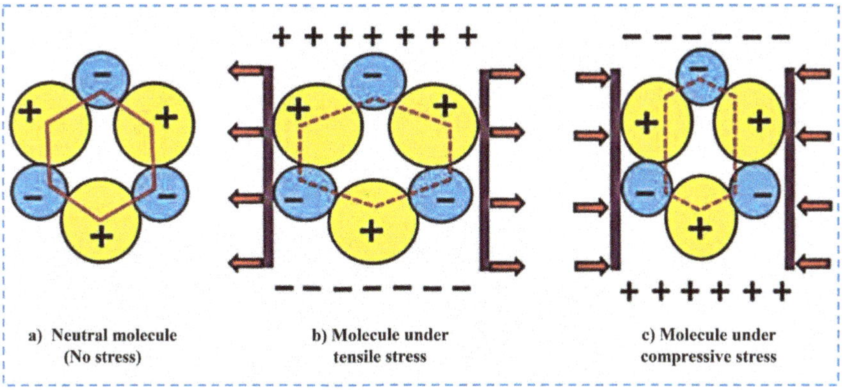

Fig. 3 Schematic representation of crystal structure distortion of piezoelectric materials under stress [58]

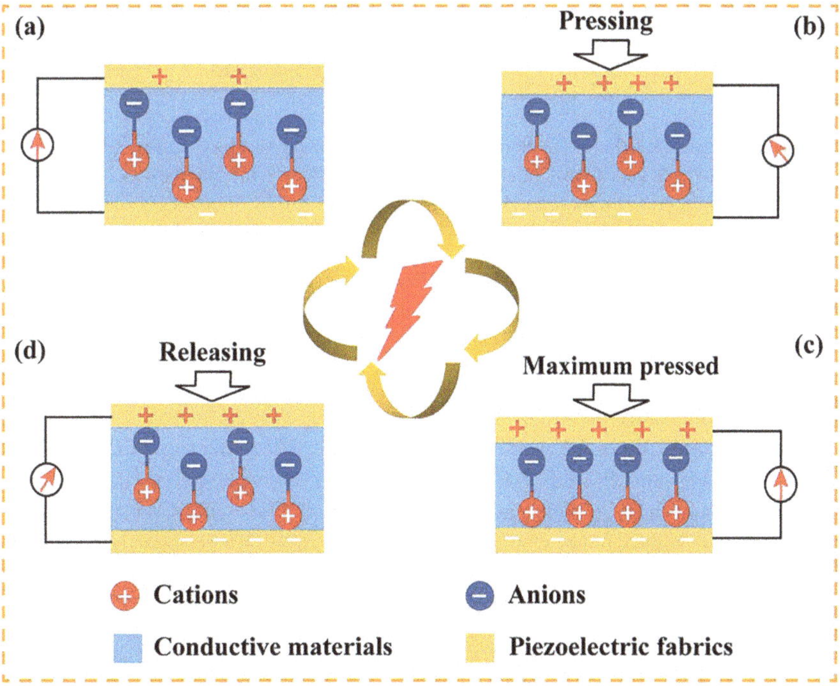

Fig. 4 Working mechanism of film-based PENGs. **a** The dipole makes placed alignment in a regular direction after poling process; **b** the piezopotential occurred after applied compressive forces and **c** reaches the maximum polarization density; **d** electronics flow back to equilibrium after the force is released [93]

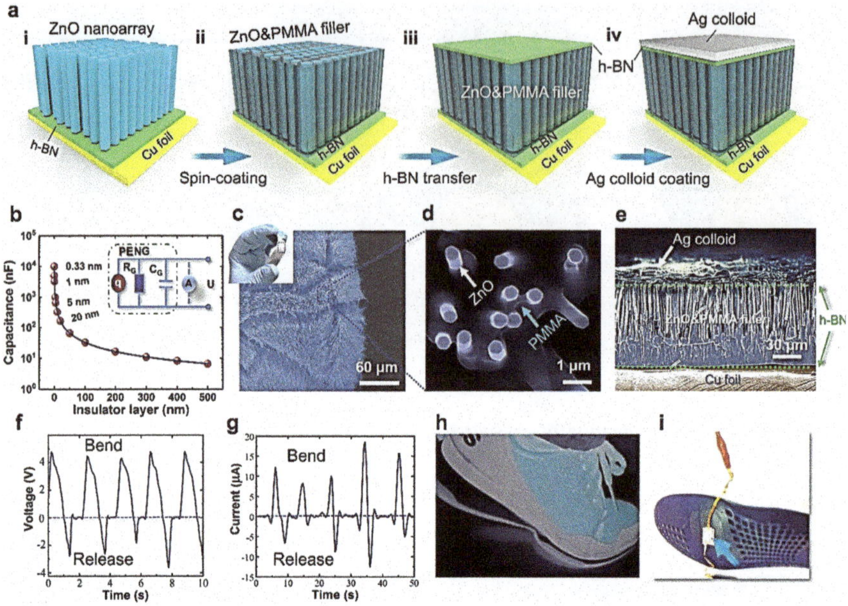

Fig. 5 **a** Schematic diagram for the fabrication of flexible nanogenerators based on ZnO nanoarray and h-BN. **b** equivalent capacitance of as-fabricated h-BN dielectrically mediated PENG in an area of 1 cm^2, inset shows equivalent circuit diagram of the PENG. **c, d** SEM images of ZnO nanoarray/h-BN/Cu paper spin-coated with PMMA. The inset is the photograph of ZnO nanoarray/h-BN/Cu paper. **e** Cross-sectional SEM image of as-fabricated nanogenerator. **f, g** Open-voltage and short-current of as-fabricated nanogenerators. **h, i** Practical wearable applications of the nanogenerators, the nanogenerators was used to collect and convert the body energy on insoles into electrical energy [113]

properties of nanopolymers. The piezoelectric nanopolymers as matrices or fillers play an important role in the performance of PENGs. Piezoelectric PVDF [4, 102], PVDF-HFP [44], and P(VDF-TrFE) [103–105] nanopolymers are widely used in PENGs. Adding binary oxides, ZnO [106, 107]; perovskites ($BaTiO_3$ [108], $KNbO_3$ [109, 110], and $NaNbO_3$ [111, 112]) nanofillers to nanopolymers enhances overall performance of PENGs.

4.3 Hybrid Nanogenerator

HNG is a new class of self-powered system that constitute of TENG and PENG, [68, 114] or TENG and Electromagnetic nanogenerator (EMG) [16], or the combination of TENG, PENG, and thermoelectric nanogenerator (ThEG) [115]. HNGs are designed to exploit the synergy of TENG, PENG, EMG, ThEG to improve the output efficiencies of energy harvesters needs for portable and self-powered wearable electronic devices. The combined electrical coupling mechanisms of TENG, PENG,

EMG, and ThEG effects enhance the electric output and energy conversion efficiency of hybrid generators, surpassing the sum of the individual contributions from each mechanism. The schematic of working mechanism of a HNG, PENG-TENG, is shown in Fig. 6. Here, Li – ZnO/PVDF/MWCNT piezocomposite acts as PENG as well as positive triboelectric material for TENG, whereas PTFE/PDMS acts as a negative triboelectric material for TENG. The top and bottom element aluminum and Ag act as electrode for the external electric circuit. The output voltage is the sum of the output voltage generated from the PENG (Li – ZnO/PVDF/MWCNT) PENG, and TENG (Li – ZnO/PVDF/MWCNT-PTFE/PDMS) under the external applied stress. An example of HNG, consists of TENG and EMG, for energy harvesting is shown in Fig. 7. In this case, HNG exploits the triboelectric and electromagnetic effect together to generate electrical energy. PTNG, a HNG, consists of PENG and TENG is shown in Fig. 8. PTNG is a mechanical energy harvesting device that uses piezoelectric and triboelectric phenomena. The HNG is used as real-time monitor for human activities, such as elbow bending (Fig. 8a), finger imparting (Fig. 8b), palm bending (Fig. 8c), and knee bending (Fig. 8d). The PTNG lit up to 40 LEDs (Fig. 8e) and power up a wrist watch (Fig. 8f) and a digital calculator (Fig. 8i).

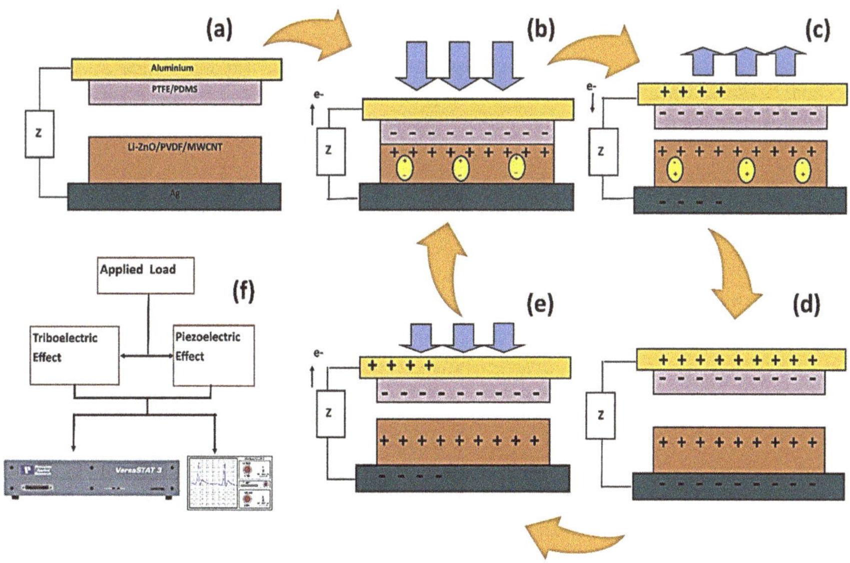

Fig. 6 Working mechanism of hybrid PENG-TENG under applied stress **a** initial condition of the PTENG (no applied stress) **b** PTENG in full contact mode (stress applied) **c** PTENG going back to its initial position (stress release) **d** PTENG gets back to its initial position (stress removed) **e** PTENG going towards full contact mode (stress applied) **f** Block diagram: Measurement of Triboelectric and Piezoelectric response of the PTENG with VERASTAT3 and Oscilloscope [116]

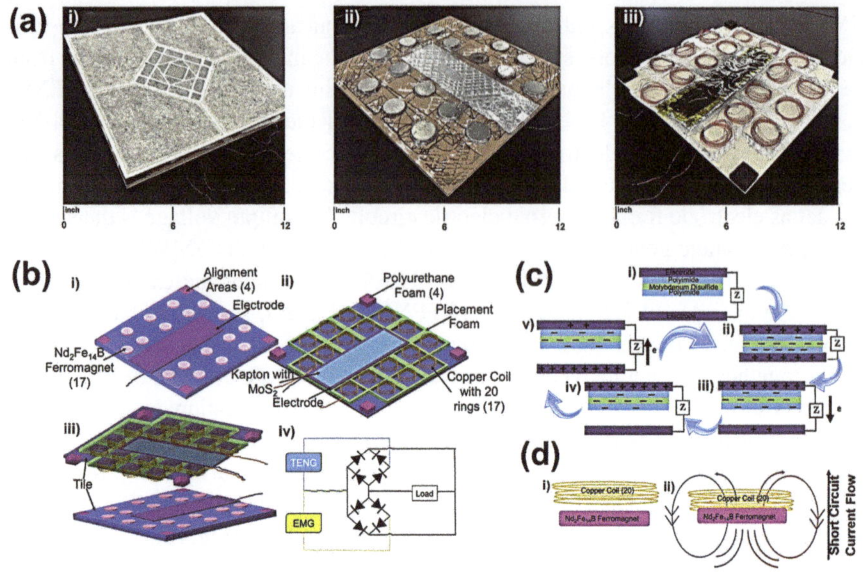

Fig. 7 Working mechanism of TENG-EMG. Optical image of the energy tile (i) complete tile, (ii) bottom part, (iii). top part **b** internal structure of the tile (i) bottom part, (ii) top part, (iii) alignment of the top and bottom part, (iv) combined rectifier circuit for combined DC output (**c**). Working mechanism of the triboelectric portion (i) initial stage, (ii) contact stage, (iii)–(iv) separation stage, (v) stage **d** working mechanism of the electromagnetic portion (i) initial stage, (ii) contact stage [16]

5 Commonly Used Nanopolymers in Energy Harvesting

Some widely used nanopolymers in energy nanogenerators are: Polydimethylsiloxane, Polyvinylidene fluoride, Poly(vinylidene fluoride-co-hexafluoropropylene), Polytetrafluoroethylene, Polyvinyl alcohol, Polyethylene oxide, Nylon 6,6, and Ecoflex 00-30.

5.1 Polydimethylsiloxane

Polydimethylsiloxane (PDMS) is a widely used polymer in nanogenerators. This is the most common silicone synthetic polymer. The backbone of PDMS is made up of repeating silicon-oxygen-silicon bonds. In addition to their links to oxygen to form polymer chains, silicon atoms are also linked to the organic groups, such as methyl ($-CH_3$), dimethyl ($-(CH_3)_2$), and trimethyl ($-(CH_3)_3$) groups. Figure 9 shows the siloxane monomer, combining methyl groups to the siloxane backbone create polydimethylsiloxane. The most common silicone, trimethylsilyloxy end block polydimethylsiloxane have the following structure, as shown in Fig. 10. These

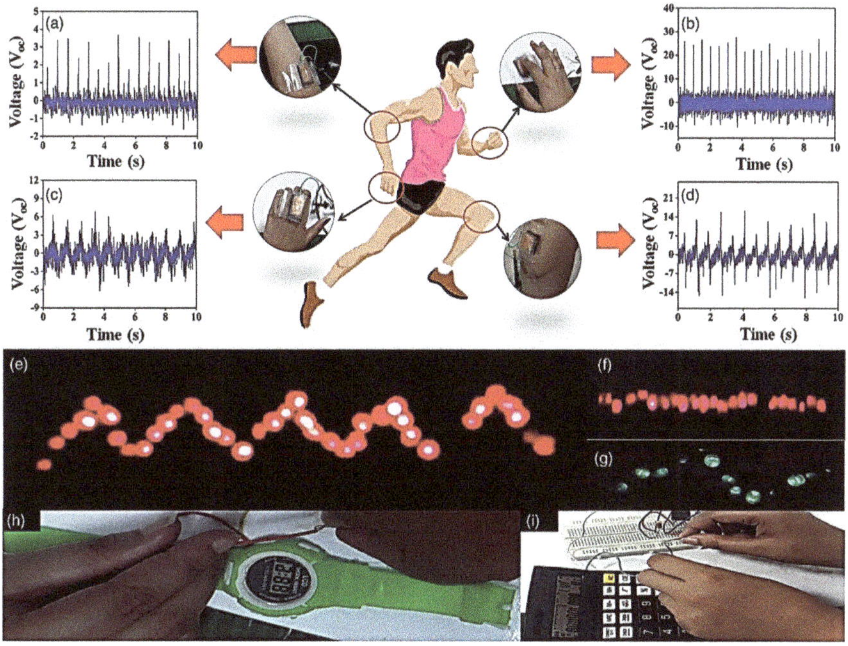

Fig. 8 Schematic representation of as-fabricated PTNG's real-time application for human activities sensing **a** elbow bending, **b** finger imparting, **c** palm bending, and **d** knee bending. With the help of fabricated PTNG lightening up **e** 40 red LEDs in series connection, **f** 22 red LEDs in parallel connection, and **g** 13 green LEDs in series connection, powering of a **h** digital wrist watch and **i** digital calculator [117]

$$(Si - O - Si). \qquad CH_3[Si(CH_3)_2O]_nSi(CH_3)_3,$$

Fig. 9 Siloxane monomer and polydimethylsiloxane

are the linear polymers and liquids, either small or large values of n. Apart from that, other groups, such as phenyl ($-C_6H_5$), vinyl ($-CH{=}CH_2$), and trifluoropropyl ($-(CH_2)_2-CF_3$) can be substituted for methyl group. The simultaneous presence of organic and inorganic groups in the backbone gives silicones the combination of unique properties. PDMS's low glass-transition temperatures and ability to maintain flexibility over wide temperature range make it durable and suitable for many applications.

PDMS has been widely used in wearable energy generators [16, 116, 118] and sensors [76, 119, 120] due to their high flexibility, stretchability, and biocompatibility.

$$CH_3[Si(CH_3)_2O]_nSi(CH_3)_3.$$

Fig. 10 Trimethylsilyloxy end block polydimethylsiloxane

A PDMS-based triboelectric nanogenerator (TENG) is shown in Fig. 11. PDMS has tunable properties and its electrical properties can be enhanced by adding other polymers [76], conductive fillers [121, 122], dielectrics [42], or by chemical treatment [35]. In TENGs, PDMS is used a negative triboelectric layer due to its position in the top of the triboelectric series [2]. On the other hand, in PENGs, PDMS is used as a matrix for piezoelectric filler materials [21, 123, 124]. High flexibility, durability, and biocompatibility are very crucial for flexible wearable nanogenerators. PDMS meets these requirements for wearable nanogenerators. Tailoring properties of PDMS will open the door for use in flexible wearable devices.

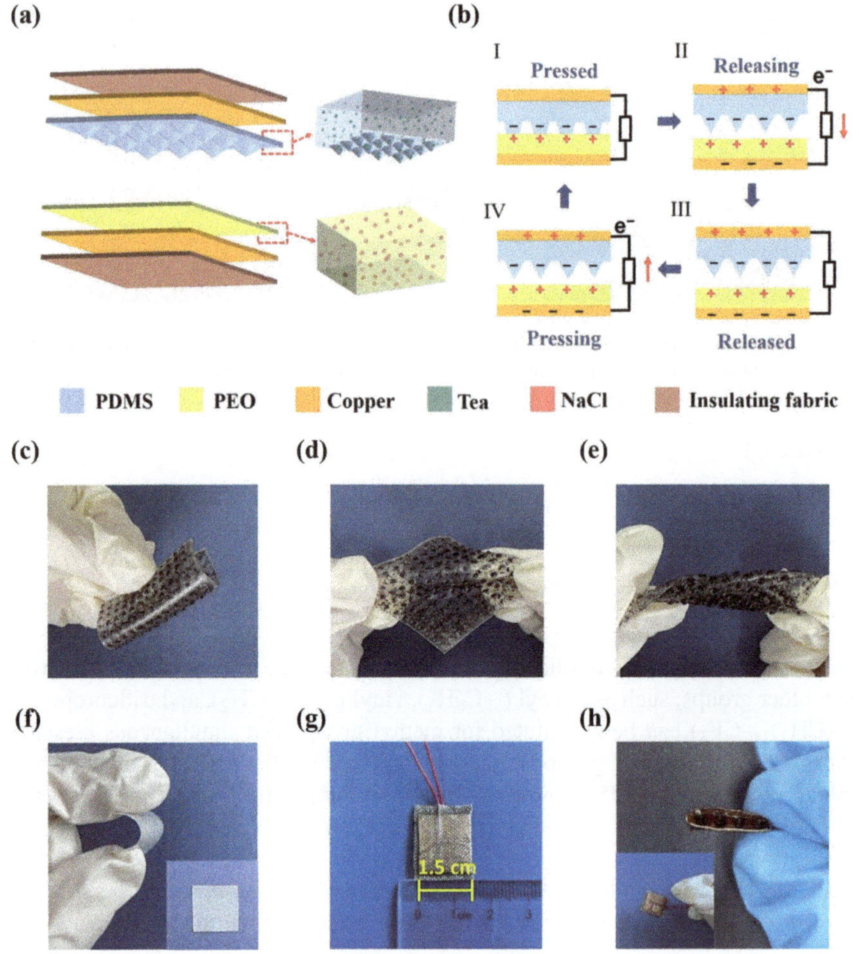

Fig. 11 **a** The internal structure of the TENG and **b** working principle of PDMS-PEO TENGs. image shows flexibility of PDMS in different deformation **c** folding, **d** stretching, and **e** twisting **f** Image of the NaCl-PEO film in the original and bending states. Images showing the **g** size, **h** hydrophobicity and cross-section of nanogenerator based sensor [125]

5.2 Polyvinylidene Fluoride

Polyvinylidene fluoride (PVDF) is a thermoplastic fluoropolymer produced by the polymerization of vinylidene difluoride ($(C_2H_2F_2)_n$). It has a relatively high melting point, mechanical strength, chemical and radiation resistance, good wear resistance, and low thermal conductivity [48, 126]. PVDF is a semicrystalline polymer. Its degree of crystallinity depends on the crystallization conditions, and it has at least four distinct phases, $\alpha-$, $\beta-$, $\gamma-$, ε—forms [3]. The two most important phases are α and β, Fig. 12; α forms under normal crystallization conditions while β forms as a result of pressure crystallization and mechanical deformation of PVDF film.

Among all crystalline phases, β-phase has spontaneous polarization and high piezoelectric properties [3, 68, 127]. Several method have been used to induce the β-phase of PVDF, such as stretching [128], annealing or poling [129, 130], adding carbon nanotubes (CNT) [131], multiwalled carbon nanotube (MWCNT) [132, 133] fillers in PVDF. Electrospinning of PVDF is a popular method of inducing β-phase [134]. This method eliminates the need for post-poling treatment of PVDF. Electrospinning of PVDF provide interesting properties, such as high porosity, pore size distribution, and high length-to-diameter ratio. PVDF has diverse application including in lithium-ion battery [135], pressure and temperature monitor, [136], tectile sensor [137, 138], exercise monitor [139], healthcare monitoring [140], and in energy harvesting nanogenerators [52, 141–144]. PVDF is a promising candidate for piezoelectric owing to its excellent piezoelectric, ferroelectric properties, and flexibility [4, 128, 145]. The application of PVDF-based piezoelectric nanogenerators are energy harvester [4, 146, 147], self-powered sensor [145], and pressure sensor [102] are a few of the examples. The PVDF-based PENG is shown in Fig. 13. PVDF has been widely used in TENGs due to their high flexibility, high output performance, and potential application in energy harvesting, flexible and wearable electronic devices, and health monitoring [148].

PVDF can be used in a variety of forms, such as films [144, 149], nanofibers [17], and nanocomposites [150, 151]. PVDF is the most significant tribonegative (electron acceptor) polymer. The polar β-phase of PVDF is responsible for high triboelectric performances in TENG. PVDF-based TENGs have applications in energy harvester [152, 153], wind energy harvester [154], and self-powered anticorrosion and antifouling [155]. An example of PVDF-based PENG physiological sensor is shown in Fig. 14.

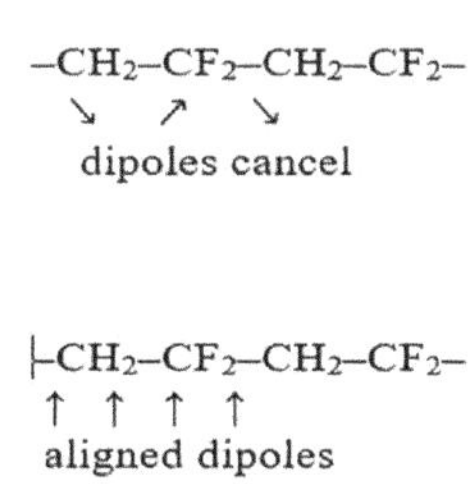

Fig. 12 α-phase and β-phase of PVDF

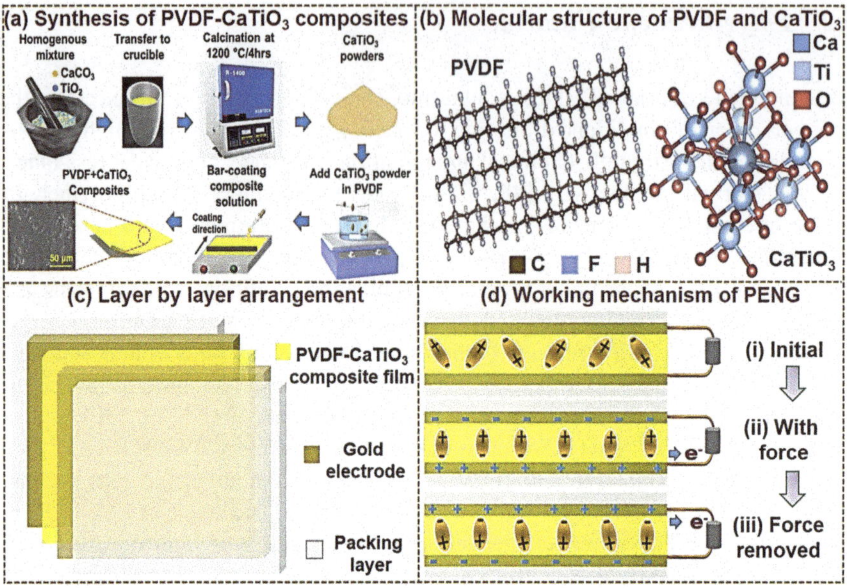

Fig. 13 **a** Synthesis methodology of CTO particles and PVDF-CTO composites; **b** molecular structure of PVDF and CTO; **c** layer-by-layer arrangement of PENG, and **d** working mechanism of fabricated PENG [139]

5.3 Poly(Vinylidene Fluoride-Co-Hexafluoropropylene)

Poly(vinylidene fluoride-co-hexafluoropropylene) (PVDF-HFP) has become a promising candidate for polymer-based energy harvesting devices due to their good chemical stability, biocompatibility, flexibility, and strong tribo-negativity associated with the fluorine counterparts [156]. PVDF-HFP is a copolymer of vinylidene fluoride with hexafluoropropylene. The polymerization of PVDF is shown in Fig. 15, and copolymerization of vinylidene fluoride (VDF) with hexafluoropropylene (HFP) to make PVDF-HFP is shown in Fig. 16. The presence of VDF enhances the electroactive process, β-phase, which is crucial for the piezoelectric performance of PVDF-HFP. On the other hand, adding HFP enhances the processability, adhesion, and hydrophobicity while decreasing the crystallinity and β-phase content [157]. PVDF-HFP exhibits better piezoelectric performance than PVDF owing to the β-crystalline phase. Compared to PVDF, PVDF-HFP has higher piezoelectric strain constant ($|d_{31}/d_{33}| > 1$) and electromechanical coupling factor ($k_{31} = 0.187$) [158]. The piezoelectric property of PVDF-HFP is associated with its crystal orientation in which fluorine atoms are bonded to carbon atoms, creating dipole moments. An increase in piezoelectricity is observed when dipoles are parallel to each other or all in the same direction under external force [159]. PVDF-HFP has several phases, such

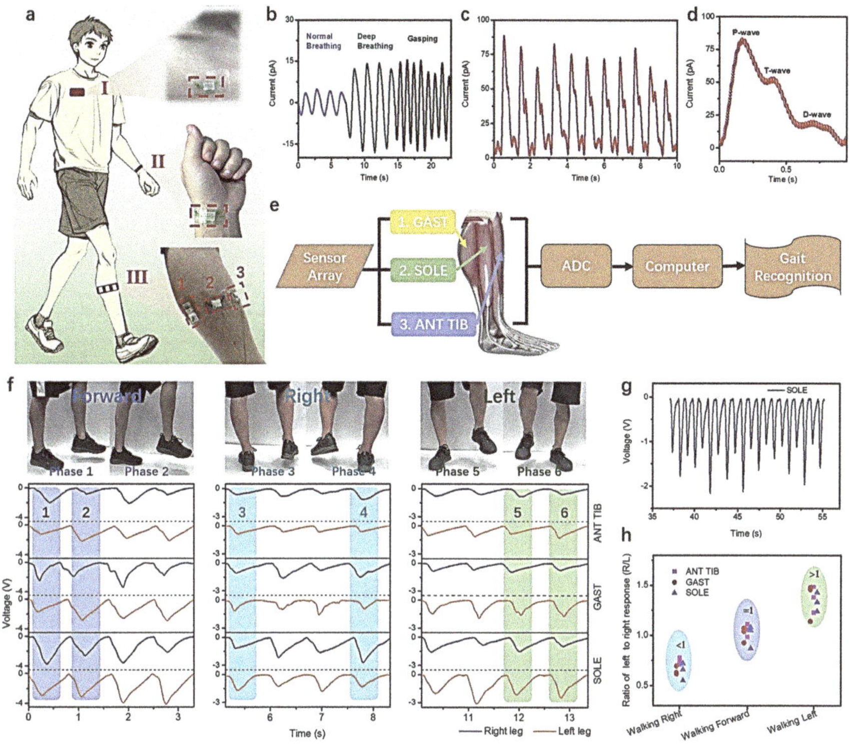

Fig. 14 Application of the PVDF-ZnO based flexible device for human physiological monitoring. **a** Schematic illustration of the pressure sensors assembled on (I) chest, (II) wrist, and (III) three calf muscles. **b** The electrical outputs of different breathing patterns. **c** The real-time signals of wrist pulses of a healthy person. **d** The expanded pulse wave containing three peaks, i.e., P-wave, T-wave, D-wave. **e** Schematic diagram of the gait recognition system. **f** Signals of the ANT TIB (anterior tibias), GAST (gastrocnemius) and SOLE (soleus) to detect gait states of walking forward, left and right, respectively. **g** Multiple signals of the left leg when walking left. **h** The ratio of calf response amplitude (Right leg) to the other calf response amplitude (Left leg) in different walking directions [82]

as α-, β-, and γ-phases [159]. Among them, β-phase possesses the piezoelectric properties. The conversion of mechanical energy to electrical energy can be efficiently achieved by increasing the β content in the PVDF-HFP [43].

The PVDF-HFP has a wide range of applications in harvesting energies. The PVDF-HFP is used in lithium batteries due to its hydrophobic nature, which helps in maintain electrode integrity and reduce charge transfer resistance in the battery [44, 160, 161]. Due to the flexibility, thermal stability, and improved mechanical properties, PVDF-HFP has been used in the piezoelectric [43, 156, 158, 159, 162] and triboelectric nanogenerators [163–167]. Figure 17 shows the PVDF-HFP nanofibers used in PENG.

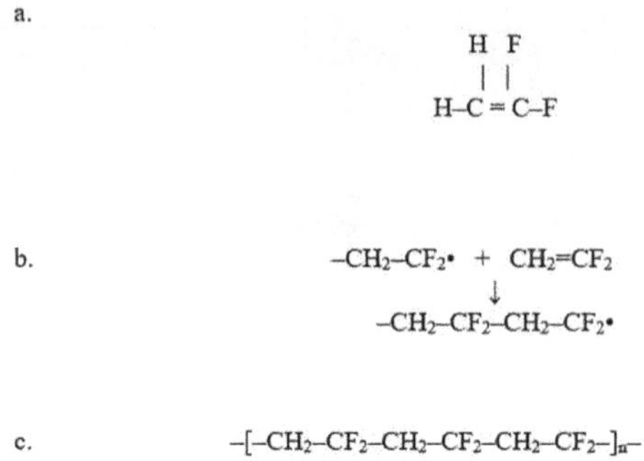

Fig. 15 Schematic of the polymerization of vinylidene fluoride for production of PVDF

a. Initiation

$$RO\text{–}OR \xrightarrow{\Delta} 2RO^{\bullet}$$

$$RO^{\bullet} + CH_2 = CF_2 \rightarrow RO\text{–}CH_2\text{–}CF_2^{\bullet}$$

b. Propagation

$$\text{-}(CH_2\text{-}CF_2)_x^{\bullet} + CF_2 = C(CF_3)F \rightarrow \text{-}(CH_2\text{-}CF_2)_x\text{-}CF_2\text{-}C(F)(CF_3)^{\bullet}$$

$$\text{-}(CF_2\text{-}C(F)(CF_3))_y^{\bullet} + CH_2 = CF_2 \rightarrow \text{-}(CF_2\text{-}C(F)(CF_3))_y\text{-}CH_2\text{-}CF_2^{\bullet}$$

C. Termination

$$P^{\bullet} + P^{\bullet} \rightarrow P\text{–}P$$

d. Overall Reaction

$$x\,CH_2 = CF_2 + y\,CF_2 = C(CF_3)F \xrightarrow[\text{Initiator}]{\text{Free-radical}} [\text{-}(CH_2\text{-}CF_2)_x\text{-}(CF_2\text{-}C(F)(CF_3))_y\text{-}]$$

Fig. 16 Schematic of the copolymerization of vinylidene fluoride with hexafluoropropylene for production of PVDF-HFP

5.4 Polytetrafluoroethylene

Polytetrafluoroethylene (PTFE) is a linear fluoropolymer. It is formed by the polymerization of tetrafluoroethylene (TFE), as shown in Fig. 18. The fundamental properties of PTFE arise from the atomic structure of fluorine and carbon, as well as their covalent bonds in the specific chemical structure. The backbone of PTFE is made of strong carbon–carbon bonds, and the pendant groups are carbon–fluorine bonds. The presence of fluorine atoms forms a uniform and continuous covering around the C–C bonds, making them chemically resistant and stable [168]. The strong electron-attractive PTFE has become a promising candidate for negative triboelectric materials. The position of PTFE is on the top of the triboelectric series [2] and the charge

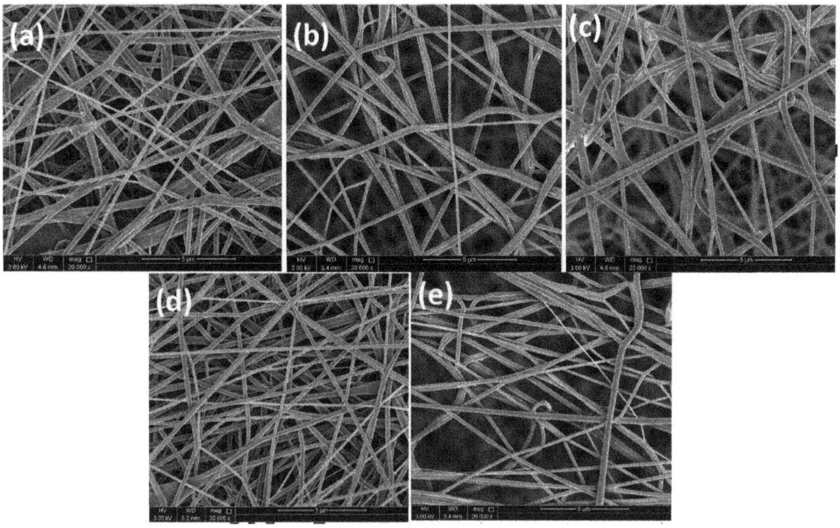

Fig. 17 SEM images of **a** neat PVDF-HFP, **b** PVDF-HFP/1 wt.% ZnO, **c** PVDF-HFP/0.5 wt.% Co-ZnO, **d** PVDF-HFP/1 wt.% Co-ZnO, **e** PVDF-HFP/2 wt.% Co-ZnO [175]

$$n\,CF_2 = CF_2 \xrightarrow[\text{Initiator}]{\text{Free-radical}} [-CF_2\text{-}CF_2\text{-}]_n$$

Fig. 18 Schematic of the polymerization of tetrafluoroethylene to make PTFE

affinity of PTFE is $-190\ \text{nCJ}^{-1}$ [169]. The PTFE has been used in TENGs as a negative triboelectric material due to their strong electron-attracting ability [170] in the presence of fluorine atom in the structure of PTFE, as shown in Fig. 19. The application of PTFE in TENGs, including high performance and flexible nanogenerator [34, 171, 172], self-powered electronics [72, 173], real-time rain sensor [174].

5.5 *Polyvinyl Alcohol*

Polyvinyl alcohol (PVA) is a semicrystalline polymer made of polymerization of vinyl acetate. The schematic of the polymerization of Vinyl alcohol to make polyvinyl alcohol is shown in Fig. 20. PVA is biocompatible and non-toxic and has good chemical and thermal stability [176]. PVA is easy to process since it is a water-soluble polymer that readily reacts with different cross-linking agents and forms a gel. These properties have led to diverse applications of PVA, including cosmetics, pharmaceuticals, food packaging industries [177, 178], and energy harvesters [179]. PVA is a promising candidate for energy-harvesting nanogenerators. It is used as a positive triboelectric material in TENG; it loses its electron during triboelectrification

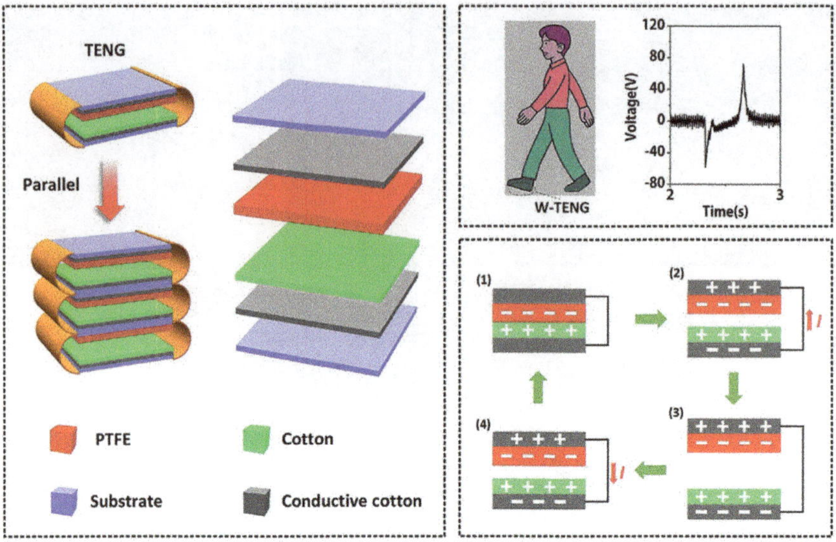

Fig. 19 The schematic illustration of W-TENG and combined W-TENG. **a** The SEM image of PTFE film surface. **b** The Schematic illustration of W-TENG operating mechanism [34]

[170]. PVA can be used as both film [180] and nanofiber form [181] in energy harvesting devices. The application of PVA in energy harvesters, including self-power electronic devices [179, 182, 183], high-performance hybrid nanogenerator [180, 184], TENG-based human–machine interaction sensor [185], and self-powered sensor [186]. The schematic of preparation of PVA for energy harvesting device, TENG, is shwon Fig. 21.

Fig. 20 Schematic of the polymerization of Vinyl alcohol to make polyvinyl alcohol

Vinyl acetate

$CH_2{=}CH{-}OCOCH_3$

↓ Free-radical polymerization

Poly (vinyl acetate)

$-[-CH_2-CH(OCOCH_3)-]_n-$

↓ Hydrolysis / Saponification

Poly (vinyl alcohol)

$-[-CH_2-CH(OH)-]_n-$

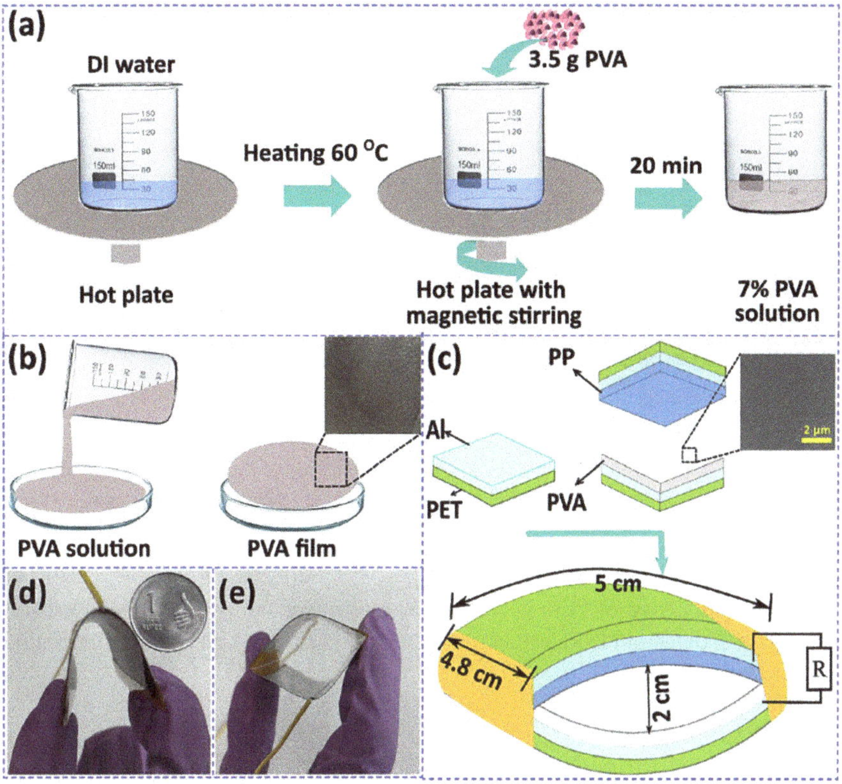

Fig. 21 **a** Schematic flow chart for preparing 7% PVA solution. **b** Illustration of PVA thin film formation. **c** Fabricating PVA-TENG. Photograph of the as-fabricated device showing **d** flexibility and **e** arch structure [73]

5.6 Polyethylene Oxide

Polyethylene oxide (PEO) is a nonionic ethylene oxide (EO) homopolymer. It is made through alkali metal hydroxide or zinc chloride catalyst-assisted heterogeneous polymerization of ethylene oxide. Figure 22 the schematic of EO polymerization to make PEO. Higher molecular weight polymers of EO are called polyethylene oxide (PEO), whereas polymers with EO molecular weight below 30, 000 g/mol are referred to as polyethylene glycol (PEG). PEO is a biocompatible polymer and is widely used for pharmaceuticals, cosmetics, medical applications, laxatives, and food additives [187]. Recently, PEO has been widely used in nanoenergy harvesting devices due to its biocompatibility and processability. In nanoenergy harvesting devices, PEO is used as a composite film [188] as well as nanofiber [189, 190]. PEO has a strong electron-donating capability; owing to this reason, it is used as a positive triboelectric material. Combining PEO with other polymers enhanced the electron donating capabilities, making it suitable for positive triboelectric materials in TENG

(a) Initiation

$$RO^- + \begin{matrix} O \\ / \backslash \\ CH_2\ CH_2 \end{matrix} \rightarrow RO\text{-}CH_2\text{-}CH_2\text{-}O^-$$

(b) Propagation

Successive ring opening of EO units:

$$RO\text{-}(CH_2\text{-}CH_2\text{-}O)_n^- + \begin{matrix} O \\ / \backslash \\ CH_2\ CH_2 \end{matrix} \rightarrow RO\text{-}(CH_2\text{-}CH_2\text{-}O)_{n+1}^-$$

(c) Termination

By protonation:

$$RO\text{-}(CH_2\text{-}CH_2\text{-}O)_n^- + H^+ \rightarrow HO\text{-}(CH_2\text{-}CH_2\text{-}O)_n\text{-}H$$

Fig. 22 Schematic of the polymerization of ethylene oxide to make polyethylene oxide, where n is the number of ethylene oxide

[191]. The application of PEO in energy harvesting, including nanogenerator-based human-health detector [192], self-powered sensor [125, 193], and green triboelectric nanogenerator [171, 194].

5.7 Nylon 6, 6

Nylon 6,6 is a semicrystalline synthetic resin and a type of polyamide or nylon. It consists of two monomers, such as adipic acid and hexamethylenediamine. The poly-condensation of adipic acid and hexamethylenediamine creates polyamide or nylon 6,6, as shown in Fig. 23. The application of nylon 6,6 includes electrical insulation components, automotive parts, medical devices, seals and gaskets, machinery parts, and pipe and hose components, to name a few. In nanogenerators, nylon 6,6 nanopolymer is used as a nanofiber. The SEM images of nylon 6,6 nanofiber, and TiO_2/nylon 6, 6/ and SiO_2/nylon 6, 6 nanofibers are shown in Fig. 24. The flexible and wearable TENG uses nylon 6,6 due to its biocompatibility, flexibility, and affordability [195, 196]. Nylon 6,6 has an electron-donating tendency as its position is on top of the triboelectric series compared to the other positive triboelectric materials. The TENG energy harvesting devices use nylon 6,6 nanofiber as a positive triboelectric material. The electrospun nylon 6,6 nanofiber is used in high-performance TENG [197–199], self-powered energy harvesting device [200], and multimode functional energy generator [201, 202].

$$\begin{array}{l} n\,HOOC\text{-}(CH_2)_4\text{-}COOH \\ +\,n\,H_2N\text{-}(CH_2)_6\text{-}NH_2 \end{array} \xrightarrow{\Delta} \boxed{[\text{-}NH\text{-}(CH_2)_6\text{-}NH\text{-}CO\text{-}(CH_2)_4\text{-}CO\text{-}]_n} + 2n\,H_2O$$

Fig. 23 Polycondensation of adipic acid and hexamethylenediamine to make polyamide 6,6 (nylon 6,6)

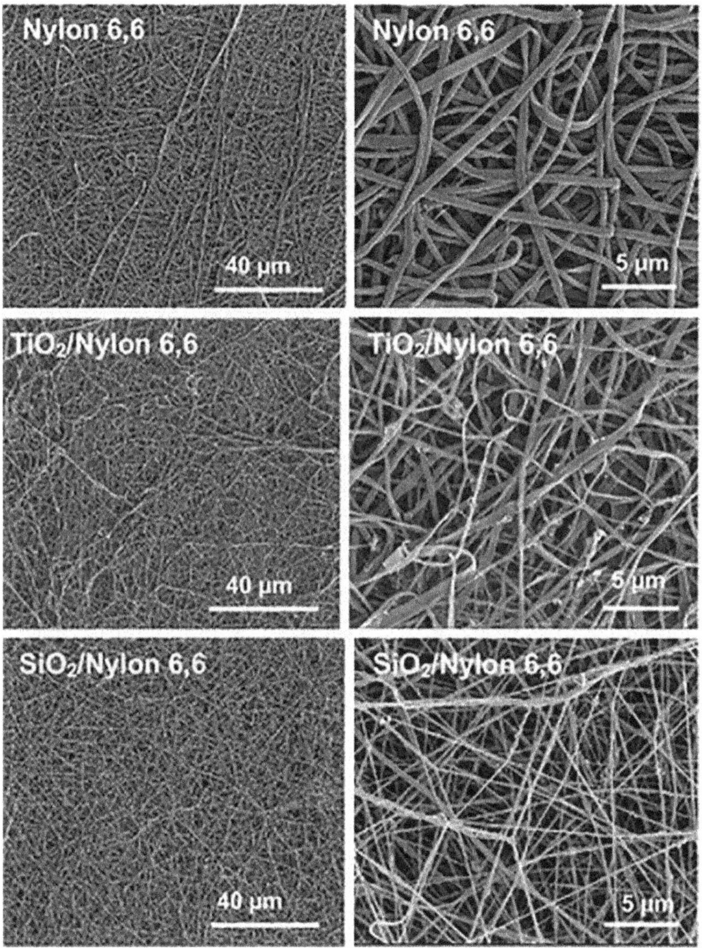

Fig. 24 SEM images of nylon 6,6, TiO2/nylon 6,6, and SiO2/nylon 6,6 nanofiber membranes [37]

5.8 *Ecoflex 00-30*

Ecoflex 00-30 is a commercially available platinum-catalyzed silicone rubber having a shore hardness of 00-30 and tensile strength of 200% [203]. Recently, ecoflex 00-30 has got significant attention in flexible energy harvesting devices due to its high durability, flexibility, and eco-friendliness [38]. It can be used in TENG as a negative

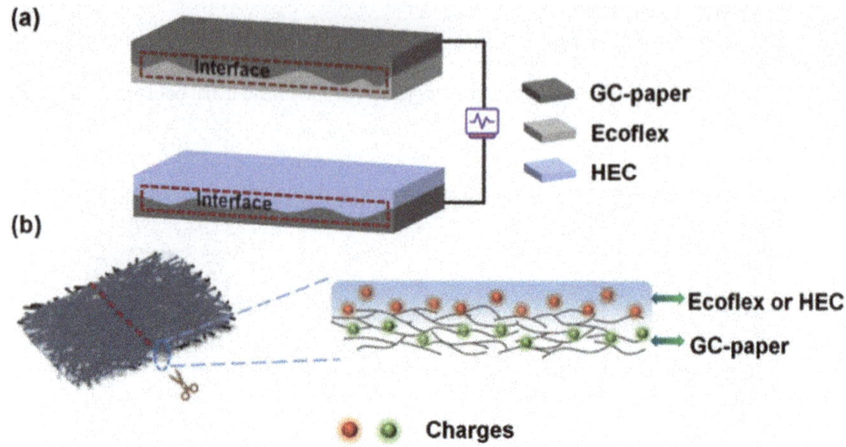

Fig. 25 **a** Schematic diagram of the paper based-TENG employing the integrated graphene composite paper (GC) as electrode while the Ecoflex and hydroxyethyl cellulose (HEC) as friction layers (FL) **b** Schematic diagram of the interface between the friction layer and graphene composite paper [212]

triboelectric material owing to its strong electron affinity [204]. The charge affinity of silicones is recorded as $-72\ nCJ^{-1}$ [169]. The high flexibility made ecoflex 00-30 is a promising candidate for flexible energy harvesting devices. The ecoflex 00-30 used in flexible TENG device [183, 205], hybrid nanogenerator [206] TENG based pressure sensor [207], self-powered human health monitoring [208], and human motion monitoringdevices [209, 210].

Ecoflex 00-30 is also used in flexible piezoelectric energy harvesters [65, 211]. The application of Ecoflex 00-30 in TENG is shown in Fig. 25.

6 Application of Nanopolymers

Nanopolymers are promising candidates for energy harvesting devices owing to its easy processability, flexibility, conductivity [7, 213, 214], piezoelectricity [215, 216], pyro-electricity, [5, 217], and ferroelectricity [26, 218]. With the development of Internet of Things (IoT), nanopolymer-based energy harvesting has become a popular choice for self-powered wearable energy harvesting devices, such as pulse sensor [219–221], pressure sensor [222, 223], and human health monitor [76, 224] due to its high flexibility and biocompatibility. Energy harvesting devices, such as TENG, PENG, and HNG scavenge energy from the ambient environment. TENG harvest energy from wind energy [154, 225, 226] ocean wave energy [227], rain-drop energy [228–230], acoustic energy [231–233], human motion [234]. The schematic of energy harvesting using TENG is shown in Fig. 26. The application of PENG is energy harvesting [235, 236], acoustic energy harvesting [237], wearable and implantable bioelectronic devices [120, 238], self-powered active sensors [239], and flexible piezoelectric sensor for human–machine interaction (HMI) [106]. Furthermore, HNGs are increasingly used to power wearable devices due to their ability to harvest energy from human motion. The integration of piezoelectric and triboelectric mechanisms allows for efficient energy conversion from body movements, making them suitable for powering small electronic. The application of high performance HNGs is wave energy harvesting [240], biomechanical and biochemical energy harvesting [241]. The in-vitro and in-vivo application of nanopolymer-based sensor is shown in Fig. 27.

Nanopolymers has great potential for a wide range of energy harvesting applications due to their versatile properties, such as easy processability, flexibility, and tunable electronic properties. Their ability to be integrated into wearable devices, smart textiles, and low-power electronics makes them promising candidates for future sustainable energy systems.

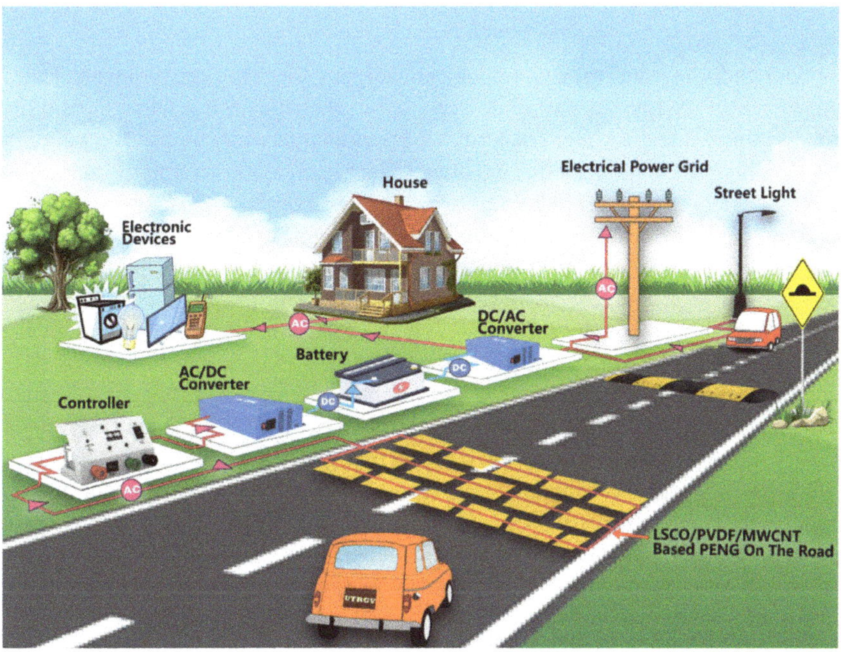

Fig. 26 Schematic diagram of the energy harvesting system using a PVDF/$La_{1-x}Sr_xCoO_3$/MWCNT-based PENG [144]

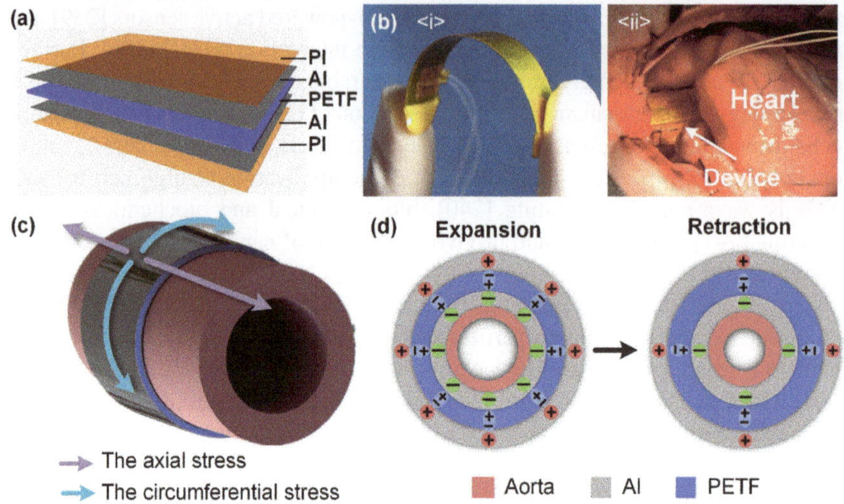

Fig. 27 **a** Exploded view of the device showing the multilayer thinfilm structure. **b** Photograph of the device showing its flexibility (i) and the device wrapping on the ascending aorta of porcine (ii). **c** The circumferential stress in expanded aorta wall and the distribution of induced charge in device. **d** The charge distributions in device of expansion and retraction state [241]

7 Conclusion

Nanopolymers have an essential role in our everyday life. They are promising candidate for clean energy harvesting devices due to their flexibility, biocompatibility, processability, low cost, and friendliness. Nanopolymers can be used in nanofilm and nanofiber formation, which broadens their application in energy harvesting devices. Nanopolymers can be conductive, dielectric, piezoelectric, and ferroelectric. Based on their properties, they can be used either in triboelectric nanogenerators or piezoelectric nanogenerators. The triboelectric nanogenerators use two nanopolymers: one with a high electron-donating tendency and another one with a high electron-accepting tendency. In contrast, a piezoelectric nanogenerator uses only one piezoelectric nanopolymer or nanopolymer with piezoelectric nanocomposites. The nanopolymer-based triboelectric and piezoelectric energy harvesting devices show outstanding performances in scavenging electrical energy from human body movement, wind energy, rain-drop energy, wave energy, and vibrational energy. It is crucial to tune the properties of nanopolymers, such as the charge density, piezoelectricity, ferroelectricity, and dielectricity, to enhance the performance of nanopolymer-based energy harvesting devices. One of the main challenges of nanopolymer-based energy harvesting is durability since nanopolymers degrade over time. Repeating friction in the case of triboelectric nanogenerators and repeating stress-relaxation in the case of piezoelectric nanogenerators degrade the nanopolymer and affect the output performance of nanogenerators. This issue needs to be addressed for the long-term application of nanopolymer-based energy harvesting.

Author Contributions Md. Arafat Hossain: writing original draft; Md. Wasikur Rahman and Victoria Padilla: writing-review & editing, Nazmul Islam: supervision, writing-review & editing, and Mohammed Jasim Uddin: supervision, writing-review & editing.

References

1. Saiyad, M., Shah, N.: Nanopolymers in drug delivery system. Mater. Today Proc. **67**, 25–30 (2022)
2. Zou, H., et al.: Quantifying the triboelectric series. Nat. Commun. **10**, 1–9 (2019)
3. Koseki, Y., Aimi, K., Ando, S.: Crystalline structure and molecular mobility of PVDF chains in PVDF/PMMA blend films analyzed by solid-state 19F MAS NMR spectroscopy. Polym. J. **44**, 757–763 (2012)
4. Chen, H., et al.: Piezoelectric nanogenerator based on in situ growth all-inorganic $CsPbBr_3$ Perovskite Nanocrystals in PVDF fibers with long-term stability. Adv. Func. Mater. **31**, 1–12 (2021)
5. Wada, Y., Hayakawa, R.: Piezoelectricity and pyroelectricity of polymers. Jpn. J. Appl. Phys. **15**, 2041–2057 (1976)
6. Park, T., et al.: Photothermally activated pyroelectric polymer films for harvesting of solar heat with a hybrid energy cell structure. ACS Nano **9**, 11830–11839 (2015)
7. Kumar, D., Sharma, R.C.: Advances in conductive polymers. Eur. Polymer J. **34**, 1053–1060 (1998)
8. Nezakati, T., Seifalian, A., Tan, A., Seifalian, A.M.: Conductive polymers: opportunities and challenges in biomedical applications. Chem. Rev. **118**, 6766–6843 (2018)
9. Wang, Z.L., Song, J.: Piezoelectric nanogenerators based on zinc oxide nanowire arrays. Science **312**, 242–246 (2006)
10. Fan, F.R., Tian, Z.Q., Lin Wang, Z.: Flexible triboelectric generator. Nano Energy **1**, 328–334 (2012)
11. Fan, F.R., Tang, W., Wang, Z.L.: Flexible nanogenerators for energy harvesting and self-powered electronics. Adv. Mater. **28**, 4283–4305 (2016)
12. Feng, H., et al.: Nanogenerator for biomedical applications. Adv. Healthcare Mater. **7**, 1–18 (2018)
13. Althagafy, K., Alotibi, E., Al-Dossari, M., Alhashmi Alamer, F.: Design and construction of a flexible conductor based on a complex conductive polymer: PEDOT: PSS/polyaniline and its application as a pressure sensor. Results Phys. **51**, 106689 (2023)
14. Mehra, A., Chaudhary, M., De Souza, F., Gupta, R.K.: Recent Advancements in Conducting Polymers for Biomedical Sensors, pp. 325–349. Springer Nature Singapore, Singapore (2024)
15. Gao, W., Ota, H., Kiriya, D., Takei, K., Javey, A.: Flexible electronics toward wearable sensing. Acc. Chem. Res. **52**, 523–533 (2019)
16. Islam, E., et al.: Electromagnetic-triboelectric-hybrid energy tile for biomechanical green energy harvesting. Nano Energy **77**, 105250 (2020)
17. Abir, S.S.H., et al.: Nanofiber-based substrate for a triboelectric nanogenerator: high-performance flexible energy fiber mats. ACS Appl. Mater. Interfaces. **13**, 60401–60412 (2021)
18. Abir, S.S.H., et al.: Piezoelectricity in chalcogenide perovskites. Nat. Commun. **15**, 5768 (2024)
19. Abdullah, A.M., et al.: Synthesis and fabrication of self-sustainable triboelectric energy case for powering smart electronic devices. Nano Energy **73**, 104774 (2020)
20. Wu, T., Wang, X., Cao, X., Wang, N.: NdFeB-based magnetic triboelectric nanogenerator for enhanced bioenergy harvesting and tactile perception. Nano Energy **128**, 109883 (2024)
21. Shawon, S.M.A.Z., et al.: Surface modified hybrid $ZnSnO_3$ nanocubes for enhanced piezoelectric power generation and wireless sensory application. Nano Energy **92**, 106653 (2022)

22. Poh, T.Y., et al.: Inhaled nanomaterials and the respiratory microbiome. Part. Fibre Toxicol. **15**, 1–16 (2018)
23. K, N., Rout, C.S.: Conducting polymers: a comprehensive review on recent advances in synthesis, properties and applications. RSC Adv. **11**, 5659–5697 (2021)
24. Smith, M., Kar-Narayan, S.: Piezoelectric polymers: theory, challenges and opportunities. Int. Mater. Rev. **67**, 65–88 (2022)
25. Zhang, X., et al.: Electrical stimulation system based on electroactive biomaterials for bone tissue engineering. Mater. Today **68**, 177–203 (2023)
26. Li, Q., Wang, Q.: Ferroelectric polymers and their energy-related applications. Macromol. Chem. Phys. **217**, 1228–1244 (2016)
27. Wang, S., Bai, Y., Zhang, T.: Materials, systems, and devices for wearable bioelectronics (2020)
28. Thakre, A., Kumar, A., Song, H.-C., Jeong, D.-Y., Ryu, J.: Pyroelectric energy conversion and its applications—flexible energy harvesters and sensors. Sensors **19**, 2170 (2019)
29. Stoyanov, O.V., Haghi, A., Zaikov, G.E.: Nanopolymers and Modern Materials: Preparation, Properties, and Applications. CRC Press (2013)
30. Shi, Q., Sun, Z., Zhang, Z., Lee, C.: Triboelectric nanogenerators and hybridized systems for enabling next-generation IoT applications. Research (2021)
31. Cai, W., Gong, X., Cao, Y.: Polymer solar cells: recent development and possible routes for improvement in the performance. Sol. Energy Mater. Sol. Cells **94**, 114–127 (2010)
32. Li, G., Zhu, R., Yang, Y.: Polymer solar cells. Nat. Photonics **6**, 153–161 (2012)
33. Choi, D., et al.: Recent advances in triboelectric nanogenerators: from technological progress to commercial applications. ACS Nano **17**, 11087–11219 (2023)
34. Zhang, Z., Cai, J.: High output triboelectric nanogenerator based on PTFE and cotton for energy harvester and human motion sensor. Curr. Appl. Phys. **22**, 1–5 (2021)
35. Menge, H.G., Kim, J.O., Park, Y.T.: Enhanced triboelectric performance of modified PDMS nanocomposite multilayered nanogenerators. Materials **13**, 4156 (2020)
36. Xu, M., et al.: An aeroelastic flutter based triboelectric nanogenerator as a self-powered active wind speed sensor in harsh environment. Extreme Mech. Lett. **15**, 122–129 (2017)
37. Yanilmaz, M., Zhu, J., Lu, Y., Ge, Y., Zhang, X.: High-strength, thermally stable nylon 6,6 composite nanofiber separators for lithium-ion batteries. J. Mater. Sci. **52**, 5232–5241 (2017)
38. Basith, S.A., Chandrasekhar, A.: Synergistic effect of dual surface modified ecoflex polymer for dynamic triboelectric nanogenerator towards sustainable battery-free tally counter. Adv. Mater. Technol. **8**, 1–14 (2023)
39. Kim, M.K., Kim, M.S., Kwon, H.B., Jo, S.E., Kim, Y.J.: Wearable triboelectric nanogenerator using a plasma-etched PDMS-CNT composite for a physical activity sensor. RSC Adv. **7**, 48368–48373 (2017)
40. Hwang, H., et al.: Metal-free, flexible triboelectric generator based on MWCNT mesh film and PDMS layers. Appl. Surf. Sci. **442**, 693–699 (2018)
41. Pandey, P., et al.: Nafion-mediated barium titanate-polymer composite nanofibers-based triboelectric nanogenerator for self-powered smart street and home control system. Nano Energy **107**, 108134 (2023)
42. Chen, J., et al.: Enhancing performance of triboelectric nanogenerator by filling high dielectric nanoparticles into sponge PDMS film. ACS Appl. Mater. Interfaces. **8**, 736–744 (2016)
43. Chen, J., et al.: Enhancing piezoelectric properties of PVDF-HFP composite nanofibers with cellulose nanocrystals. Mater. Today Commun. **39**, 108872 (2024)
44. Carena, E., et al.: PVDF-HFP based, quasi-solid nanocomposite electrolytes for lithium metal batteries. Small **20**, 1–12 (2024)
45. L¨otters, J.C., Olthuis, W., Veltink, P.H., Bergveld, P.: The mechanical properties of the rubber elastic polymer polydimethylsiloxane for sensor applications. J. Micromech. Microeng. **7**, 145–147 (1997)
46. Mata, A., Fleischman, A.J., Roy, S.: Characterization of Polydimethylsiloxane (PDMS) properties for biomedical micro/nanosystems. Biomed. Microdevice **7**, 281–293 (2005)

47. Hu, D., et al.: Strategies to achieve high performance piezoelectric nanogenerators. Nano Energy **55**, 288–304 (2019)
48. Dallaev, R., et al.: Brief review of PVDF properties and applications potential. Polymers **14**, 4793 (2022)
49. Gregorio, R.: Determination of the α, β, and γ crystalline phases of poly(vinylidene fluoride) films prepared at different conditions. J. Appl. Polym. Sci. **100**, 3272–3279 (2006)
50. Mohammadi, B., Yousefi, A.A., Bellah, S.M.: Effect of tensile strain rate and elongation on crystalline structure and piezoelectric properties of PVDF thin films. Polym. Testing **26**, 42–50 (2007)
51. Kabir, E., Khatun, M., Nasrin, L., Raihan, M.J., Rahman, M.: Pure β-phase formation in polyvinylidene fluoride (PVDF)-carbon nanotube composites. J. Phys. D: Appl. Phys. 50 (2017)
52. Joshi, B., et al.: Supersonically sprayed PVDF and ZnO flowers with built-in nanocuboids for wearable piezoelectric nanogenerators. Nano Energy **112**, 108447 (2023)
53. Park, C., et al.: Enhancing the piezoelectric coefficient of $SrTiO_3$ nanocubes and PVDF film deposited by supersonic spraying for energy-harvesting nanogenerators. Chem. Eng. J. **491**, 151945 (2024)
54. Shi, X. et al.: Electrospinning of nanofibers and their applications for energy devices. J. Nanomaterials 2015 (2015); He, Z., Rault, F., Lewandowski, M., Mohsenzadeh, E., Salau¨n, F.: Electrospun PVDF nanofibers for piezoelectric applications: A review of the influence of electrospinning parameters on the β phase and crystallinity enhancement. Polymers **13**, 1–23 (2021)
55. Ichangi, A. et al.: Electrospun $BiFeO_3$ nanofibers for vibrational energy harvesting application. Adv. Eng. Mater. 24 (2022)
56. Nilsson, E., Lund, A., Jonasson, C., Johansson, C., Hagstr¨om, B.: Poling and characterization of piezoelectric polymer fibers for use in textile sensors. Sensors Actuat. A: Phys. **201**, 477–486 (2013)
57. Mishra, S., Unnikrishnan, L., Nayak, S.K., Mohanty, S.: Advances in piezo-electric polymer composites for energy harvesting applications: a systematic review. Macromol. Mater. Eng. **304**, 1–25 (2019)
58. Mohammadpourfazeli, S., et al.: Future prospects and recent developments of polyvinylidene fluoride (PVDF) piezoelectric polymer; fabrication methods, structure, and electro-mechanical properties. RSC Adv. **13**, 370–387 (2023)
59. Bouhamed, A. et al.: A hybrid piezoelectric composite flexible film based on PVDF-HFP for boosting power generation. Compos. Sci. Technol. 208 (2021)
60. Abdolmaleki, H., Haugen, A.B., Buhl, K.B., Daasbjerg, K., Agarwala, S.: Interfacial engineering of PVDF-TrFE toward higher piezoelectric, ferroelectric, and dielectric performance for sensing and energy harvesting applications. Adva. Sci. 10 (2023)
61. Huang, Z.-X., et al.: Self-poled piezoelectric polymer composites via melt-state energy implantation. Nat. Commun. **15**, 819 (2024)
62. Lipsa, L., Parida, S., Behura, S.K., Choudhary, A., Roy, A.: Scalable piezoelectric energy harvesting from pvdf-hfp/bzt-bct/go hybrid composite film. Appl. Phys. Lett. **125**, 043902 (2024)
63. Kim, M., Wu, Y.S., Kan, E.C., Fan, J.: Breathable and flexible piezoelectric ZnO@PVDF fibrous nanogenerator for wearable applications. Polymers **10**, 745 (2018)
64. Choudhry, I., Khalid, H.R., Lee, H.K.: Flexible piezoelectric transducers for energy harvesting and sensing from human kinematics. ACS Appl. Electron. Mater. **2**, 3346–3357 (2020)
65. Wen, D.-L., et al.: High-performance hybrid nanogenerator for self-powered wireless multi-sensing microsystems. Microsyst. Nanoeng. **9**, 94 (2023)
66. Guo, X., et al.: Doping engineering of conductive polymers and their application in physical sensors for healthcare monitoring. Macromol. Rapid Commun. **45**, 2300246 (2024)
67. Abdullah, A.M., et al.: KNN based piezo-triboelectric lead-free hybrid energy films. Nano Energy **86**, 106133 (2021)

68. Jiang, T., et al.: Spring-assisted triboelectric nanogenerator for efficiently harvesting water wave energy. Nano Energy **31**, 560–567 (2017)
69. Wang, S., et al.: Sliding-Triboelectric nanogenerators based on in-plane charge-separation mechanism. Nano Lett. **13**, 1869–2316 (2013)
70. Wang, S., Xie, Y., Niu, S., Lin, L., Wang, Z.L.: Freestanding triboelectric-layer-based nanogenerators for harvesting energy from a moving object or human motion in contact and non-contact modes. Adv. Mater. **26**, 2818–2824 (2014)
71. Yang, Y., et al.: Single-electrode-based sliding triboelectric nanogenerator for self-powered displacement vector sensor system. ACS Nano **7**, 7342–7351 (2013)
72. Amini, S., Muktar Ahmed, R.F.S., Ankanathappa, S.M., Sannatham-megowda, K.: Polyvinyl alcohol-based economical triboelectric nanogenerator for self-powered energy harvesting applications. Nanotechnology **35**, 035403 (2024)
73. Oh, W., et al.: Contact electrification of porous PDMS-nickel ferrite composites for effective energy harvesting. Mater. Sci. Eng. B **292**, 116397 (2023)
74. Sun, Q., et al.: Charge dispersion strategy for high-performance and rain-proof triboelectric nanogenerator. Adv. Mater. **36**, 1–10 (2024)
75. Chowdhury, A.R., et al.: Decentralized triboelectric electronic health monitoring flexible microdevice. Medical Devices & Sensors **3**, 1–12 (2020)
76. Cheng, J., et al.: Triboelectric microplasma powered by mechanical stimuli. Nat. Commun. **9**, 1–11 (2018)
77. Bai, Y. et al.: Washable multilayer triboelectric air filter for efficient particulate matter PM2.5 removal. Adv. Funct. Mater. **28**, 1–8 (2018)
78. He, W., et al.: Ultrahigh output piezoelectric and triboelectric hybrid nanogenerators based on ZnO nanoflakes/polydimethylsiloxane composite films. ACS Appl. Mater. Interfaces. **10**, 44415–44420 (2018)
79. Qian, Y., Kang, D.J.: Poly(dimethylsiloxane)/ZnO nanoflakes/three-dimensional graphene heterostructures for high-performance flexible energy harvesters with simultaneous piezoelectric and triboelectric generation. ACS Appl. Mater. Interfaces. **10**, 32281–32288 (2018)
80. Manchi, P., Graham, S.A., Patnam, H., Paranjape, M.V., Yu, J.S.: ZnO nanoflakes embedded polymer matrix for high-performance mechanical energy harvesting. Adv. Mater. Technol. **7**, 1–11 (2022)
81. Yang, T., et al.: Hierarchically structured PVDF/ZnO core-shell nanofibers for self-powered physiological monitoring electronics. Nano Energy **72**, 104706 (2020)
82. He, Q., Li, X., Zhang, J., Zhang, H., Briscoe, J.: P-N junction-based ZnO wearable textile nanogenerator for biomechanical energy harvesting. Nano Energy **85**, 105938 (2021)
83. Chu, Y.L., et al.: Enhanced Nanogenerator performances of 1-D Al-doped ZnO nanorod arrays through ultrasonic wave systems. ACS Appl. Electron. Mater. **5**, 1277–1285 (2023)
84. Choi, M., et al.: Mechanical and electrical characterization of PVDF-ZnO hybrid structure for application to nanogenerator. Nano Energy **33**, 462–468 (2017)
85. Yang, F., et al.: The high-speed ultraviolet photodetector of ZnO nanowire Schottky barrier based on the triboelectric-nanogenerator-powered surface-ionic-gate. Nano Energy **60**, 680–688 (2019)
86. Pu, X., et al.: Flexible PVDF/nylon-11 electrospun fibrous membranes with aligned ZnO nanowires as potential triboelectric nanogenerators. Chem. Eng. J. **398**, 125526 (2020)
87. Chen, Y., et al.: Dynamic electronic doping for correlated oxides by a triboelectric nanogenerator. Adv. Mater. **30**, 1–8 (2018)
88. Chen, X., et al.: Electron trapping & blocking effect enabled by MXene/TiO_2 intermediate layer for charge regulation of triboelectric nanogenerators. Nano Energy **98**, 107236 (2022)
89. Huang, Y.J., Ke, K.H., Chung, C.K.: Effect of commercial P25 TiO_2 doping in polydimethylsiloxane tribo-film on the output performance of triboelectric nanogenerator and its application. Surf. Coat. Technol. **482**, 130721 (2024)
90. Du, X., et al.: Improved triboelectric nanogenerator output performance through polymer nanocomposites filled with core–shell-structured particles. ACS Appl. Mater. Interfaces. **10**, 25683–25688 (2018)

91. Tantraviwat, D., et al.: Tuning the dielectric constant and surface engineering of a $BaTiO_3$/ porous PDMS composite film for enhanced triboelectric nanogenerator output performance. ACS Omega **6**, 29765–29773 (2021)
92. Zhang, C., et al.: Recent progress of wearable piezoelectric nanogenerators. ACS Appl. Electron. Mater. **3**, 2449–2467 (2021)
93. Jeong, C.K., Baek, C., Kingon, A.I., Park, K.-I., Kim, S.-H.: Lead-free perovskite nanowire-employed piezopolymer for highly efficient flexible nanocomposite energy harvester. Small **14**, 1–8 (2018)
94. Manchi, P., Graham, S.A., Dudem, B., Patnam, H., Yu, J.S.: Improved performance of nanogenerator via synergetic piezo/triboelectric effects of lithium niobate microparticles embedded composite films. Compos. Sci. Technol. **201**, 108540 (2021)
95. Hajra, S., et al.: Lead-free flexible Bismuth Titanate-PDMS composites: a multifunctional colossal dielectric material for hybrid piezo-triboelectric nano-generator to sustainably power portable electronics. Nano Energy **89**, 106316 (2021)
96. Shehata, N., et al.: Stretchable nanofibers of polyvinylidenefluoride (PVDF)/thermoplastic polyurethane (TPU) nanocomposite to support piezoelectric response via mechanical elasticity. Sci. Rep. **12**, 1–11 (2022)
97. Mirjalali, S., et al.: Enhanced piezoelectricity of PVDF-TrFE nanofibers by intercalating with electrosprayed $BaTiO_3$. ACS Appl. Mater. Interfaces. **15**, 41806–41816 (2023)
98. Pratihar, S., Chandran, A.M., Bhat, A.R., Mural, P.K.S.: Piezoelectric nanogenerators based on poly(vinylidene fluoride) doped with high entropy oxide nanoparticles for sensitive pressure sensors. ACS Appl. Nano Mater. **7**, 20553–20568 (2024)
99. Yang, J., et al.: Eardrum-inspired active sensors for self-powered cardiovascular system characterization and throat-attached anti-interference voice recognition. Adv. Mater. **27**, 1316–1326 (2015)
100. Yamaguchi, T., Yamamoto, D., Arie, T., Akita, S., Takei, K.: Wrist flexible heart pulse sensor integrated with a soft pump and a pneumatic balloon membrane. RSC Adv. **10**, 17353–17358 (2020)
101. Liu, G., et al.: Vertically aligned ZnO nanoarray directly orientated on Cu paper by h-BN monolayer for flexible and transparent piezoelectric nanogenerator. Nano Energy **109**, 108265 (2023)
102. Li, J., et al.: Enhanced piezoelectric output of the PVDF-TrFE/ZnO flexible piezoelectric nanogenerator by surface modification. Appl. Surf. Sci. **463**, 626–634 (2019)
103. Cho, Y., et al.: $BaTiO_3$@PVDF-TrFE nanocomposites with efficient orientation prepared via phase separation nano-coating method for piezoelectric performance improvement and application to 3D-PENG. Chem. Eng. J. **427**, 131030 (2022)
104. Peng, R., et al.: Enhanced piezoelectric energy harvester by employing free-standing single-crystal $BaTiO_3$ films in PVDF-TrFE based composites. Adv. Func. Mater. **34**, 1–9 (2024)
105. Deng, W., et al.: Cowpea-structured PVDF/ZnO nanofibers based flexible self-powered piezoelectric bending motion sensor towards remote control of gestures. Nano Energy **55**, 516–525 (2019)
106. Borazan, S., Celik Bedeloglu, A.: Self-standing piezoelectric nanogenerator fabrics from ZnO-doped PVDF nanofiber yarns. J. Appl. Polymer Sci. **141**, 1–10 (2024)
107. Jiang, H., et al.: Synthesis of lead-free piezoelectric potassium sodium niobates for the preparation of anti-fouling KNN porous membrane. Ceram. Int. **50**, 27255–27264 (2024)
108. Aleksandrova, M., Tsanev, T., Pandiev, I., Dobrikov, G.: Study of piezoelectric behaviour of sputtered KNbO3 nanocoatings for flexible energy harvesting. Energy **205**, 118068 (2020)
109. Paria, S., et al.: Approach for enhancement in output performance of randomly oriented $ZnSnO_3$ nanorod-based piezoelectric nanogenerator via p–n Hetero-junction and Surface passivation layer. ACS Appl. Electron. Mater. **2**, 2565–2578 (2020)
110. D′Ambrogio, G. et al.: Piezoelectric biosensor for smart cardiovascular grafts based on $NaNbO_3$ fibers/PDMS structured composite. Mater. Design **223**, 111195 (2022)
111. Paranjape, M.V., et al.: Perovskite V –$NaNbO_3$ embedded PDMS composite film-based robust hybrid nanogenerator for efficient mechanical energy harvesting. Compos. Sci. Technol. **242**, 110195 (2023)

112. Ukasi, S., et al.: Gamma glycine enhances efficiency of organic hybrid piezoelectric-triboelectric nanogenerators. Nano Energy **119**, 109045 (2024)
113. Zhao, Q., et al.: Highly sensitive, reliable and flexible pressure sensor based on piezoelectric PVDF hybrid film using MXene nanosheet reinforcement. J. Alloy. Compd. **886**, 161069 (2021)
114. Wang, X., Wang, Z.L., Yang, Y.: Hybridized nanogenerator for simultaneously scavenging mechanical and thermal energies by electromagnetic-triboelectric-thermoelectric effects. Nano Energy **26**, 164–171 (2016)
115. Chowdhury, A.R., et al.: Lithium doped zinc oxide based flexible piezoelectric-triboelectric hybrid nanogenerator. Nano Energy **61**, 327–336 (2019)
116. Ojha, S., Maity, P., Bera, S., Kumar Si, S., Bhusan Khatua, B.: Metal–organic framework–derived ZnO-assisted β -phase-stabilized high-performance PVDF/ZnO-PDMS/rGO Nanocomposites as Piezo–Tribo Hybrid Nanogenerator. Energy Technol. **11**, 1–12 (2023)
117. Ajimsha, R., Mahapatra, A., Das, A., Sahu, V., Misra, P.: High output power density owing to enhanced charge transfer in ZnO-based triboelectric nanogenerator. Energy **263**, 125646 (2023)
118. Jao, Y.T., et al.: A textile-based triboelectric nanogenerator with humidity-resistant output characteristic and its applications in self-powered healthcare sensors. Nano Energy **50**, 513–520 (2018)
119. Das, N.K., Veeralingam, S., Badhulika, S.: Zinc ferrite nanoparticle-based wearable piezoelectric nanogenerators as self-powered sensors to monitor human motion. ACS Applied Nano Materials **6**, 13431–13442 (2023)
120. Luo, F., Chen, B., Ran, X., Ouyang, W., Shang, L.: PEO-PDMS-based triboelectric nanogenerators as self-powered sensors for driver status monitoring. Chem. Eng. J. **451**, 138961 (2023)
121. Feng, S., et al.: Synergistic effects of $BaTiO_3$/multiwall carbon nanotube as fillers on the electrical performance of triboelectric nanogenerator based on polydimethylsiloxane composite films. Energ. Technol. **7**, 1–8 (2019)
122. Park, H., Hyeon, D.Y., Jung, M., Park, K.I., Park, J.: Piezoelectric BaTiO3 microclusters and embossed ZnSnO3 microspheres-based monolayer for highly-efficient and flexible composite generator. Compos. Part B Eng., 108476 (2020)
123. Jeronimo, K., Koutsos, V., Cheung, R., Mastropaolo, E.: PDMS-ZnO piezoelectric nanocomposites for pressure sensors. Sensors **21**, 5873 (2021)
124. Baskin, Z.L., Shabalin, D.A., Vyrazheikin, E.S., Dedov, S.A.: Product range, properties, and application of fluoropolymers manufactured in Kirovo-Chepetsk chemical industrial complex. Russ. J. Gen. Chem. **79**, 527–538 (2009)
125. Yun, J., et al.: Piezo-triboelectric hybridized nanogenerator embedding MXene based bifunctional conductive filler in polymer matrix for boosting electrical power. Nano Energy **105**, 108018 (2023)
126. Salimi, A., Yousefi, A.: FTIR studies of β-phase crystal formation in stretched PVDF films. Polym. Testing **22**, 699–704 (2003)
127. Bera, B., Sarkar, M.D.: Piezoelectricity in PVDF and PVDF based piezoelectric nanogenerator: a Concept. IOSR J. Appl. Phys. **09**, 95–99 (2017)
128. Lovinger, A.J.: Annealing of Poly(vinylidene fluoride) and formation of a fifth phase. Macromolecules **15**, 40–44 (1982)
129. Ruan, L., et al.: Properties and applications of the β phase Poly(vinylidene fluoride). Polymers **10**, 228 (2018)
130. Yu, S., et al.: Formation mechanism of β-phase in PVDF/CNT composite prepared by the sonication method. Macromolecules **42**, 8870–8874 (2009)
131. Yu, H., et al.: Enhanced power output of an electrospun PVDF/MWCNTs-based nanogenerator by tuning its conductivity. Nanotechnology **24**, 405401 (2013)
132. Eun, J.H., Sung, S.M., Kim, M.S., Choi, B.K., Lee, J.S.: Effect of MWCNT content on the mechanical and piezoelectric properties of PVDF nanofibers. Mater. Des. **206**, 109785 (2021)

133. Fakhri, P., et al.: Flexible hybrid structure piezoelectric nanogenerator based on ZnO nanorod/ PVDF nanofibers with improved output. RSC Adv. **9**, 10117–10123 (2019)
134. He, J. et al: Electrospun PVDF-based polymers for lithium-ion battery separators: a review. Polymers 16 (2024)
135. Lee, J.S., Shin, K.Y., Cheong, O.J., Kim, J.H., Jang, J.: Highly sensitive and multifunctional tactile sensor using free-standing ZnO/PVDF thin film with graphene electrodes for pressure and temperature monitoring. Sci. Rep. **5**, 1–8 (2015)
136. Alam, M.M., Sultana, A., Sarkar, D., Mandal, D.: Electroactive β -crystalline phase inclusion and photoluminescence response of a heat-controlled spin-coated PV DF/TiO_2 free-standing nanocomposite film for a nanogenerator and an active nanosensor. Nanotechnology **28**, 365401 (2017)
137. Rahman, M.A., Rubaiya, F., Islam, N., Lozano, K., Ashraf, A.: Graphene-coated PVDF/PAni fiber mats and their applications in sensing and nanogeneration. ACS Appl. Mater. Interfaces. **14**, 38162–38171 (2022)
138. Panda, S., et al.: Biocompatible $CaTiO_3$−PV DF composite-based piezoelectric nanogenerator for exercise evaluation and energy harvesting. Nano Energy **102**, 107682 (2022)
139. Sasmal, A., Seetharaman, A., Maiti, P., Sudhakar, S., Arockiarajan, A.: Charge-defect tuned pvdf based ternary biocompatible composite as skin touch actuated single electrode triboelectric nanogenerator for wireless healthcare monitoring. Chem. Eng. J. **487**, 150573 (2024)
140. Singh, H.H., Khare, N.: Flexible ZnO-PVDF/PTFE based piezo-tribo hybrid nanogenerator. Nano Energy **51**, 216–222 (2018)
141. Lu, L., Ding, W., Liu, J., Yang, B.: Flexible PVDF based piezoelectric nanogenerators. Nano Energy **78**, 105251 (2020)
142. Zhu, Q., et al.: A high performance nanocellulose-PVDF based piezoelectric nano-generator based on the highly active CNF@ZnO via electrospinning technology. Nano Energy **127**, 109741 (2024)
143. Shohag, S.M.A.U., et al.: Enhanced piezoelectric nanogenerators with Sr-doped lanthanum cobaltite ($La_{1-x}Sr_xCoO_3$) and multiwalled carbon nanotubes for energy harvesting. ACS Appl. Eng. Mater. **2**, 2842–2855 (2024)
144. Wang, S., et al.: Core-shell structured silk Fibroin/PVDF piezoelectric nanofibers for energy harvesting and self-powered sensing. Nano Materials Sci. **4**, 126–132 (2022)
145. Zeyrek Ongun, M., Oguzlar, S., Kartal, U., Yurddaskal, M., Cihanbegendi, O.: Energy harvesting nanogenerators: Electrospun β-PVDF nanofibers accompanying ZnO NPs and ZnO@Ag NPs. Solid State Sci. **122**, 106772 (2021)
146. Hasanzadeh, M., Ghahhari, M.R., Bidoki, S.M.: Enhanced piezoelectric performance of PVDF-based electrospun nanofibers by utilizing in situ synthesized graphene-ZnO nanocomposites. J. Mater. Sci.: Mater. Electron. **32**, 15789–15800 (2021)
147. Bindhu, A., Arun, A.P., Pathak, M.: Review on Polyvinylidene fluoride-based triboelectric nanogenerators for applications in health monitoring and energy harvesting. ACS Appl. Electron. Mater. **6**, 47–72 (2024)
148. Bai, P. et al.: Dipole-moment-induced effect on contact electrification for triboelectric nanogenerators. Nano Res., 990–997 (2014)
149. Singh, H.H., Singh, S., Khare, N.: Enhanced β -phase in PVDF polymer nanocomposite and its application for nanogenerator. Polym. Adv. Technol. **29**, 143–150 (2018)
150. Tayyab, M., et al.: Enhanced output in polyvinylidene fluoride nanofibers based triboelectric nanogenerator by using printer ink as nano-fillers. Nano Energy **77**, 105178 (2020)
151. Singh, S.K., et al.: Seed power: natural seed and electrospun Poly(vinyl difluoride) (PVDF) Nanofiber Based Triboelectric Nanogenerators with high output power density. ACS Appl. Bio Mater. **2**, 3164–3170 (2019)
152. Luo, C., et al.: Preparation and application of high performance PVDF/PS electrospinning film-based triboelectric nanogenerator. Chem. Phys. Lett. **813**, 140276 (2023)
153. Feng, Y., et al.: Leaves based triboelectric nanogenerator (TENG) and TENG tree for wind energy harvesting. Nano Energy **55**, 260–268 (2019)

154. Feng, Y., et al.: Paper-based triboelectric nanogenerators and their application in self-powered anticorrosion and antifouling. J. Mater. Chem. A **4**, 18022–18030 (2016)
155. Gajula, P., Yoon, J.U., Woo, I., Bae, J.W.: Development of turbocharging-ability hybrid nanogenerators comprising bipolar PVDF-HFP/MXene electrospun composites. Small **2407001**, 1–13 (2024)
156. Domalanta, M.R.B., Ali, M.R.R., Maalihan, R.D., Caldona, E.B.: Mechanistic effects of HFP content on the surface structure and protective action of PVDF-HFP coatings. Prog. Org. Coat. **198**, 108879 (2025)
157. Chinya, I., Pal, A., Sen, S.: Flexible, hybrid nanogenerator based on Zinc Ferrite nanorods incorporated poly(vinylidene fluoride-co-hexafluoropropylene) nanocomposite for versatile mechanical energy harvesting. Mater. Res. Bull. **118**, 110515 (2019)
158. Polat, K.: Energy harvesting from a thin polymeric film based on PVDF-HFP and PMMA blend. Appl. Phys. A **126**, 497 (2020)
159. Liu, T., et al.: The PVDF-HFP gel polymer electrolyte for $Li-O_2$ battery. Solid State Ionics **318**, 88–94 (2018)
160. Roy, J., et al.: Room temperature spin injection study across semi-crystalline PVDF-HFP thin films in PVDF-HFP/NiFe bilayers and Ag/(NiFe or Co)/PVDF-HFP/NiFe spin valves. J. Phys.: Condens. Matter **36**, 505803 (2024)
161. Mitra, R., Sheetal Priyadarshini, B., Ramadoss, A., Manju, U.: Stretchable polymer-modulated PVDF-HFP/TiO_2 nanoparticles-based piezoelectric nanogenerators for energy harvesting and sensing applications. Mater. Sci. Eng., B **286**, 116029 (2022)
162. Vu, D.L., Vo, C.P., Le, C.D., Ahn, K.K.: Enhancing the output performance of fluid-based triboelectric nanogenerator by using poly(vinylidene fluoride-co-hexafluoropropylene)/ionic liquid nanoporous membrane. Int. J. Energy Res. **45**, 8960–8970 (2021)
163. Saikh, M.M. et al.: Self-Polarized ZrO_2/Poly(vinylidene fluoride-co-hexafluoropropylene) nanocomposite-based piezoelectric nanogenerator and single-electrode triboelectric nanogenerator for sustainable energy harvesting from human movements. Physica Status Solidi (A) Appl. Mater. Sci. **218**, 1–10 (2021)
164. He, Y., et al.: Enhancing output performance of PVDF-HFP fiber-based nanogenerator by hybridizing silver nanowires and perovskite oxide nanocrystals. Nano Energy **98**, 107343 (2022)
165. Lin, M.-F., Chang, K.-W., Lee, C.-H., Wu, X.-X., Huang, Y.-C.: Electrospun P3HT/PVDF-HFP semiconductive nanofibers for triboelectric nanogenerators. Sci. Rep. **12**, 14842 (2022)
166. Amrutha, B., et al.: Performance optimization of MoS_2-doped PVDF-HFP nanofiber triboelectric nanogenerator as sensing technology for smart cities. Appl. Mater. Today **41**, 102503 (2024)
167. Parangusan, H., Ponnamma, D., Al-Maadeed, M.A.A.: Stretchable Electrospun PVDF-HFP/Co-ZnO Nanofibers as Piezoelectric Nanogenerators. Sci. Rep. **8**, 1–11 (2018)
168. Khandelwal, G., Maria Joseph Raj, N.P., Kim, S.: Materials beyond conventional triboelectric series for fabrication and applications of triboelectric nanogenerators. Adv. Energy Mater. **11**, 1–32 (2021)
169. Zhang, R., Olin, H.: Material choices for triboelectric nanogenerators: a critical review. EcoMat **2**, 1–13 (2020)
170. Zhao, P., et al.: Emulsion electrospinning of polytetrafluoroethylene (PTFE) nanofibrous membranes for high-performance triboelectric nanogenerators. ACS Appl. Mater. Interfaces. **10**, 5880–5891 (2018)
171. Li, G.Z., et al.: High-performance transparent and flexible triboelectric nanogenerators based on PDMS-PTFE composite Films. Adv. Electron. Mater. **5**, 1–8 (2019)
172. Wang, M., et al.: Single-electrode triboelectric nanogenerators based on sponge-like porous PTFE thin films for mechanical energy harvesting and self-powered electronics. J. Mater. Chem. A **5**, 12252–12257 (2017)
173. Liu, H., et al.: Real-time acid rain sensor based on a Triboelectric nanogenerator made of a PTFE–PDMS composite film. ACS Appl. Electron. Mater. **3**, 4162–4171 (2021)

174. Koski, A., Yim, K., Shivkumar, S.: Effect of molecular weight on fibrous PVA produced by electrospinning. Mater. Lett. **58**, 493–497 (2004)
175. McKeen, L.W.: Fluoropolymers (2012)
176. Ben Halima, N.: Poly(vinyl alcohol): review of its promising applications and insights into biodegradation. RSC Adv. **6**, 39823–39832 (2016)
177. Kwon, H.-J., Hong, S.-M., Park, S.-M., Lee, C.W.: Characterization of acid-modified polyvinyl alcohol and its application to barrier-coated paper for eco-friendly food packaging. Food Packag. Shelf Life **43**, 101271 (2024)
178. Patnam, H., Dudem, B., Graham, S.A., Yu, J.S.: High-performance and robust triboelectric nanogenerators based on optimal microstructured poly(vinyl alcohol) and poly(vinylidene fluoride) polymers for self-powered electronic applications. Energy **223**, 120031 (2021)
179. Wang, Y., Zhang, L., Lu, A.: Highly stretchable, transparent cellulose/PVA composite hydrogel for multiple sensing and triboelectric nanogenerators. J. Mater. Chem. A **8**, 13935–13941 (2020)
180. Sardana, S., Saddi, R., Mahajan, A.: MXene-functionalized KNN dielectric nanofillers incorporated in PVA nanofibers for high-performance triboelectric nanogenerator. Appl. Phys. Lett. **122**, 162902 (2023)
181. Shi, L., et al.: Enhanced performance triboelectric nanogenerators based on solid polymer electrolytes with different concentrations of cations. Nano Energy **64**, 103960 (2019)
182. Luo, X., et al.: A flexible multifunctional triboelectric nanogenerator based on MXene/PVA hydrogel. Adv. Func. Mater. **31**, 1–9 (2021)
183. Liu, J. et al.: A high-performance, biocompatible, and fully biodegradable piezo-triboelectric hybrid nanogenerator based on PVA/Glycine/PVA heterostructured piezoelectric film. Nano Energy, 109310 (2024)
184. Yang, J., et al.: Transparent self-powered triboelectric sensor based on PVA/PA hydrogel for promoting human-machine interaction in nursing and patient safety. Nano Energy **97**, 107199 (2022)
185. Jaiswal, M., et al.: Sodium niobate nanowires embedded PVA-hydrogel-based triboelectric nanogenerator for versatile energy harvesting and self-powered CO gas sensor. Small **2403699**, 1–10 (2024)
186. Herzberger, J., et al.: Polymerization of ethylene oxide, propylene oxide, and other alkylene oxides: synthesis, novel polymer architectures, and bioconjugation. Chem. Rev. **116**, 2170–2243 (2016)
187. Bo, X., et al.: A Stretchable solid ionic electrode-based triboelectric nanogenerator for biomechanical energy harvesting and self-powered sensors. Small **19**, 2303415 (2023)
188. Maccaferri, E., et al.: New application field of polyethylene oxide: PEO nanofibers as epoxy toughener for effective CFRP delamination resistance improvement. ACS Omega **7**, 23189–23200 (2022)
189. Tao, D. et al.: Electro-spun nanofibers-based triboelectric nanogenerators in wearable electronics: status and perspectives. npj Flexible Electron. **9**, 4 (2025)
190. Hao, Y., et al.: PEO/cysteine composite nanofiber-based triboelectric nanogenerators for harvesting tiny mechanical energy. J. Mater. Chem. A **13**, 1853–1862 (2025)
191. Lin, C., Zhao, H., Huang, H., Ma, X., Cao, S.: PEO/cellulose composite paper based triboelectric nanogenerator and its application in human-health detection. Int. J. Biol. Macromol. **228**, 251–260 (2023)
192. Lee, C., Park, S., Cho, C., Oh, J.H.: Improvement in the electrical performance of Poly(ethylene oxide) nanofiber-based triboelectric nanogenerators by introducing sodium chloride. ACS Appl. Electron. Mater. **5**, 6047–6055 (2023)
193. Johnson, A.D., et al.: High performance flexible green triboelectric nanogenerator with polyethylene oxide/mica tribo-positive composite material. Appl. Mater. Today **39**, 102321 (2024)
194. Winnacker, M., et al.: Polyamide / PEG blends as biocompatible biomaterials for the convenient regulation of cell adhesion and growth. Macromol. Rapid Commun. **40**, 1900091 (2019)

195. Mathew, D.T., et al.: Surface area enhanced Nylon-6,6 nanofiber engineered triboelectric nanogenerator for self-powered seat monitoring applications. ACS Sustain. Chem. Eng. **10**, 14126–14135 (2022)
196. Bhatta, T., et al.: High-performance triboelectric nanogenerator based on MXene functionalized polyvinylidene fluoride composite nanofibers. Nano Energy **81**, 105670 (2021)
197. Bairagi, S., et al.: High-performance triboelectric nanogenerators based on commercial textiles: electrospun Nylon 66 nanofibers on Silk and PVDF on polyester. ACS Appl. Mater. Interfaces. **14**, 44591–44603 (2022)
198. Okbaz, A.: GnPs/PVDF decorated thermoplastic veils to boost the triboelectric nanogenerator output performance toward highly efficient energy harvesting. Energy Convers. Manage. **270**, 116204 (2022)
199. Bhatta, T., et al.: Siloxene/PVDF composite nanofibrous membrane for high-performance triboelectric nanogenerator and self-powered static and dynamic pressure sensing applications. Adv. Func. Mater. **32**, 1–15 (2022)
200. Graham, S.A., Patnam, H., Manchi, P., Paranjape, M.V., Yu, J.S.: Superficial modification of polyamide-66 film for enhanced electrical performance and multimode functional triboelectric nanogenerators. Int. J. Energy Res. **46**, 17391–17403 (2022)
201. Liu, J., et al.: A PA66 lamellar crystal film with excellent triboelectric performance in vertical contact separation mode. J. Mater. Chem. C **11**, 12222–12233 (2023)
202. ecoflex^{TM}00-30 product information. https://www.smooth-on.com/products/ecoflex-00-30/?quicksearch. Accessed 28 January 2025
203. Shrestha, K., et al.: A siloxene/ecoflex nanocomposite-based triboelectric nano-generator with enhanced charge retention by MoS_2/LIG for self-powered touchless sensor applications. Adv. Func. Mater. **32**, 1–12 (2022)
204. Sarkar, D., Das, N., Sau, S., Basu, R., Das, S.: Micro-patterned $BaTiO_3$@Ecoflex nanocomposite-assisted self-powered and wearable triboelectric nanogenerator with improved charge retention by 2D $MoTe_2$/PVDF nanofibrous layer. J. Mater. Chem. C **12**, 984–1001 (2023)
205. Lim, Y.P., et al.: Modulating ZnO growth structures for maximum power output of hybrid piezo/triboelectric nanogenerator. Adv. Func. Mater. **32**, 1–11 (2022)
206. Vivekananthan, V., Chandrasekhar, A., Alluri, N.R., Purusothaman, Y., Kim, S.J.: A highly reliable, impervious and sustainable triboelectric nanogenerator as a zero-power consuming active pressure sensor. Nanoscale Adv. **2**, 746–754 (2020)
207. Salvadores Fernandez, C. et al.: A Triboelectric nanocomposite for sterile sensing, energy harvesting, and haptic diagnostics in interventional procedures from surgical gloves. Adv. Healthcare Mater. **12**, 1–10 (2023)
208. Zhao, L., et al.: Triboelectric gait sensing analysis system for self-powered IoT-based human motion monitoring. InfoMat **6**, 1–13 (2024)
209. Zhang, Y., et al.: Ultra-stretchable monofilament flexible sensor with low hysteresis and linearity based on MWCNTs/ecoflex composite materials. Macromol. Mater. Eng. **306**, 1–9 (2021)
210. Yu, J., et al.: Skin-conformal $BaTiO_3$/ecoflex-based piezoelectric nanogenerator for self-powered human motion monitoring. Mater. Lett. **269**, 127686 (2020)
211. Liu, H., et al.: Sustainable paper-based triboelectric nanogenerator with improved output performance through a general strategy of interface fusion. Nano Energy **124**, 109464 (2024)
212. Le, T.H., Kim, Y., Yoon, H.: Electrical and electrochemical properties of conducting polymers. Polymers 9 (2017)
213. Awuzie, C.I.: Conducting polymers. Mater. Today. Proc. **4**, 5721–5726 (2017)
214. Scheffler, S., Poulin, P.: Piezoelectric fibers: processing and challenges. ACS Appl. Mater. Interfaces. **14**, 16961–16982 (2022)
215. Tang, T. et al.: Stretchable polymer composites with ultrahigh piezoelectric performance. Nat. Sci. Rev. 10 (2023)
216. Furukawa, T.: Piezoelectricity and pyroelectricity in polymers. IEEE Trans. Electr. Insul. **24**, 375–394 (1989)

217. Kepler, R.G., Anderson, R.A.: Ferroelectric polymers. Adv. Phys. **41**, 1–57 (1992)
218. Chen, G., Au, C., Chen, J.: Textile triboelectric nanogenerators for wearable pulse wave monitoring. Trends Biotechnol. **39**, 1078–1092 (2021)
219. Du, M., et al.: Hybrid nanogenerator for biomechanical energy harvesting, motion state detection, and pulse sensing. Adv. Mater. Technol. **7**, 1–8 (2022)
220. Mathew, A.A., Vivekanandan, S.: Design and simulation of single-electrode mode triboelectric nanogenerator-based pulse sensor for healthcare applications using COMSOL multiphysics. Energy Technol. 10 (2022)
221. Lin, M.F., Xiong, J., Wang, J., Parida, K., Lee, P.S.: Core-shell nanofiber mats for tactile pressure sensor and nanogenerator applications. Nano Energy **44**, 248–255 (2018)
222. Rasel, M.S., et al.: An impedance tunable and highly efficient triboelectric nanogenerator for large-scale, ultra-sensitive pressure sensing applications. Nano Energy **49**, 603–613 (2018)
223. Yi, F., Zhang, Z., Kang, Z., Liao, Q. & Zhang, Y.: Recent Advances in Triboelectric Nanogenerator-Based Health Monitoring. Adv. Funct. Mater. 29 (2019)
224. Lin, Z., et al.: Super-robust and frequency-multiplied triboelectric nanogenerator for efficient harvesting water and wind energy. Nano Energy **64**, 103908 (2019)
225. Ahmed, A., et al.: Farms of triboelectric nanogenerators for harvesting wind energy: a potential approach towards green energy. Nano Energy **36**, 21–29 (2017)
226. Rodrigues, C., et al.: Emerging triboelectric nanogenerators for ocean wave energy harvesting: state of the art and future perspectives. Energy Environ. Sci. **13**, 2657–2683 (2020)
227. Liang, Q., Yan, X., Liao, X., Zhang, Y.: Integrated multi-unit transparent triboelectric nanogenerator harvesting rain power for driving electronics. Nano Energy **25**, 18–25 (2016)
228. Lai, Y., Hsiao, Y., Wu, H., Wang, Z.L.: Waterproof fabric-based multi-functional triboelectric nanogenerator for universally harvesting energy from raindrops, wind, and human motions and as self-powered sensors. Adv. Sci. 6 (2019)
229. Zhang, Y., et al.: A flexible droplet-based triboelectric-electromagnetic hybrid generator for raindrop energy harvesting. Nano Energy **121**, 109253 (2024)
230. Yang, J., et al.: Triboelectrification-based organic film nanogenerator for acoustic energy harvesting and self-powered active acoustic sensing. ACS Nano **8**, 2649–2657 (2014)
231. Fan, X., Chen, J., Yang, J., Peng Bai, Z.L., Wang, Z.L.: Triboelectric Nano-generator for Acoustic Energy Harvesting and a self powered sounf recording. ACS Nano **9**, 4236–4243 (2015).
232. Chen, F., et al.: A novel triboelectric nanogenerator based on electrospun polyvinylidene fluoride nanofibers for effective acoustic energy harvesting and self-powered multifunctional sensing. Nano Energy **56**, 241–251 (2019)
233. Xia, K., et al.: Milk-based triboelectric nanogenerator on paper for harvesting energy from human body motion. Nano Energy **56**, 400–410 (2019)
234. Ren, X., Fan, H., Zhao, Y., Liu, Z.: Flexible lead-free $BiFeO_3$/PDMS-based nanogenerator as piezoelectric energy harvester. ACS Appl. Mater. Interfaces. **8**, 26190–26197 (2016)
235. Guo, R., Guo, Y., Duan, H., Li, H., Liu, H.: Synthesis of orthorhombic perovskite-type $ZnSnO_3$ single-crystal nanoplates and their application in energy harvesting. ACS Appl. Mater. Interfaces. **9**, 8271–8279 (2017)
236. Jean, F., Khan, M.U., Alazzam, A., Mohammad, B.: Advancement in piezoelectric nanogenerators for acoustic energy harvesting. Microsystems and Nanoengineering 10 (2024)
237. Das, K.K., Basu, B., Maiti, P., Dubey, A.K.: Piezoelectric nanogenerators for self-powered wearable and implantable bioelectronic devices. Acta Biomater. **171**, 85–113 (2023)
238. Hu, Y., Wang, Z.L.: Recent progress in piezoelectric nanogenerators as a sustainable power source in self-powered systems and active sensors. Nano Energy **14**, 3–14 (2015)
239. Panda, S., et al.: Hybrid nanogenerators for ocean energy harvesting: mechanisms, designs, and applications. Small **19**, 1–21 (2023)
240. Hansen, B.J., Liu, Y., Yang, R., Wang, Z.L.: Hybrid nanogenerator for concurrently harvesting biomechanical and biochemical energy. ACS Nano **4**, 3647–3652 (2010)
241. Cheng, X., et al.: Implantable and self-powered blood pressure monitoring based on a piezoelectric thinfilm: simulated, in vitro and in vivo studies. Nano Energy **22**, 453–460 (2016)

Advanced Printing Techniques for Nanopolymers

Niranjan Patra, Pradeep Kumar Panda, and Ştefan Ţălu

Abstract Advanced materials for 3D and 4D printing have been developed through the integration of nanotechnology with additive manufacturing. The use of nano polymers in printed structures allows for increased mechanical, thermal, electrical, and stimuli-responsive qualities, as well as environmental adaptability. Nano polymers are polymeric matrices that contain fillers on the nanoscale. Nanocomposites, biopolymers, and stimuli-responsive materials are some of the important classes covered, and their significance in cutting-edge fields including aerospace, biomedicine, smart clothing, and environmentally friendly packaging is emphasised. Adding nano polymers to 4D printing makes it far easier to change shapes and have them self-actuate in response to external stimuli. This chapter highlights the revolutionary possibilities of nano polymers for nextgen additive manufacturing in terms of intelligent, environmentally sustainable, and application-specific solutions.

Keywords Additive manufacturing · Nano polymers · Nanocomposites · Smart materials · Functional materials

1 Introduction

3D printing, or additive manufacturing (AM), is a revolutionary fabrication technique that uses a layer-by-layer deposition method to create three-dimensional objects from digital models. Several industries, including electronics, aerospace, healthcare, and

N. Patra (✉)
Department of Chemistry, Koneru Lakshmaiah Education Foundation, Greenfield, Vaddeswaram, Andhra Pradesh, India
e-mail: patraji@gmail.com

P. K. Panda
Department of Chemical Engineering and Materials Science, Yuan Ze University, Taoyuan, Taiwan

Ş. Ţălu
Development and Innovation Management (DMCDI), The Directorate of Research, Technical University of Cluj-Napoca, Cluj-Napoca, Romania

B. Sharma et al. (eds.), *Nanostructured Polymers*, Engineering Materials,
https://doi.org/10.1007/978-981-95-3581-1_12

the automotive sector, have seen profound paradigm shifts in manufacturing since AM's inception in the 1980s. Digital manufacturing and Industry 4.0 projects rely on 3D printing because it can create lightweight, complex geometric pieces with little material waste [1]. The three main materials used in traditional 3D printing are metals, ceramics, and polymers. Polymers have several advantages, such as being inexpensive, easy to process, and versatile. Having said that, the structures and qualities that these materials usually produce are static with the introduction of time as a functional dimension, four-dimensional (4D) printing has marked a significant progress. Shape morphing, self-healing, and functional reconfiguration are just a few examples of the dynamic transformations that can occur when materials are created to respond to external stimuli like light, moisture, temperature, or pH. The utilisation of smart materials like shape-memory polymers (SMPs), hydrogels, and liquid crystal elastomers (LCEs) has created new possibilities in areas like biomedical devices, soft robotics, and deployable aircraft systems [2–4].

A key obstacle to attaining more functional and industrial integration, despite the fast growth of AM technologies, is the material restrictions. For high-performance or adaptive applications, traditional polymers frequently fall short in terms of mechanical strength, thermal stability, and responsiveness. One potential solution to these problems is the use of nanotechnology in polymer systems. To improve structural, functional, and environmental qualities beyond what conventional polymer materials can achieve, scientists have developed nano polymers, which are essentially polymer matrices mixed with nanoscale fillers like nanoparticles, nanofibers, or nanoclusters [4–6].

Ever since Charles Hull introduced stereolithography (SLA) in 1986—a method for precise production using photopolymer resins—the area of polymer-based AM has come a long way [7, 8]. More sophisticated processes, such as fused deposition modelling (FDM) and selective laser sintering (SLS), have made it possible to employ thermoplastic filaments and polymer powders in a wide variety of items [9, 10]. The use of polylactic acid (PLA), acrylonitrile butadiene styrene (ABS), and polyamide (PA) in early 3D printing of polymers has given way to a preference for high-performance polymers like polyether ether ketone (PEEK) and biocompatible resins, which are opening up new possibilities for use in biomedical and industrial fields [11, 12]. As the printing technology has progressed from 3 to 4D, new material properties have been required. Nanoscale engineering of stimuli-responsive materials improves their qualities, making them more resilient mechanically, faster to actuate, and with a customisable modulus. Combining mechanical performance with multifunctionality is made possible through the incorporation of nanostructured components. For instance, nano polymer composites are well-suited for structural and load-bearing applications in the aerospace and automotive industries due to the increased tensile strength, toughness, and fatigue resistance brought about by nanofillers such nanocellulose, carbon nanotubes (CNTs), and graphene [13]. Alumina, silica, and boron nitride are thermal additives that enhance high-temperature performance, which is essential in modern manufacturing environments [14, 15].

Nanotechnology adds smart capabilities to 4D printing through the use of stimuli-responsive nanomaterials including quantum dots, magnetite, and gold nanoparticles.

By facilitating environmental responsiveness, these materials allow printed buildings to self-heal, rearrange, or morph in response to actual environmental conditions [4, 5, 16]. From autonomous robotic components to wearable biomedical devices, this capability is crucial for the development of adaptive systems [12]. Nano polymers applicability to printed electronics is further enhanced by the use of conductive nanomaterials, such as silver nanoparticles, carbon nanotubes, and graphene. Sensors, antennas, and energy storage systems can be made from these materials since they are lightweight, flexible, and functionally integrated. New developments in soft robotics and individualised healthcare stand to gain the most from the combination of electrical capabilities with mechanical adaptability [17, 18]. Concern for the environment is also rising to the forefront of AM studies. A lessening of additive manufacturing's impact on the environment is driving the development of biodegradable and bio-derived nano polymers. Examples of such materials include those generated from starch, nanocellulose, or chitosan. There is a growing need for eco-friendly manufacturing methods, and these green materials can help with that [19, 20].

Finally, a giant leap forward in material innovation has occurred with the incorporation of nanotechnology into additive manufacturing processes based on polymers. Nano polymers are leading the way in next-gen 3D and 4D printing technology because of their improved mechanical, thermal, electrical, and adaptable capabilities. Improving printability across multiple AM platforms, increasing their functional response profiles, and perfecting the synthesis of nano polymers will be the focus of future efforts as research advances. Materials science, nanotechnology, and digital manufacturing are coming together in a way that could change the way printed materials work in a lot of different industries [21].

2 Fundamentals of 3D and 4D Printing

3D printing builds products layer by layer straight from a digital model which is also called additive manufacturing (AM). 3D printing creates structures by depositing material precisely where it is required, as opposed to the traditional method of subtractive manufacturing which entails removing material from a solid block. Benefits like as mass customisation, reduced material waste, and the ability to create complex geometries are made possible by this paradigm change. 3D printing has the potential to revolutionise numerous industries due to these advantages [7].

A digital model, usually made with CAD software, is the starting point for the 3D printing process. Digitally slicing the model into several horizontal layers follows its completion, followed by conversion to a print-ready format like STL or AMF. The printer follows the cross-sectional patterns represented by the layers as they fabricate the object. The G-code is used to convert the sliced data (STL file) into instructions for the printer to deposit material for each layer. The printing process involves carefully depositing, curing, or fusing layers of the chosen material—be it thermoplastic, resin, metal powder, or composite—in order to construct the object from the bottom up. Traditional methods would have a hard time, if not impossible, reproducing the complex shapes made feasible by the layer-by-layer deposition process.

After printing, the majority of components are subjected to post-processing in order to improve their performance or aesthetics. Thermal curing, chemical sintering, support removal, and surface polishing are some of the extra processes that may be required depending on the material and its intended use. To get the mechanical strength, dimensional correctness, and aesthetic quality you want in the end, these stages are essential.

2.1 3D Printing Techniques

Multiple methods have emerged as a result of 3D printing's growth; these methods differ in their underlying concepts, benefits, and limits. These techniques allow for the production of complicated structures with fine-grained control over mechanical characteristics, resolution, and material deposition.

Stereolithography (SLA) is a UV laser procedure that selectively cures liquid photopolymer resin; it is one of the oldest and most used technologies. Laser scanning of the resin's surface in accordance with a digital model initiates a photopolymerization reaction, which causes the substance to harden in layers. SLA printing is perfect for complex dental models and prosthetics due to its high resolution and flawless surface treatments. The brittleness of SLA resins need additional curing and solvent cleaning as post-processing processes to increase durability [22].

Fused deposition modelling (FDM) is another popular method; it uses a heated nozzle to extrude thermoplastic filaments. As it cools and comes into touch with the build platform, the molten polymer is deposited in layers, eventually hardening. Both academic research and industrial prototypes make heavy use of FDM because of its versatility and low cost. A few downsides to consider are the decreased resolution, the visibility of layer lines, and the impact of anisotropic mechanical characteristics on part strength [9].

Selective laser sintering (SLS) is an option worth considering for applications that require materials with better mechanical characteristics and more design flexibility. A dense, solid structure can be produced using this method by selectively fusing polymer or metal powders using a powerful laser, eliminating the need for supplementary support materials. Although SLS enables the creation of long-lasting parts with intricate geometries, it is an expensive process that demands cautious manipulation of powder to avoid flaws like porosity and distortion [5].

Another state-of-the-art 3D printing method is PolyJet printing, which uses inkjet technology to deposit liquid photopolymer droplets that are quickly cured by ultraviolet light. The technology allows for the printing of functioning prototypes with various attributes in a single build, thanks to its support for multi-material and multicolor printing. Digital light processing (DLP) is similar to SLA in that it cures many layers at once using a digital projector screen, but it uses vat photopolymerization instead. This allows for more rapid printing without sacrificing precision.

Direct ink writing (DIW) is a relatively new but increasingly significant method for depositing inks with high viscosity, such as formulations based on nanomaterials or polymer composites. Biomedical engineers, flexible electronics, and soft robots can all benefit from DIW printing's ability to create conductive, biocompatible, and stimuli-responsive structures. Ongoing research aims to optimise process settings and material formulations for these 3D printing techniques, with the goal of improving performance and expanding their utility. Each of these methods has its own set of advantages and constraints.

2.2 *Material Considerations in 3D Printing*

The functionality of 3D-printed objects are greatly affected by the materials that are chosen. With their varied mechanical, thermal, and chemical characteristics, additive manufacturing has investigated a wide range of polymers. Many 3D printing processes, such as FDM and SLS, rely on thermoplastics as a medium. In particular, the biodegradability and printability of polylactic acid (PLA) made from sustainable sources make it a desirable material. Not appropriate for high-performance applications due to its weak thermal stability and brittleness. While acrylonitrile butadiene styrene (ABS) prints with better toughness and heat resistance than PLA, it also produces fumes that could be harmful to the operator. As a result of its combination of strength, flexibility, and chemical resistance, polyethylene terephthalate glycol (PETG) is the material of choice for both practical prototypes and uses involving food. Premium engineering-grade polymers, including PEEK and Polyetherimide, are highly resistant to heat and chemicals, making them perfect for use in aircraft, medical implants, and high-performance engineered parts [23].

Photopolymers are also important in SLA, DLP, and PolyJet printing, in addition to traditional thermoplastics. These techniques prints with good resolution and no rough spots because these resins are crosslinked by heat and light. Newer, more resilient formulations with enhanced impact resistance were developed in response to the fact that conventional photopolymer resins have fragility. Furthermore, biocompatible resins have recently attracted interest from the dentistry and medical communities for use in creating custom implants and prosthesis for individual patients [24].

Nanocomposite materials have recently advanced, greatly expanding the potential of 3D printing. Digitally printed parts benefit from increased mechanical strength and stiffness and decreased weight when carbon fibre is mixed with polymer matrices. Similar to how graphene and carbon nanotubes enhance electrical and thermal conductivity, these materials are well-suited for use in functional electronics and sensors. The use of silica nanoparticles has been shown to improve wear resistance and dimensional stability, while nanocellulose-based composites are gaining traction as eco-friendly additive manufacturing options [4].

The use of hydrogels and bioinks in 3D bioprinting has dramatically altered its trajectory in the biomedical field, opening the door to the creation of organ models and tissue scaffolds. Biocompatible and supportive of cell adhesion and proliferation, natural biopolymers including collagen, gelatin, and alginate are great options. In 4D bioprinting, printed materials can change shape over time in reaction to outside triggers, and now researchers are looking at smart hydrogels that can react to surroundings [25].

An additional important group is functional and conductive polymers, which are necessary for 3D-printed electrical components. An abundance of flexible circuits and energy storage devices have been manufactured using conductive polymers like PEDOT: PSS and polypyrrole (PPy). Soft robotics, artificial muscles, and bioelectronic interfaces have all been given new life by ionic conducting polymers, which may transport ions in reaction to an electric field.

Materials research and development for 3D printing still faces a number of obstacles, even with recent improvements. Because of the issue of weak interlayer bonding that can arise from layer-by-layer deposition, one of the main issues is that printed parts have limited mechanical strength, especially in the Z-direction. Since many polymers break down at high processing temperatures, thermal stability is another issue that limits their utility in harsh settings. Material designers face additional challenges when attempting to strike a balance between biodegradability and durability. Also, not all materials work well with current printing methods; some may necessitate tweaks to the formulation, print settings, or post-processing procedures for the best results.

In order to overcome these constraints, scientists are investigating novel polymer formulations that may expand additive manufacturing's capabilities. These include materials with nano-enhancements, shape-memory polymers, and self-healing properties. The incorporation of nanotechnology into 3D printing materials is anticipated to have a game-changing impact on various industries, enhancing functionality, efficiency, and sustainability, as the field progresses.

2.3 4D Printing: Concept, Stimuli-Responsive Materials, and Self-actuation

4D printing expands the capabilities of 3D printing by incorporating time-dependent transformations, which can produce even more complex structures. This innovation has the potential to completely change the manufacturing industry. Printing objects with the ability to respond to environmental factors like heat, light, pH, or moisture is the goal of four-dimensional printing (4D printing). Use of stimuli-responsive materials with shape-memory effects, self-healing capabilities, or programmed actuation allows for this dynamic behavior [26, 27].

Smart materials like shape-memory polymers (SMPs), hydrogels, liquid crystal elastomers (LCEs), and electroactive polymers (EAPs) are the main drivers behind the concept of 4D printing [28–31]. One example is SMPs, which can restore a previously determined shape when heated, thanks to a process called thermally induced shape transition. In contrast, hydrogels are well-suited for use in biological contexts including drug delivery and tissue engineering due to their ability to expand or contract in reaction to changes in humidity or pH [28, 29, 32].

A key component of 4D-printed materials' self-actuation mechanism is the controlled programming of material properties that takes place during printing. Objects that bend, twist, or expand in a pre-determined manner upon stimulation can be designed by strategically integrating anisotropic arrangements of responsive materials within a 3D-printed structure. Significant breakthroughs in biomedical implants, deployable aircraft structures, and soft robotics—areas where adaptability and utility are paramount—have been made possible by this technique [2, 31, 33]. 4D printing has the promise to revolutionise several industries, but it isn't without its problems, such as a lack of readily available materials, complicated computational modelling, and poor actuation speeds. The responsiveness, mechanical durability, and biocompatibility of 4D-printed materials are being improved through current research in nano-enhanced polymers, which is addressing these issues [34, 35].

2.4 Key Differences Between 3 and 4D Printing

While additive manufacturing is the same for both 3D and 4D printing, the two technologies couldn't be more different in how they work and how materials behave. The main difference between 3D-printed and 4D-printed structures is that the former are static and the latter are dynamically adaptable. Unless handled differently during post-processing, the created objects from 3D printing will keep their original shape and characteristics. Objects printed with 4D printing, on the other hand, can gain intelligence and change over time in response to targeted environmental stimuli such as heat, light, moisture etc. Biomedicine, aerospace, and soft robots are just a few of the many areas that can benefit from this change made possible by the material's intrinsic responsiveness. Thirdly, 4D printing and 3D printing use very different standards for choosing their materials. 4D printing requires stimuli-responsive polymers that may experience regulated morphological changes, in contrast to 3D printing which mostly uses static thermoplastics, thermosets, and composites. This differentiation emphasises the growing requirement for cutting-edge nano-engineered materials that can offer 4D-printed systems enhanced mechanical, electrical, and actuation capabilities [10, 15, 28, 29, 36].

3 Nano Polymers: Composition and Properties

Polymers that incorporate nanoscale components, usually ranging from 1 to 100 nm, within their structure or as individual inclusions are called nano polymers. These materials stand out from regular polymers due to their exceptional physicochemical characteristics, which include increased chemical resistance, electrical conductivity, thermal stability, and mechanical strength [15]. Advanced additive manufacturing relies on new functionalities made possible by changing polymer interactions with nanoparticles or nanostructures. Different types of nanopolymers are defined by different combinations of chemical elements, molecular architecture, and functional properties (Fig. 1). Characterised by unique structural features and adaptable physicochemical qualities, each class serves a unique purpose.

3.1 Polymer Nanocomposites

Nanoscale fillers, such as carbon-based nanomaterials (such as carbon nanotubes, graphene, and carbon dots), ceramic nanoparticles (such as silica, titania, and zirconia), and metallic nanoparticles (such as silver, gold, and copper), are integrated into typical polymer matrices in polymer nanocomposites. Through controlling interfacial adhesion and nanoscale dispersion, these nanofillers greatly improve the mechanical strength, electrical conductivity, thermal stability, and chemical resistance of the polymer. Final nanocomposite characteristics, including toughness, optical transparency, and rheological behaviour, are determined in large part by the homogeneous dispersion of nanofillers. Composites made of polymers reinforced with carbon nanotubes (CNTs) are perfect for use in aircraft and flexible electronics due to their high mechanical durability and exceptional electrical conductivity. Biopolymers treated with graphene oxide (GO) had improved water barrier and

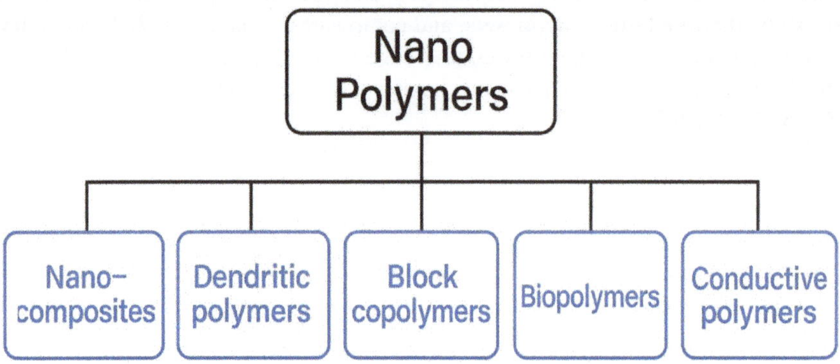

Fig. 1 Classification of nano polymers based on their molecular architecture and functional attributes

thermal conductivity, making them beneficial for applications such as food packaging and wearable electronics [16].

3.2 Dendritic Polymers

A highly branched, tree-like molecular structure is a defining feature of dendritic polymers, which include dendrimers and hyperbranched polymers. Their many functional groups, microscopic size, and accurate molecular weight distribution make them useful in nanoelectronics, drug delivery, and catalysis. Because of their carefully regulated structure, dendritic polymers have excellent solubility and host–guest interactions. Tunable surface characteristics and regulated release mechanisms in biological applications are made possible by their terminal functional groups, which allow for further chemical alterations.

For instance, PAMAM dendrimers' biocompatibility, controlled release capabilities, and capacity to encapsulate hydrophobic pharmaceuticals have made them a promising candidate for targeted drug delivery [37]. Their focused therapeutic applications are made possible by the enhanced solubility in aqueous solutions caused by the presence of exterior functional groups and internal voids.

3.3 Block Copolymer-Based Nano Polymers

Block copolymers are nanoscale phase-separable self-assembling structures made of chemically different polymer segments. Membrane technology, nanoscale lithography, and energy storage rely on their capacity to generate nanostructured domains including micelles, lamellae, and cylinders. Mechanical strength, flexibility, and permeability can be fine-tuned by selectively functionalising the block segments [38, 39]. In emulsification, controlled drug administration, and nanofiltration membranes, for example, amphiphilic block copolymers with hydrophobic and hydrophilic domains are widely utilised. For microelectronic nanopatterning applications, block copolymers of polystyrene and poly(methyl methacrylate) (PS-PMMA) provide excellent control over the microphase separation [40, 41].

3.4 Biopolymer-Based Nano Polymers

Biopolymer-based nano polymers create eco-friendly and biodegradable nanomaterials by utilising polymers found in nature, including cellulose, chitosan, alginate, and silk fibroin. Medical implants, tissue scaffolding, and biodegradable packaging are just a few examples of the many uses for materials that undergo nanoscale engineering to enhance mechanical strength, thermal stability, and functional compatibility. One example is hydrogels made of nanocellulose, which are great for tissue engineering and wound healing because of their high water retention capacity and mechanical flexibility. Because of their bioactivity and capacity to regulate cellular interactions, chitosan nanoparticles are extensively investigated for use in antibacterial applications and medication delivery [42].

3.5 Conductive and Smart Nano Polymers

Nanopolymers with conductive and stimuli-responsive properties are becoming increasingly popular in new areas such wearable sensors, flexible electronics, and soft robotics. In response to environmental cues like light, temperature, or pH, these polymers display optical responsiveness, shape memory, and controllable electrical conductivity. Reversible redox behaviour, processability, and stability are three key characteristics of materials like poly(3,4-ethylenedioxythiophene) (PEDOT), polypyrrole (PPy), and polyaniline (PANI) that make them especially useful as sensors. The incorporation of nanofillers such as carbon nanotubes and silver nanoparticles into smart polymers improves their responsiveness and conductivity, leading to the development of next-generation optoelectronic devices [43]. All of these different kinds of nanopolymers are essential for making 3D and 4D printing materials more versatile and useful.

3.6 Structural Characteristics Influencing Printability

The printability of nano polymers in additive manufacturing is highly affected by their molecular and nanoscale structural properties. The overall resolution, and mechanical integrity of the printed object are determined by these parameters, which impact the material's flow, adhesion, and solidification during layer-by-layer deposition. A crucial consideration is the polymer's molecular weight. The mechanical strength and toughness of polymers are usually better when the molecular weight is higher. But, in processes like fused deposition modelling (FDM) and direct ink writing (DIW), a material with a high molecular weight may not extrude smoothly due to severe chain entanglement. A striking a balance between processability and mechanical performance requires an ideal molecular weight distribution.

Important factors also include the interactions of the nanofillers incorporated in the polymer matrix. Nanoparticles improve mechanical performance and structural stability when they are evenly dispersed among polymer chains and have strong interfacial bonding. These interactions, however, can change the material's rheological properties, especially its shear-thinning and viscosity. To guarantee consistent and predictable flow behaviour during printing, precise formulation procedures are required. Another important characteristic in photo-curable systems like stereolithography (SLA) and digital light processing (DLP) is the degree of crosslinking within the polymer network. To get the ideal combination of stiffness and flexibility in the printed material, it is necessary to use an optimal crosslinking density. If the crosslinking is not done properly, the printed structure will be fragile and easy to break, and if it is done too much, it will lose its shape and be inaccurate in its dimensions. In applications involving multi-material or multi-layer assemblies, surface attributes like interfacial adhesion and surface energy have an even greater impact on print quality. Cohesive layering and robust mechanical interlocks are impossible without adequate wettability and interlayer bonding. For better adhesion and compatibility between printed layers, recipes sometimes include surface-modified nanofillers. To achieve accurate material deposition and layer resolution, it is necessary to modify thermal and rheological characteristics such viscosity, melting point, and glass transition temperature (Tg). Consistent performance across various additive manufacturing techniques is made possible through the fine-tuning of these thermal parameters made possible by nano polymer compositions. When printing at high speeds or with high resolution, this degree of control becomes crucial. How successfully different 3D and 4D printing systems process nano polymers is determined by their structural and physicochemical features taken together. Creating high-performance materials that can handle the intricate needs of contemporary additive manufacturing requires an understanding of these qualities and an optimisation of them.

3.7 *Mechanical, Thermal, and Rheological Properties of Nano Polymers*

Materials used in 3D and 4D printing have seen significant improvements in mechanical, thermal, and rheological properties by the inclusion of nanoscale fillers into polymer matrices. To satisfy the demanding performance standards of functional prototypes and final components made using additive manufacturing, these property improvements are crucial.

Nano polymers outperform traditional polymer systems in terms of mechanical properties such as tensile strength, Young's modulus, and fracture toughness. By enhancing the distribution of loads within the polymer matrix, nanomaterials like carbon nanotubes (CNTs), graphene, or nanoclays make materials more resistant to deformation and the development of cracks when subjected to stress. Particularly for structural uses and components intended to withstand cyclic loads, these mechanical improvements are crucial. Nanoscale reinforcements also improve stimuli-responsive polymers' elastic recovery and shape memory, which are important for 4D printing because the materials used in the process need to be able to undergo controlled, reversible transformations [15].

Nanomaterial integration also improves thermal behaviour, another crucial performance parameter. By increasing heat conductivity and delaying thermal deterioration, nanofillers including silica, alumina, and boron nitride broaden the temperature range in which printed components can be used. Stabilising the printing process under high thermal loads and producing heat-resistant components for demanding settings like aerospace and automotive are both made possible by improved crystallinity and heat dissipation properties. Nano polymers incorporated with nanostructures can respond to thermal stimuli and display programmable behaviours such as self-healing and shape recovery. These properties make them valuable components of intelligent materials systems [44, 45].

For material extrusion or photopolymerization to go smoothly and produce accurate prints, rheological characteristics are just as important. Nanomaterial dispersion in polymer inks changes thixotropy, shear-thinning behaviour, and viscosity, which in turn changes the dynamics of flow and deposition. Printing artefacts like nozzle clogging and flow instabilities can be minimised with a balanced rheological profile, which also allows for defect-free layer stacking and enhanced interlayer adhesion. Rheology optimisation controls the printed structure's resolution, dimensional correctness, and structural integrity, especially in extrusion-based methods as DIW and fused deposition modelling (FDM) [4]. When combined, nano polymers' improved mechanical, thermal, and rheological properties make them a promising new material system for HPAM. Because of their modifiable characteristics, 3D and 4D printed items have more possible uses than ever before, and new possibilities for design, usefulness, and sustainability are opening up in many different industries.

3.8 Functionalization and Enhancement Using Nanofillers

Nano polymers are incredibly adaptable for advanced printing applications due to the addition of designed nanofillers, which enhance their functional qualities. Nanomaterial integration allows for application-specific tuning of mechanical strength, electrical conductivity, thermal stability, and biocompatibility. Carbon nanotubes (CNTs) have a stellar reputation for being very strong, highly conductive electrically, and having an unusually high aspect ratio. Their incorporation into nano polymer matrices greatly improves electrical performance and mechanical reinforcement, rendering them perfect for printed electronics and flexible conductive materials [46]. In a similar vein, graphene and graphene oxide all help keep structures lightweight while improving electrical responsiveness, mechanical stability, and thermal conductivity (Fig. 2). Because of these characteristics, they are ideal for applications requiring mechanical strength and functional flexibility, such as biomedical scaffolds, energy storage devices, and flexible electronics [5, 35]. Another potential nanofiller with improved biodegradability, pliability, and water resistance is nanocellulose, which comes from sustainable sources like plant fibres. Due to its biocompatibility and natural origin, it has potential uses in tissue engineering and medical scaffolding; furthermore, it is highly valuable in the creation of sustainable 3D printing materials [47, 48].

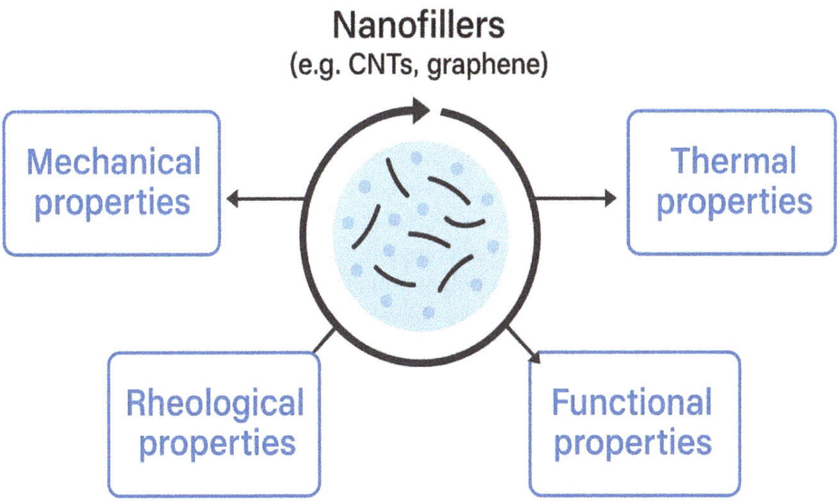

Fig. 2 Schematic representation of the structure–property relationship in nano polymers. Nanofillers such as carbon nanotubes (CNTs) and graphene influence the overall performance of polymer matrices by enhancing mechanical strength, thermal stability, rheological behavior, and functional responsiveness—critical attributes for high-performance 3D and 4D printing applications

In addition, nanoparticles of metal oxides (NiO, ZnO, TiO_2, and SiO_2) provide functional improvements such resistance to ultraviolet light, antibacterial characteristics, and photocatalytic activity. These qualities pave the way for the creation of biomedical implants, self-cleaning surfaces, and protective coatings that can withstand the test of time and the elements. By integrating various reinforcing agents into a single polymer matrix, hybrid nanocomposites provide an advantage over individual nanofillers. Improved electrical conductivity, mechanical strength, and chemical reactivity can all be achieved simultaneously, for instance, in polymer-ceramic composites or CNT-graphene hybrids. For high-tech additive manufacturing, these customised nanocomposites provide new opportunities, especially in the aerospace, medicinal, and energy sectors, where the need for multifunctional performance is paramount.

4 Types of Nano Polymers for 3D and 4D Printing

3D and 4D printing using nano polymers has increased material reactivity, broadened additive manufacturing's applicability, and improved material characteristics. Nanocomposite polymers are an important subset of nano polymers because they combine the processing ease of regular polymers with the enhanced mechanical, thermal, and electrical properties of nanofillers.

4.1 Nanocomposite Polymers

Incorporating nanoparticles or nanostructured reinforcements into a polymer matrix, nanocomposite polymers vastly enhance performance parameters like electrical conductivity, thermal stability, response to external stimuli, and strength. When using sophisticated additive manufacturing procedures that need functionality, precision, and long-term durability, these materials become crucial. The design of polymer-nanoparticle hybrid systems involves combining the best features of both components by adding nanoscale fillers to polymer matrices. Some of the benefits of these hybrid materials include higher stiffness, lower weight, better electrical and thermal conductivities, and better resistance to wear. In determining the end characteristics of the printed parts, the choice of nanoparticles and polymer matrices is crucial.

For instance, graphene and carbon nanotubes (CNTs) are two examples of carbon-based nanofillers that can greatly improve the electrical and mechanical characteristics of polymer matrices. This makes them attractive for use in fields like biomedical engineering, electronics, and aerospace. For biosensors and conductive ink applications, metallic nanoparticles like silver (Ag), gold (Au), and copper (Cu) offer antibacterial characteristics and increased electrical conductivity [49, 50]. Additionally, nanoparticles derived from ceramics, such as silica (SiO_2), alumina (Al_2O_3), and titanium dioxide (TiO_2), are perfect for use as protective coatings or structural components because they increase resistance to wear, thermal stability, and ultraviolet light [4, 8, 51].

The development of polymer nanocomposites for additive manufacturing has led to the formulation of several promising material systems with specific enhancements: A biodegradable polymer commonly used in 3D printing, polylactic acid (PLA) is sometimes reinforced with carbon nanotubes (CNTs) to improve electrical conductivity, mechanical strength, and thermal stability. PLA is well-suited for use in smart packaging, wearable electronics, and biomedical implants due to the enhanced tensile modulus and thermal conductivity that result from CNT integration [13, 52]. Thermoplastic acrylonitrile butadiene styrene (ABS) enhanced with graphene improves ABS's tensile strength, impact resistance, and electrical conductivity; it is a popular thermoplastic in fused deposition modelling (FDM). The lightweight and long-lasting properties of graphene-reinforced ABS make it a popular choice for component manufacturing in the automotive, aerospace, and consumer electronics industries [53]. An environmentally preferable substitute for additive manufacturing is nanocellulose, which is produced from renewable resources like wood pulp and food scraps. Nanocellulose improves the biodegradability, mechanical strength, and barrier qualities of polymer resins while keeping the structure lightweight. These materials have great potential as biomedical scaffolding, packaging, and environmentally friendly consumer products [54, 55]. By using nanocomposite polymers into 3D and 4D printing, the performance of additive manufacturing is improved and its potential uses are expanded. Utilising the distinct characteristics of nanoscale fillers, experts in the field are consistently creating ground-breaking materials that challenge the limits of contemporary production methods.

4.2 *Stimuli-Responsive Nano Polymers for 4D Printing*

4D printing has changed the game for additive manufacturing since it allows for the creation of structures that can change themselves in response to outside forces. In contrast to traditional 3D-printed materials, 4D-printed items can mechanically react to changes in environmental factors like light, humidity, temperature, or pH by changing their shape or function over time (Fig. 3). The main drivers of this behaviour are nano polymers that are sensitive to external stimuli and can change their state in either a reversible or irreversible way. The addition of nanoparticles to these polymers increases their mechanical strength, responsiveness, and functional diversity. As a result, they are ideal for use in biomedical devices, aerospace, smart textiles, and soft robotics.

Direct ink writing (DIW) hydrogels based on multi-responsive nanocellulose are an outstanding example. Composites made of cellulose nanofibers (CNF), poly(N-isopropylacrylamide)/poly(acrylic acid) and dual networks of these reinforcing agents show modulation of temperature, pH, and ionic strength. An excellent example of a programmable material system, they can bend in both directions and carry out logical operations in response to consecutive stimuli [56].

4.2.1 Shape-Memory Polymers (SMPs) in 4D Printing

Due to their extraordinary capacity to undergo reversible shape transformations, shape-memory polymers (SMPs) play a crucial role as stimuli-responsive materials that are extensively used in 4D printing. Excessive heat, light, or moisture can cause certain materials to momentarily deform before causing them to go back to their normal shape. Contacts between the molecular architecture of the polymer and its

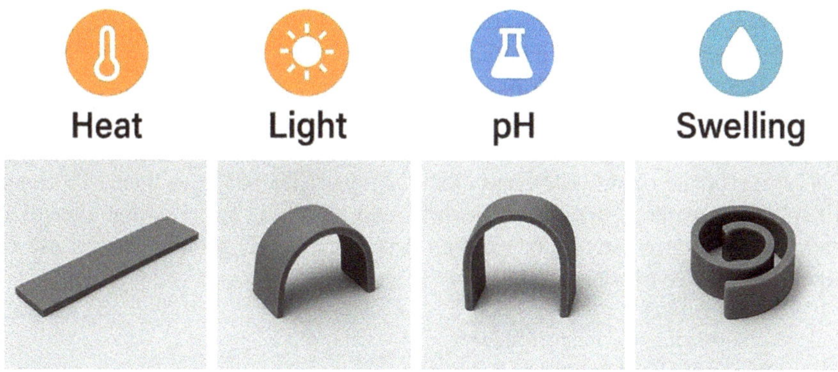

Fig. 3 Examples of 4D-printed shape-responsive structures exhibiting transformations triggered by external stimuli such as heat, light, pH, and swelling

network of reversible contacts determine the material's responsiveness and mechanical behaviour. This interaction is what gives rise to the shape-memory effect (SME) in SMPs [57]. The shape-memory technique has two primary steps when applied in real-world scenarios. At the outset, during programming, the polymer is subjected to controlled environmental conditions—typically heat or mechanical stress—in order to distort it into a temporary configuration. This transient shape is retained by the substance until reactivation occurs, even after the external stimulus has been eliminated. The recovery phase involves restoring the programmed form by relaxing the polymer chains to their original structure in response to a specific trigger, like thermal energy.

Carbon nanotubes (CNTs), graphene, and metallic nanoparticles are common nanofillers that researchers use to improve the 4D printing performance of SMPs. More accurate and dependable shape recovery is made possible by these nanostructures, which also greatly enhance the actuation speed, thermal conductivity, and mechanical strength of SMPs. Additionally, the incorporation of nanoparticles can expand the spectrum of stimuli capable of inducing the shape-memory response, hence enhancing the material's adaptability and versatility for intricate uses.

It is possible to classify SMPs into several types of stimuli-responsive nanopolymers based on the kind of outside input that triggers the shape change. This categorisation makes it easier to create smart materials that are specialised for certain functional requirements in fields including aerospace engineering, biomedical devices, and soft robotics [4, 5, 58].

4.2.2 Thermo-Responsive Nano Polymers

Nano polymers with thermoresponsive characteristics can alter their mechanical characteristics, solubility, or form in response to changes in temperature. Usually, a structural change in the polymer, like the glass transition temperature (Tg) or melting point (Tm), is what triggers the reaction. When heated beyond a specific point, polymers with a lower critical solution temperature (LCST) like poly(N-isopropylacrylamide) (PNIPAAm) change from a hydrophilic to a hydrophobic state [4, 5, 58]. Drug delivery, biological scaffolds, and soft actuators are just a few of the many potential uses for polymers that have had nanomaterials like carbon nanotubes (CNTs), graphene oxide (GO), and silica nanoparticles added to them. To create self-folding structures, deployable devices, and morphing surfaces that alter their shape in reaction to internal or external heat sources, 4D printing makes use of thermo-responsive polymers.

4.2.3 Photo-Responsive Nano Polymers

UV, visible, and near-infrared (NIR) light are only a few examples of the wavelengths that cause photo-responsive nano polymers to alter their structure or conformation. Incorporating photochromic monomers like azobenzene or spiropyran into these polymers allows for light-induced isomerisation, which in turn causes changes in shape, colour, or mechanical actuation [59, 60]. To make photo-responsive polymers even more sensitive to light and allow for fine-grained control over their reactions, nanomaterials like graphene derivatives, quantum dots, and gold nanoparticles are frequently added [61, 62]. Photonic sensors, self-healing coatings, and optomechanical actuators all make extensive use of these materials. Remotely controlled biomedical devices, soft robotics, and reconfigurable electronics are just a few of the many potential uses for 4D printed photo-responsive polymers, which allow for the creation of shape-shifting structures that can be activated by regulated light exposure.

4.2.4 pH-Responsive Nano Polymers

Nano polymers that are sensitive to changes in pH can alter their solubility or structure. Biomedical settings (such as acidic tumour microenvironments or intestinal pH gradients) are ideal for the natural occurrence of pH swings, making these materials ideal for such situations. Chitosan and polyacrylic acid (PAA) are two examples of polymers that can change their form in relation to variations in pH through protonation or deprotonation processes [63, 64].

These polymers are made more responsive and stable with the addition of nanofillers such as pH-sensitive metal oxides, layered double hydroxides (LDHs), and mesoporous silica nanoparticles. Smart drug delivery systems, controlled-release capsules, and bioresponsive scaffolds that alter their shape in response to particular biological conditions can all be made using 4D printing and pH-responsive nano polymers.

4.3 Biodegradable and Sustainable Nano Polymers

Additive manufacturing has put a lot of effort into developing biodegradable and sustainable nano polymers due to the growing concern for the environment and the need to reduce plastic waste. Because of how slowly they break down, traditional polymers made of petroleum add to environmental contamination over the long run. Biodegradable nano polymers, on the other hand, are perfect for green uses since they naturally degrade under certain environmental circumstances. Polyhydroxyalkanoates (PHAs), cellulose, chitosan, starch, and polylactic acid (PLA) are among the renewable sources from which these materials are mainly sourced. They are ideal for environmentally friendly 3D and 4D printing because the incorporation of nanotechnology increases their processability, thermal stability, and mechanical strength [20].

4.3.1 Nanocellulose-Based Materials

Among eco-friendly nanomaterials, nanocellulose has recently risen to the top of the additive manufacturing food chain for polymers. The most useful biopolymer in plants, algae, and bacteria is cellulose, and nanocellulose is a byproduct of this. Nanocellulose is prized for its mechanical strength, low density, huge specific surface area, and easily adjustable surface chemistry. Its biodegradability and environmental friendliness make it a great material for 3D and 4D printed nano polymer composites [54, 55]. Commonly used nanocellulose comes in three primary types, each with its own set of structural and functional advantages. The mechanical strength of polymer matrices can be greatly enhanced by the incorporation of cellulose nanocrystals (CNCs), which are hard, rod-shaped particles with a high degree of crystallinity and great reinforcing potential. Nanofibrils made of cellulose, on the other hand, enhance resilience, toughness, and flexibility by forming lengthy networks of intertangled fibrils. Microbes like *Komagataeibacter xylinus* produce bacterial nanocellulose (BNC), which is highly regarded in the biomedical field due to its excellent water-holding capacity, biocompatibility, and high purity [65].

Biodegradable polymers such as chitosan, polyhydroxyalkanoates (PHA), or polylactic acid (PLA) can benefit from the addition of nanocellulose, which increases printability, processability, and thermal characteristics without sacrificing compostability. Optimal properties for high-performance additive manufacturing include improved functional responsiveness, dimensional precision, and layer adhesion in these composites. A variety of uses have prompted research into materials derived from nanocellulose. In eco-friendly packaging, PLA/nanocellulose sheets show better barrier characteristics and increased tensile strength without compromising biodegradability. The extracellular matrix-like structure and moisture retention of BNC-based scaffolds are encouraging their use in tissue engineering and wound healing in the biomedical domain. In addition, nanocellulose has been used to create biodegradable electronics, such as smart textile components and wearable sensors, by combining it with conductive hydrogels and inks [48, 66].

4.3.2 Bio-Derived Nano Polymers for Eco-Friendly Additive Manufacturing

Sustainable additive manufacturing is being advanced by a wide range of bio-derived nano polymers, not limited to nanocellulose. One of the most well-known is polylactic acid (PLA), which is made from sustainable sources including sugarcane and maize starch. 3D printing has used PLA, a biodegradable polymer, to a large extent. Enhancing its mechanical and thermal performance with nanoscale fillers like graphene oxide, metal oxide nanoparticles, or nanocellulose doesn't compromise its environmental credentials. Biomedical implants, packaging, and one-time use items are among the many uses for the PLA nanocomposites that are created [67]. Polyhydroxyalkanoate (PHA) is a microbial polyester that breaks down in soil and compost on its own, making it another bio-based polymer that is attracting interest. Nanostructured PHAs are advantageous for environmentally friendly medical devices and food packaging solutions due to their enhanced flexibility, strength, and processing ease [68].

The polysaccharide chitosan, sourced from crab shells, possesses natural antibacterial action and is biodegradable. Thermoplastic chitosan-based nanocomposites can be 3D-printed into bioactive coatings, wound dressings, and tissue engineering scaffolds when reinforced with nanoparticles like silver, graphene oxide, or carboxymethyl cellulose (CNCs). These additives improve mechanical and biological performance. Another common biodegradable polymer is starch, which may be transformed into thermoplastics that facilitate additive manufacturing. Researchers have enhanced starch's mechanical strength and barrier performance by adding nanomaterials such nano-silica, CNCs, or montmorillonite clays. This makes it an attractive material for biodegradable containers and one-time-use cutlery.

5 Challenges and Future Prospects

Nano polymers for 3D and 4D printing have come a long way, but there are still a number of obstacles that make them hard to use and accept. Achieving homogeneous dispersion of nanofillers inside polymer matrices is a major challenge. Nanoparticle agglomeration can cause printability issues, poor interfacial bonding, and uneven characteristics. A significant technical challenge is finding the optimal nanofiller dispersion that does not compromise the rheological characteristics. Getting the printability and functionality right is another big hurdle. Although the incorporation of nanoparticles enhances mechanical and responsive characteristics, it has the potential to change the ink's flow behaviour and viscosity, which in turn impacts print stability, layer adhesion, and resolution. It is critical to develop formulations that can be adjusted rheologically while keeping print quality high and improving functionality.

Another consideration is the material's compatibility with currently used printing processes. A lot of nano polymers can't be used in regular additive manufacturing processes since they need specific tools or certain post-processing steps. Nano polymer synthesis must also be cost-effective and scalable for widespread use in industry. Complex and time-consuming processes are typically required for the synthesis of high-performance nanocomposites, making their large-scale commercial viability an uncertain proposition. Concerning safety and regulation, our understanding of the effects of nanoparticles on the environment and human health over the long term is limited. For nanotechnology to be used responsibly in production, there must be thorough toxicological assessments and environmentally friendly disposal methods.

Nano polymers show a lot of promise for additive manufacturing in the future. Biodegradable nanocomposites made of cellulose, chitosan, or starch are one example of the innovative green nanomaterials that provide a greener alternative to conventional plastics. The creation of tailor-made nano polymer formulations that are both printable and highly effective is anticipated to be expedited by the use of machine learning and AI-driven materials design. Advances in biofabrication, soft robotics, and flexible electronics, along with 3D/4D printing, will pave the way for novel applications in areas including adaptive devices, multifunctional systems, and personalised medicine. Nano polymers have the potential to revolutionise additive manufacturing through ongoing multidisciplinary research and innovation. This will pave the way for printed materials that are smarter, stronger, and more sustainable, which will have far-reaching implications in engineering and biomedicine.

6 Conclusions

The combination of nanotechnology with 3D and 4D printing has brought about a new age in modern manufacturing. Materials are now engineered systems' active, responsive, and multifunctional components, rather than passive elements. Engineered by incorporating nanoscale fillers into polymer matrices, nano polymers have greatly increased the performance of materials for additive manufacturing. For printing needs that are both dynamic and application-specific, they are ideal because of their increased mechanical strength, thermal stability, electrical conductivity, and, most significantly, response to environmental stimuli.

In this chapter, we have covered the basics of nano polymer, printability, and types. We have also looked at their widespread uses in fields such as biomedicine, aerospace, wearable technology, and sustainable product development. Particular focus was placed on stimuli-responsive systems employed in 4D printing, which permit printed structures to undergo transformations upon application of external stimuli. In the coming years, additive manufacturing will be greatly influenced by the advancement of nano polymers that are intelligent, biocompatible, and environmentally safe. Thanks to nano polymers, 3D and 4D printing have the potential to revolutionise material functionality and make a significant impact on healthcare, sustainability, and intelligent system design.

References

1. Vaidya, S., Ambad, P., Bhosle, S.: Industry 4.0—A Glimpse, Procedia Manuf. 20, 233–238 (2018). https://doi.org/10.1016/j.promfg.2018.02.034
2. Arif, Z.U., Khalid, M.Y., Zolfagharian, A., Bodaghi, M.: 4D bioprinting of smart polymers for biomedical applications: recent progress, challenges, and future perspectives. React. Funct. Polym. **179**, 105374 (2022). https://doi.org/10.1016/j.reactfunctpolym.2022.105374
3. Chen, J., Virrueta, C., Zhang, S., Mao, C., Wang, J.: 4D printing: the spotlight for 3D printed smart materials. Mater. Today **77**, 66–91 (2024). https://doi.org/10.1016/j.mattod.2024.06.004

4. Falahati, M., Ahmadvand, P., Safaee, S., Chang, Y.C., Lyu, Z., Chen, R., Li, L., Lin, Y.: Smart polymers and nanocomposites for 3D and 4D printing. Mater. Today **40**, 215–245 (2020). https://doi.org/10.1016/j.mattod.2020.06.001
5. Mehta, A., Vasudev, H., Prakash, C., Pramanik, A., Basak, A., Shankar, S.: An overview of 3D printing for modeling polymers, bio-nanocomposite, and nanocomposites materials, LTD (2025). https://doi.org/10.1016/B978-0-12-824329-9.00019-X
6. Muhammad, A., Valero, C., De Angelis, P., Koutroumanis, N., Semitekolos, D., Jiménez, B., Rivera, R., Sáenz Ezquerro, C., Srivastava, R., Pappas, P.N., Galiotis, C., Charitidis, C.A., Chiavazzo, E., Asinari, P., Laspalas, M., Chiminelli, A., Fasano, M.: From nanoscale to printed products: Multiscale modeling and experimental characterization of graphene-enhanced polylactic acid composites for 3D printing. Compos. Part B Eng. **298**, 112354 (2025). https://doi.org/10.1016/j.compositesb.2025.112354
7. Farid, M.I., Wu, W., Liu, X., Wang, P.P.: Additive manufacturing landscape and materials perspective in 4D printing. Int. J. Adv. Manuf. Technol. **115**, 2973–2988 (2021). https://doi.org/10.1007/s00170-021-07233-w
8. He, T., Yip, W.S., Yan, E.H., Tang, J., Rehan, M., Teng, L., Wong, C.H., Sun, L., Zhang, B., Guo, F., Zhang, S., To, S.: 3D printing for ultra-precision machining: current status, opportunities, and future perspectives. Front. Mech. Eng. 19 (2024). https://doi.org/10.1007/s11465-024-0792-4
9. Ali, S., Deiab, I., Pervaiz, S.: State-of-the-art review on fused deposition modeling (FDM) for 3D printing of polymer blends and composites: innovations, challenges, and applications. Int. J. Adv. Manuf. Technol. **135**, 5085–5113 (2024). https://doi.org/10.1007/s00170-024-14820-0
10. Fang, R., Zhang, X., Song, B., Zhang, Z., Zhang, L., Song, J., Yao, Y., Gao, M., Zhou, K., Wang, P., Lu, J., Shi, Y.: 3D and 4D printing of electromagnetic metamaterials. Engineering (2024). https://doi.org/10.1016/j.eng.2024.10.017
11. Cheng, P., Li, S., Peng, Y., Le Duigou, A., Wang, K., Ahzi, S.: 3D/4D printed functional continuous fiber-reinforced polymer composites: progress and perspectives. Chinese. J. Mech. Eng. Addit. Manuf. Front. **2**, 100090 (2023). https://doi.org/10.1016/j.cjmeam.2023.100090
12. Lan, W., Huang, X., Huang, D., Wei, X., Chen, W.: Progress in 3D printing for bone tissue engineering: a review. J. Mater. Sci. **57**, 12685–12709 (2022). https://doi.org/10.1007/s10853-022-07361-y
13. Agarwala, S., Goh, G.L., Goh, G.D., Dikshit, V., Yeong, W.Y.: 3D and 4D Printing of Polymer/CNTs-Based Conductive Composites. Elsevier Inc. (2019). https://doi.org/10.1016/B978-0-12-816805-9.00010-7
14. Liu, H., Wang, F., Wu, W., Dong, X., Sang, L.: 4D printing of mechanically robust PLA/TPU/Fe_3O_4 magneto-responsive shape memory polymers for smart structures. Compos. Part B Eng. **248**, 110382 (2023). https://doi.org/10.1016/j.compositesb.2022.110382
15. Xu, W., Jambhulkar, S., Zhu, Y., Ravichandran, D., Kakarla, M., Vernon, B., Lott, D.G., Cornella, J.L., Shefi, O., Miquelard-Garnier, G., Yang, Y., Song, K.: 3D printing for polymer/particle-based processing: a review. Compos. Part B Eng. **223**, 109102 (2021). https://doi.org/10.1016/j.compositesb.2021.109102
16. Burela, R.G., Kamineni, J.N., Harursampath, D.: Multifunctional Polymer Composites for 3D and 4D Printing. Elsevier Inc. (2019). https://doi.org/10.1016/B978-0-12-816805-9.00008-9
17. Raies, M., Peng, Z., Ahmad, F., Sohail, S., Shah, A., Wasy, A., Ali, A., Qiao, G., Shehzad, K.: Comprehensive review of 3D/4D printing of soft materials, methods and applications. Appl. Mater. Today **43**, 102667 (2025). https://doi.org/10.1016/j.apmt.2025.102667
18. Vashishtha, G., Chauhan, S., Yadav, N., Chhabra, D., Gupta, M.K., Kumar, R., Li, Z.: Shaping the future: latest developments in 3D printing stimuli-responsive soft polymers. Int. J. Adv. Manuf. Technol. **136**, 4215–4237 (2025). https://doi.org/10.1007/s00170-025-15095-9
19. Dananjaya, S.A.V., Chevali, V.S., Dear, J.P., Potluri, P., Abeykoon, C.: 3D printing of biodegradable polymers and their composites—current state-of-the-art, properties, applications, and machine learning for potential future applications. Prog. Mater. Sci. **146**, 101336 (2024). https://doi.org/10.1016/j.pmatsci.2024.101336

20. Siddiqui, M.A.S., Rabbi, M.S., Ahmed, R.U., Billah, M.M.: Biodegradable natural polymers and fibers for 3D printing: a holistic perspective on processing, characterization, and advanced applications. Clean. Mater. **14**, 100275 (2024). https://doi.org/10.1016/j.clema.2024.100275
21. Monfared, V., Ramakrishna, S., Nasajpour-Esfahani, N., Toghraie, D., Hekmatifar, M., Rahmati, S.: Science and technology of additive manufacturing progress: processes, materials, and applications. Met. Mater. Int. **29**, 3442–3470 (2023). https://doi.org/10.1007/s12540-023-01467-x
22. Brinckmann, S.A., Patra, N., Yao, J., Ware, T.H., Frick, C.P., Fertig, R.S.: Stereolithography of SiOC polymer-derived ceramics filled with SiC micronwhiskers. Adv. Eng. Mater. **20**, 1–10 (2018). https://doi.org/10.1002/adem.201800593
23. Khan, T., Ali, M., Riaz, Z., Butt, H., Abu Al-Rub, R.K., Dong, Y., Umer, R.: Recent developments in improving the fracture toughness of 3D-printed fiber-reinforced polymer composites. Compos. Part B Eng. **283**, 111622 (2024). https://doi.org/10.1016/j.compositesb.2024.111622
24. Generalova, A.N., Vikhrov, A.A., Prostyakova, A.I., Apresyan, S.V., Stepanov, A.G., Myasoedov, M.S., Oleinikov, V.A.: Polymers in 3D printing of external maxillofacial prostheses and in their retention systems. Int. J. Pharm. **657**, 124181 (2024). https://doi.org/10.1016/j.ijpharm.2024.124181
25. Yoon, J., Han, H., Jang, J.: Nanomaterials-incorporated hydrogels for 3D bioprinting technology. Nano Converg. 10 (2023). https://doi.org/10.1186/s40580-023-00402-5
26. Han, X., Saiding, Q., Cai, X., Xiao, Y., Wang, P., Cai, Z., Gong, X., Gong, W., Zhang, X., Cui, W.: Intelligent Vascularized 3D/4D/5D/6D-Printed Tissue Scaffolds. Springer Nature Singapore (2023). https://doi.org/10.1007/s40820-023-01187-2
27. Strutynski, C., Evrard, M., Désévédavy, F., Gadret, G., Jules, J.C., Brachais, C.H., Kibler, B., Smektala, F.: 4D Optical fibers based on shape-memory polymers. Nat. Commun. **14**, 1–11 (2023). https://doi.org/10.1038/s41467-023-42355-7
28. Naghib, S.M., Hosseini, S.N., Beigi, A.: 3D 4D printing of chitosan-based scaffolds for wound dressing applications. Carbohydr. Polym. Technol. Appl. 8 (2024). https://doi.org/10.1016/j.carpta.2024.100594
29. Soleymani, S., Naghib, S.M.: 3D and 4D printing hydroxyapatite-based scaffolds for bone tissue engineering and regeneration. Heliyon. **9**, e19363 (2023). https://doi.org/10.1016/j.heliyon.2023.e19363
30. Shinde, S., Mane, R., Vardikar, A., Dhumal, A., Rajput, A.: 4D printing: from emergence to innovation over 3D printing. Eur. Polym. J. **197**, 112356 (2023). https://doi.org/10.1016/j.eurpolymj.2023.112356
31. Miao, S., Castro, N., Nowicki, M., Xia, L., Cui, H., Zhou, X., Zhu, W., Jun Lee, S., Sarkar, K., Vozzi, G., Tabata, Y., Fisher, J., Zhang, L.G.: 4D printing of polymeric materials for tissue and organ regeneration. Mater. Today. **20**, 577–591 (2017). https://doi.org/10.1016/j.mattod.2017.06.005
32. Grira, S., Khalifeh, H.A., Alkhedher, M., Ramadan, M.: 3D printing algae-based materials: pathway towards 4D bioprinting. Bioprinting. **33**, e00291 (2023). https://doi.org/10.1016/j.bprint.2023.e00291
33. Subash, A., Kandasubramanian, B.: 4D printing of shape memory polymers. Eur. Polym. J. **134**, 109771 (2020). https://doi.org/10.1016/j.eurpolymj.2020.109771
34. Khalid, M.Y., Arif, Z.U., Noroozi, R., Zolfagharian, A., Bodaghi, M.: 4D printing of shape memory polymer composites: a review on fabrication techniques, applications, and future perspectives. J. Manuf. Process. **81**, 759–797 (2022). https://doi.org/10.1016/j.jmapro.2022.07.035
35. Sajjad, R., Chauhdary, S.T., Anwar, M.T., Zahid, A., Khosa, A.A., Imran, M., Sajjad, M.H.: A review of 4D printing—technologies, shape shifting, smart polymer based materials, and biomedical applications. Adv. Ind. Eng. Polym. Res. **7**, 20–36 (2024). https://doi.org/10.1016/j.aiepr.2023.08.002
36. Han, X., Yang, Y., Xu, Y., Hong, X., Tang, Z., Zhang, H., Liu, N., Li, M., Wang, Z., Zheng, A.: 3D micro-nano printing technology as a transformative tool apply for microneedle drug delivery. J. Drug Deliv. Sci. Technol. **96**, 105709 (2024). https://doi.org/10.1016/j.jddst.2024.105709

37. Ouyang, W., Xu, X., Lu, W., Zhao, N., Han, F., Chen, S.C.: Ultrafast 3D nanofabrication via digital holography. Nat. Commun. **14**, 1–9 (2023). https://doi.org/10.1038/s41467-023-37163-y
38. Pourhajrezaei, S., Abbas, Z., Khalili, M.A., Madineh, H., Jooya, H., Babaeizad, A., Gross, J.D., Samadi, A.: Bioactive polymers: a comprehensive review on bone grafting biomaterials. Int. J. Biol. Macromol. **278**, 134615 (2024). https://doi.org/10.1016/j.ijbiomac.2024.134615
39. Subramanian, B., Das, P., Biswas, S., Roy, A., Basak, P.: Polymers for additive manufacturing and 4D-printing for tissue regenerative applications, INC (2022). https://doi.org/10.1016/B978-0-323-88524-9.00001-2
40. Dada, M., Popoola, P.: The design of hybrid polymeric materials for tissue engineering: 3D bioprinting, polymeric protheses, smart polymers, and shape memory polymeric biomaterials, LTD (2024).https://doi.org/10.1016/B978-0-443-15564-2.00014-7
41. Nyabadza, A., Kane, J., Vázquez, M., Sreenilayam, S., Brabazon, D.: Multi-material production of 4D shape memory polymer composites. Encycl. Mater. Compos. **2**, 879–894 (2021). https://doi.org/10.1016/B978-0-12-819724-0.00057-4
42. Sarkar, Z., Singh, H., Iqubal, M.K., Baboota, S., Khan, S., Parveen, R., Ali, J.: Involvement of macromolecules in 3D printing for wound healing management: A narrative review. Int. J. Biol. Macromol. 282 (2024). https://doi.org/10.1016/j.ijbiomac.2024.136991
43. Subedi, S., Liu, S., Wang, W., Shovon, S.M.A.N., Chen, X., Ware, H.O.T.: Multi-material vat photopolymerization 3D printing : a review of mechanisms and applications. Npj Adv. Manuf., 1–17 (2024). https://doi.org/10.1038/s44334-024-00005-w
44. Wang, Z., Ma, D., Liu, J., Xu, S., Qiu, F., Hu, L., Liu, Y., Ke, C., Ruan, C.: 4D printing polymeric biomaterials for adaptive tissue regeneration. Bioact. Mater. **48**, 370–399 (2025). https://doi.org/10.1016/j.bioactmat.2025.01.033
45. Alshebly, Y.S., Nafea, M., Mohamed Ali, M.S., Almurib, H.A.F.: Review on recent advances in 4D printing of shape memory polymers. Eur. Polym. J. **159**, 110708 (2021). https://doi.org/10.1016/j.eurpolymj.2021.110708
46. Solanki, A., Ranganath, M.S., Singholi, A.K.S.: Review on advancements in 3D/4D printing for enhancing efficiency, cost-effectiveness, and quality. Int. J. Interact. Des. Manuf. (2024). https://doi.org/10.1007/s12008-024-02029-0
47. Noroozi, R., Arif, Z.U., Taghvaei, H., Khalid, M.Y., Sahbafar, H., Hadi, A., Sadeghianmaryan, A., Chen, X.: 3D and 4D bioprinting technologies: a game changer for the biomedical sector? Ann. Biomed. Eng. **51**, 1683–1712 (2023). https://doi.org/10.1007/s10439-023-03243-9
48. Hwang, S., Han, Y., Gardner, D.J.: Melt compounding of spray-dried cellulose nanofibrils/polypropylene and their application in 3D printing. Cellulose **31**, 7531–7552 (2024). https://doi.org/10.1007/s10570-024-06038-w
49. Zhan, L., Zhou, Y., Liu, R., Sun, R., Li, Y., Tian, Y., Fan, B.: Advances in growth factor-containing 3D printed scaffolds in orthopaedics. Biomed. Eng. Online. 24 (2025). https://doi.org/10.1186/s12938-025-01346-z
50. Ul Hassan, R., Sharipov, M., Ryu, W.H.: Electrohydrodynamic (EHD) printing of nanomaterial composite inks and their applications. Micro Nano Syst. Lett. 12 (2024). https://doi.org/10.1186/s40486-023-00194-7
51. Harun-Ur-Rashid, M., Bin Imran, A., Susan, M.A.B.H.: Emerging 3D Printed Polymers and Composites for Water Quality Preservation. Elsevier (2024). https://doi.org/10.1016/b978-0-323-95486-0.00010-7
52. Cieślik, M., Susik, A., Banasiak, M., Bogdanowicz, R., Formela, K., Ryl, J.: Tailoring diamondised nanocarbon-loaded poly(lactic acid) composites for highly electroactive surfaces: extrusion and characterisation of filaments for improved 3D-printed surfaces. Microchim. Acta **190**, 1–17 (2023). https://doi.org/10.1007/s00604-023-05940-7
53. Kabir, A., Chowdhury, M.A., Hossain, N., Islam, M.A., Aoyon, H., Masud Rana, M.: Graphene reinforced 3D printed polymer nanocomposites for diverse applications. Results Chem. **6**, 101197 (2023). https://doi.org/10.1016/j.rechem.2023.101197
54. Sajab, M.S., Mohan, D., Ling, T.T., Kaco, H., Ishak, W.H.W., Ding, G.: Nanocellulose-based composites for 3D printed stimuli-responsive swelling hydrogel. Ind. Crops Prod. **222**, 119541 (2024). https://doi.org/10.1016/j.indcrop.2024.119541

55. Dalloul, F., Mietner, J.B., Navarro, J.R.G.: Production and 3D printing of a nanocellulose-based composite filament composed of polymer-modified cellulose nanofibrils and high-density polyethylene (HDPE) for the fabrication of 3D complex shapes. Fibers **10** (2022). https://doi.org/10.3390/fib10100091
56. Lee, Y.W., Ceylan, H., Yasa, I.C., Kilic, U., Sitti, M.: 3D-printed multi-stimuli-responsive mobile micromachines. ACS Appl. Mater. Interfaces **13**, 12759–12766 (2021). https://doi.org/10.1021/acsami.0c18221
57. Thakur, V., Singh, R., Kumar, R., Gehlot, A.: 4D printing of thermoresponsive materials: a state-of-the-art review and prospective applications. Int. J. Interact. Des. Manuf. **17**, 2075–2094 (2023). https://doi.org/10.1007/s12008-022-01018-5
58. Ali, F., Kalva, S.N., Koc, M.: Advancements in 3D printing techniques for biomedical applications: a comprehensive review of materials consideration, post processing, applications, and challenges. Springer International Publishing (2024). https://doi.org/10.1007/s43939-024-00115-4
59. Szukalski, A., Korbut, A., Ortyl, E.: Structural and light driven molecular engineering in photochromic polymers. Polymer (Guildf). **192**, 122311 (2020). https://doi.org/10.1016/j.polymer.2020.122311
60. Jiang, Z.C., Liu, Q., Xiao, Y.Y., Zhao, Y.: Liquid crystal elastomers for actuation: a perspective on structure-property-function relation. Prog. Polym. Sci. **153**, 101829 (2024). https://doi.org/10.1016/j.progpolymsci.2024.101829
61. Dananjaya, V., Marimuthu, S., (Chunhui) Yang, R., Grace, A.N., Abeykoon, C.: Synthesis, properties, applications, 3D printing and machine learning of graphene quantum dots in polymer nanocomposites. Prog. Mater. Sci. **144**, 101282 (2024). https://doi.org/10.1016/j.pmatsci.2024.101282
62. Tavan, M., Yousefian, Z., Bakhtiar, Z., Rahmandoust, M., Mirjalili, M.H.: Carbon quantum dots: multifunctional fluorescent nanomaterials for sustainable advances in biomedicine and agriculture. Ind. Crops Prod. **231**, 121207 (2025). https://doi.org/10.1016/j.indcrop.2025.121207
63. Huh, K.M., Kang, H.C., Lee, Y.J., Bae, Y.H.: PH-sensitive polymers for drug delivery. Macromol. Res. **20**, 224–233 (2012). https://doi.org/10.1007/s13233-012-0059-5
64. Chu, S., Shi, X., Tian, Y., Gao, F.: PH-Responsive polymer nanomaterials for tumor therapy, front. Oncol. **12**, 1–22 (2022). https://doi.org/10.3389/fonc.2022.855019
65. Sumini, M., Andrade, G.J.S., Tischer, C.A., Kobayashi, R.K.T., Nakazato, G.: Production of bacterial cellulose by Komagataeibacter xylinus: biochemistry, synthesis and applications. Cellulose **32**, 115–131 (2024). https://doi.org/10.1007/s10570-024-06179-y
66. Shabestani, N., Jafari-Gharabaghlou, D., Gholami, S., Zarghami, N.: An overview of the various nanofiber scaffolds techniques with a focus on the 3D nanofiber-based scaffolds application in medicine. J. Inorg. Organomet. Polym. Mater. **33**, 3355–3371 (2023). https://doi.org/10.1007/s10904-023-02560-9
67. Giner-Grau, S., Lazaro-Hdez, C., Pascual, J., Fenollar, O., Boronat, T.: Enhancing polylactic acid properties with graphene nanoplatelets and carbon black nanoparticles: a study of the electrical and mechanical characterization of 3D-printed and injection-molded samples. Polymers (Basel). **16** (2024). https://doi.org/10.3390/polym16172449
68. Yousefi, A.M., Wnek, G.E.: Poly(hydroxyalkanoates): emerging biopolymers in biomedical fields and packaging industries for a circular economy. Biomed. Mater. Devices. **3**, 19–44 (2024). https://doi.org/10.1007/s44174-024-00166-4

Nano Polymer Composites for 3D and 4D Printing Applications

Chitte Prem Kumar, N. Sivanagaraju, D. Loknath, G. Naga Mallikarjun Rao, and M. R. K. Vakkalagadda

Abstract Polymer nanocomposites (nanoparticles integrated into a polymer matrix) have been recognized as a transformative category of materials in the domains of 3D and 4D printing, thereby facilitating novel avenues for advanced manufacturing and functional applications. The integration of nanoparticles within polymer matrices enhances mechanical strength, thermal durability, electrical conductivity, and functional responsiveness, thereby enabling the fabrication of high-performance components with tailored attributes. Nanopolymers have significantly improved printability in 3D printing, surface resolution, and structural integrity, rendering them applicable in sectors such as aerospace, biomedical devices, and electronics. In the swiftly evolving domain of 4D printing, nanopolymer composites facilitate dynamic, stimulus-responsive modifications in printed entities that are activated by external stimuli including heat, light, or moisture. This versatility encourages the creation of intelligent materials that may find use in deployable infrastructure, self-repairing systems, and soft robotics. This chapter provides an overview of nanoparticles for 3D and 4D printing. The structural, biological, and aeronautical applications of nano polymers are also discussed.

Keywords Polymers · 3D printing · 4D printing · Polymer composites · Graphene · CNTs · Nano-reinforcements · Nanocomposites · Stimuli · Biomedical applications

1 Introduction

Nanopolymer composites are a class of polymers that contain nanoscale reinforcements such as nanoparticles, nanotubes, and nanofibers. These reinforcements or fillers improve the required properties such as mechanical, thermal, electrical, and stimuli responsiveness (Crosby and Lee [1]). These type of polymers exhibits better

C. Prem Kumar · N. Sivanagaraju · D. Loknath · G. Naga Mallikarjun Rao ·
M. R. K. Vakkalagadda (✉)
School of Mechanical Engineering, VIT-AP University, Amaravati, India
e-mail: ravindra.btw@gmail.com

B. Sharma et al. (eds.), *Nanostructured Polymers*, Engineering Materials,
https://doi.org/10.1007/978-981-95-3581-1_13

properties than the conventional polymers. Nanoscale polymers are generally in the range of 1–100 nm. These materials belong to a class of nanomaterials and have drawn significant interest because of their special qualities, which include strength, flexibility, resistance to chemicals, and customized functionality (Khan et al. [2]). There are key types of nanopolymer composites as shown in Fig. 1, including

- **Nano polymer composites**: These types of polymers are blended with organic or inorganic nanoparticles, which exhibit superior strength, flexibility, and stiffness compared to other polymeric materials.
- **Conductive nano polymers**: Polymers that have conductive nanomaterials like carbon nanotubes or graphene incorporated within the polymer matrix. Sensors, energy storage, and printed electronics all benefit greatly from these materials.
- **Polymer nanofibers:** These fibers are created using techniques such as electrospinning. They have superior mechanical qualities, tunable porosity, and high surface area qualities of which the fibers have become ideal components for tissue engineering, filtration, and ultra-light, tubular structures.
- **Biodegradable nano polymers:** Polymers are designed with interventions at the microscopic level, which help in sustaining lives, while enabling enhanced performance, without harming the environment (Rashid et al. [3]).

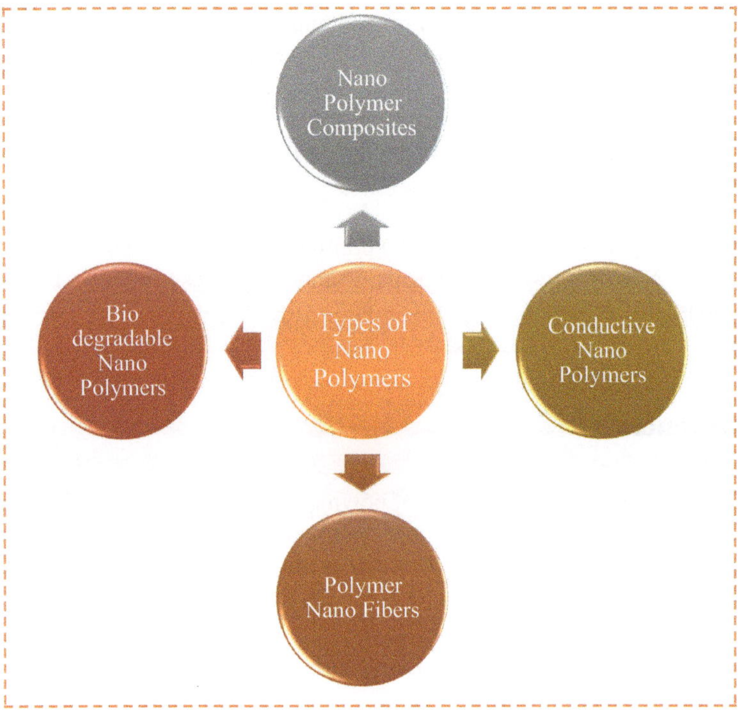

Fig. 1 Types of nano polymers

3D printing, or additive manufacturing, is a fabrication technique that is used to create a three-dimensional object (3D) by printing and joining several layers upon one another as shown in Fig. 2. 3D printing uses slicing software and creates the required model of the object (Lakkala et al. [4]). First, designing a product using computer-aided design (CAD) software defines product part particulars. Slicing software slices the model into multiple thin, horizontal layers with specific G-code settings to create a print file to provide instructions to the 3D printer. The specification directs the printer to prepare the object as the material is deposited layer by layer starting from the bottom layer up using those guidelines. It may contain plastics, metals, ceramics, resins, and even biological materials (Lee et al. [5]). The intended application decides the material along with the corresponding print material. Postprocessing is possible after printing for attaining the requisite quality and utility-from the cutting away of the supports, to smoothing, to finishes. It is possible to realize intricate geometries, which may be difficult to produce by conventional manufacturing means, if not completely impossible (Monteiro et al. [6]).

4D printing advances 3D printing by including the extra dimension of time. This might represent material-made structures with respect to smart materials like shape memory polymers and hydrogels, which could change behaviour or shape or even function with time intervals because of stimuli like heat, light, or humidity from the outside (Momeni et al. [7]). This enables the designing of dynamic structures that shape themselves according to their surrounding conditions. These benefits make the concept very interesting in fields such as medical devices, aerospace, construction, and wearable technologies. For instance, 4D-produced stents can self-expand in the body, and self-assembling components those with an improved thermal response,

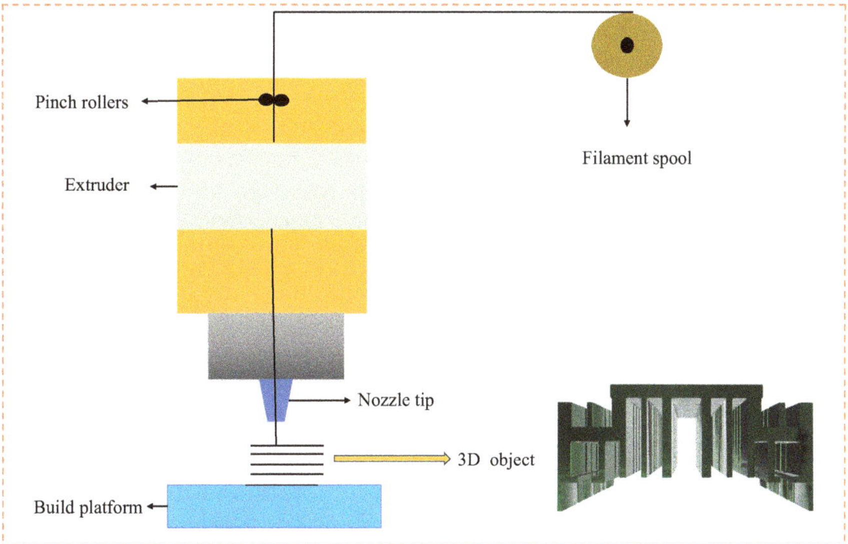

Fig. 2 3D printing using FDM

enabling adaptive transformation. The exploration of 4D printing has surged dramatically across the globe, captivating the imaginations of many (Kuang et al. [8]). A key characteristic of 4D printing is its dynamic nature; rather than being a static form, it may change and grow over time using computer instructions that have been preprogrammed. The primary focus of 4D printing is the material's ability to change shape. These 4D printed materials are exceptionally suitable for the fabrication of robots, lifters, microtubes, actuators, and other engineering components (Chen et al. [9]).

Polymers are of paramount importance in 3D and 4D printing technologies due to the very vast array of qualities and mainly functional versatility. Commonly used are thermoplastics such as ABS, PLA, and PETG in 3D printing because they are straightforward to mold and have mechanical strength, while photopolymers and elastomers are well suited in the development of parts where high levels of detail or elasticity is needed (Urquiza [10]). A 4D-printed mechanism uses newly developed materials like shape-memory polymers (SMPs), hydrogels, and liquid crystal elastomers (LCEs) that allow dynamic stimuli such as heat, light, or moisture to act on them. These polymers support the creation of efficient techniques, adaptable designs, and novelty in many industrial segments from aerospace to health with highly functional designs (Wan et al. [11]).

Nano-reinforcements have a key role to play in enhancing the properties of polymers by integrating select nanomaterials like carbon nanotubes, graphene, nanoclays, or metal nanoparticles into the polymer matrix. The important aspect is the introduction of unique characteristics, like flame retardancy, electrical conductivity, or self-healing capabilities, through modified polymers, making them ideal for high-performance applications in aviation, automobile electronics, and biomedicine fields. This composite characterizes a synergistic interaction that is established between the polymer and the nanomaterial in the accomplishment of superior performance quite impossible from conventional reinforcements (Li et al. [12]).

Different types of reinforcements, for example fibrous, particulate, and nanomaterial-based reinforcements may be incorporated in the material to enhance its mechanical, thermal, and functional properties. Examples of reinforcing fibers are glass, carbon, and aramid fibers. Such fibers have high stiffness and strength, which make them suitable fibrous reinforcement in composite materials used in the automotive and aerospace industries. Particulate reinforcing materials such as metal powders, ceramic particles, and microspheres improve some physical properties such as hardness, dimensional stability, and wear resistance of metals or polymers. Nanoreinforcement materials such as carbon nanotubes, graphene, and nanoclays, enhance specific structural qualities, including electrical conductivity, thermal stability, and barrier performance. Generally, the motive behind the selection of each type of reinforcement is governed by the application and requirements for the intended aspects to be added to the final product composite. Table 1 shows various reinforcements of nano polymers.

Table 1 Various types of reinforcements used in polymers

Reinforcement type	Examples	Key properties	Applications	References
Fibrous reinforcements	Glass fibers, carbon fibers, aramid fibers	High strength and stiffness	Aerospace, automotive, wind turbines, sports equipment	Liang et al. [13]
Particulate reinforcements	Ceramic particles, metal powders, microspheres	Improved wear resistance and hardness	Cutting tools, coatings, construction materials	Zhang et al. [14] and Ghasempour-Mouziraji et al. [15]
Nanomaterial-based reinforcements	Carbon nanotubes, Graphene, nanoclays	Enhanced thermal, electrical, and barrier properties	Electronics, aerospace, biomedical devices, packaging	Kishore et al. [16]
Hybrid reinforcements	Combination of fibers and particles	Tailored multiproperty enhancement	High-performance composites, advanced structural components	Mashayekhi et al. [17]

2 Nanopolymer Composites for 3D Printing Applications

Nano polymers are innovative materials designed for 3D printing that include nanoscale reinforcements including carbon nanotubes, graphene, nanoclays, and silica nanoparticles. These reinforcements greatly increase the attributes of base polymers, including mechanical strength, thermal stability, and electrical conductivity. PLA and ABS based nanocomposites, for example, improve the stiffness and longevity of structural and automotive parts, whilst conductive nanopolymers allow for the manufacture of electronic components such as sensors and circuits. Elastomeric nano polymers with increased elasticity and durability are well suited for soft robotics and flexible sealing (Thakur et al. [18]).

2.1 Nanopolymer Composites for Improving Structural or Functional Properties in 3D Printing

The advent of nanocomposites has revolutionized 3D printing by enabling the production of components with remarkably improved structural properties. By integrating nanomaterials such as carbon nanotubes (CNTs), graphene, nanoclays, and metal oxide nanoparticles into polymer matrices, nanocomposites exhibit superior mechanical strength, stiffness, durability, and exceptional strength-to-weight ratios compared to traditional materials (Shahbazi et al. [19]; Tundwal et al. [20]). These advancements not only enhance the performance of 3D-printed components but also expand the possibilities for high-performance applications across various industries, including aerospace, automotive, construction, and biomedical engineering. The inherent design freedom offered by 3D printing further amplifies the potential of nanocomposites, allowing the precise fabrication of intricate geometries that leverage their unique properties. This capability makes nanocomposites indispensable for applications requiring lightweight, robust, and highly customizable structures.

Numerous studies have investigated the impact of various nanomaterials on enhancing the structural and functional performance of cementitious composites. These investigations have consistently demonstrated a range of benefits, including improved durability, refined pore structures, and increased resistance to environmental degradation. Carbon Nanotubes (CNTs) significantly reduce permeability, enhancing resistance to aggressive environments (Ramezani et al. [21]). Nano Calcium Carbonate (NCC) refines the microstructure, accelerates hydration, and improves early-age strength and long-term durability (Camiletti et al. [22]). Nano-Silica (NS) enhances durability by refining the pore structure and densifying the matrix, leading to reduced permeability (Althoey et al. [23]). Nanoclay (NC) achieves higher pore refinement and enhanced thermal properties, making it suitable for extreme temperature applications (Khatoon et al. [24]). Nano Titanium (NTI) exhibits high photocatalytic activity, enabling self-disinfection and self-cleaning properties, reducing maintenance requirements (Khatoon et al. [24]).

Xu et al. [25] investigated the use of 3D printed polymeric octet lattice structures as reinforcements for cementitious composites to enhance their ductility and reduce the need for high reinforcing ratios as shown in Fig. 3. Experimental and numerical studies were conducted on mortar specimens reinforced with these structures, demonstrating significantly improved flexural ductility compared to plain mortar. They demonstrated the potential of additive manufacturing in developing innovative reinforcement solutions for cementitious materials, leading to improved performance and more sustainable construction practices.

Zhang et al. [26] investigated the development of 3D core-framework lattice cementitious composites (CS-LCCs) to address the challenge of balancing mechanical properties and thermal conductivity in building materials. CS-LCCs consist of

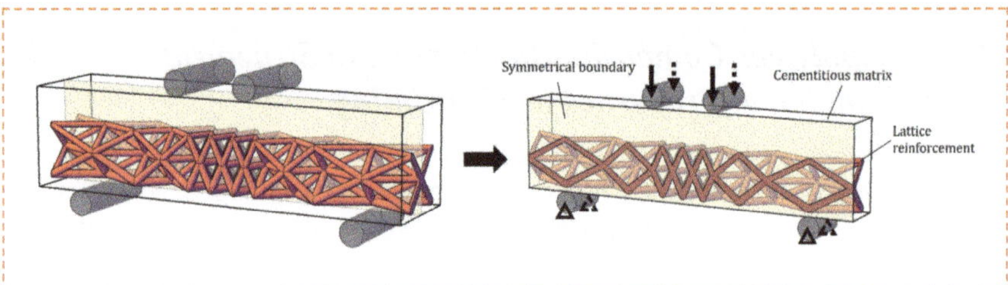

Fig. 3 Schematics of flexural loading conditions in numerical simulation for polymer lattice with cementitious matrix [25]

a 3D printed polymeric framework filled with cement mortar and the results showed that while the average compressive strength of CS-LCCs was lower than that of solid cement mortar cubes, their specific strength was comparable. Notably, CS-LCCs exhibited significantly enhanced ductility due to the presence of the polymeric framework. A hinging and stretching coupled deformation mechanism was identified through testing and finite element analysis to explain the deformation behavior of CS-LCCs. Furthermore, CS-LCCs demonstrated low thermal conductivity across various temperatures due to the ordered porous nature of the lattice structure. This study highlights the potential of lattice-inspired 3D printing in developing energy-efficient building materials with improved mechanical and thermal properties.

Significant advancements in the fabrication and assembly of electrodes and electrolytes have been pivotal in enhancing the performance of electrochemical energy storage (EES) devices, such as batteries and supercapacitors. Traditional manufacturing methods often face limitations in precisely controlling the intricate geometries, architectures, and porosity of these critical components, which can significantly compromise overall efficiency. 3D printing has emerged as a transformative technology, offering unprecedented control over device geometry, enabling precise control over dimensions, porosity, and morphology to optimize energy and power densities. For instance, inkjet printing (IJP) of carbon-coated $LiFePO_4$ (LFP) electrodes has demonstrated remarkable improvements in conductivity and rate capability due to the high porosity and thin-film characteristics of the printed structures (Delannoy et al. [27]). Similarly, Zhu et al. [28] demonstrated the potential of direct-ink writing techniques to fabricate graphene aerogels using graphene-based inks, showcasing their applicability in high-performance supercapacitors. This approach allows for the use of high-concentration ink formulations, leading to significant enhancements in the structural and functional properties of the printed materials.

Further, Sun et al. [29] developed a 3D interdigitated microbattery architecture (3D-IMA) by directly writing $Li_4Ti_5O_{12}$ (LTO) as the anode and $LiFePO_4$ (LFP) as the cathode material. This innovative design achieves significantly higher areal energy and power densities compared to conventional rechargeable batteries, attributed to the precise fabrication of high-aspect-ratio structures within a compact footprint. These advancements underscore the transformative potential of 3D printing in revolutionizing the design and functionality of EES devices, paving the way for scalable, high-performance solutions across a wide range of applications.

Kim et al. [30] demonstrated the fabrication of efficient piezoelectric nanoparticle-polymer composite materials using 3D optical printing. The 3D microstructures were created by incorporating barium titanate ($BaTiO_3$, BTO) nanoparticles into photocurable polymer solutions and exposing them to dynamically altered digital optical masks. To improve mechanical-to-electrical conversion efficiency, BTO nanoparticles were chemically modified with acrylate surface groups, forming covalent linkages with the polymer matrix during light exposure. Composites with 10% mass loading of chemically modified BTO nanoparticles exhibited piezoelectric coefficients of approximately 40 pC/N, which were more than times higher than those of composites with unmodified BTO nanoparticles and more than twice as high as those of composites containing unmodified BTO nanoparticles and carbon nanotubes.

These findings present a novel approach to 3D printing piezoelectric polymers and lay the foundation for creating highly efficient piezoelectric polymer materials through nano interfacial engineering.

2.2 *Nanopolymer Composites for Improving Mechanical Properties in 3D Printing*

In 3D printing, nano polymers have emerged as revolutionary materials significantly improving the mechanical properties of produced components. Nanoparticles like graphene, carbon nanotubes, or nanoclays are incorporated into polymer matrices due to their remarkable strength, stiffness, and durable structures of the materials produced. This increased surface-area-to-volume ratio of nanoparticles that will directly contribute to enhanced tensile strength, thermal stability, and interlayer adhesion simultaneously addresses some drawbacks usually associated with conventional 3D parts. Moreover, progress is being made in nano polymers equipped with self-lubricating capabilities. Materials are intended to be manufactured for advanced applications in commercial and consumer markets, automotive, medical, and aerospace sectors.

Silva et al. [31] assessed the mechanical characteristics of graphene oxide (GO)-reinforced low-density polyethylene (LDPE) nanocomposites intended for 3D printing. They are composed of between 0.5% and 1.0% GO. The tensile test revealed a significant difference between LDPE-1.5%wt GO and pure LDPE, raising Young's modulus from 276 to 384 MPa (39.13%) as shown in Fig. 4. The composite loses its ductility as the amount of GO in the polymer matrix increases, making it suitable for use in 3D printing methods and so promoting further research on printed materials for a range of uses. Also, they performed an impact test and revealed that the materials' toughness remained relatively constant. The observed decline in additive materials, however, could be connected to the ductility loss brought on by the additional fillers.

Singh et al. [32] used the micro-compounding extrusion method to assess the mechanical performance of Multiwall Carbon Nanotubes (MWCNTs) distributed in Acrylonitrile Butadiene Styrene (ABS). Tensile characteristics were shown to be significantly improved by the addition of 1, 2, and 3 weight percent of MWCNTs in comparison to pure ABS. They concluded it may greatly enhance the mechanical characteristics of items that are 3D printed. Ranaifar et al. [33] analyzed the mechanical properties of the 3D printed honeycomb structure of ABS with nanoscale fillers of CNTs and chopped carbon fibers as shown in Fig. 5a. They investigated compositions containing 5% CNTs and CFs. According to their findings, specimens of ABS, CF-ABS, and CNT-ABS honeycombs with different hex sizes and print-layer heights underwent compression testing. Smaller hex sizes often showed higher compressive strength than bigger hex sizes for a constant print-layer height as shown in Fig. 5b. Across materials and print-layer heights of 0.2 mm and 0.3 mm, this pattern held true, with smaller hex-size specimens exhibiting higher stresses before failure. These

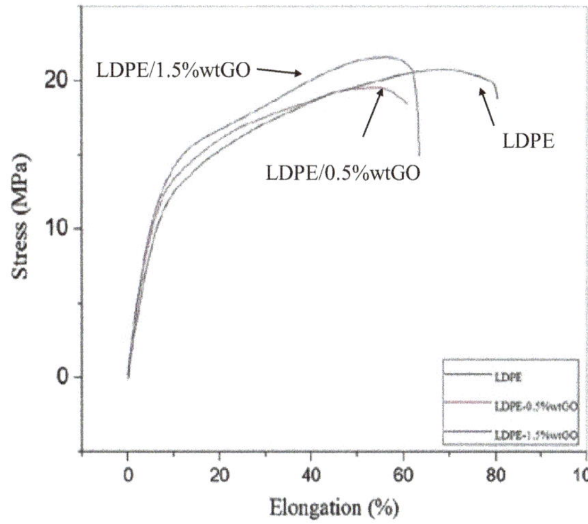

Fig. 4 Tensile tests graph for pure LDPE sample, LDPE/GO −0.5%, LDPE/GO −1.5% [31]

results demonstrate the negative correlation between compressive strength and hex size, which may be further validated by additional studies.

Kim et al. [34] created 3D printed specimens using fused filament fabrication (FFF) of thermoplastic polyurethane (TPU) and fillers of MWCNTs to assess the mechanical properties. For the tensile test, TPU/MWCNT 5%, 10%, and 15% specimens were produced. Their results showed adding 5% MWCNT to TPU increases strain tolerance despite large stiffness variations. While MWCNTs typically improve mechanical qualities, excessive content produces particle aggregation and internal flaws, resulting in decreased strength, elongation, and adhesion forces. Low MWCNT

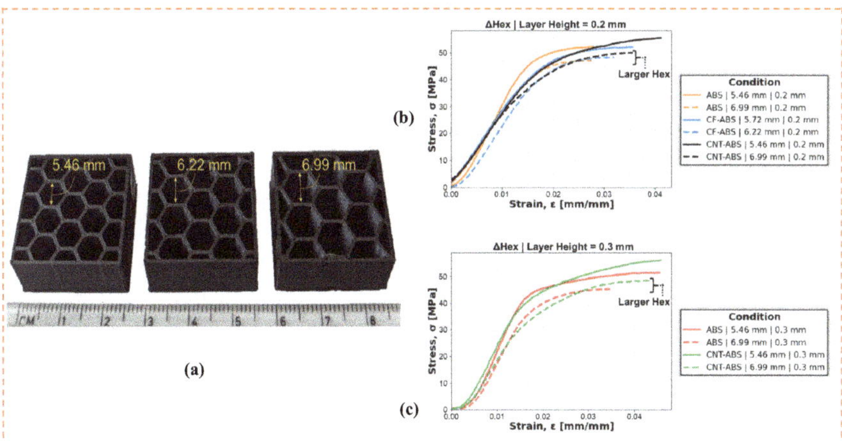

Fig. 5 **a** Honeycomb structures of ABS, CNTs-ABS, and CFs-ABS and **b** Stress–strain of compression testing of honeycomb structure with different layers height [33]

concentrations improve tensile strength, whereas greater quantities compromise composite performance owing to increased stiffness and cohesion.

Farajian et al. [35] investigated the mechanical properties (tensile, impact, pull-off test, and microhardness) of polylactic acid (PLA) coated with PU and MWCNTs with varying concentrations. The samples were immersed in NaCl 3.5 wt.% for 10, 20, and 30 days to assess their mechanical performance in corrosive environments. MWCNTs significantly improve the coating performance and durability of 3D-printed PLA samples. The tensile and impact strength of PU/MWCNTs-coated samples dropped marginally with longer immersion in NaCl 3.5 wt.% solution but remained much greater than uncoated PLA samples. The pull-off test shows that PU/MWCNT coatings increase adherence to PLA substrates, with an initial adhesion strength of 2.8 MPa, but this strength declines with prolonged immersion in NaCl solution due to PLA degradation and hydrolysis. Corrosion accelerates over time owing to chloride ions and acidic byproducts, which degrade the substrate and reduce coating adherence. MWCNTs may exacerbate galvanic corrosion, resulting in reduced interfacial bonding after extended exposure. Pure PLA (77 Shore D) has a microhardness drop of up to 17% after 30 days of immersion in NaCl due to surface corrosion and layer gaps. PU and PU/MWCNT coatings improve hardness to 80 and 82 Shore D, respectively, by reinforcing the PU matrix, improving interfacial adhesion, and filling defects. However, prolonged NaCl exposure diminishes microhardness by surface deterioration. Table 2 shows various nano polymers for improving mechanical properties.

2.3 *Nanopolymer Composites for Improving Thermal and Electrical Conductivity in 3D Printing*

In 3D printing applications, the nanopolymers could successfully substitute other materials to enhance electrical and thermal conductivity. In the research conducted, through the addition of nanoparticles such as metal oxides, carbon nanotubes, or graphene into polymer matrices, nanocomposites of exceptional conductivity have been produced (Hisham et al. [46]). These can greatly improve attributes of the traditional polymers as their electrical conductivity surpasses limits and as their heat dissipation is weak; both are the main obstacles to higher application value. Increased thermal conductivity enables superior heat control during the 3D print process, thereby further increasing the print's precision. Such material is highly conducive to 3D procedures, as greater thermal conductivity usually makes the best print, enabling structures that don't break when subjected to thermal stresses. Likewise, it enables improved electrical conductivity to carry out useful printings such as circuits, sensors, and electromagnetic absorption plates (Tuli et al. [47]). Most importantly, such an innovative combination of multiple benefits makes these composites the most versatile composite material applicable to different markets, sectors from electronics to aerospace.

Table 2 Various nanopolymers for improving mechanical properties

Matrix material	Reinforcements	Enhanced mechanical properties	References
Polyethylene terephthalate glycol (PETG)	CNTs, graphene, and carbon black	Young's modulus	Crapnell et al. [36]
PLA	CNTs	Tensile strength increased by 22.8% and modulus increased by 7.41%	Sharifi et al. [37]
Polyetheretherketone (PEEK)	CNTs and GNPs	Enhanced modulus strength and compressive strength	Schneider et al. [38]
ABS	CNTs	Young's modulus varies with the concentration of CNTs	Rafiee et al. [39]
PLA	MXene	Adding 0.25–0.5 weight percent MXene to PLA increased stiffness, fracture strength, and energy absorption capacity	Srinivas et al. [40]
PLA and Polycaprolactone (PCL)	Chitosan (CS) and MWCNTs	Increased load bearing capacity, compression modulus increased compared to pure, and ultimate strength improved	Kallivokas et al. [41]
Polyamide-imide (PAI)	Aramid nanofibers and carbon nanotubes	Enhanced tensile strength up to 105 MPa and functional properties	Kim et al. [42]
Acrylic photosensitive resin	GO	Improved mechanical strength as compared to pure resin	Garcia et al. [43]

(continued)

Table 2 (continued)

Matrix material	Reinforcements	Enhanced mechanical properties	References
Polypropylene (PP)	Titanium nitride (TiN)	Improved tensile (41.5%), flexural (33.7%), and impact (18.0%) strength	Michailidis et al. [44]
PP	Halloysite nanotubes (HNTs)	Exhibited high storage modulus and impact strength with varied percentages	Patiño-Almanza et al. [45]

Hossain et al. [48] evaluated the thermal and electrical properties of PLA and high-density *polyethylene* (HDPE) by introducing Titanium Dioxide *nanoparticles* (TiO_2) using 3D printed parts. They prepared various compositions with varying concentrations of TiO_2 and samples named M1, M2, M3, and M4. In thermogravimetric analysis, their results showed the thermal decomposition occurred in the range of 467.51 °C to 472.37 °C which is greater than that of M3 sample as shown in Fig. 6a. However, the results also indicated limited thermal stability. In contrast to previous composite filaments for 3D printing, nanoparticles, and 60% recycled plastics broke down at a temperature that was suitable. The test assessed the 3D-printed filaments' electrical resistance, which is intended for use in high-voltage insulation applications. High-performance insulators were found in non-conductive plastic filters, including recycled and nanoparticle-based types. The material's great purity was demonstrated by its average nonconductivity, which was 48.86 GΩ as shown in Fig. 6b.

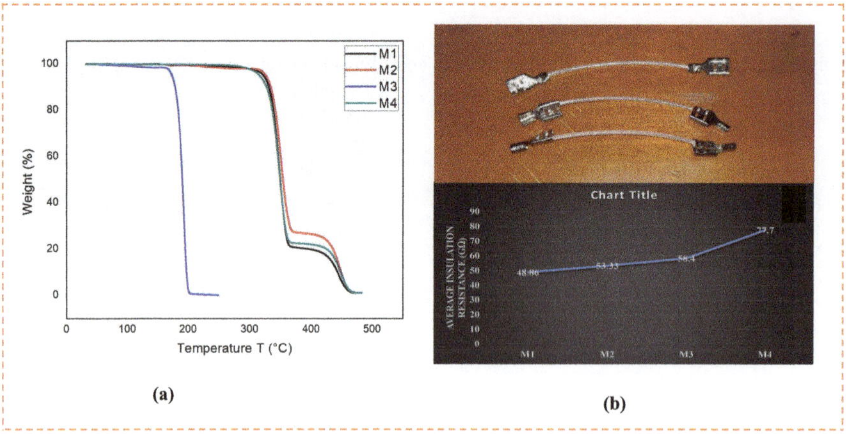

Fig. 6 **a** TGA analysis of filaments and **b** Insulation resistance test of the filaments [48]

Wang et al. [49] created nanocomposites of isotactic PP and GNPs using fused deposition modeling to test the thermal conductivity as shown in Fig. 7. In this work, ball milling and melt extrusion processes were combined to create isotactic polypropylene/GNP (iPP/GNP) composite filament that was optimized for FDM. The printed components demonstrated a thermal conductivity of 0.64 W/m· K with a 5-weight percent GNP loading, significantly exceeding the 0.42 W/m·K thermal conductivity of FP samples by making use of the FDM-induced orientation of GNPs as shown in Fig. 7a. In contrast to the 3.83×10^{-7} m^2K/W found in FP samples, the interfacial thermal resistance of VP printed components was remarkably lowered to 1.97×10^{-7} $m^2 \cdot K/W$, demonstrating the efficacy of filler orientation in increasing thermal conductivity. Also, they observed thermal behavior and revealed strong α and mild β endothermic peaks were seen in neat iPP, however, β-crystals were not present in composites because of β/α transitions that occurred during heating. Initially, the addition of GNP increased crystallinity; however, by limiting the mobility of the iPP molecular chain, a larger GNP concentration somewhat decreased it.

Łapińska et al. [50] analyzed the electrical and thermal properties of PLA/graphite/molybdenum disulfide composites using the FFF method. The DSC experiments indicated overlapping exothermic and endothermic peaks during the glass transition (~60 °C), but no crystallization during cooling since PLA crystallizes slowly. Because PLA crystallizes into both α and α' forms, it becomes difficult to calculate initial crystallinity because the α' crystallites reorganize into the α form when heated as shown in Fig. 8a. The thermal conductivity test results revealed that the G10 and G10/M1 samples showed improvements of up to 400%, in thermal conductivity, with the maximum conductivity occurring at 100% infill as shown in Fig. 8b. Additionally, notable improvements were noted at 80% infill, with the best G10/M1 sample achieving around 350%. Depending on the structure and infill density, adding 1 weight percent MoS_2 to 10 weight percent graphite/PLA increased heat conductivity by 1.27–1.41 times, indicating a synergistic outcome.

Li et al. [51] created a 3D printed nanocomposite by incorporating silver and graphene nanoparticles into a curable resin. Their findings indicated that electrical conductivity improved with varying nanoparticle concentrations, which may be used in wearable electronic devices and smart fabrics. Wei et al. [52] found that adding 5.6wt percent graphene to ABS enhanced the electrical conductivity by four orders of magnitude in polymer nanocomposite as shown in Fig. 9. Bedi et al. [53] used thermogravimetric analysis (TGA) to demonstrate that adding graphene increased the thermal stability and resistance of a composite. Several nanopolymers are shown in Table 3 to improve the electrical and thermal conductivity of different polymeric materials.

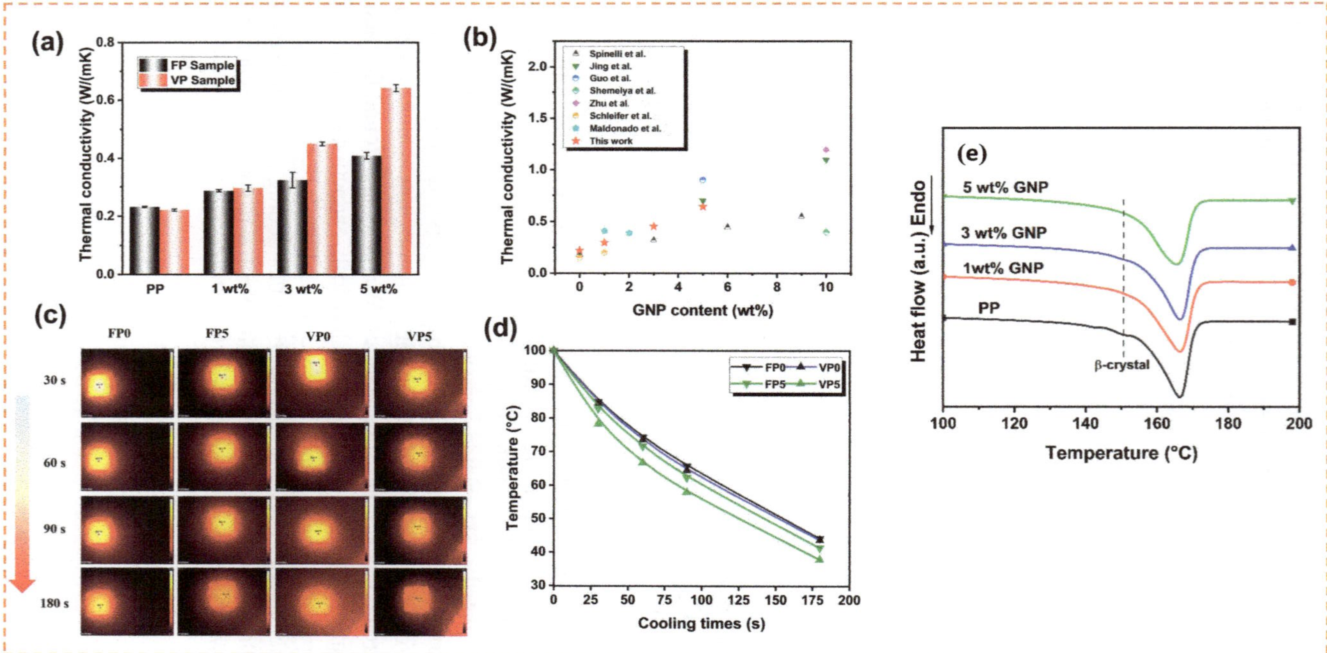

Fig. 7 **a** Thermal conductivity of FP and VP printed parts; **b** Comparison of thermal conductivity of the as-printed composites with those of previously reported works **c** Thermal imaging test results of different printed parts for FP0, FP5, VP0, and VP5 samples; **d** Temperature changes curve of various FDM printed parts. **e** DSC test of samples [49]

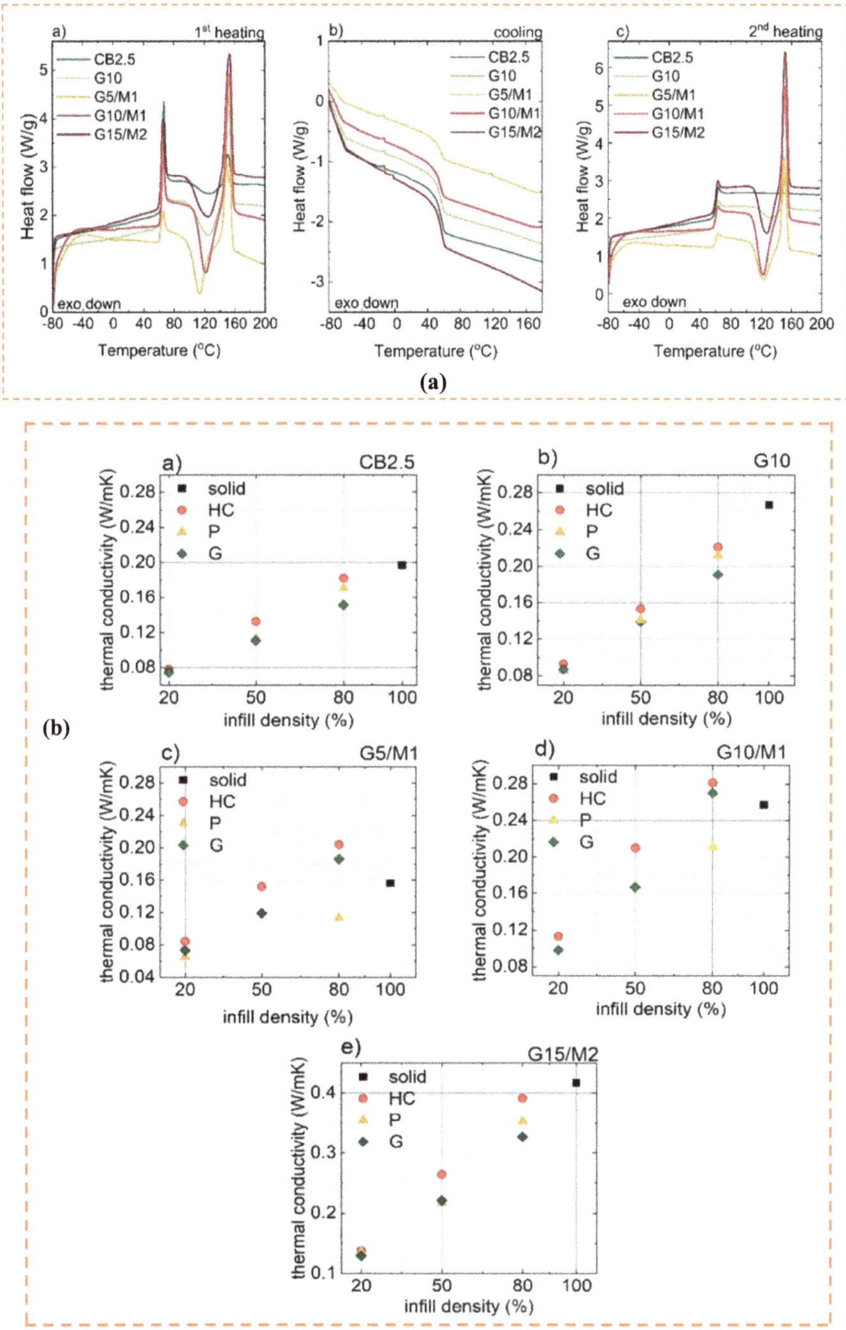

Fig. 8 **a** DSC graphs of filaments and **b** Thermal conductivity of filaments of various samples [50]

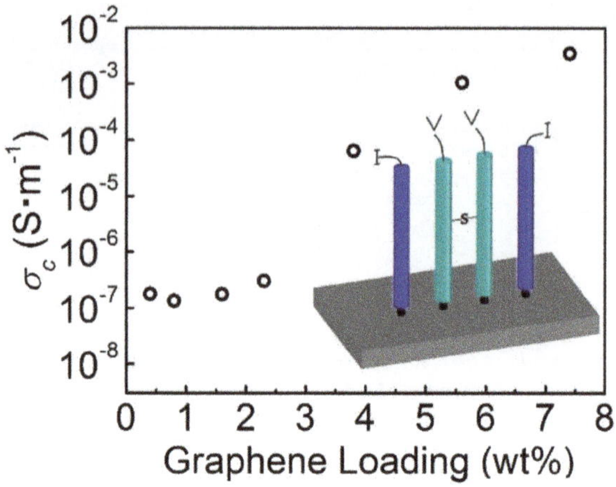

Fig. 9 Electrical conductivity (σ_C) of G-ABS composites varies with graphene loading [52]

Table 3 Various nano polymers for improving the thermal and electrical properties

Matrix	Reinforcements	Improved thermal and electrical properties	References
PLA	Boron nitride (BN) and Graphene (GR)	Thermal conductivity of 3.82 W/mK and tensile strength	Fang et al. [54]
PLA	CNTs	CNTs improved thermal stability and conductivity, enhanced electrical conductivity of 1.8×10^{-9} S m^{-1}	Sharifi et al. [37]
ABS	MWCNTs	For 1, 2, and 3 wt.% of ABS/ MWCNTs thermal properties enhanced significantly	Singh et al. [32]
PLA and HDPE	CNTs	Increased electrical conductivity for various concentrations and applicable for electro-conductive composites	Mora et al. [55]
Polyamide 12 (PA12)	MWCNTs and CB	5 wt.% of fillers improved thermal and electrical properties	Vidakis et al. [56]
PLA	CNFs and Iron oxide (Fe_2O_3)	Adding conductive fillers enhanced conductivity of the samples	Alam et al. [57]
TPU	MWCNTs	All samples showed anisotropic electrical conductivity; highest conductivity obtained for 5 wt.% of MWCNTs	Tzounis et al. [58]
Poly (vinyl alcohol) (PVA)	Graphene nanoplate (GNP)	Among all samples, 10 wt.% of GNP showed high electrical conductivity	Yang et al. [59]
PLA	GNPs	Enhanced thermal and electrical properties and potential for biomaterials	Y. Wang et al. [60]

2.4 Nanopolymer Composites for Biomedical Applications in 3D Printing

The usage of nanopolymers in 3D printing for biomedical applications is growing because of their remarkable qualities, which include high accuracy, biocompatibility, and adjustable functionality (Silva et al. [61]). These materials make it possible to precisely fabricate complex structures at the nanoscale, such as drug delivery systems, tissue scaffolds, and medical implants as shown in Fig. 10. Nanoparticles are well suited for use in regenerative medicine, customized healthcare, and sophisticated diagnostics because of their improved mechanical strength, thermal stability, and bioactive qualities. Their flexibility and versatility continue to spur biomedical engineering innovation (Murugesan and Scheibel [62]).

Supporting structures called scaffolds are used in biomedical applications to replicate the extracellular environment and enable cell growth and survival inside a certain volume. Depending on the study's target application, 3D printing was shown to be a viable method for creating scaffolds with unique geometry (Hollister et al. [64]). For usage in biological applications, Khodaei et al. [65] created scaffolds of PLA and akermanite (Ak) as shown in Fig. 11. The study's objective is to examine ultrasonic vibration in scaffold 3D printing. The stages and morphology of the scaffolds were found to be unaffected. When compared to traditional printing techniques, this work demonstrates how well ultrasonic-assisted 3D printing produces nanocomposite scaffolds with noticeably larger nanoparticle loadings. To build the customized complicated structure for bone tissue scaffolds that have been strengthened by the filling of

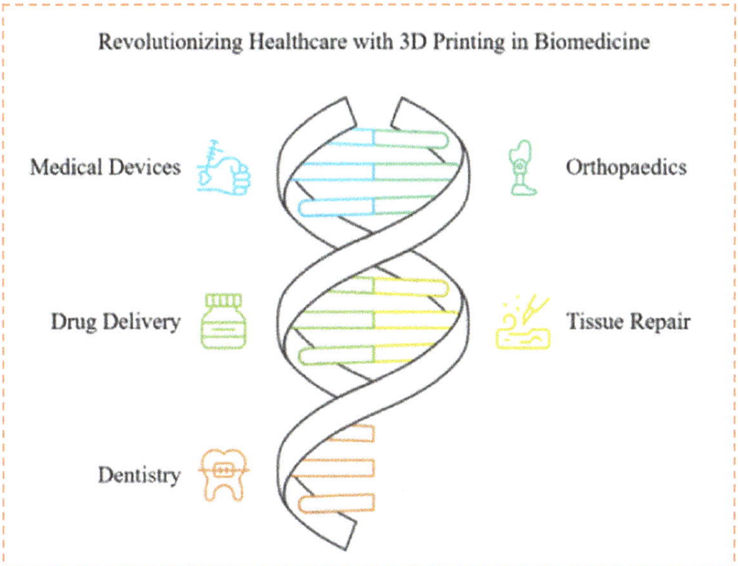

Fig. 10 Biomedical applications of 3D printing (Tyagi et al. [63])

graphene layers, Feng et al. [66] developed a biodegradable UV-cured resin via SLA. Trimethylolpropane trimethacrylate (TEGDMA), an easily accessible polyurethane resin, and phenylbis (2, 4, 6-trimethylbenzoyl)-phosphine oxide (Irgacure 819) as an oligomer, a reactive diluent, and a photoinitiator constitute the composite. When compared to conventional direct casting methods, the mechanical performance of the components produced by SLA was enhanced due to the incorporation and effective dispersion of graphene fillers. Images of jawbones and gyroid scaffolds made of graphene-reinforced nanocomposite and pure UV-cured resin are displayed in Fig. 12.

By providing precise surgical templates for precise guidance and better anatomical vision, 3D printing technology improves surgical planning. It is extensively utilized in several surgical specialties, including neurosurgery, orthopedics, and vascular surgery, where it improves patient safety, decreases mistakes, and produces better results. It also supports medical education and is still being used and developed

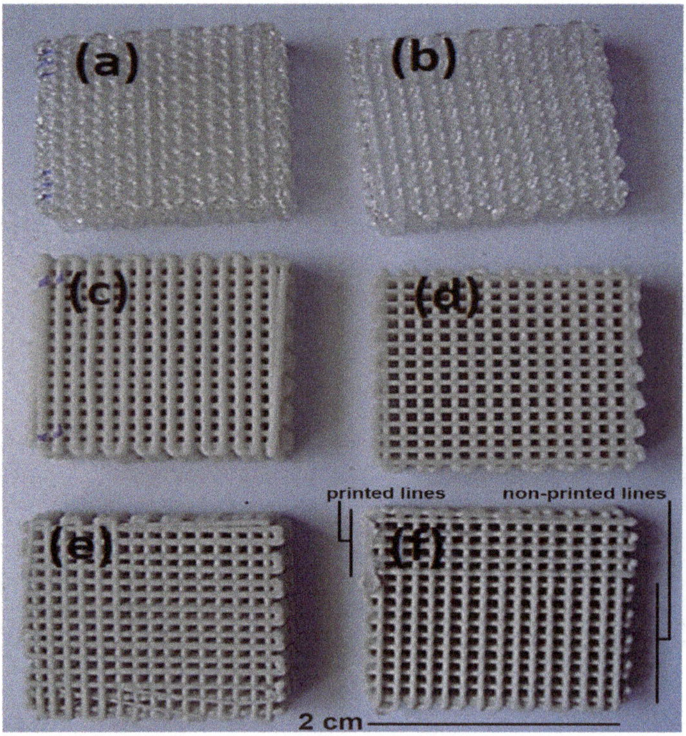

Fig. 11 **a**, **b** PLA scaffolds made using both conventional and ultrasonic-assisted 3D printing; **c**, **d** PLA-20Ak scaffolds made using both conventional and ultrasonic-assisted 3D printing; **e**, **f** PLA-40Ak scaffolds made using both conventional and ultrasonic-assisted 3D printing. The samples are 20 × 20 × 5 mm in size [65]

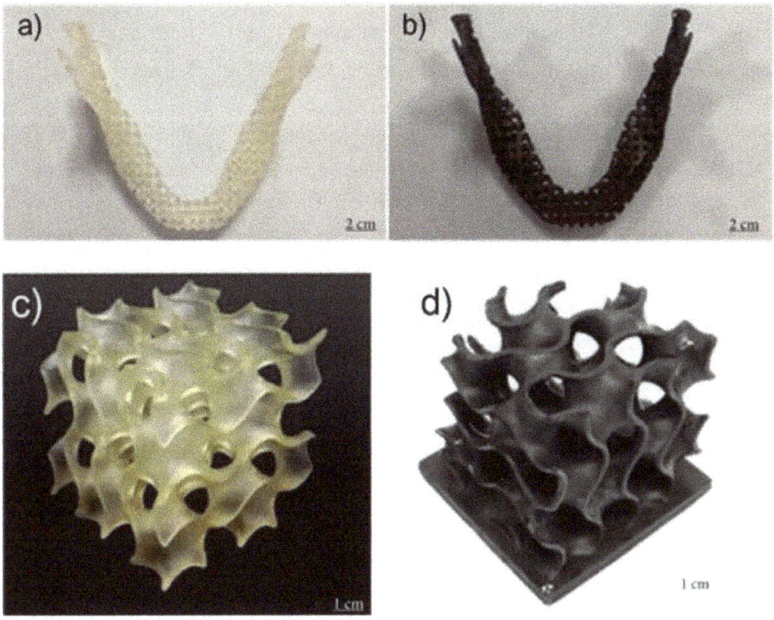

Fig. 12 Pictures of **a**, **c** monolithic UV-cured resin and **b**, **d** graphene-reinforced nanocomposite jawbone with a square architecture and gyroid scaffold for bone tissue engineering [66]

at an increasing rate (Qiu et al. [67]). Also, for medical devices 3D printing has significant importance and is used to produce components that are difficult to produce by conventional techniques. Because of this, 3D printing is used to create many patient-specific hearing aids, including those that need to fit the patient's ear structure (Dodziuk [68] and Auricchio and Marconi [69]). Much more is generated and used now than this figure. The benefits of this approach allow for the creation of hearing aids in a significantly shorter amount of time than is typically required. This technique has promise for a variety of healthcare devices. In addition to hearing aids, the visually handicapped can also benefit from specialized eye lenses, stethoscopes, and spectacles (Mertz [70]). Applications of nanopolymers in the biomedical field are displayed in Table 4.

2.5 Nanopolymer Composites for Other Specific Applications in 3D Printing

Through 3D printing, nanopolymers have made a wide range of uses possible in several industries. They are employed to produce flexible circuits, electromagnetic

Table 4 Biomedical applications of nanopolymers in 3D printing

Materials	Printing method	Application	References
PLA and silver nanoparticles	FDM	Biomedical	Katseli et al. [71]
PU	FDM	Tissue engineering	Hendrikson et al. [72]
SWCNTs/Agarose	Extrusion printing	Tissue engineering, biosensors, organ Printing	Nadernezhad et al. [73]
nHA Polycaprolactone, alginate, gelatine	FDM	Bone tissue engineering	Hernandez et al. [74]
Nanocellulose Alginate and gelatin	Extrusion printing	Tissue regeneration	Sultan and Mathew [75]
Silver nanoparticles/ Alginate	SLA	Bionic ears	Mannoor et al. [76]
Graphene Nano-58S bioactive	SLS	Bone repair and regeneration	Gao et al. [77]
MWCNTs/ Polycaprolactone	Extrusion printing	Cardiac tissue engineering	Ho et al. [78]
nHA Polylactic acid	FDM	Vascularized bone tissue repair	Holmes et al. [79]
nHA Polyethylene glycol diacrylate and polyethylene glycol	FDM	Osteochondral tissue regeneration	Castro et al. [80]

shielding materials, and conductive components for electronics. Nano polymers are useful to the automotive and aerospace sectors because they may be used to create lightweight, robust parts with exceptional mechanical and thermal qualities. These materials are also being used in energy storage devices that offer better energy density and performance, like supercapacitors and 3D-printed batteries. Applications in the environment include the development of sophisticated filtration systems and sensors for pollutant detection and water purification, demonstrating their ability to address global issues (Ryan et al. [81]). Nanopolymer composites have specific uses in 3D printing, as shown in Table 5

3 Nanopolymer Composites for 4D Printing Applications

4D printing is a cutting-edge additive manufacturing technology that employs smart materials capable of undergoing transformation over time in response to environmental triggers such as heat, light, moisture, pH, or magnetic fields. Among these materials, SMPs and their composites (SMPCs) have gained significant attention due to their rapid and intelligent responsiveness. SMPCs have been utilized to

Table 5 Specific applications of nanopolymer composites in 3D printing

Matrix	Fillers	Applications	References
Polyaniline	graphene quantum dots (GQDs)	Supercapacitor	Kuzhandaivel et al. [82]
poly (ethylene oxide)	GQDs	Lithium-ion batteries	H. Liu et al. [83]
Silica molecular imprinted polymer	GQDs	Nanosensor	Mehrzad-Samarin et al. [84]
polypyrene	GQDs	Supercapacitors	M et al. [85]
memory Polymers	GQDs	Electronic memory devices	Fan et al. [86]
polyvinyl alcohol (PVA)	–	Flexible electronic display devices	Luo et al. [87]

enhance the response speed to thermal stimuli [88] or to introduce responsiveness to thermo-magnetic triggers [89]. Additionally, they offer improved mechanical properties while retaining the inherent characteristics of the polymer matrix [90, 91]. These advanced materials are applied across various industries, including biomedical devices for drug delivery and minimally invasive implants, soft robotics for adaptive movement, aerospace and automotive components with self-healing or shape-changing capabilities, and wearable smart textiles.

3.1 Nanopolymer Composites to Enable Various External Stimuli in 4D Printing

Shape memory polymers typically possess glass transition temperatures (T_g) higher than their operational temperatures. These polymers adopt a permanent shape when heated above their T_g and revert to a temporary shape upon cooling. The shape transformation effect primarily occurs in bilayer structures, where significant differences in their thermal expansion coefficients enable the behavior. In thermo-responsive materials, the shared interface between the layers creates a stress field, allowing the structure to bend. The stimulus-responsive behaviours relevant to 4D printing are illustrated in Fig. 13 [92].

Temperature is the most widely used stimulus in 4D printing [93, 94], as it allows for precise control of structural adjustments, mechanical manipulations, and straightforward operation. This versatility has made temperature-responsive applications a prominent area of research and development in 4D printing technology [95–97]. Light has become the second most commonly studied stimulus in the field, attracting growing interest from researchers [98]. Its use offers numerous advantages, including rapid switching [99], precise focusing [100], biocompatibility, sustainability as an energy source [101], wireless and remote controllability [102], the ability to induce color changes [103], and the regulation of mechanical properties.

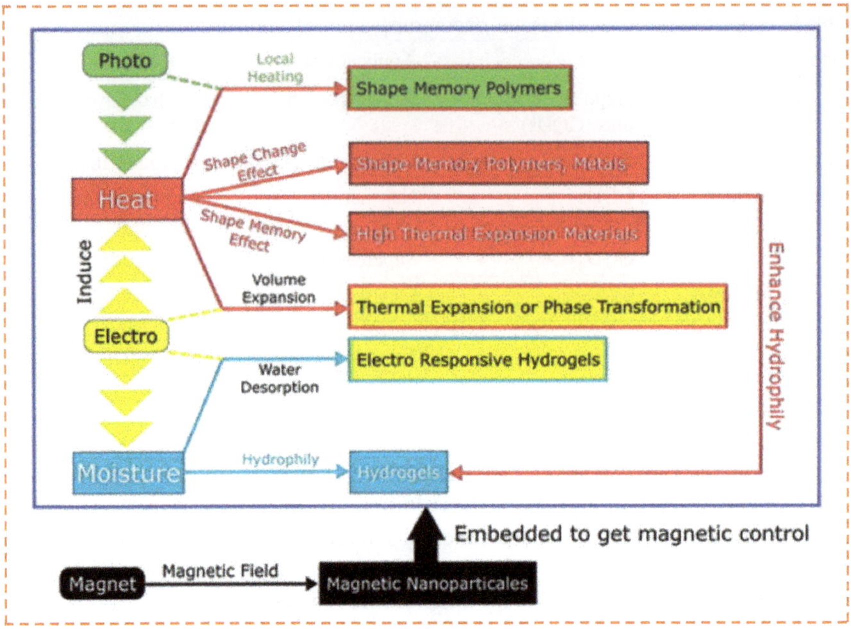

Fig. 13 Response of stimuli effect on smart material [92]

Magnetic fields serve as an alternative stimulus that has been investigated extensively in various studies [104–106]. Using magnetic fields as actuators provides several benefits, including fast and efficient responses [107, 108], minimal safety risks [94], remote controllability [109, 110], and the ability to accelerate movement in 4D-printed objects with precise, target-oriented trajectories [111]. In chemistry, pH refers to the potential or power of hydrogen, making it a significant stimulus for 4D-printed objects. Its ability to trigger reactions such as swelling, shrinking, dissociation, or degradation enhances its utility [95]. Moreover, pH can drive functions like expansion, contraction, and torsion [112]. It also plays a key role in response modulation, controlled drug release [113], self-healing [114], color transitions [115, 116], biodegradation, and biocompatibility. Electric fields have been investigated as stimuli in numerous studies, demonstrating characteristics like rapid response and remote controllability [117, 118]. However, their use as a stimulus can pose challenges, including localized heating, membrane damage, and potential cell death [119].

3.2 Nanopolymer Composites for Improving Mechanical Properties in 4D Printing

SMPs have notable limitations, including low mechanical strength and shape recovery stress. The mechanical strength of SMPs typically ranges between 5 and 100 MPa, depending on their type and composition. Similarly, their recovery stress usually falls within a few tenths to several tens of megapascals. A common approach to enhance the properties of pristine SMPs is the incorporation of high-modulus inorganic or organic fillers [120]. SMPCs have been utilized to enhance the response speed to thermal stimuli [88] or to introduce responsiveness to thermo-magnetic triggers [89]. Additionally, they offer improved mechanical properties while retaining the inherent characteristics of the polymer matrix [90, 91].

Choong et al. [121] found that incorporating nanosilica into SMPs significantly improved their mechanical strength by an order of magnitude and increased elongation by 85.2% compared to the unmodified polymers. In a similar vein, Kausar et al. [122] investigated the use of graphene nanofillers in polymer matrices, showing enhanced mechanical performance for 4D printing applications. Yousefi et al. [123] demonstrated that adding Fe_3O_4 nanoparticles to SMPs boosted their mechanical strength and facilitated rapid shape memory responses when activated by thermal and magnetic stimuli. Namathoti et al. [124] improved the mechanical properties of SMPs by incorporating MWCNTs and HNTs via 4D printing techniques. Patel et al. [125] enhanced the mechanical properties of microwave-induced SMPU/GNP composites. Prasad et al. [126] increased the mechanical properties of SMPU by incorporating MWCNTs into its matrix. Hashmi et al. [127] also improved the mechanical strength of SMPU samples by adding MWCNTs. Additionally, Patel et al. [128] enhanced the mechanical strength of a polyurethane matrix by integrating both GNPs and MWCNTs. Finally, Bi et al. [129] showed improvements in the mechanical properties of 4D printed thermoreversible cross-linked network composites.

3.3 Nanopolymer Composites for Improving Shape Recovery in 4D Printing

SMPs are known for their ability to undergo significant shape changes in response to external stimuli. However, one of the challenges associated with SMPs is their relatively low recovery stress and limited cycle durability compared to other smart materials like SMAs [130, 131]. These limitations can hinder their practical applications, particularly in high-performance settings or those that require repeated use over time. However, the integration of nanofillers into SMPs not only enhances the mechanical properties but also accelerates the actuation process, enabling more efficient and versatile 4D printing applications.

Choong et al. [121] proved that the inclusion of nanosilica in SMPs notably enhanced their shape recovery capabilities. Similarly, Kausar et al. [122] improved the shape recovery properties of SMPs by integrating graphene nanofillers into the matrices for 4D printing applications. Yousefi et al. [123] achieved rapid shape memory responses in SMPs by introducing Fe_3O_4 nanoparticles, which were activated by thermal and magnetic stimuli. Namathoti et al. [124] boosted the shape recovery performance of SMPs by incorporating MWCNTs and HNTs through 4D printing techniques. Patel et al. [125] enhanced shape recovery properties in microwave-induced SMPU/GNP composites. Prasad et al. [126] also improved the shape recovery of SMPU by incorporating MWCNTs into the polymer matrix. Hashmi et al. [127] further enhanced the shape recovery of SMPU samples with the addition of MWCNTs. Moreover, Patel et al. [128] improved the shape recovery characteristics of a polyurethane matrix by integrating both GNPs and MWCNTs. Lastly, Bi et al. [129] reported significant improvements in the shape recovery properties of 4D printed thermoreversible cross-linked network composites.

3.4 Other Applications of Nanopolymer Composites in 4D Printing

SMP nanocomposites offer a wide range of possibilities for enhancing 4D printing technologies across diverse applications. Whether improving mechanical properties, enabling self-healing, enhancing sensing and actuation, or creating smart coatings, the integration of nanocomposites into 4D printed objects opens up new opportunities for innovation in fields like robotics, medicine, energy, and environmental sustainability [132, 133]. The adaptability of nanocomposites ensures that they can be tailored to meet the specific requirements of different industries, making them a key component of the next generation of smart materials [134–137]. Table 6 shows various applications of nanopolymer composites in 4D printing.

4 Summary

This chapter presents polymer nanocomposites and explains how they play a role in 3D and 4D printing technologies. This chapter explains polymers through the addition of nanoparticles, nanotubes, and nanofibers, which significantly improve the mechanical, thermal, electrical, and functional properties of a polymer matrix, enabling the development of advanced, tailored parts. The chapter describes various types of nanopolymer composites and their ability to improve structural strength, and responsiveness to external factors and discusses their use in aerospace, biomedical engineering, electronics, energy harvesting, and environmental engineering applications. Particularly, importance of 4D printing, in which nanopolymer composites can

Table 6 Various applications of nanopolymer composites in 4D printing

Material Used	4D printing method	Applications	References
SMP & ABS	FDM	Hyper-redundant robotic arm	Mostafaie et al. [138]
SMPU	FDM	Mechanical gripper	Gay et al. [139]
SMPU	FDM	Reversible energy absorbing	Shafranek et al. [140]
SMP	FDM	Energy absorbing devices	Zang et al. [141]
Electroactive polymer	DIW	Actuators	Bar et al. [142]
Silicone elastomer	DIW	Soft actuators	Miriyev et al. [143]
Thermoresponsive SMP	FDM	Shape morphing structures	Yu et al. [144]
Methyl acrylate	SLA	Flexible electronic devices	Zarek et al. [145]

act for responsive, active materials for smart devices, soft robotics, and self-healing materials.

References

1. Crosby, A.J., Lee, J.: Polymer nanocomposites: the "Nano" effect on mechanical properties. Polym. Rev. **47**(2), 217–229 (2007). https://doi.org/10.1080/15583720701271278
2. Khan, I., Saeed, K., Khan, I.: Nanoparticles: properties, applications and toxicities. Arab. J. Chem. **12**(7), 908–931 (2017). https://doi.org/10.1016/j.arabjc.2017.05.011
3. Rashid, A.B., Haque, M., Islam, S.M.M., Labib, K.R.U.: Nanotechnology-enhanced fiber-reinforced polymer composites: Recent advancements on processing techniques and applications. Heliyon **10**(2), e24692 (2024). https://doi.org/10.1016/j.heliyon.2024.e24692
4. Lakkala, P., Munnangi, S.R., Bandari, S., Repka, M.: Additive manufacturing technologies with emphasis on stereolithography 3D printing in pharmaceutical and medical applications: a review. Int. J. Pharmaceutics X **5**, 100159 (2023). https://doi.org/10.1016/j.ijpx.2023.100159
5. Lee, J., An, J., Chua, C.K.: Fundamentals and applications of 3D printing for novel materials. Appl. Mater. Today **7**, 120–133 (2017). https://doi.org/10.1016/j.apmt.2017.02.004
6. Monteiro, S.A., Scheid, C., Deon, M., Merib, J.: Fundamentals, recent applications, and perspectives of 3D printing in sample preparation approaches. Microchem. J. **195**, 109385 (2023). https://doi.org/10.1016/j.microc.2023.109385
7. Momeni, F., HassaniN, S.M., Liu, X., Ni, J.: A review of 4D printing. Mater. Des. **122**, 42–79 (2017). https://doi.org/10.1016/j.matdes.2017.02.068
8. Kuang, X., Roach, D.J., Wu, J., Hamel, C.M., Ding, Z., Wang, T., Dunn, M.L., Qi, H.J.: Advances in 4D printing: materials and applications. Adv. Functi. Mater. **29**(2) (2018). https://doi.org/10.1002/adfm.201805290
9. Chen, Y., Zhang, M., Sun, Y., Phuhongsung, P.: Improving 3D/4D printing characteristics of natural food gels by novel additives: a review. Food Hydrocolloids **123**, 107160 (2021). https://doi.org/10.1016/j.foodhyd.2021.107160

10. Urquiza, E.A.F.: Advances in additive manufacturing of polymer-fused deposition modeling on textiles: from 3D printing to innovative 4D printing—a review. Polymers **16**(5), 700 (2024). https://doi.org/10.3390/polym16050700
11. Wan, X., He, Y., Liu, Y., Leng, J.: 4D printing of multiple shape memory polymer and nanocomposites with biocompatible, programmable and selectively actuated properties. Addit. Manuf. **53**, 102689 (2022). https://doi.org/10.1016/j.addma.2022.102689
12. Li, Z., Liu, M., Young, R.J.: Dependence of the reinforcement of polymer-based nanocomposites upon the nanofiller geometry. Nano Mater. Sci. (2024).https://doi.org/10.1016/j.nanoms.2024.04.014
13. Liang, J., Wang, L., Liu, B., He, X., Guo, J., Zhang, T., Li, X., Ma, Y., Tian, W.: Experimental study of mechanical properties of 3D braided aramid/carbon fiber composites. J. Text. Inst. **115**(11), 2224–2232 (2023). https://doi.org/10.1080/00405000.2023.2283823
14. Zhang, L., Li, Z., Lu, T., He, F., Ye, J.: Preparation and properties of porous calcium phosphate ceramic microspheres modified with magnesium phosphate surface coating for bone defect repair. Ceram. Int. **50**(5), 7514–7527 (2023). https://doi.org/10.1016/j.ceramint.2023.12.055
15. Ghasempour-Mouziraji, M., Lagarinhos, J., Afonso, D., De Sousa, R.A.: A review study on metal powder materials and processing parameters in laser metal deposition. Opt. Laser Technol. **170**, 110226 (2023). https://doi.org/10.1016/j.optlastec.2023.110226
16. Kishore, K., Sheikh, M.N., Hadi, M.N.: Doped multi-walled carbon nanotubes and nanoclay based-geopolymer concrete: an overview of current knowledge and future research challenges. Cement Concr. Compos. **105774**,(2024). https://doi.org/10.1016/j.cemconcomp.2024.105774
17. Mashayekhi, A., Hassanli, R., Zhuge, Y., Ma, X., Chow, C.W.: Synergistic effects of fiber hybridization on the mechanical performance of seawater sea-sand concrete. Constr. Build. Mater. **416**, 135087 (2024). https://doi.org/10.1016/j.conbuildmat.2024.135087
18. Thakur, V., Singh, R., Kumar, R., Sharma, S., Singh, S., Li, C., Zhang, Y., Eldin, S.M., Alqarni, S.A.: Mechanically sustainable and primary recycled thermo-responsive ABS–PLA polymer composites for 4D printing applications: Fabrication and studies. Rev. Adv. Mater. Sci. **63**(1) (2024). https://doi.org/10.1515/rams-2023-0149
19. Shahbazi, M., Jäger, H., Ettelaie, R., Chen, J., Kashi, P.A., Mohammadi, A.: Dispersion strategies of nanomaterials in polymeric inks for efficient 3D printing of soft and smart 3D structures: a systematic review. Adv. Coll. Interface. Sci. **333**, 103285 (2024). https://doi.org/10.1016/j.cis.2024.103285
20. Tundwal, A., Kumar, H., Binoj, B.J., Sharma, R., Kumar, G., Kumari, R., Dhayal, A., Yadav, A., Singh, D., Kumar, P.: Developments in conducting polymer-, metal oxide-, and carbon nanotube-based composite electrode materials for supercapacitors: a review. RSC Adv. **14**(14), 9406–9439 (2024). https://doi.org/10.1039/d3ra08312h
21. Ramezani, M., Dehghani, A., Sherif, M.M.: Carbon nanotube reinforced cementitious composites: a comprehensive review. Constr. Build. Mater. **315**, 125100 (2021). https://doi.org/10.1016/j.conbuildmat.2021.125100
22. Camiletti, J., Soliman, A.M., Nehdi, M.L.: Effect of nano-calcium carbonate on early-age properties of ultra-high-performance concrete. Mag. Concr. Res. **65**(5), 297–307 (2013). https://doi.org/10.1680/macr.12.00015
23. Althoey, F., Zaid, O., Martínez-García, R., Alsharari, F., Ahmed, M., Arbili, M.M.: Impact of Nano-silica on the hydration, strength, durability, and microstructural properties of concrete: a state-of-the-art review. Case Studies Constr. Mater. **18**, e01997 (2023). https://doi.org/10.1016/j.cscm.2023.e01997
24. Khatoon, N., Chu, M.Q., Zhou, C.H.: Nanoclay-based drug delivery systems and their therapeutic potentials. J. Mater. Chem. B **8**(33), 7335–7351 (2020). https://doi.org/10.1039/d0tb01031f
25. Xu, Y., Zhang, H., Gan, Y., Šavija, B.: Cementitious composites reinforced with 3D printed functionally graded polymeric lattice structures: experiments and modelling. Addit. Manuf. **39**, 101887 (2021). https://doi.org/10.1016/j.addma.2021.101887
26. Song, J., Cao, M., Cai, L., Zhou, Y., Chen, J., Liu, S., Zhou, B., Lu, Y., Zhang, J., Long, W., Li, L.: 3D printed polymeric formwork for lattice cementitious composites. J. Build. Eng. **43**, 103074 (2021). https://doi.org/10.1016/j.jobe.2021.103074

27. Delannoy, P., Riou, B., Brousse, T., Bideau, J.L., Guyomard, D., Lestriez, B.: Ink-jet printed porous composite LiFePO4 electrode from aqueous suspension for microbatteries. J. Power Sources **287**, 261–268 (2015). https://doi.org/10.1016/j.jpowsour.2015.04.067
28. Zhu, C., Liu, T., Qian, F., Han, T.Y., Duoss, E.B., Kuntz, J.D., Spadaccini, C.M., Worsley, M.A., Li, Y.: Supercapacitors based on three-dimensional hierarchical graphene aerogels with periodic macropores. Nano Lett. **16**(6), 3448–3456 (2016). https://doi.org/10.1021/acs.nanolett.5b04965
29. Sun, K., Wei, T., Ahn, B.Y., Seo, J.Y., Dillon, S.J., Lewis, J.A.: 3D printing of interdigitated Li-Ion microbattery architectures. Adv. Mater. **25**(33), 4539–4543 (2013). https://doi.org/10.1002/adma.201301036
30. Kim, K., Zhu, W., Qu, X., Aaronson, C., McCall, W.R., Chen, S., Sirbuly, D.J.: 3D optical printing of piezoelectric nanoparticle-polymer composite materials. ACS Nano **8**(10), 9799–9806 (2014). https://doi.org/10.1021/nn503268f
31. Silva, B.M.M., Tegon, C.C., Soares, M.M., Gonçalves, R.L.P., Oliveira, A.A., Couto, A.A., Massi, M.: Development and performance of graphene oxide reinforced LDPE nanocomposite polymers for 3D printing. REM—Int. Eng. J. **77**(2) (2024). https://doi.org/10.1590/0370-44672023770059
32. Singh, P., Singari, R.M., Mishra, R.S.: Enhanced mechanical properties of MWCNT reinforced ABS nanocomposites fabricated through additive manufacturing process. Polymers Adv. Technol. **35**(2) (2024). https://doi.org/10.1002/pat.6308
33. Ranaiefar, M., Singh, M., Halbig, M.C.: Additively manufactured carbon-reinforced ABS honeycomb composite structures and property prediction by machine learning. Molecules **29**(12), 2736 (2024). https://doi.org/10.3390/molecules29122736
34. Kim, H., Jeon, J., Lee, S., Cho, J., Lee, I.: Printing characteristics of TPU/MWCNT conductive composite using FFF. Int. J. Precis. Eng. Manuf. **25**(6), 1303–1309 (2024). https://doi.org/10.1007/s12541-024-00974-0
35. Farajian, J., Hatami, O., Bakhtiari, M., Darabinajand, B., Mahboubkhah, M.: Investigation of mechanical properties of 3D-printed PLA coated with PU/MWCNTs in a corrosive environment. Arab. J. Sci. Eng. **49**(8), 11181–11193 (2024). https://doi.org/10.1007/s13369-023-08632-9
36. Crapnell, R.D., Bernalte, E., Sigley, E., Banks, C.E.: Recycled PETg embedded with graphene, multi-walled carbon nanotubes and carbon black for high-performance conductive additive manufacturing feedstock. RSC Adv. **14**(12), 8108–8115 (2024). https://doi.org/10.1039/d3ra08524d
37. Sharifi, J., Rizvi, G., Fayazfar, H.: Sustainable 3D printing of enhanced carbon nanotube-based polymeric nanocomposites: green solvent-based casting for eco-friendly electrochemical sensing applications. Int. J. Adv. Manuf. Technol. **131**(9–10), 4825–4837 (2024). https://doi.org/10.1007/s00170-024-13337-w
38. Schneider, J., Basak, S., Hou, Y., Koo, J.H., Wardle, B.L., Gadegaard, N., Kumar, S.: 3D-Printed Polyetheretherketone Smart Polymer Nanocomposite scaffolds: Mechanical, Self-Sensing, and Biological attributes. Adv. Eng. Mater. **26**(4) (2024). https://doi.org/10.1002/adem.202301659
39. Rafiee, R., Zehtabzadeh, H., Amini, M.R.: Predicting mechanical properties of 3D printed nanocomposites using multi-scale modeling. Addit. Manuf. **83**, 104055 (2024). https://doi.org/10.1016/j.addma.2024.104055
40. Srinivas, P., Jacob, L., C, M. S., Butt, H., Barsoum, I., Al-Rub, R.K.A., Zaki, W.: Mechanical properties, energy absorption, and shape memory behavior of 3D printed PLA-MXene nanocomposites and gyroid lattices. Adv. Eng. Mater. **26**(12) (2024). https://doi.org/10.1002/adem.202301698
41. Kallivokas, S.V., Kontaxis, L.C., Psarras, S., Roumpi, M., Ntousi, O., Kakkos, I., Deligianni, D., Matsopoulos, G.K., Fotiadis, D.I., Kostopoulos, V.: A combined computational and experimental analysis of PLA and PCL hybrid nanocomposites 3D printed scaffolds for bone regeneration. Biomedicines **12**(2), 261 (2024). https://doi.org/10.3390/biomedicines12020261

42. Kim, H.C., Tu, R., Sodano, H.A.: Room temperature 3D printing of high-temperature engineering polymer and its nanocomposites with porosity control for multifunctional structures. Compos. B Eng. **279**, 111444 (2024). https://doi.org/10.1016/j.compositesb.2024.111444
43. Garcia, G.E.S., De Sousa, R.R., Junior, Gouveia, J.R., Santos, D.J.D.: Graphene Oxide-based nanocomposites for stereolithography (SLA) 3D printing: comprehensive mechanical characterization under combined loading modes. Polymers **16**(9), 1261 (2024). https://doi.org/10.3390/polym16091261
44. Michailidis, N., Petousis, M., Moutsopoulou, A., Argyros, A., Ntintakis, I., Papadakis, V., Nasikas, N.K., Vidakis, N.: Engineering response of biomedical grade isotactic polypropylene reinforced with titanium nitride nanoparticles for material extrusion three-dimensional printing. Europ. J. Mater. **4**(1) (2024b). https://doi.org/10.1080/26889277.2024.2340944
45. Patiño-Almanza, R., García-Méndez, R.F., Rivera-Armenta, J.L., Strachota, A., Almendarez-Camarillo, A.: 3D printing polypropylene composites reinforced with functionalized halloysite: Balance between stiffness and impact resistance. Polym. Compos. **45**(7), 6439–6452 (2024). https://doi.org/10.1002/pc.28208
46. Hisham, M., C, M. S., C, D., Jacob, L., Butt, H.: Additive manufacturing of carbon nanocomposites for structural applications. J. Mater. Res. Technol. **28**, 4674–4693 (2024). https://doi.org/10.1016/j.jmrt.2024.01.049
47. Tuli, N.T., Khatun, S., Rashid, A.B.: Unlocking the future of precision manufacturing: a comprehensive exploration of 3D printing with fiber-reinforced composites in aerospace, automotive, medical, and consumer industries. Heliyon **10**(5), e27328 (2024). https://doi.org/10.1016/j.heliyon.2024.e27328
48. Hossain, M., Chowdhury, M.A., Mahamud, S., Saha, R.K., Zahid, M., Ferdous, J., Hossain, N., Mobarak, M.H.: Electro-mechanical analysis of nanostructured polymer matrix composite materials for 3D printing using machine learning. Chem. Eng. J. Adv. **19**, 100626 (2024). https://doi.org/10.1016/j.ceja.2024.100626
49. Wang, Z., Yang, Q., Zheng, X., Zhang, S., He, P., Han, R., Chen, G.: Fused Deposition modeling of isotactic polypropylene/graphene nanoplatelets composites: achieving enhanced thermal conductivity through filler orientation. Polymers **16**(6), 772 (2024). https://doi.org/10.3390/polym16060772
50. Łapińska, A., Grochowska, N., Cieplak, K., Płatek, P., Wood, P., Deuszkiewicz, P., Dużyńska, A., Sztorch, B., Głowcka, J., Przekop, R., Jóźwik, P., Gołofit, T., Plichta, A.: Architecture influence on acoustic performance, EMI shielding, electrical and thermal, properties of 3D printed PLA/graphite/molybdenum disulfide composites. Mater. Des. **245**, 113241 (2024). https://doi.org/10.1016/j.matdes.2024.113241
51. Li, J., Chen, H., Liu, Y., Wang, J., Lu, M., Chiu, C.: Photocurable 3D-printed AgNPs/Graphene/polymer nanocomposites with high flexibility and stretchability for ECG and EMG smart clothing. Chem. Eng. J. **484**, 149452 (2024). https://doi.org/10.1016/j.cej.2024.149452
52. Wei, X., Li, D., Jiang, W., Gu, Z., Wang, X., Zhang, Z., Sun, Z.: 3D Printable Graphene Composite. Scientific Reports **5**(1) (2015). https://doi.org/10.1038/srep11181
53. Bedi, S.S., Mallesha, V., Mahesh, V., Mahesh, V., Mukunda, S., Negi, S., Ponnusami, S.A.: Thermal characterization of 3D printable multifunctional graphene-reinforced polyethylene terephthalate glycol (PETG) composite filaments enabled for smart structural applications. Polym. Eng. Sci. **63**(9), 2841–2856 (2023). https://doi.org/10.1002/pen.26409
54. Fang, B., Zhang, G., Zou, F., Wang, Y., Dou, B., Zhang, H., Guo, J., Xu, J.: 3D printing interconnected segregated composites to simultaneously enhance thermal conductivity and mechanical properties. ACS Appl. Polymer Mater. **6**(8), 4904–4911 (2024). https://doi.org/10.1021/acsapm.4c00654
55. Mora, A., Verma, P., Kumar, S.: Electrical conductivity of CNT/polymer composites: 3D printing, measurements and modeling. Compos. B Eng. **183**, 107600 (2019). https://doi.org/10.1016/j.compositesb.2019.107600
56. Vidakis, N., Petousis, M., Tzounis, L., Velidakis, E., Mountakis, N., Grammatikos, S.A.: Polyamide 12/Multiwalled Carbon nanotube and Carbon Black nanocomposites manufactured by 3D printing Fused Filament Fabrication: A comparison of the electrical, thermoelectric,

and mechanical properties. C—J. Carbon Res. **7**(2), 38 (2021). https://doi.org/10.3390/c7020038
57. Alam, F., Varadarajan, K., Kumar, S.: 3D printed polylactic acid nanocomposite scaffolds for tissue engineering applications. Polym. Testing **81**, 106203 (2019). https://doi.org/10.1016/j.polymertesting.2019.106203
58. Tzounis, L., Petousis, M., Grammatikos, S., Vidakis, N.: 3D printed thermoelectric polyurethane/multiwalled carbon nanotube nanocomposites: a novel approach towards the fabrication of flexible and stretchable organic thermoelectrics. Materials **13**(12), 2879 (2020). https://doi.org/10.3390/ma13122879
59. Yang, L., Chen, Y., Wang, M., Shi, S., Jing, J.: Fused deposition modeling 3D printing of novel poly(vinyl alcohol)/graphene nanocomposite with enhanced mechanical and electromagnetic interference shielding properties. Ind. Eng. Chem. Res. **59**(16), 8066–8077 (2020). https://doi.org/10.1021/acs.iecr.0c00074
60. Wang, Y., Lei, M., Wei, Q., Wang, Y., Zhang, J., Guo, Y., Saroia, J.: 3D printing biocompatible l-Arg/GNPs/PLA nanocomposites with enhanced mechanical property and thermal stability. J. Mater. Sci. **55**(12), 5064–5078 (2020). https://doi.org/10.1007/s10853-020-04353-8
61. Silva, M., Pinho, I.S., Covas, J.A., Alves, N.M., Paiva, M.C.: 3D printing of graphene-based polymeric nanocomposites for biomedical applications. Functional Compos. Mater. **2**(1) (2021). https://doi.org/10.1186/s42252-021-00020-6
62. Murugesan, S., Scheibel, T.: Copolymer/Clay nanocomposites for biomedical applications. Adv. Funct. Mater. **30**(17) (2020). https://doi.org/10.1002/adfm.201908101
63. Tyagi, N., Bhardwaj, V., Sharma, D., Tomar, R., Chaudhary, V., Khanuja, M., Singh, M.K., Sharma, G.: 3D printing technology in the pharmaceutical and biomedical applications: a critical review **2**(1), 178–190 (2023). https://doi.org/10.1007/s44174-023-00105-9
64. Hollister, S., Maddox, R., Taboas, J.: Optimal design and fabrication of scaffolds to mimic tissue properties and satisfy biological constraints. Biomaterials **23**(20), 4095–4103 (2002). https://doi.org/10.1016/s0142-9612(02)00148-5
65. Khodaei, M., Razavi, H., Nosrati, H.: Effects of ultrasonic vibration on 3D printing of poly-lactic Acid/Akermanite nanocomposite scaffolds. Heliyon **10**(20), e39240 (2024). https://doi.org/10.1016/j.heliyon.2024.e39240
66. Feng, Z., Li, Y., Hao, L., Yang, Y., Tang, T., Tang, D., Xiong, W.: Graphene-Reinforced biodegradable resin composites for stereolithographic 3D printing of bone structure scaffolds. J. Nanomater. **2019**, 1–13 (2019). https://doi.org/10.1155/2019/9710264
67. Qiu, K., Haghiashtiani, G., McAlpine, M.C.: 3D printed organ models for surgical applications. Annu. Rev. Anal. Chem. **11**(1), 287–306 (2018). https://doi.org/10.1146/annurev-anchem-061417-125935
68. Dodziuk, H.: Applications of 3D printing in healthcare. Polish J. Cardio-Thoracic Surgery **3**, 283–293 (2016). https://doi.org/10.5114/kitp.2016.62625
69. Auricchio, F., Marconi, S.: 3D printing: clinical applications in orthopaedics and traumatology. EFORT Open Rev. **1**(5), 121–127 (2016). https://doi.org/10.1302/2058-5241.1.000012
70. Mertz, L.: Dream it, design it, print it in 3-D: What can 3-D printing do for you? IEEE Pulse **4**(6), 15–21 (2013). https://doi.org/10.1109/mpul.2013.2279616
71. Katseli, V., Economou, A., Kokkinos, C.: Single-step fabrication of an integrated 3D-printed device for electrochemical sensing applications. Electrochem. Commun. **103**, 100–103 (2019). https://doi.org/10.1016/j.elecom.2019.05.008
72. Hendrikson, W.J., Rouwkema, J., Clementi, F., Van Blitterswijk, C.A., Farè, S., Moroni, L.: Towards 4D printed scaffolds for tissue engineering: exploiting 3D shape memory polymers to deliver time-controlled stimulus on cultured cells. Biofabrication **9**(3), 031001 (2017). https://doi.org/10.1088/1758-5090/aa8114
73. Nadernezhad, A., Khani, N., Skvortsov, G.A., Toprakhisar, B., Bakirci, E., Menceloglu, Y., Unal, S., Koc, B.: Multifunctional 3D printing of heterogeneous hydrogel structures. Scientific Reports **6**(1) (2016). https://doi.org/10.1038/srep33178
74. Hernandez, I., Kumar, A., Joddar, B.: A bioactive hydrogel and 3D printed polycaprolactone system for bone tissue engineering. Gels **3**(3), 26 (2017). https://doi.org/10.3390/gels3030026

75. Sultan, S., Mathew, A.P.: 3D printed scaffolds with gradient porosity based on a cellulose nanocrystal hydrogel. Nanoscale **10**(9), 4421–4431 (2018). https://doi.org/10.1039/c7nr08966j
76. Mannoor, M.S., Jiang, Z., James, T., Kong, Y.L., Malatesta, K.A., Soboyejo, W.O., Verma, N., Gracias, D.H., McAlpine, M.C.: 3D printed Bionic ears. Nano Lett. **13**(6), 2634–2639 (2013). https://doi.org/10.1021/nl4007744
77. Gao, C., Liu, T., Shuai, C., Peng, S.: Enhancement mechanisms of graphene in nano-58S bioactive glass scaffold: mechanical and biological performance. Scientific Reports **4**(1) (2014). https://doi.org/10.1038/srep04712
78. Ho, C.M.B., Mishra, A., Lin, P.T.P., Ng, S.H., Yeong, W.Y., Kim, Y., Yoon, Y.: 3D printed polycaprolactone carbon nanotube composite scaffolds for cardiac tissue engineering. Macromolecular Bioscience **17**(4) (2016). https://doi.org/10.1002/mabi.201600250
79. Holmes, B., Bulusu, K., Plesniak, M., Zhang, L.G.: A synergistic approach to the design, fabrication and evaluation of 3D printed micro and nano featured scaffolds for vascularized bone tissue repair. Nanotechnology **27**(6), 064001 (2016). https://doi.org/10.1088/0957-4484/27/6/064001
80. Castro, N.J., Patel, R., Zhang, L.G.: Design of a novel 3D printed bioactive nanocomposite scaffold for improved osteochondral regeneration. Cell. Mol. Bioeng. **8**(3), 416–432 (2015). https://doi.org/10.1007/s12195-015-0389-4
81. Ryan, K.R., Down, M.P., Hurst, N.J., Keefe, E.M., Banks, C.E.: Additive manufacturing (3D printing) of electrically conductive polymers and polymer nanocomposites and their applications. eScience **2**(4), 365–381 (2022). https://doi.org/10.1016/j.esci.2022.07.003
82. Kuzhandaivel, H., Manickam, S., Balasingam, S.K., Franklin, M.C., Kim, H., Nallathambi, K.S.: Sulfur and nitrogen-doped graphene quantum dots/PANI nanocomposites for supercapacitors. New J. Chem. **45**(8), 4101–4110 (2021). https://doi.org/10.1039/d1nj00038a
83. Liu, H., Xu, L., Tu, H., Luo, Z., Zhu, F., Deng, W., Zou, G., Hou, H., Ji, X.: Interfacial interaction of multifunctional GQDs reinforcing polymer electrolytes for All-Solid-State LI battery. Small **19**(33) (2023). https://doi.org/10.1002/smll.202301275
84. Mehrzad-Samarin, M., Faridbod, F., Ganjali, M.R.: A luminescence nanosensor for Ornidazole detection using graphene quantum dots entrapped in silica molecular imprinted polymer. Spectrochim. Acta Part A Mol. Biomol. Spectrosc. **206**, 430–436 (2018). https://doi.org/10.1016/j.saa.2018.08.026
85. M, V., H, V., P, A. S., H, D.: Effect of different gel electrolytes on conjugated polymer—graphene quantum dots based electrode for solid state hybrid supercapacitors. Polymer-Plastics Technol. Mater. **59**(18), 2068–2075 (2020). https://doi.org/10.1080/25740881.2020.1784221
86. Fan, F., Zhang, B., Cao, Y., Yang, X., Gu, J., Chen, Y.: Conjugated polymer covalently modified graphene oxide quantum dots for ternary electronic memory devices. Nanoscale **9**(30), 10610–10618 (2017). https://doi.org/10.1039/c7nr02809a
87. Luo, X., Wan, R., Zhang, Z., Song, M., Yan, L., Xu, J., Yang, H., Lu, B.: 3D-Printed Hydrogel-Based flexible electrochromic device for wearable displays. Adv. Sci. **11**(38) (2024b). https://doi.org/10.1002/advs.202404679
88. Rahmatabadi, D., Khajepour, M., Bayati, A., Mirasadi, K., Yousefi, M.A., Shegeft, A., Ghasemi, I., Baniassadi, M., Abrinia, K., Bodaghi, M., Baghani, M.: Advancing sustainable shape memory polymers through 4D printing of polylactic acid-polybutylene adipate terephthalate blends. Eur. Polymer J. **216**, 113289 (2024). https://doi.org/10.1016/j.eurpolymj.2024.113289
89. Rahmatabadi, D., Mirasadi, K., Bayati, A., Khajepour, M., Ghasemi, I., Baniassadi, M., Abrinia, K., Bodaghi, M., Baghani, M.: 4D printing thermo-magneto-responsive PETG-Fe_3O_4 nanocomposites with enhanced shape memory effects. Appl. Mater. Today **40**, 102361 (2024). https://doi.org/10.1016/j.apmt.2024.102361
90. Zhang, R., Tian, J., Wu, Y., Chou, W., Yang, J., Xue, P.: An investigation on shape memory behaviors of UHMWPE-based nanocomposites reinforced by graphene nanoplatelets. Polym. Testing **99**, 107217 (2021). https://doi.org/10.1016/j.polymertesting.2021.107217

91. Thostenson, E., Li, C., Chou, T.: Nanocomposites in context. Compos. Sci. Technol. **65**(3–4), 491–516 (2004). https://doi.org/10.1016/j.compscitech.2004.11.003
92. Zhang, Z., Demir, K.G., Gu, G.X.: Developments in 4D-printing: a review on current smart materials, technologies, and applications. Int. J. Smart Nano Mater. **10**(3), 205–224 (2019). https://doi.org/10.1080/19475411.2019.1591541
93. Pinho, A., Buga, C., Piedade, A.: The chemistry behind 4D printing. Appl. Mater. Today **19**, 100611 (2020). https://doi.org/10.1016/j.apmt.2020.100611
94. Lui, Y.S., Sow, W.T., Tan, L.P., Wu, Y., Lai, Y., Li, H.: 4D printing and stimuli-responsive materials in biomedical aspects. Acta Biomater. **92**, 19–36 (2019). https://doi.org/10.1016/j.actbio.2019.05.005
95. Ge, Q., Dunn, C.K., Qi, H.J., Dunn, M.L.: Active origami by 4D printing. Smart Mater. Struct. **23**(9), 094007 (2014). https://doi.org/10.1088/0964-1726/23/9/094007
96. Mao, Y., Yu, K., Isakov, M.S., Wu, J., Dunn, M.L., Qi, H.J.: Sequential Self-Folding structures by 3D printed digital shape memory polymers. Scientific Reports **5**(1) (2015). https://doi.org/10.1038/srep13616
97. Tao, R., Ji, L., Li, Y., Wan, Z., Hu, W., Wu, W., Liao, B., Ma, L., Fang, D.: 4D printed origami metamaterials with tunable compression twist behavior and stress-strain curves. Compos. B Eng. **201**, 108344 (2020). https://doi.org/10.1016/j.compositesb.2020.108344
98. Roppolo, I., Chiappone, A., Angelini, A., Stassi, S., Frascella, F., Pirri, C.F., Ricciardi, C., Descrovi, E.: 3D printable light-responsive polymers. Mater. Horiz. **4**(3), 396–401 (2017). https://doi.org/10.1039/c7mh00072c
99. Jeong, H.Y., Lee, E., Ha, S., Kim, N., Jun, Y.C.: Multistable thermal actuators via multimaterial 4D printing. Adv. Mater. Technol. **4**(3) (2018). https://doi.org/10.1002/admt.201800495
100. Ali, M., Alam, F., Fah, Y.F., Shiryayev, O., Vahdati, N., Butt, H.: 4D printed thermochromic Fresnel lenses for sensing applications. Compos. B Eng. **230**, 109514 (2021). https://doi.org/10.1016/j.compositesb.2021.109514
101. Chen, H., Noirbent, G., Liu, S., Brunel, D., Graff, B., Gigmes, D., Zhang, Y., Sun, K., Morlet-Savary, F., Xiao, P., Dumur, F., Lalevée, J.: Bis-chalcone derivatives derived from natural products as near-UV/visible light sensitive photoinitiators for 3D/4D printing. Mater. Chem. Frontiers **5**(2), 901–916 (2020). https://doi.org/10.1039/d0qm00755b
102. Hagaman, D.E., Leist, S., Zhou, J., Ji, H.: Photoactivated polymeric bilayer actuators fabricated via 3D printing. ACS Appl. Mater. Interfaces. **10**(32), 27308–27315 (2018). https://doi.org/10.1021/acsami.8b08503
103. Choi, J., Kwon, O., Jo, W., Lee, H.J., Moon, M.: 4D Printing Technology: a review. 3D Printing Additive Manuf. **2**(4), 159–167. https://doi.org/10.1089/3dp.2015.0039
104. Dorishetty, P., Balu, R., Sreekumar, A., De Campo, L., Mata, J.P., Choudhury, N.R., Dutta, N.K.: Robust and tunable hybrid hydrogels from photo-cross-linked soy protein isolate and regenerated silk fibroin. ACS Sustain. Chem. Eng. **7**(10), 9257–9271 (2019). https://doi.org/10.1021/acssuschemeng.9b00147
105. Zhu, P., Yang, W., Wang, R., Gao, S., Li, B., Li, Q.: 4D printing of complex structures with a fast response time to magnetic stimulus. ACS Appl. Mater. Interfaces. **10**(42), 36435–36442 (2018). https://doi.org/10.1021/acsami.8b12853
106. Zhao, W., Zhang, F., Leng, J., Liu, Y.: Personalized 4D printing of bioinspired tracheal scaffold concept based on magnetic stimulated shape memory composites. Compos. Sci. Technol. **184**, 107866 (2019). https://doi.org/10.1016/j.compscitech.2019.107866
107. Ashammakhi, N., Ahadian, S., Zengjie, F., Suthiwanich, K., Lorestani, F., Orive, G., Ostrovidov, S., Khademhosseini, A.: Advances and future perspectives in 4D bioprinting. Biotechnol. J. **13**(12) (2018). https://doi.org/10.1002/biot.201800148
108. Wu, H., Zhang, X., Ma, Z., Zhang, C., Ai, J., Chen, P., Yan, C., Su, B., Shi, Y.: A material combination concept to realize 4D printed products with newly emerging property/functionality. Adv. Sci. **7**(9) (2020). https://doi.org/10.1002/advs.201903208
109. Adam, G., Benouhiba, A., Rabenorosoa, K., Clévy, C., Cappelleri, D.J.: 4D Printing: Enabling technology for microrobotics applications. Adv. Intell.Syst. **3**(5) (2021). https://doi.org/10.1002/aisy.202000216

110. Breger, J.C., Yoon, C., Xiao, R., Kwag, H.R., Wang, M.O., Fisher, J.P., Nguyen, T.D., Gracias, D.H.: Self-folding thermo-magnetically responsive soft microgrippers. ACS Appl. Mater. Interfaces. **7**(5), 3398–3405 (2015). https://doi.org/10.1021/am508621s
111. Li, T., Li, G., Liang, Y., Cheng, T., Dai, J., Yang, X., Liu, B., Zeng, Z., Huang, Z., Luo, Y., Xie, T., Yang, W.: Fast-moving soft electronic fish. Sci. Adv. **3**(4) (2017). https://doi.org/10.1126/sciadv.1602045
112. Hu, Y., Wang, Z., Jin, D., Zhang, C., Sun, R., Li, Z., Hu, K., Ni, J., Cai, Z., Pan, D., Wang, X., Zhu, W., Li, J., Wu, D., Zhang, L., Chu, J.: Botanical-Inspired 4D printing of hydrogel at the microscale. Adv. Functi. Mater. **30**(4) (2019). https://doi.org/10.1002/adfm.201907377
113. Falahati, M., Ahmadvand, P., Safaee, S., Chang, Y., Lyu, Z., Chen, R., Li, L., Lin, Y.: Smart polymers and nanocomposites for 3D and 4D printing. Mater. Today **40**, 215–245 (2020). https://doi.org/10.1016/j.mattod.2020.06.001
114. Mitchell, A., Lafont, U., Hołyńska, M., Semprimoschnig, C.: Additive manufacturing—a review of 4D printing and future applications. Addit. Manuf. **24**, 606–626 (2018). https://doi.org/10.1016/j.addma.2018.10.038
115. Chen, C., Zhang, M., Mujumdar, A.S., Phuhongsung, P.: Investigation of 4D printing of lotus root-compound pigment gel: effect of pH on rapid colour change. Food Res. Int. **148**, 110630 (2021). https://doi.org/10.1016/j.foodres.2021.110630
116. Shanthamma, S., Preethi, R., Moses, J.A., Anandharamakrishnan, C.: 4D Printing of sago starch with turmeric blends: a study on pH-triggered spontaneous color transformation. ACS Food Sci. Technol. **1**(4), 669–679 (2021). https://doi.org/10.1021/acsfoodscitech.0c00151
117. Chu, H., Yang, W., Sun, L., Cai, S., Yang, R., Liang, W., Yu, H., Liu, L.: 4D printing: a review on recent progresses. Micromachines **11**(9), 796 (2020). https://doi.org/10.3390/mi11090796
118. Shao, L., Zhao, B., Zhang, Q., Xing, Y., Zhang, K.: 4D printing composite with electrically controlled local deformation. Extreme Mech. Lett. **39**, 100793 (2020). https://doi.org/10.1016/j.eml.2020.100793
119. Goya, G.F., Asín, L., Ibarra, M.R.: Cell death induced by AC magnetic fields and magnetic nanoparticles: current state and perspectives. Int. J. Hyperth. **29**(8), 810–818 (2013). https://doi.org/10.3109/02656736.2013.838646
120. Meng, Q., Hu, J.: A review of shape memory polymer composites and blends. Compos. A Appl. Sci. Manuf. **40**(11), 1661–1672 (2009). https://doi.org/10.1016/j.compositesa.2009.08.011
121. Choong, Y.Y.C., Maleksaeedi, S., Eng, H., Yu, S., Wei, J., Su, P.: High speed 4D printing of shape memory polymers with nanosilica. Appl. Mater. Today **18**, 100515 (2019). https://doi.org/10.1016/j.apmt.2019.100515
122. Kausar, A., Ahmad, I., Zhao, T., Aldaghri, O., Eisa, M.H.: Polymer/graphene nanocomposites via 3D and 4D printing—design and technical potential. Processes **11**(3), 868 (2023). https://doi.org/10.3390/pr11030868
123. Yousefi, M.A., Rahmatabadi, D., Baniassadi, M., Bodaghi, M., Baghani, M.: 4D Printing of multifunctional and biodegradable PLA-PBAT-Fe_3O_4 nanocomposites with supreme mechanical and shape memory properties. Macromolecular Rapid Commun. (2024). https://doi.org/10.1002/marc.202400661
124. Namathoti, S., Vakkalagadda, M.R.K.: Development of multiwalled carbon nanotubes/halloysite nanotubes reinforced thermal responsive shape memory polymer nanocomposites for enhanced mechanical and shape recovery characteristics in 4D printing applications. Polymers **15**(6), 1371 (2023). https://doi.org/10.3390/polym15061371
125. Patel, K.K., Purohit, R.: Improved shape memory and mechanical properties of microwave-induced thermoplastic polyurethane/graphene nanoplatelets composites. Sens. Actuators A **285**, 17–24 (2018). https://doi.org/10.1016/j.sna.2018.10.049
126. Prasad, H.C., Hashmi, S.A.R., Naik, A., Bhargaw, H.N.: Improved shape memory effects in multiwalled-carbon-nano-tube reinforced thermosetting polyurethane composites. J. Appl. Polymer Sci. **134**(7) (2016). https://doi.org/10.1002/app.44389
127. Hashmi, S., Prasad, H.C., Abishera, R., Bhargaw, H.N., Naik, A.: Improved recovery stress in multi-walled-carbon-nanotubes reinforced polyurethane. Mater. Des. **1980–2015**(67), 492–500 (2014). https://doi.org/10.1016/j.matdes.2014.10.062

128. Patel, K.K., Purohit, R., Hashmi, S.A.R.: Microwave-triggered thermo-responsive shape memory polymer (Polyurethane)/GNPS/MWCNTS composites. Strength Mater. **53**(6), 998–1010 (2021). https://doi.org/10.1007/s11223-022-00367-6
129. Bi, H., Ye, G., Sun, H., Ren, Z., Gu, T., Xu, M.: Mechanically robust, shape memory, self-healing and 3D printable thermoreversible cross-linked polymer composites toward conductive and biomimetic skin devices applications. Addit. Manuf. **49**, 102487 (2021). https://doi.org/10.1016/j.addma.2021.102487
130. Rousseau, I.A.: Challenges of shape memory polymers: a review of the progress toward overcoming SMP's limitations. Polym. Eng. Sci. **48**(11), 2075–2089 (2008). https://doi.org/10.1002/pen.21213
131. Mu, T., Liu, L., Lan, X., Liu, Y., Leng, J.: Shape memory polymers for composites. Compos. Sci. Technol. **160**, 169–198 (2018). https://doi.org/10.1016/j.compscitech.2018.03.018
132. Duan, H., Gu, J., Sun, H., Zeng, H., Rodriguez-Morales, J.A.: A thermodynamic constitutive model based on uncoupled physical mechanisms for polymer-based shape memory composites and its application in 4D printing. Appl. Math. Model. **115926**,(2025). https://doi.org/10.1016/j.apm.2025.115926
133. Liu, Y., Wang, L., Liu, Y., Zhang, F., Leng, J.: Recent progress in shape memory polymer composites: driving modes, forming technologies, and applications. Compos. Commun. **51**, 102062 (2024). https://doi.org/10.1016/j.coco.2024.102062
134. Kumar, C.P., Sivanagaraju, N., Vakkalagadda, M.R.K.: Shape memory polymers and composites in 4D printing: techniques and applications. Prog. Addit. Manuf. (2025). https://doi.org/10.1007/s40964-025-01099-0
135. Prem Kumar, C., Sivanagaraju, N., Loknath, D., Mallikarjun Rao, G.N., Vakkalagadda, M.R.K.: Shape memory polymers, blends, and composites: processing, properties, and applications. Polymer-Plastics Technol. Mater. **1–29**,(2025). https://doi.org/10.1080/25740881.2025.2460063
136. Namathoti, S., Vakkalagadda, M.R.K.: Mechanical and shape recovery characterization of MWCNTs/HNTs-reinforced thermal-responsive shape-memory polymer nanocomposites. Polymers **15**(3), 710 (2023). https://doi.org/10.3390/polym15030710
137. Namathoti, S., Ps, R.S., Vakkalagadda, M.R.K.: Characterization of 4D Mprinted shape memory and conventional polymer blends for thermal responsive applications. J. Polymer Res. **31**(8) (2024). https://doi.org/10.1007/s10965-024-04074-5
138. Mostafaei, A., Elliott, A.M., Barnes, J.E., Li, F., Tan, W., Cramer, C.L., Nandwana, P., Chmielus, M.: Binder jet 3D printing—Process parameters, materials, properties, modeling, and challenges. Prog. Mater Sci. **119**, 100707 (2020). https://doi.org/10.1016/j.pmatsci.2020.100707
139. Gay, D.: Composite materials. In CRC Press eBooks (2014).https://doi.org/10.1201/b17106
140. Shafranek, R.T., Millik, S.C., Smith, P.T., Lee, C., Boydston, A.J., Nelson, A.: Stimuli-responsive materials in additive manufacturing. Prog. Polym. Sci. **93**, 36–67 (2019). https://doi.org/10.1016/j.progpolymsci.2019.03.002
141. Zhang, X., Han, Y., Zhu, M., Chu, Y., Li, W., Zhang, Y., Zhang, Y., Luo, J., Tao, R., Qi, J.: Bio-inspired 4D printed intelligent lattice metamaterials with tunable mechanical property. Int. J. Mech. Sci. **272**, 109198 (2024). https://doi.org/10.1016/j.ijmecsci.2024.109198
142. Bar-Cohen, Y., Anderson, I.A.: Electroactive polymer (EAP) actuators—background review. Mech. Soft Mater. **1**(1) (2019). https://doi.org/10.1007/s42558-019-0005-1
143. Miriyev, A., Stack, K., Lipson, H.: Soft material for soft actuators. Nature Commun. **8**(1) (2017). https://doi.org/10.1038/s41467-017-00685-3
144. Yu, K., Ritchie, A., Mao, Y., Dunn, M.L., Qi, H.J.: Controlled sequential shape changing components by 3D printing of shape memory polymer multimaterials. Procedia IUTAM **12**, 193–203 (2015). https://doi.org/10.1016/j.piutam.2014.12.021
145. Zarek, M., Layani, M., Cooperstein, I., Sachyani, E., Cohn, D., Magdassi, S.: 3D printing of shape memory polymers for flexible electronic devices. Adv. Mater. **28**(22), 4449–4454 (2015). https://doi.org/10.1002/adma.201503132

Nano Polymers Towards Advanced Energy Storage Devices

Barnika Chakraborty

Abstract The global energy demand has been steadily increasing in recent years, driven by the rapid growth of industries, urbanization (Vo et al. 2024), and population expansion (Jones and Warner 2016). This surge in energy consumption has highlighted the pressing need for the development of energy storage systems that are not only efficient and capable of storing high amounts of energy but also sustainable (Schainker) and environmentally friendly. Energy storage plays a critical role in balancing the supply and demand for electricity, especially in renewable energy systems where energy production is intermittent and variable. However, despite significant advancements in energy storage technologies, there remain substantial challenges related to energy density, stability, cost, and sustainability.

Keywords Energy storage devices · Next-generation batteries · Energy density · Renewable energy systems · Electrochemical stability

1 Introduction

The global energy demand has been steadily increasing in recent years, driven by the rapid growth of industries, urbanization (Vo et al. 2024), and population expansion (Jones and Warner 2016). This surge in energy consumption has highlighted the pressing need for the development of energy storage systems that are not only efficient and capable of storing high amounts of energy but also sustainable (Schainker) and environmentally friendly. Energy storage plays a critical role in balancing the supply and demand for electricity, especially in renewable energy systems where energy production is intermittent and variable. However, despite significant advancements in energy storage technologies, there remain substantial challenges related to energy density, stability, cost, and sustainability.

Traditional energy storage devices, such as lithium-ion batteries (LIBs), supercapacitors, and fuel cells, have made substantial strides in recent years (Amir et al.

B. Chakraborty (✉)
Institute for Materials Science, Kiel, Germany
e-mail: bach@tf.uni-kiel.de

B. Sharma et al. (eds.), *Nanostructured Polymers*, Engineering Materials,
https://doi.org/10.1007/978-981-95-3581-1_14

2023). They have provided reliable solutions for a variety of applications ranging from portable electronics to electric vehicles. Yet, these technologies still face limitations, including issues with energy capacity, lifespan, charging times, and environmental impact. For instance, while LIBs are widely used due to their relatively high energy density, concerns about their long-term stability, safety (Lain and Kendrick 2021), and the availability of raw materials for their production continue to pose challenges for their widespread adoption in large-scale energy storage applications [1]. Additionally, the use of fossil-fuel-derived energy sources and the demand for batteries containing rare minerals further heightens the need for sustainable alternatives.

In this context, nanopolymers (Hossain and Hoque 2018)—an emerging class of materials with remarkable physicochemical properties—have garnered considerable attention. These materials are composed of nanostructured polymers that exhibit unique characteristics, such as an extraordinarily high surface area, excellent electrical conductivity, and enhanced mechanical stability. These attributes make nanopolymers highly suitable for integration into energy storage systems, as they can significantly improve the performance of devices by increasing energy density, charge/discharge efficiency, and overall cycle life. Moreover, the adaptability of nanopolymers allows for the fine-tuning of their properties, enabling their use in a wide range of energy storage applications, from batteries to supercapacitors and beyond [2].

This chapter delves into the transformative role of nanopolymers in advancing energy storage technologies. It will explore the various synthesis methods used to fabricate nanopolymers, highlight their key properties that make them ideal candidates for energy storage applications, and review the current and potential future applications of these materials in enhancing the performance of energy storage devices. Furthermore, the chapter will address the future prospects of nanopolymers in energy storage, discussing the challenges and opportunities that lie ahead as researchers continue to explore new avenues for making energy storage more efficient, sustainable, and scalable. Through a comprehensive examination of these materials, we aim to provide insights into how nanopolymers could play a pivotal role in the future of energy storage technologies, helping to meet the growing global energy demand in a more sustainable manner.

2 Fundamentals of Nanopolymers

Nanopolymers are a category of materials with structures and properties at the nanoscale, typically between 1 and 100 nanometers. At this scale, the properties of the materials are substantially different from those at the bulk level, leading to enhanced performance in various applications, particularly in energy storage systems. They exhibit unique mechanical, electrical, and chemical properties due to their high surface area, molecular organization, and other quantum-scale phenomena, which make them ideal candidates for improving energy storage technologies (2018).

2.1 Definition and Classification

Nanopolymers can be defined as polymers with nanometer-scale structural features that enhance or enable specific functionalities critical to energy storage devices. Depending on their composition and structure, nanopolymers can be classified into several categories, each contributing differently to energy storage performance.

1. **Conductive Polymers**: Conductive polymers are organic polymers that possess intrinsic electrical conductivity. Normally, polymers are insulators, but conductive polymers achieve electrical conductivity when doped with particular ions or by incorporating conductive nanomaterials such as carbon nanotubes or graphene. These polymers can conduct electricity without requiring metallic components, offering advantages such as lightweight, flexibility, and processability (Nezakati et al. 2018). For energy storage systems, conductive polymers are integral to the development of electrodes and other components in batteries, supercapacitors, and fuel cells. Examples of conductive polymers include polypyrrole (PPy), polyaniline (PANI), and PEDOT:PSS (Poly (3,4-ethylenedioxythiophene) polystyrene sulfonate). These materials are especially relevant in supercapacitors, where rapid charge and discharge cycles are required.
2. **Polymer Nanocomposites**: These are materials in which a polymer matrix is reinforced with nanoparticles or nanofillers. The nanofillers are typically inorganic materials like metal oxide nanoparticles, carbon nanotubes (CNTs), graphene, or quantum dots. The addition of nanofillers into polymers enhances various properties such as electrical conductivity, mechanical strength, thermal stability, and chemical resistance. In energy storage devices, polymer nanocomposites are often used in the development of advanced electrodes and electrolytes (Tofighy and Mohammadi 2020). The combination of the polymer matrix and nanoparticles can lead to highly efficient, durable, and cost-effective materials for energy storage applications.
3. **Polymer Electrolytes**: Polymer electrolytes are polymers that conduct ions rather than electrons and are essential in energy storage devices like batteries, capacitors, and fuel cells. Polymer electrolytes are typically used in lithium-ion batteries (LIBs), solid-state batteries, and supercapacitors. These electrolytes are flexible, lightweight, and have the ability to be tailored for high ionic conductivity. Solid polymer electrolytes (SPEs), gel polymer electrolytes (GPEs), and ionic liquid-based polymer electrolytes are used in a variety of energy storage applications (Asthana and Pal 2020). One of the significant advantages of polymer electrolytes is their ability to form solid or semi-solid electrolytic media that can replace liquid electrolytes, thus reducing leakage and enhancing safety.

The classification of nanopolymers allows engineers and scientists to select the appropriate material for specific applications, depending on the desired properties (conductivity, strength, flexibility, etc.) and the challenges of the energy storage device in question.

2.2 Synthesis Methods

The synthesis of nanopolymers involves various techniques that are designed to achieve specific properties, such as high surface area, conductivity, porosity, and enhanced mechanical stability (Getahun et al. 2024). These methods provide control over the material's morphology, which is essential for their use in energy storage applications. The key synthesis methods include:

1. **Electrospinning**: Electrospinning is a method for producing polymer nanofibers by applying a high-voltage electric field to a polymer solution. This process pulls the polymer solution into fine fibers that can range from a few nanometers to several microns in diameter, depending on the parameters of the electrospinning process. The resulting nanofibers exhibit extremely high surface area and porosity, which are crucial for applications in energy storage devices like supercapacitors and lithium-ion batteries. The high surface area facilitates greater ion exchange and improved charge/discharge efficiency. Furthermore, electrospun nanofibers are also mechanically strong and flexible, which is advantageous for wearable electronics and flexible energy storage devices.
2. **Sol-Gel Processing**: Sol-gel processing is a technique used to produce nanostructured materials through the transition of a sol (a colloidal suspension of solid particles in a liquid) to a gel and ultimately to a solid. In the sol-gel process, metal alkoxides or other precursors are dissolved in a solvent and then undergo hydrolysis and polycondensation reactions to form a gel. The gel undergoes thermal treatment to create a solid, nanoporous polymer matrix. Sol-gel processing allows for fine control over the porosity, structure, and composition of the nanomaterials, making it ideal for producing high-surface-area polymer electrolytes and electrodes used in energy storage devices. This method is frequently used for producing coatings, thin films, and composite materials.
3. **Chemical Vapor Deposition (CVD)**: Chemical Vapor Deposition (CVD) is a widely used technique for synthesizing thin films of polymers by chemically reacting gaseous precursors to form solid materials on a substrate. The CVD process allows for the precise control of the film's thickness, composition, and properties, and it is especially effective for producing conductive polymer films used in energy storage devices. These films are often used as electrode materials in batteries and supercapacitors because of their ability to provide high conductivity and stability. CVD is particularly useful in the fabrication of nanopolymer coatings or layers that improve energy storage efficiency by enabling rapid electron flow.
4. **In-Situ Polymerization**: In-situ polymerization involves the polymerization of monomers in the presence of nanoparticles or nanomaterials, allowing the formation of composite nanomaterials. This method enables the nanoparticles to be uniformly dispersed within the polymer matrix during the polymerization process, leading to improved material properties. This approach is commonly

used to integrate conductive fillers, such as CNTs, graphene, or metal nanoparticles, into a polymer matrix to enhance the overall conductivity and mechanical strength of the resulting material. In energy storage devices, this technique can improve the performance of electrodes and electrolytes by improving their electrochemical properties, mechanical strength, and charge/discharge rates.

Advancements in **3D printing** and **self-assembly methods** have further expanded the potential applications of nanopolymers. 3D printing enables the fabrication of complex nanopolymer structures with precise control over the geometry, which is particularly useful for prototyping energy storage devices. Self-assembly methods rely on the inherent chemical interactions between molecules or particles to organize themselves into specific structures, which can also be applied to the synthesis of nanopolymers for energy storage applications.

2.3 Structural and Functional Properties

Nanopolymers possess unique structural and functional properties due to their nanoscale dimensions, which make them highly efficient and effective for use in energy storage applications (Anbusagar et al. 2018). These properties include:

1. **High Conductivity**: One of the defining characteristics of nanopolymers is their ability to conduct electricity. This is especially important for energy storage devices like batteries, supercapacitors, and fuel cells. In conductive polymers, the electrical conductivity arises due to the presence of conjugated double bonds within the polymer chain, which allows for the flow of charge carriers. In nanocomposite materials, the incorporation of conductive nanomaterials such as carbon nanotubes or graphene enhances the overall conductivity of the polymer matrix. The high conductivity enables faster charge and discharge cycles in energy storage devices and ensures efficient energy storage and release.
2. **Enhanced Mechanical Strength**: Nanopolymers often exhibit superior mechanical properties compared to bulk polymers, thanks to the incorporation of nanoparticles that reinforce the polymer matrix. The incorporation of nanoparticles such as CNTs or graphene can significantly improve the tensile strength, modulus, and durability of the material. This is crucial in energy storage systems where repeated mechanical stress (e.g., bending, compression) could otherwise lead to degradation of the material. The enhanced mechanical strength increases the lifespan and reliability of energy storage devices, especially in applications like flexible electronics or wearable devices.
3. **Flexibility and Lightweight Nature**: Many nanopolymers are inherently flexible and lightweight, which makes them ideal for applications in flexible electronics, wearable devices, and portable energy storage systems. The lightweight nature reduces the overall weight of energy storage devices, which is crucial in applications like electric vehicles or mobile electronics. The flexibility allows for the development of stretchable batteries or supercapacitors that can conform

to various shapes, such as curved surfaces or human skin, which is essential in wearable energy storage devices.

4. **Electrochemical Stability**: Electrochemical stability refers to the ability of the material to maintain its performance over many charge/discharge cycles without significant degradation. This is a critical property for energy storage devices that are designed for long-term use. Nanopolymers exhibit excellent electrochemical stability, thanks to their strong molecular structure and resistance to chemical degradation. This ensures that energy storage devices using nanopolymers can operate efficiently for an extended period, reducing the need for frequent replacements or maintenance.
5. **High Surface-to-Volume Ratio**: Due to their nanoscale size, nanopolymers have an extremely high surface-to-volume ratio, which increases the material's interaction with ions and other electrochemical species in energy storage systems. This feature facilitates faster ion diffusion and more efficient charge/discharge cycles. In applications such as supercapacitors and lithium-ion batteries, the increased surface area leads to greater storage capacity and faster energy transfer, improving the overall performance of the device.

In conclusion, nanopolymers are rapidly gaining attention as a transformative material for improving the efficiency, sustainability, and performance of energy storage devices. Their unique properties, including high conductivity, mechanical strength, flexibility, and electrochemical stability, make them highly suitable for the next generation of energy storage technologies. The ability to synthesize these materials using advanced methods such as electrospinning, sol-gel processing, CVD, and in-situ polymerization allows for precise control over their properties, enabling their application across a wide range of energy storage systems. As research and development in this field continue to evolve, nanopolymers are expected to play a pivotal role in shaping the future of energy storage technologies.

3 Nanopolymer-Based Energy Storage Devices

Nanopolymers are playing an increasingly important role in enhancing the performance of various energy storage technologies. By incorporating nanopolymeric materials into the components of these devices, energy storage systems can benefit from improved efficiency, faster charging and discharging capabilities, and better stability over time (Sonika et al. 2022). Below is a detailed exploration of how nanopolymers are contributing to the development and optimization of four major energy storage devices: **Lithium-Ion Batteries (LIBs), Sodium-Ion Batteries (SIBs), Supercapacitors, and Fuel Cells**.

3.1 Lithium-Ion Batteries (LIBs)

Lithium-Ion Batteries (LIBs) are one of the most common and widely used energy storage devices, particularly in portable electronics and electric vehicles (EVs), due to their high energy density, long cycle life, and relatively low self-discharge rates (Kebede et al. 2022). However, their performance can be further improved with the incorporation of nanopolymers. Nanopolymers enhance the different components of LIBs, including the electrolytes, binders, and electrodes.

1. **Polymer Electrolytes: Improving Ionic Conductivity and Safety** Polymer electrolytes are used in LIBs to facilitate the movement of lithium ions between the cathode and anode. Traditional liquid electrolytes, though widely used, are often flammable and can pose safety risks. Nanopolymers, particularly solid or gel polymer electrolytes (GPEs), are being researched as alternatives that offer improved ionic conductivity, flexibility, and better safety features. By incorporating nanoparticles such as ceramic fillers, these nanopolymers help achieve high ionic conductivity while maintaining electrochemical stability. This reduces the risk of leakage or short-circuiting, making the battery safer while improving its performance by enhancing ion diffusion and increasing cycle stability.
2. **Electrode Binders: Enhancing Adhesion and Mechanical Integrity** In LIBs, binders are crucial for maintaining the mechanical integrity of the electrode materials. Nanopolymers serve as advanced binders that improve the adhesion between active materials (such as graphite or lithium iron phosphate) and the current collectors. The high surface area and mechanical properties of nanopolymers enhance the bond between the electrode materials, improving the overall strength and stability of the electrode. This ensures that the electrode materials stay intact during charge and discharge cycles, reducing the risk of capacity degradation and improving the overall lifespan of the battery.
3. **Nanocomposite Cathodes: Increasing Energy Density and Stability** Nanopolymers also contribute to the enhancement of the cathode materials in LIBs. By integrating nanomaterials (such as carbon nanotubes, graphene, or metal oxide nanoparticles) with the polymer matrix, the resulting nanocomposite cathodes exhibit higher energy density and enhanced stability. The high surface area and improved conductivity of nanocomposites help improve the overall electrochemical performance of the battery, enabling faster electron flow and increasing the charge storage capacity. These nanocomposite cathodes can also offer better cycling stability, reducing capacity loss over time and increasing the longevity of the battery.

Improved Fast Charging Capabilities: Nanopolymers also contribute to the enhancement of fast charging capabilities in LIBs. By improving the ionic conductivity and structural integrity of the electrodes, electrolytes, and binders, nanopolymers facilitate faster ion diffusion, allowing for quicker charge and discharge cycles. This improves the practical applicability of LIBs in devices that require quick recharging times, such as electric vehicles and portable electronics.

3.2 Sodium-Ion Batteries (SIBs)

Sodium-Ion Batteries (SIBs) are gaining traction as an alternative to lithium-ion batteries, particularly due to the abundance and lower cost of sodium. While SIBs offer similar performance to LIBs, they also face challenges, such as lower energy density and reduced cycling stability (Wu et al. 2024). Nanopolymers offer a promising approach to overcoming these challenges and improving the performance of SIBs in several key areas.

1. **Conductive Polymer Coatings: Facilitating Charge Transfer** One of the main limitations of SIBs is the relatively low conductivity of sodium ions compared to lithium ions, which can lead to slower charge and discharge rates. Nanopolymers, particularly conductive polymer coatings, can enhance the conductivity of the electrodes in SIBs by improving the overall charge transfer between the electrode materials and the electrolyte. By coating electrode materials like sodium-based oxides or carbon, conductive polymers create a conductive network that facilitates more efficient electron flow and ion movement. This helps improve the power density and overall performance of the battery.
2. **Gel Polymer Electrolytes: Enhancing Ionic Conductivity and Safety** Like LIBs, SIBs require electrolytes to allow the flow of ions between the cathode and anode. Gel polymer electrolytes (GPEs) are emerging as an attractive alternative to traditional liquid electrolytes, offering improved safety and stability. GPEs, when combined with sodium salts and nanoparticles, exhibit high ionic conductivity and are more resistant to leakage and thermal instability. These nanopolymer-based gel electrolytes enhance the overall performance of the SIB by improving cycling stability, safety, and the battery's capacity retention over extended use.
3. **Stabilization of Electrode Interfaces: Enhancing Cycling Performance and Capacity Retention** The performance of SIBs is often limited by the stability of the electrode interfaces, particularly during repeated charge and discharge cycles. Nanopolymers help stabilize the interfaces between the electrode materials and the electrolyte, reducing the formation of unwanted byproducts and minimizing degradation. This stabilization improves the cycling performance of the battery and enhances its capacity retention. By controlling the interfacial interactions at the nanoscale, nanopolymers can significantly increase the lifespan and efficiency of SIBs.

3.3 Supercapacitors

Supercapacitors, or ultracapacitors, are energy storage devices known for their ability to store and release energy very quickly. They have high power density but typically lower energy density compared to batteries (Olabi et al. 2022). Nanopolymers are playing a key role in enhancing the performance of supercapacitors, particularly in terms of capacitance, power density, and cycling stability.

1. **Forming Conductive Polymer Networks: Increasing Capacitance and Charge Storage** Supercapacitors rely on the electrostatic storage of energy in the electric field between two electrodes. Nanopolymers, particularly conductive polymers such as polyaniline and polypyrrole, can be used to form high-conductivity networks in the electrode materials. The incorporation of nanofillers like carbon nanotubes or graphene within the polymer matrix can significantly increase the overall surface area of the electrodes, allowing for more efficient energy storage. The larger surface area provides more sites for charge accumulation, resulting in increased capacitance and overall energy storage capacity.
2. **Enhancing Electrode Stability: Reducing Degradation Over Cycles** One of the key challenges in supercapacitors is the degradation of electrode materials over repeated charge and discharge cycles. Nanopolymers improve the mechanical properties of the electrodes, reducing the likelihood of cracking or breaking during cycling. By increasing the mechanical integrity and stability of the electrodes, nanopolymers ensure that supercapacitors maintain their performance over a large number of cycles, which is essential for applications in which long cycle life is required, such as in electric vehicles, renewable energy systems, and portable devices.
3. **Providing High Power Density: Facilitating Rapid Charge-Discharge Cycles Without Compromising Energy Efficiency** Supercapacitors are often preferred for applications that require rapid charge-discharge cycles, and nanopolymers can significantly enhance their power density. By incorporating conductive polymers and nanomaterials into the electrode matrix, nanopolymers allow for faster electron and ion movement, reducing the internal resistance of the supercapacitor and facilitating high-power performance. This is essential for applications in which quick bursts of energy are needed, such as in power backup systems, regenerative braking systems in electric vehicles, and consumer electronics.

3.4 Fuel Cells

Fuel cells are devices that convert chemical energy into electrical energy through an electrochemical reaction, typically involving hydrogen or methanol. Proton exchange membrane fuel cells (PEMFCs) are among the most common types, and nanopolymers are vital for improving their performance (Smith 2000), particularly the proton exchange membranes (PEMs), which are responsible for conducting protons between the anode and cathode while preventing the passage of electrons.

1. **Improving Proton Conductivity: Enhancing Overall Efficiency** Proton exchange membranes in fuel cells need to exhibit high proton conductivity to ensure efficient energy conversion. Nanopolymers, particularly those based on sulfonated polymers, can improve the proton conductivity of the membrane, allowing for better fuel cell performance. The incorporation of nanomaterials such as silica, graphene oxide, or carbon nanotubes into the polymer matrix

enhances the proton transport pathways, resulting in a more efficient fuel cell system. This improvement helps to increase the overall efficiency of the fuel cell by reducing the resistance to proton movement.

2. **Enhancing Mechanical Durability: Prolonging Operational Lifespan** PEMs must maintain their mechanical integrity over the course of many cycles. Nanopolymers provide enhanced mechanical strength to the membranes, preventing them from becoming brittle or degrading over time. By incorporating nanofillers into the polymer matrix, the mechanical properties of the PEMs can be enhanced, improving their resistance to cracking, swelling, and other forms of degradation. This prolongs the operational lifespan of fuel cells and ensures they can function efficiently for extended periods.
3. **Reducing Catalyst Degradation: Ensuring Long-Term Performance with Minimal Efficiency Losses** In fuel cells, catalysts are responsible for facilitating the chemical reactions that generate electricity. However, catalyst degradation can occur over time, reducing the efficiency and lifespan of the fuel cell. Nanopolymers can help reduce catalyst degradation by providing a stable and protective environment for the catalyst particles. The nanopolymer matrix acts as a barrier, preventing the catalyst from leaching or undergoing deactivation, ensuring long-term performance and minimal loss of efficiency.

In summary, nanopolymers play a crucial role in enhancing the performance, safety, and longevity of energy storage devices, from **lithium-ion batteries** to **fuel cells**. By improving various components like electrolytes, electrodes, binders, and membranes, nanopolymers enable faster charging, higher capacity, and better durability across these technologies, paving the way for more efficient and sustainable energy storage solutions in the future.

4 Recent Advances and Innovations

Nanopolymers have seen significant advancements in recent years, pushing the boundaries of energy storage technologies. These innovations span various areas, from the development of high-performance conductive nanocomposites to bio-inspired polymers that offer sustainable solutions. Below, we explore some of the most exciting advances in nanopolymer-based energy storage systems.

4.1 Conductive Polymer Nanocomposites

One of the most promising areas of nanopolymer development is the creation of **conductive polymer nanocomposites** (Idumah 2021). These materials have demonstrated considerable improvements in conductivity, capacitance, and overall performance in energy storage devices such as batteries and supercapacitors.

1. **Polyaniline (PANI) and Polypyrrole (PPy) Nanocomposites**: Among the conductive polymers, **polyaniline (PANI)** and **polypyrrole (PPy)** have been extensively researched due to their excellent conductivity and electrochemical properties. These polymers have been used in a variety of energy storage applications, including supercapacitors and batteries, where their high capacitance is a critical factor for enhancing energy storage. PANI and PPy, when combined with nanomaterials, exhibit significantly enhanced charge storage capacities and provide an avenue for increasing the energy density of supercapacitors and batteries.
2. **Hybrid Nanocomposites**: Recent advancements have focused on the development of **hybrid nanocomposites** that combine conductive polymers with high-performance nanomaterials like **graphene**, **carbon nanotubes (CNTs)**, and **metal oxides**. These hybrid materials leverage the unique properties of the individual components to enhance the overall performance. For example, **graphene** and **carbon nanotubes** offer high conductivity and mechanical strength, while **metal oxides** can contribute to higher energy storage capacities and stability. By integrating these materials into a polymer matrix, the resulting composites demonstrate superior **charge storage capacity**, **mechanical resilience**, and **electrochemical stability** compared to their pure counterparts.
3. **Flexibility and Tunability**: One of the most exciting features of conductive polymer nanocomposites is their **flexibility** and **tunability**. Unlike traditional inorganic materials used in energy storage devices, these hybrid nanocomposites can be easily molded and adapted to various form factors, making them ideal for applications in **wearable electronics** and **flexible energy storage devices**. The **tunable electrical properties** of these materials also allow for custom designs that can meet the specific demands of different energy storage systems, from high power-density supercapacitors to high energy-density batteries.
4. As a result of these advances, conductive polymer nanocomposites are now considered **highly attractive candidates** for next-generation energy storage devices that require **lightweight**, **flexible**, and **efficient** materials capable of sustaining high performance over long periods.

4.2 Polymer Electrolytes for Solid-State Batteries

Solid-state batteries represent one of the most significant advancements in energy storage technologies, offering the potential for improved safety, higher energy density, and better overall performance compared to traditional liquid electrolyte-based batteries. Polymer electrolytes, which are a key component of solid-state batteries (Song et al. 2023), have seen substantial innovation in recent years.

1. **Enhanced Safety and Energy Density**: The primary advantage of solid-state batteries is their ability to operate at higher voltages without the risks associated with liquid electrolytes, such as leakage or thermal runaway. Solid-state

batteries using **polymer electrolytes** prevent the issues related to thermal instability, making them much safer for use in various applications, particularly in **electric vehicles** and **portable electronics**. By eliminating the flammable liquid electrolyte, solid-state batteries significantly reduce the risk of short circuits, fires, or even explosions.
2. **Cross-Linked Polymer Matrices**: To optimize the ionic conductivity of polymer electrolytes while maintaining their **mechanical integrity**, researchers have explored the use of **cross-linked polymer matrices**. These matrices enhance the mechanical properties of the polymer electrolyte, providing improved **stability** and **durability**. Cross-linking, which involves chemically bonding polymer chains together, helps create a more robust network that resists deformation during the cycling process of a battery. This is especially critical in **solid-state lithium-ion batteries**, where maintaining structural integrity over thousands of cycles is necessary for long-term reliability.
3. **Plasticized Polymer Blends**: Another innovative approach in solid-state polymer electrolytes is the use of **plasticized polymer blends**. By incorporating plasticizers, the flexibility and ionic conductivity of the polymer electrolyte are improved, allowing for better performance in real-world conditions. These plasticized blends are especially valuable for enabling high ionic transport in solid-state batteries, which is often a limitation in conventional solid-state systems. The plasticizers help optimize the interaction between lithium ions and the polymer chains, enhancing conductivity without compromising the mechanical properties of the electrolyte.
4. **High-Voltage Operation**: Recent advancements have also made it possible for solid-state batteries to operate at higher voltages, which is crucial for improving the **energy density** of the battery. New designs and formulations of polymer electrolytes have enabled better ion migration, even at higher voltages, allowing these solid-state batteries to achieve higher overall energy storage. This is a critical step toward the development of batteries that can power electric vehicles for longer distances, charge faster, and provide more efficient energy storage in portable electronics.

4.3 Self-Healing and Flexible Nanopolymers

As energy storage devices become more advanced and are used in more diverse applications, **self-healing** and **flexible** materials have emerged as innovative solutions (Tong et al. 2021) to improve device longevity, performance, and resilience.

1. **Self-Healing Nanopolymers**: **Self-healing polymers** are materials that can repair themselves after being damaged, which is a revolutionary advancement in the field of energy storage. In the context of batteries and supercapacitors, these self-healing properties can prevent the degradation of key components, such as the electrodes or electrolytes, and reduce material failure over time. This could

significantly increase the **lifetime** of energy storage devices, making them more cost-effective and reliable over extended periods.

2. Self-healing polymers can be incorporated into various components of an energy storage device, such as **electrolytes**, **electrodes**, and **membranes**, to mitigate the effects of wear and tear. When the material is damaged (for example, by cracking or puncturing), the polymer molecules re-bond and repair the damaged area autonomously. This capability is particularly useful in **lithium-ion batteries (LIBs)**, where repeated cycles of charging and discharging can cause cracks in the electrodes, leading to capacity loss.
3. **Flexible Nanopolymers**: The growing demand for flexible and portable electronics has driven the development of **flexible energy storage devices**. Nanopolymers play a crucial role in this area due to their **mechanical flexibility**, enabling energy storage systems to be integrated into **wearable electronics**, **smart textiles**, and **flexible displays**. Flexible supercapacitors, for example, are highly desirable for applications that require lightweight, thin, and bendable energy storage solutions.
4. Flexible nanopolymers not only enhance the mechanical properties of energy storage devices but also contribute to their **electrochemical performance**. They allow for efficient **charge storage** and **fast charge/discharge cycles**, while maintaining flexibility over a wide range of mechanical deformations. This makes them ideal for applications where the device will be subjected to bending, stretching, or other mechanical stresses.

4.4 Bio-Inspired and Green Nanopolymers

The shift toward **sustainability** and **green chemistry** has led to the development of **bio-inspired nanopolymers** that are derived from renewable resources (Ahmed et al. 2023). These materials aim to replace traditional petrochemical-based polymers, offering a more sustainable alternative while still providing the necessary performance for energy storage applications.

1. **Sustainable Sourcing**: Recent research has focused on **bio-based nanopolymers** derived from renewable sources such as **cellulose**, **chitosan**, and **lignin**. These natural polymers offer several advantages, including biodegradability, renewability, and lower environmental impact during production. For example, **cellulose-based polymers** are derived from plant fibers and have been explored as potential candidates for use in both **electrolytes** and **electrodes** in energy storage devices. Chitosan, a polymer derived from chitin (found in crustacean shells), has been studied for its electrochemical stability and potential for use in supercapacitors and batteries.
2. **Electrochemical Performance**: Bio-inspired polymers can provide high **electrochemical performance**, including high **ionic conductivity** and **energy storage capacity**, similar to their synthetic counterparts. By modifying the chemical structure of natural polymers and incorporating nanomaterials, researchers have

developed bio-based nanopolymers with **improved electrochemical stability**, **faster charge/discharge cycles**, and **longer cycle life**. These polymers can contribute to more environmentally friendly energy storage systems without compromising performance.

3. **Green Chemistry**: The development of green nanopolymers is in line with global efforts to reduce the environmental impact of materials used in energy storage. By utilizing bio-renewable resources and reducing reliance on petroleum-based products, bio-inspired nanopolymers are a key step toward creating more sustainable and **eco-friendly energy storage technologies**.

4.5 Computational Modeling and AI-Driven Material Design

The integration of **artificial intelligence (AI)** and **machine learning** with **computational modeling** is revolutionizing the way nanopolymers are designed for energy storage applications (Huang et al. 2024). By leveraging advanced computational techniques, researchers can predict the properties and behaviors of nanopolymeric materials before they are physically synthesized, accelerating the material development process.

1. **AI-Driven Material Design**: AI and machine learning algorithms can analyze vast amounts of data from previous experiments to predict which combinations of polymers, nanoparticles, and processing techniques will result in the best performance for a given application. This enables the design of nanopolymers with **optimized electrical conductivity**, **enhanced mechanical properties**, and **superior electrochemical performance**. Through AI-driven optimization, researchers can explore a much larger parameter space and identify novel nanopolymeric materials that may have otherwise been overlooked.
2. **Computational Modeling for Nanopolymer Properties**: Computational modeling allows researchers to simulate the **structural properties**, **ionic conductivity**, and **thermal stability** of nanopolymers under different conditions. These models can predict how nanopolymers will behave in real-world energy storage systems, reducing the need for costly and time-consuming experimental work. By simulating interactions at the molecular level, researchers can design nanopolymers that are tailored to specific applications, improving the overall efficiency and performance of energy storage devices.
3. **Accelerating Innovation**: The combination of AI, machine learning, and computational modeling is expected to drastically accelerate the pace of innovation in nanopolymer-based energy storage devices. As these tools become more advanced, they will enable researchers to rapidly prototype and optimize new materials, potentially leading to breakthroughs in energy storage technology.

These **recent advances** in nanopolymers have the potential to revolutionize the **energy storage industry**, enhancing performance, safety, sustainability, and cost-effectiveness across a range of technologies. The continued development and optimization of these materials will undoubtedly play a critical role in advancing the next generation of energy storage devices.

5 Challenges and Future Prospects

While nanopolymers have shown immense potential in improving the performance and efficiency of energy storage devices, there remain several challenges that must be addressed to fully realize their commercial potential. The challenges span various aspects, including stability, cost-effectiveness, and environmental sustainability. However, ongoing research is paving the way for overcoming these barriers, promising a bright future for nanopolymers in energy storage applications.

5.1 Stability and Scalability Issues

One of the key challenges in the development of nanopolymer-based energy storage devices is ensuring **stability** and **scalability**. While nanopolymers offer superior properties such as high conductivity, flexibility, and mechanical resilience, maintaining their **thermal** and **electrochemical stability** over extended use is a significant hurdle (Mahmud 2023).

1. **Thermal Stability**: Nanopolymers, particularly those based on organic materials, are often susceptible to **thermal degradation** at high temperatures. In energy storage systems, especially those used in electric vehicles or portable electronics, temperature fluctuations are common, and materials must withstand these conditions without losing performance. The degradation of nanopolymers at elevated temperatures can lead to **capacity fading**, **electrolyte breakdown**, or **structural damage**, thus reducing the overall life cycle of the device. Researchers are working on enhancing the **thermal stability** of nanopolymers by incorporating heat-resistant components or by modifying their chemical structure to withstand higher operational temperatures.
2. **Electrochemical Stability**: For energy storage devices like **batteries** and **supercapacitors**, the electrochemical stability of nanopolymers is equally crucial. Nanopolymers need to maintain their properties over thousands of **charge-discharge cycles** without significant degradation. Electrochemical stability is often compromised by **side reactions** at the electrode-electrolyte interface, leading to the formation of unwanted byproducts that can impede charge transport. This is especially important in **solid-state batteries**, where **polymer electrolytes** must remain stable at high voltages and current densities. To improve

electrochemical stability, researchers are exploring the modification of polymer backbones, incorporating **nanoparticles** to enhance conductivity, or designing **multifunctional** materials that can mitigate degradation.

3. **Scalability**: Another significant challenge for nanopolymers is their **scalability**. While nanopolymer-based materials may show excellent performance at the laboratory scale, scaling up the synthesis process for large-scale production presents several challenges. The current methods for synthesizing nanopolymers, such as **electrospinning**, **chemical vapor deposition** (**CVD**), and **sol-gel processing**, often require precise control over conditions such as temperature, pressure, and precursor concentration, making them difficult to scale up efficiently. Furthermore, **high-performance nanopolymers** may require **specialized equipment** and expensive raw materials, which increases production costs and complicates the commercialization process. As a result, scaling up the production of nanopolymer-based materials while maintaining their high performance remains a key challenge for their widespread use in energy storage applications.

Efforts are underway to develop more **efficient and cost-effective manufacturing processes** that can produce high-quality nanopolymers on a large scale. Innovations such as **continuous flow processing**, **automated fabrication**, and **low-cost raw materials** could help improve scalability and make nanopolymer-based energy storage devices more commercially viable.

5.2 Cost-Effectiveness

The **cost-effectiveness** of nanopolymers remains a significant challenge for their widespread adoption in energy storage technologies (Ahmed et al. 2024). While the performance benefits of nanopolymers are clear, the cost of their synthesis often limits large-scale commercialization. Several factors contribute to this issue:

1. **High Synthesis Costs**: The synthesis of high-performance nanopolymers typically involves expensive raw materials, specialized chemicals, and complex fabrication techniques. For instance, **conductive polymer nanocomposites** often require the integration of high-cost nanomaterials such as **graphene**, **carbon nanotubes (CNTs)**, or **metal oxides**. These materials are not only costly but also require precise handling and fabrication, which further drives up production costs. Additionally, some synthesis techniques used to produce nanopolymers, such as **chemical vapor deposition** (**CVD**) and **sol-gel processing**, require specialized equipment and energy-intensive processes, making them economically challenging for large-scale production.
2. **Research into Cost-Effective Techniques**: In response to this challenge, significant efforts are being made to identify and develop **cost-effective manufacturing methods** that can lower production costs without compromising the performance of nanopolymer-based energy storage devices. Some of these efforts include:

- **Green synthesis techniques**: These involve the use of **environmentally friendly methods** to synthesize nanopolymers, such as **bio-inspired processes, solvent-free reactions**, or using **renewable resources** as raw materials. These approaches can help reduce the overall cost of production while minimizing the environmental impact.
- **Scalable fabrication methods**: Advances in **3D printing** and **roll-to-roll processing** could provide **scalable** and **cost-effective** ways to produce nanopolymer-based energy storage devices. These techniques are capable of producing materials in large quantities at relatively low costs.
- **Use of cheaper raw materials**: Researchers are investigating the use of **lower-cost alternative materials** for producing nanopolymers that still maintain high performance. For instance, substituting **graphene** with **cheaper carbon-based materials** or replacing **precious metal catalysts** with **abundant materials** could significantly reduce material costs.

3. **Economic Viability**: Achieving cost-effectiveness will require balancing performance improvements with **manufacturing cost**. As the demand for energy storage devices increases, manufacturers will seek ways to optimize production processes and reduce costs to make nanopolymer-based devices more affordable for mass adoption. This will likely involve **optimizing supply chains**, leveraging economies of scale, and increasing production efficiency.

5.3 Environmental Impact and Sustainability

As the global demand for energy storage devices increases, there is growing pressure to develop materials that are both high-performing and **environmentally sustainable**. Traditional polymers used in energy storage devices, such as those based on **petrochemicals**, can have significant environmental impacts, from resource extraction to disposal. Nanopolymers, on the other hand, offer the potential for more **sustainable energy storage solutions**, but their environmental impact remains a concern (Gulati et al. 2024).

1. **Development of Biodegradable Nanopolymers**: One of the most promising trends in nanopolymer research is the development of **biodegradable nanopolymers** derived from renewable and sustainable resources. Materials like **cellulose**, **chitosan**, and **lignin**, which are sourced from **plants** and **biomass**, can be used to create biodegradable polymers that perform similarly to traditional synthetic polymers. these bio-based nanopolymers can break down naturally in the environment, reducing the risk of pollution and minimizing waste in landfills or oceans

   The development of **bio-based nanopolymers** for energy storage applications holds significant promise for reducing the overall environmental footprint of energy storage technologies. For instance, **cellulose-derived nanocomposites** could be used in batteries and supercapacitors, offering high performance while being more environmentally friendly. Moreover, the production of these bio-based materials typically involves less **energy-intensive processing**, further reducing their environmental impact.

2. **Recycling and Reusability**: In addition to biodegradability, efforts are also being made to develop **recyclable nanopolymers** that can be reused at the end of the life cycle of energy storage devices. Researchers are investigating **closed-loop recycling systems** for nanopolymers, where materials from used devices can be recovered and reincorporated into the production of new devices. This could greatly reduce the demand for new raw materials and minimize waste, contributing to a **circular economy** in energy storage technologies.
3. **Sustainable Sourcing**: To ensure the long-term sustainability of nanopolymers, research is also focused on sourcing materials in a **sustainable** and **ethical manner**. The extraction of raw materials for nanopolymers, such as rare earth metals or certain carbon-based nanomaterials, can have environmental and social impacts. By prioritizing **responsible sourcing** and considering the **life cycle** of nanopolymer materials, manufacturers can minimize the environmental impact of energy storage devices and align with global sustainability goals.
4. **Green Chemistry and Eco-Friendly Manufacturing**: **Green chemistry principles** are being incorporated into the design of nanopolymers and their synthesis processes to reduce their environmental impact. These approaches involve the use of non-toxic solvents, renewable feedstocks, and **low-energy synthesis methods** that minimize the consumption of natural resources and reduce harmful emissions. The future of nanopolymers in energy storage will likely depend on the ability to integrate these eco-friendly approaches to create both high-performance and sustainable materials.

5.4 14.6 Conclusion

Nanopolymers have emerged as a transformative class of materials with exceptional promise in revolutionizing energy storage technologies. Their ability to enhance key properties such as conductivity, mechanical stability, and sustainability positions them at the forefront of next-generation solutions for energy storage systems. With their versatile range of applications, nanopolymers are playing a crucial role in advancing **batteries**, **supercapacitors**, and **fuel cells**, offering enhanced performance, increased energy density, and longer lifespan compared to conventional materials. Recent advancements in **conductive polymer nanocomposites**, **polymer electrolytes**, and the integration of **artificial intelligence (AI)** in material design are expanding the capabilities of nanopolymer-based energy storage devices. These innovations allow for the fine-tuning of electrochemical properties, higher efficiency, and better overall system performance, paving the way for faster, more reliable, and safer energy storage solutions. Additionally, efforts in the development of **bio-inspired** and **self-healing nanopolymers** further underscore the potential for nanopolymers to not only enhance performance but also contribute to environmentally sustainable energy solutions.

However, despite the remarkable progress, there remain significant challenges that must be overcome to fully realize the widespread adoption of nanopolymers in energy storage applications. **Cost-effectiveness**, **scalability**, and **long-term stability** are crucial barriers that need to be addressed to ensure these materials can be produced at a commercially viable scale and maintain their performance over time. While the advanced fabrication methods and new synthetic strategies offer promising solutions, finding ways to make these technologies accessible and economically feasible for large-scale production will be key to their success.

Looking forward, **future research** must focus on developing **sustainable**, **high-performance nanopolymers** that strike a balance between **efficiency** and **environmental impact**. By exploring greener synthesis techniques, recyclable materials, and biodegradable polymers, researchers can create energy storage solutions that not only meet the growing demands of the energy sector but also align with global sustainability goals. Additionally, leveraging AI-driven tools to predict material properties and optimize designs could significantly accelerate the development of next-generation nanopolymer-based devices.

In conclusion, nanopolymers stand poised to play an essential role in the future of energy storage, offering solutions that promise both high performance and environmental responsibility. As research progresses and challenges are addressed, nanopolymers have the potential to revolutionize how we store and use energy, ultimately contributing to a more sustainable and energy-efficient future.

References

1. Goodenough, J.B., Park, K.S.: The li-ion rechargeable battery: a perspective. J. Am. Chem. Soc. **135**(4), 1167–1176 (2013)
2. Zhang, Q., et al.: Polymer nanocomposites for next-generation batteries. J. Mater. Chem. A. **7**(35), 20139–20167 (2019)
3. Aricò, A.S., et al.: Nanostructured materials for advanced energy conversion and storage devices. Nat. Nanotechnol. **10**(7), 611–620 (2015)
4. Cui, J., et al.: Self-healing polymers for advanced energy storage applications. Adv. Mater. **33**(12), 2007355 (2021)
5. Gao, X., et al.: Recent advances in sodium-ion battery materials and technology. Chem. Soc. Rev. **49**(2), 898–935 (2020)
6. Janek, J., Zeier, W.G.: A solid future for battery development. Nat. Energy. **1**(9), 16141 (2016)
7. Liew, K.M., et al.: Nanopolymers: properties, synthesis, and applications in energy storage. Prog. Polym. Sci. **105**, 101245 (2020)
8. Liu, Y., et al.: Sustainable nanopolymer materials for energy applications. ACS Sustain. Chem. Eng. **10**(3), 1834–1850 (2022)
9. Lu, Y., et al.: Electrospun nanopolymers for energy storage and conversion. Adv. Funct. Mater. **31**(23), 2101245 (2021)
10. Manthiram, A.: A reflection on lithium-ion battery cathode chemistry. Nat. Commun. **11**(1), 1550 (2020)
11. Shi, C., et al.: Conductive polymer nanocomposites for energy storage. Nano Energy. **90**, 106558 (2022)

12. Sun, Y., et al.: Advances in lithium-ion battery materials. Chem. Rev. **117**(5), 2647–2690 (2017)
13. Wang, J., et al.: Nanopolymers for electrochemical energy storage applications. Mater. Today. **21**(5), 501–516 (2018)
14. Zhou, X., et al.: Advances in polymer-based supercapacitors. Energy Environ. Sci. **12**(8), 2283–2301 (2019)

Nanopolymers Applications in Medical Imaging, Diagnostics, Drug Delivery, Cell Activity Regulation, Sensors and Therapy

Ankita Rajput

Abstract Nanopolymers represent one of the most dynamic areas of nanotechnology, offering new opportunities to advance biomedical science and healthcare. This chapter explores the fundamental principles, design strategies, and wide-ranging applications of nanopolymeric systems in medicine. In particular, the chapter explores their applications in drug and gene delivery, diagnostics, medical imaging, vaccine development, cellular regulation and biosensing. Due to their compatibility with biological systems, tunable surface chemistry, and responsiveness to various stimuli, polymer-based nanomaterials possess unique characteristics that facilitate precise targeting, regulated release, and multifunctional activity. These characteristics have been key to progressing both therapeutic and diagnostic methods. Case studies highlight applications in cancer treatment, cardiovascular and metabolic disorders, infectious diseases, and neurodegenerative conditions, illustrating their translational promise. The chapter also discusses emerging multimodal and theranostic approaches that integrate treatment with real-time monitoring. Finally, the challenges of clinical implementation, including toxicity, immunogenicity, large scale production, and regulatory considerations, are addressed along with perspectives on future directions. By bridging nanoscale science with biomedical needs, nanopolymers are poised to play a pivotal role in shaping the future of precision medicine and next generation healthcare.

Keywords Nanopolymers · Targeted drug delivery · Medical imaging · Polymeric nanoparticles · Dendrimers · Cancer theranostics · Gene therapy · Biosensors · Cell activity regulation · Vaccine delivery

---

A. Rajput (✉)
Dr. B. R. Ambedkar Center for Biomedical Research, University of Delhi, Delhi, India
e-mail: ankitarajput1998@gmail.com

B. Sharma et al. (eds.), *Nanostructured Polymers*, Engineering Materials,
https://doi.org/10.1007/978-981-95-3581-1_15

1 Introduction

Nanotechnology deals at the incredibly minute scale of atomic and molecular dimensions. The latter has revolutionized health care by enhancing diagnostic techniques, strategies for treatment, and methods of drug delivery [1]. This promising field has allowed fantastic advances in drug delivery systems, therapies, and diagnostics because of the exact control of material features that is possible at the nanoscale. The distinctive features of nanomaterials allow for directed drug delivery, tailored treatments, and enhanced methods for medical imaging. Nanoparticle-based systems enhance effectiveness while minimizing toxicities. Advanced nanosensors will enable the detection of diseases much earlier and accurately. Nanoparticles and nanoscale technologies may enable active targeting of therapeutic delivery, disease diagnosis, and/or therapeutic regimens and may enable revolutionary treatments by improving patient outcomes. It enhanced treatment modalities related to cardiovascular disease, oncology, musculoskeletal disorders, neurological and psychiatric illnesses, infectious diseases, and metabolic disorders like diabetes [2]. As research progresses, nanotechnology is leading the way in medical innovation. It pushes the limits of genetic therapies, tissue engineering, and regenerative medicine and further creates new treatments by bridging the gap between molecular discovery and complex treatments [3]. Enhancing medication delivery and efficacy requires the use of a variety of nanoparticle types, such as polymeric, lipid-based, and inorganic ones. The adaptability, precise targeting, and controlled drug release of polymeric nanoparticles (PNPs) make them special. PNPs are crucial for therapeutic innovation because they improve the stability, bioavailability, and efficient distribution of active compounds. To maximize drug encapsulation, optimize release patterns, and guarantee therapeutic efficacy, the appropriate polymer system selection is essential. PNP formulations are now more stable and useful thanks to recent developments including better lyophilization methods, which have cemented their status as the go-to option for drug delivery systems [4–6] (Table 1).

2 Applications of Nanopolymers

A significant breakthrough in medical technology, nanopolymers enable targeted gene therapy, better medical imaging, and precise drug delivery. These nanoscale materials are revolutionizing the way complicated diseases like cancer are treated by lowering side effects and offering regulated release mechanisms. Their enormous potential in modern medicine is highlighted by their capacity to improve vaccine delivery, serve as advanced biosensors, and facilitate genetic alterations. However, to address safety, regulatory, and environmental concerns, more research is needed (Figs. 1 and 2).

Table 1 Advancements in nanocarriers for drug delivery and therapeutics

Type	Advantages	Applications	Examples
Nanocrystals	Improves drug dissolution rate, enhances saturation solubility, safe composition, high bioavailability and suitable for intravenous injection	Drug delivery, cancer treatment, triglyceride and cholesterol control and hyperthermic chemotherapy	Paclitaxel Nanocrystals (PTX-NCs) [7]
Polymeric NPs	Multifunctional cargo carriers, thermodynamic stability and deep tissue penetration	Drug delivery and gene delivery	polymeric micelles [8], polymeric nanoparticles, dendrimers [9]
Liposomes	Passive targeting, reduces cargo toxicity and highly efficient delivery	Drug delivery and gene therapy	cationic liposomes [10], anionic liposomes [11], neutral liposomes [12], PEGylated liposomes [13]
Dendrimers	Defined molecular weight, host–guest entrapment, uniform shape and low polydispersity	Drug delivery and gene therapy	Poly(propylene imine) dendrimer [14]
Nanocomposites	Enhanced imaging and diagnostic capabilities	Enhanced imaging and diagnostic capabilities	quantum dot-polymeric nanocomposites [15]
Nanocapsules	Sustained drug release and protective encapsulation	Drug delivery and controlled release	polymeric nanocapsules [16]
Nanoparticles with Ligands	Target specificity and enhanced drug efficacy	Cancer therapy and targeted drug delivery	folate-conjugated nanoparticles [17]

2.1 *Imaging and Diagnostic Applications*

Early cancer detection in nanomedicine depends on in vivo imaging, which uses a variety of technologies in preclinical and clinical contexts. While PET (Positron Emission Tomography), MRI (Magnetic Resonance Imaging), and CT (Computed Tomography) are widely used modalities and are most regularly used in clinical settings, near-infrared fluorescence (NIRF) imaging is often employed in preclinical research. PET is extremely sensitive to metabolic changes, MRI excels at differentiating soft tissues, and CT allows for rapid structural visualization. Every modality has advantages of its own. Nuclear imaging techniques have higher costs and radiation risks despite their exceptional accuracy.

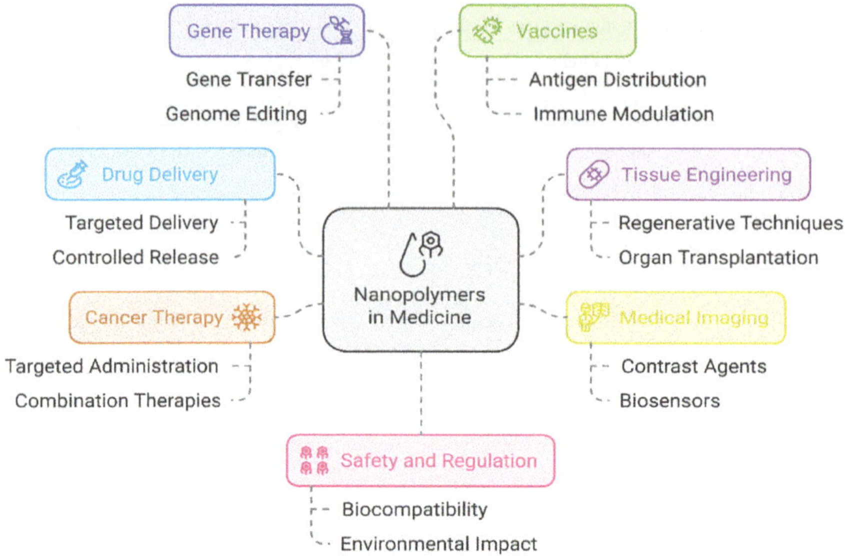

Fig. 1 Nanopolymers in medicine: applications and innovations

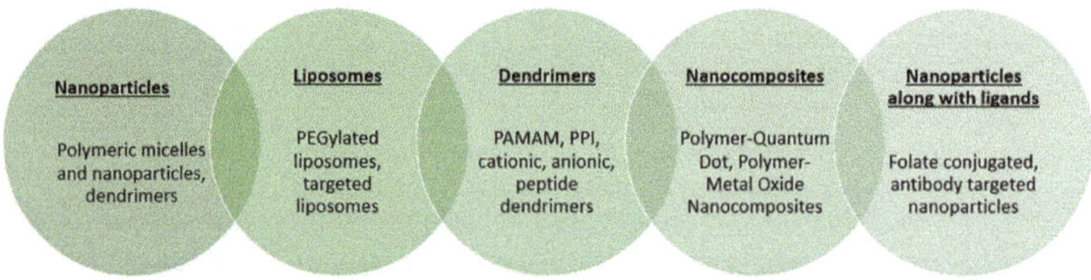

Fig. 2 Overview of nanopolymer types and their examples

However, the cost-effectiveness and radiation-free characteristics of optical imaging methods such as NIRF are limited by the depth of tissue penetration. MRI is noteworthy for offering thorough anatomical imaging without the risk of radiation exposure, but being costlier and less sensitive than alternative techniques [18, 19]. However, the combination of molecular imaging and nanotechnology has totally changed diagnostic capabilities. In terms of signal strength, stability, and biodistribution characteristics, imaging agents based on nanoparticles perform better than traditional small-molecule contrast agents. The majority of contrast materials used in medical imaging are composed of small-molecule compounds that degrade quickly and show non-specific biodistribution, which could lead to unintended hazardous effects. Functionalizing these nanoparticles with specific agents can therefore improve the sensitivity and precision of diagnostic imaging by enhancing signals, identifying analytes at low concentrations, and pinpointing the precise anatomical location of diseased alterations in specific tissue structures and disease biomarkers

[20]. Furthermore, the development of multimodal nanoparticle imaging agents allows for the integration of various imaging techniques, including optical and photoacoustic methods, PET Positron Emission Tomography) and MRI (Magnetic Resonance Imaging), to identify diseases and provide comprehensive diagnostic information [21, 22].

Furthermore, the cooperation of chemists, molecular biologists, physicists, and imaging specialists is essential to the development of molecular imaging. Significant legal, regulatory, and financial obstacles also confront the industry. Both the range of modalities and the range of molecular probes being used in molecular imaging are currently changing.

Furthermore, polymer-based engineered nanoparticles offer a viable way to advance medical imaging and diagnostic methods. By modifying these nanopolymers to identify specific biological markers, they could help identify a number of illnesses sooner. These nanostructures are ideal for safe and efficient circulation throughout the body because of their versatile surface properties and compatibility with living tissues. Additionally, the creation of contrast agents made from nanopolymers has the potential to transform medical imaging technology and improve disease monitoring and diagnosis [19]. Perfluorocarbon (PFC) gas-containing microbubbles are being used as carriers for targeted drug and gene delivery to particular organs as well as as adaptable contrast agents to enhance ultrasound imaging [23–25]. In 2018, scientists developed an ultrasound-responsive nanodroplet system that included a PFC core, Fe3O4 for imaging and HCPT and folic acid for cancer treatment. The PFC evaporated upon ultrasonic activation, releasing HCPT [26]. Although ferrotherapy shows promise for treating tumors with iron-based nanomaterials, tumors' inadequate Fenton reaction activity limits its efficacy (Fig. 3).

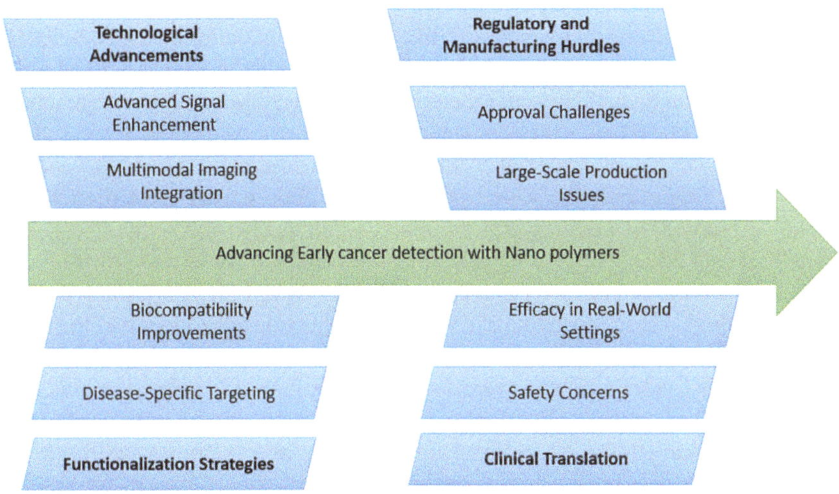

Fig. 3 Challenges and innovations in nanopolymer cancer detection

In order to get over this restriction, scientists created NIR light-activated Fe/PPy-RGD nanopolymers, which enhance cancer treatment by combining ferrotherapy and photothermal therapy. With the help of RGD ligands, these minuscule nanopolymers are engineered to precisely bind integrin αvβ3 receptors, guaranteeing precise tumor localization. These nanopolymers both accelerate the release of iron ions and produce heat for photothermal effects when exposed to near-infrared (NIR) light. This dual action causes cancer cells to undergo ferroptosis and intensifies the Fenton reaction. The Fe/PPy-RGD nanopolymers have shown effective in both lab and biological settings. They are a promising multipurpose tool for cancer therapy that combines targeted delivery, imaging, and synergistic treatment benefits because they also have magnetic resonance (MR) imaging capabilities [27]. For application in drug administration, optical imaging, and magnetic resonance imaging (MRI), the researchers successfully created multifunctional PEGylated Gd2O3 nanoparticles (NPs) that contain the fluorescent anti-cancer medication doxorubicin (dox). They used a variety of spectroscopic and microscopic techniques to validate the optical properties of these NPs. Studies of cellular absorption and in vitro cytotoxicity assays on numerous cancer cell lines (U-87, A549, and PANC-1) have shown effective drug delivery and anti-cancer effects. The line broadening in the H NMR spectrum of nearby heavy water protons showed how well the nanoparticles worked as MRI contrast agents. According to their research, Gd-PEG-dox NPs have a lot of potential as both imaging agents and dual-purpose drug delivery devices. The group anticipates that additional improvements to these nanomaterials could lead to major advancements in therapeutic and diagnostic approaches for a variety of biomedical applications [28]. Multifunctional nanostructures can be produced through strategic polymer coatings, and quantum dots (QDs) have a lot of potential as molecular imaging agents. In order to develop multimodal imaging probes that combine MRI and optical imaging capabilities, researchers have combined quantum dots (QDs) with paramagnetic materials like Fe2O3 and gadolinium chelates. These state-of-the-art nanostructures provide accurate tumor identification, surgical guidance, and simultaneous therapeutic therapy monitoring. By enabling simultaneous disease imaging and therapy, the conjugation of QDs with medicinal molecules offers up intriguing theranostic potential. Although further research is needed to address any potential limitations, the varied properties of QDs suggest great promise for enhancing biomedical imaging and customized therapeutic approaches [29]. In a groundbreaking 2002 study, Lidke and colleagues coupled epidermal growth factor to red-light emitting (CdSe) ZnS quantum dots (QDs) in order to develop a novel imaging method. These bioconjugates were specifically designed to target the erbB/HER membrane receptor. When the researchers introduced receptor-bound QDs to cultured human cancer cells, they were able to identify them as single molecules. One characteristic that set these QDs apart was their characteristic intermittent fluorescence, or "blinking." The exceptional light stability and brightness of these QD sensors made it possible to continuously and in real time monitor the kinetics of protein diffusion on the cell membrane. Remarkably, this tracking capability persisted upon internalization of the receptor-QD complexes. This innovative technique opened up new avenues for studying protein behaviour and cellular processes at previously unheard-of resolution and time scales. This work demonstrated the promise of QDs as powerful tools for long-term, high-resolution imaging of cellular dynamics, offering insights into the behaviour of intracellular trafficking networks and membrane proteins [30].

By fusing fluorescence and magnetic resonance imaging (MRI), conjugated polymer nanoparticles (CPNs) loaded with metallic nanoparticles, such IONPs, improve imaging capabilities and enable better tumor tracking. These nanoparticles enable real-time tumor detection using fluorescence imaging and provide exceptional contrast in MRI, especially in T2-weighted imaging. CPNs demonstrated efficient tumor accumulation in glioma model experiments, effectively overcoming barriers such as the blood–brain barrier. They are crucial for both therapeutic and diagnostic applications due to their dual imaging capabilities, which enable precise tracking of distribution within the body and the efficacy of therapies [31].

The development of a novel CT contrast agent employing PEG-coated gold nanoparticles (GNPs) addressed the drawbacks of iodine-based CT contrast agents, which, due to their higher X-ray attenuation, demonstrated 5.7 times greater contrast than iodine-based agents such as Ultravist. The PEG coating prolonged blood circulation time (>4 h) and enhanced imaging of hepatoma, great vessels, and cardiac ventricles in rats. The findings indicate that PEG-coated GNPs are a promising, biocompatible CT contrast agent designed of Camptothecin Derivatives for hepatoma and blood pool imaging [32]. In order to detect microtumors in mice, researchers created Au-PEG nanoparticles (15 and 30 nm) that had twice the contrast of CT imaging when compared to conventional agents. The study demonstrated the gradual buildup of nanoparticles in the liver and spleen and provided visible proof of the EPR effect. Versions coupled with antibodies (Au-PEG-HER2ab) were created to improve tumor targeting; nevertheless, because to the diverse makeup of tumors, it was difficult to compare distributions using CT imaging [33].The researchers also reported new results showing that 10 nm PEG-coated gold nanoparticles (AuNPs) improve heart function in a number of ways. According to their research, these nanoparticles suppress inflammatory reactions, lessen collagen deposition, and decrease the size of infarctions. The potential use of AuNPs in the therapy of cardiac disorders, especially myocardial infarction, is empirically supported by this ground-breaking study. New directions for investigating nanoparticle-based treatments in cardiovascular medicine are made possible by the researchers' study [34]. The biocompatibility and strong magnetic characteristics of PEG-coated iron oxide nanoparticles (IONPs) make them intriguing agents for medical imaging. These nanoparticles, which are produced by co-precipitation and described using a variety of spectroscopic and microscopic methods, provide improved contrast for imaging. They are useful for diagnostics since they are excellent at imaging malignant tissues and bacterial infections. Because of their special qualities, PEG-coated IONPs have the potential to revolutionize clinical imaging and improve methods for both illness detection and therapy [35]. By adding indocyanine green (ICG) to mesoporous silica nanoparticles (MSNs), researchers have produced a novel nanocomposite that has particles with a diameter of about 50 nm. This probe is very useful for imaging tumors in naked mice because of its remarkable optical stability, biocompatibility, and intense near-infrared II (NIR-II) fluorescence. The nanocomposite efficiently accumulates in tumor tissues and raises tumor-to-muscle ratio (TMR) values by specifically targeting xenograft tumors that express the PD-L1 receptor. A low-cost and low-toxicity alternative for accurate breast cancer diagnosis, this ICG@MSNs-PEG-PD antibody NIR-II probe may open up new therapeutic applications in cancer detection and treatment [36]. BiO_3-Dex nanoparticles (NPs), which have excellent X-ray attenuation, great biocompatibility, and ultra-small size, were created for targeted CT imaging of IBD. Following oral treatment, they were able to target colitis locations and see the GI system. These NPs have a lot of promise as a clinical CT contrast agent due to their easy production and high sensitivity [37]. A novel nanocomposite called Bi_2O_3@PCD-CUR-Glu was created by researchers for use in cancer treatment with CT imaging-guided chemotherapy and radiation. This multipurpose approach incorporates glucose for targeted administration to tumors and polycyclodextrin to improve biocompatibility with curcumin-loaded BiO_3 nanoparticles. The nanocomposite outperformed Omnipaque by a factor of 1.52 in CT contrast and showed remarkable radiosensitization, increasing efficacy by 270%. It reduced adverse effects while offering a precision therapeutic approach by specifically targeting cancer cells that overexpress HER-2 [38]. To improve imaging and therapeutic applications, researchers synthesized a series of functional polymer nanoparticles with varying amounts of the GRGDS peptide, which selectively binds integrin $\alpha v\beta 3$. Because the nanoparticles were created in a modular fashion, their structure and functionality could be precisely altered. Higher RGD loading increased targeting efficiency in vitro using $\alpha v\beta 3$-positive U87MG glioma cells, although there was a reported upper limit to this effectiveness. According to in vivo research, higher RGD loading resulted in shorter blood circulation times and increased liver absorption. The study emphasized the critical role that ligand loading plays in maximizing nanoparticle performance. Furthermore, the modular synthesis method made it possible to create nanoparticles with different concentrations of RGD peptides and 64Cu-DOTA diagnostic units, providing important new information about the relationship between ligand density, nanoparticle size, and biological behaviour for certain applications [39]. A state-of-the-art imaging system that makes use of the fluorescent properties of the PFO conjugated polymer is the PFO/PG-Dox combination. Dox initially suppresses PFO's fluorescence, but after being absorbed by cancer cells, poly (l-glutamic acid) (PG) hydrolyzes, releasing Dox and restoring PFO's fluorescence. Through fluorescence recovery, this "turn-on" process makes it easier to observe drug release in real time. PFO has exceptional properties for cellular imaging, including a propensity to concentrate in the cytoplasm of A549 cells, low cytotoxicity, and great photostability. The PFO/PG-Dox combination is perfect for monitoring drug release in imaging studies since it aggregates into nanoparticles with a diameter of about 50 nm. Researchers may visually evaluate medication transport and release processes with this setup, providing vital information about the cellular effectiveness of cancer treatments [40].

2.2 Drug Delivery

Nanopolymers have revolutionized drug delivery systems due to their distinct properties. These microscopic carriers improve the stability, solubility, and controlled release of pharmaceuticals, hence boosting their efficacy. Because of their small size, they can penetrate deeper into tissues and be better absorbed by cells, enabling more accurate and focused therapies. Drugs are released in precise locations by certain nanopolymers in response to particular triggers. Additionally, they can carry several drugs simultaneously, which promotes synergistic effects in combination treatments. These developments result in safer and more effective drug delivery, which eventually leads to better patient outcomes and fewer adverse effects. When it comes to healthcare innovation and personalized medicine, nanopolymers are setting the standard [1, 41].

Extended bloodstream circulation, improved stability, and tailored distribution to tumor locations with minimal harm to healthy cells are just a few benefits of using nanopolymers for drug delivery. The constraints of traditional chemotherapy are addressed by this nanoparticle-driven method, which allows the simultaneous delivery of numerous medications. More research is done on the use of nanobiopolymers in cancer treatment, with an emphasis on their microbial, synthetic, and natural forms [42]. Because smart nanopolymers may react to particular stimuli like pH, redox agents, light, and temperature, they are exceptionally useful in regulated medication delivery. These triggers allow for precise medication release at the intended place and are frequently found in sick tissues or externally triggered. Their responsiveness and accuracy are further improved by the addition of redox-sensitive links and sophisticated functionalization processes, which makes them an effective tool for focused and focused medicinal applications [43]. For targeted drug delivery, researchers synthesized dendrimers containing mannose alterations, and they used a variety of spectroscopic techniques to confirm the conjugation. Because of their low polydispersity index and narrow size distribution, dendrimers are perfect for drug delivery applications. The characteristic development pattern of these nanostructures, which begins at a central core and branches outward in a tree-like fashion, defines them. Dendrimers get bigger and heavier as the number of branching generations rises, which enhances their ability to transport and encapsulate medicinal molecules. Because of its scalable structure, the dendrimer's characteristics may be precisely controlled, allowing for customisation to meet specific drug delivery needs [44, 45].

The altered dendrimers improved the encapsulation of rifampicin, exhibited pH-responsive release of the drug, and indicated lower levels of cytotoxicity. Enhanced uptake by macrophages through lectin receptor-mediated endocytosis underscores their promise for targeted drug delivery to the lungs [46] (Fig. 4).

Camptothecin derivatives, including 10-hydroxycamptothecin (10HCPT) and 7-butyl-10-amino-camptothecin (BACPT), were encapsulated within a dendrimer, which is biocompatible and derived from natural metabolites such as glycerol and succinic acid. The effectiveness of medicine distribution was significantly increased as a result of this innovative method. The use of these dendrimers as carriers

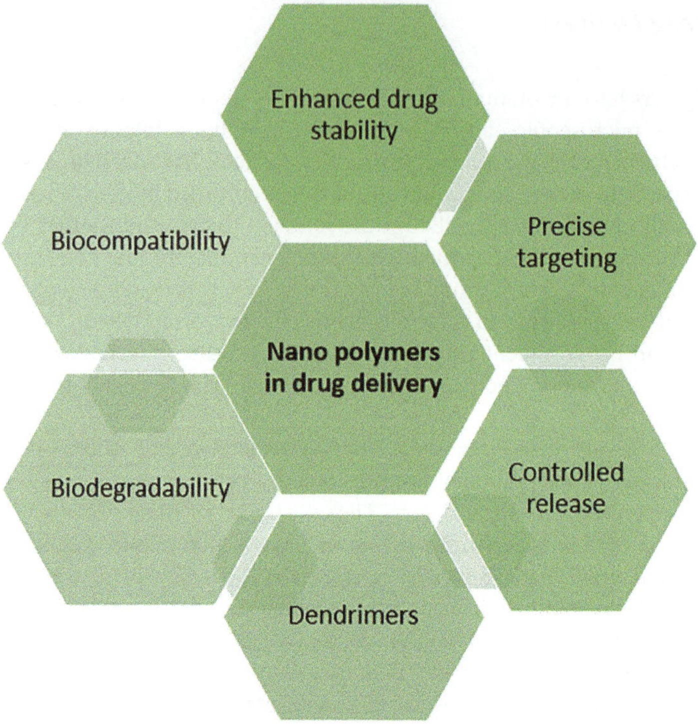

Fig. 4 Advancements in nanopolymer drug delivery

resulted in a remarkable 16-fold increase in the medicines' cellular absorption when compared to traditional techniques. Additionally, the encapsulated materials were better retained inside the cells using the dendrimer-based delivery system. These findings highlight the encouraging possibility that dendrimer-based drug delivery systems could greatly improve the efficacy of medicines developed from camptothecin [47]. Kono and a group of scientists used fourth-generation (G4) PAMAM dendrimers to develop a unique drug delivery method. Polyethylene glycol (PEG) was used to modify these dendrimers and attach them to the chemotherapeutic drug adriamycin. By joining PEG to the amino group of glutamic acid (Glu) and joining adriamycin with either an amide or hydrazone bond, the researchers employed Glu as a linker. According to their research, the amide bond was less stable and effective than the hydrazone bond arrangement. Crucially, even against multidrug-resistant (MDR) cancer cells, which are frequently challenging to treat because of their ability to evade conventional chemotherapy drugs, this formulation produced promising outcomes. This technique represents a significant advancement in dendrimer-based drug delivery systems and may offer improved cancer therapy options for malignancies that are resistant to current treatments [48].

The effectiveness of a single subconjunctival injection of dendrimer-dexamethasone (D-Dex) in treating autoimmune dacryoadenitis was investigated by

Lin and his team. In animal models, the study contrasted the effects of D-Dex with those of saline and free dexamethasone (Free-Dex). Results showed that compared to the Free- dexamethasone and saline groups, the D-Dex group showed a significant decrease in pro-inflammatory gene expression, specifically IL6, IL8, MMP9, and TNFα. This notable reduction in inflammatory markers suggests that D-Dex could be a viable treatment option for a number of eye conditions [49, 50].

Although 75% of patients with ovarian cancer achieve initial success with chemotherapy, many eventually relapse as a result of treatment resistance, which is caused by higher levels of drug transporters such ABCG2 and P-glycoprotein. The Bola/IM nanoformulation, which encapsulates imatinib within a bola-amphiphilic dendrimer, is the focus of this study. The formulation increases apoptosis and decreases drug efflux by downregulating the expression of these transporters. Additionally, Bola/IM increases the susceptibility of cancer stem cells to cisplatin and paclitaxel when combined with them, offering a viable strategy to treat medication resistance in ovarian cancer [51, 52]. In a hydrogel formulation, vinyl sulfone-functionalized PAMAM dendrimers and thiolated PEG improve the solubility and prolonged release of hydrophobic anti-cancer medications. By increasing their solubility, 30% to 80% of silibinin, camptothecin, and methotrexate are released over the course of one to four days. Cytotoxicity assays on J82 and MCF7 cancer cell lines reveal up to 95% cell death. This method shows promise for the effective intratumoral administration of hydrophobic medications in the treatment of cancer [53]. The potential of PAMAM dendrimers to enhance the pulmonary administration of the anticoagulant drug enoxaparin has been studied. G2, G2.5, and G3 PAMAM dendrimers were examined in the study in order to determine how they affected enoxaparin absorption through antifactor Xa activity and prevented DVT in animal models. The results showed that amine-terminated G2 and G3 dendrimers significantly increased enoxaparin's pulmonary absorption by 40% without seriously harming the lungs. By forming ionic connections with enoxaparin, these positively charged dendrimers prevented deep vein thrombosis with an efficacy similar to that of subcutaneous enoxaparin. In contrast, enoxaparin's bioavailability was unaffected by negatively charged G2.5 dendrimers having carboxyl terminal groups. These findings suggest that cationic PAMAM dendrimers are viable options for enoxaparin's pulmonary administration, which could increase the medication's therapeutic impact while maintaining safety [54]. PAMAM dendrimers coated with FA-conjugated MCS were created as a dual-responsive nanocarrier for regulated DOX release. In order to replicate the tumor environment, the nanocarrier (20–50 nm) demonstrated a high drug loading capacity and sustained release at pH 5.6. The Peppas-Sahlin model with non-Fickian diffusion was used to describe drug release. Because of its delayed release, the nanocarrier demonstrated less cytotoxicity in MTT assays than free DOX [55]. As possible drug carriers for glioblastoma resistant to temozolomide (TMZ), biotinylated and glycidylated PAMAM G4 dendrimers (G4Bgl) are attracting interest due to their exceptional biocompatibility, notable cellular accumulation, and blood–brain barrier crossing capabilities. Additionally, they have demonstrated efficient transport in C. elegans, suggesting that they may be used to treat nematodes [56].

In order to target tumors that express MET, this study presents a unique DPC system that combines MET-targeting peptides with G7 dendrimers, improving multivalent binding and imaging capabilities. When compared to free peptides, the system exhibits a much longer plasma half-life and a significantly higher binding affinity for MET-expressing cells, which leads to improved tumor accumulation. The DPC system opens the door for theranostic applications in cancer treatment by serving as a dependable radio imaging agent for MET-positive tumors when labeled with 89Zr. These findings make the DPC system a viable choice for the detection and management of malignancies with MET overexpression [57].

In order to isolate exosomes from urine samples and perform untargeted metabolomics analysis, this study presents a novel pH-responsive bifunctional affinity nanopolymer (pHBiAN). By modifying the pH-sensitive polymer P4VP with DSPE and CD63 aptamer, the researchers were able to separate exosomes from urine in an effective and non-destructive manner. Using real urine samples, the scientists used a novel metabolomics technique based on HPLC–MS. This approach effectively differentiated between patients with early-stage osteoarthritis (OA) and healthy persons when paired with machine learning algorithms. According to the study, catechol in urine exosomes may be a biomarker for OA. This non-invasive method demonstrated its significant clinical promise for early OA diagnosis by successfully identifying aberrant metabolic pathways. Additionally, the method has potential for wider use in the diagnosis of different illnesses [58].

A hallmark of a number of neurological disorders, most notably Alzheimer's disease and prion illnesses, is the build-up of amyloid fibrils. Certain positively charged dendrimers have been shown to have the ability to disrupt the development of amyloid fibrils or degrade already-formed mature fibrils. Additionally, several dendrimers have demonstrated potential in reducing the levels of toxic amyloid oligomers, which are thought to be detrimental intermediaries in the fibril production process. According to these results, dendrimers may one day be used as treatment for neurodegenerative diseases linked to amyloid build-up. Dendrimers offer a multifaceted strategy to treat the intricate pathology of these disorders by focusing on distinct stages of amyloid production and aggregation [59].

A cisplatin nanoparticle formulation called NC-6004 was created to improve the drug's efficacy and safety when used to treat head and neck squamous cell cancer. In animal experiments, NC-6004 shown a significant decrease in nephrotoxicity, even though it demonstrated comparable anti-tumor efficacy as cisplatin in vivo. In a tongue cancer model, it also shown more efficacy in reducing lymph node metastases. According to these findings, NC-6004 may enhance platinum-based treatments by enhancing drug delivery and lowering adverse effects, particularly nephrotoxicity [60].

Researchers have developed a novel siRNA delivery method that makes use of a biodegradable nanocarrier based on dextran. This carrier was created by using acid-sensitive acetal linkages to join amine groups to the dextran structure. These connections degrade in mildly acidic settings but remain stable in slightly alkaline ones (pH 7.4). By facilitating faster siRNA release and improving the safety profile of the cationic dextran siRNA carriers, this clever innovation raises the transfection efficiency. Experiments demonstrated that this nanocarrier was capable of efficiently delivering ChoK siRNA and lowering ChoK expression in MDA-MB-231 cells. The rapid degradation of the chemical and the efficient release of siRNA were confirmed by cellular imaging techniques. Given these encouraging findings, the researchers recommend that further research be done on this degradable dextran nanocarrier as a potential safe, reproducible, and biocompatible siRNA delivery platform for upcoming preclinical studies [61].

This report discusses the therapeutic benefits of folic acid-conjugated amphiphilic copolymers, which target cancer cells by interfering with oncogenic proteins and promoting apoptosis. The effective nanopolymer also concentrates on the targeted delivery of hydrophobic chemotherapy medicines to cancer cells in order to minimize side effects atoms. Furthermore, FA-DABA-SMA is a powerful delivery method for metabolic cancer cells because of its large, sheet-like structure, which allows it to change protein function and induce death. Its interaction with folate receptors, which are overexpressed in different cancer types, requires more study [62].

Researchers have developed a novel drug delivery method that uses magnetic-core silibinin nanopolymeric carriers (SLB-MPNPs) to treat renal cell carcinoma. The double emulsion solvent evaporation method was used to create these carriers, which showed good encapsulation efficiency, zeta potential, and suitable particle size. Crucially, compared to free silibinin, SLB-MPNPs demonstrated a stronger inhibitory effect on the development of A-498 kidney cancer cells. According to stability tests, the formulation remained stable for up to three months when refrigerated. Safety was confirmed by acute toxicity tests in mice, which showed no negative effects at a dose of 50 mg/kg. These magnetic-core silibinin nanopolymeric carriers offer a promising, efficient, and targeted approach for the treatment of renal cell carcinoma, warranting additional research because silibinin is a recognized herbal drug with few adverse effects [63].

Acetazolamide and G5 PPI dendrimers were combined to generate a compound that successfully lowers intraocular pressure. 5662.3% was the maximum drug entrapment efficiency. During a 24-h period, $83.5 \pm 1.8\%$ and $80.4 \pm 1.6\%$ of acetazolamide, respectively, were released in phosphate-buffered saline (pH 7.4) and simulated tear fluid (pH 7.4), indicating a sustained release of the medication in in vitro studies. The dendrimer formulations were tested for their ability to lower intraocular pressure, cause ocular irritation, and cause hemolytic toxicity. Results indicated that mild ocular discomfort was generated by low concentrations of the dendrimer-acetazolamide conjugate in aqueous solutions. The study found that dendrimer-based acetazolamide formulations improve the drug's duration of residence in the eye and had better results in reducing intraocular pressure for the treatment of glaucoma [64] (Fig. 5).

The application of nanopolymers in the delivery of vaccines holds significant potential for boosting immune responses. By inhibiting the degradation of antigens, enhancing interactions with immune cells, and ensuring a gradual release, these carriers improve both the effectiveness and memory of vaccines. Their diminutive size and surface properties are crucial for vaccines and immunotherapies, as they help regulate immune cell activity and cytokine production. Nanopolymers act as

Fig. 5 Enhancing drug delivery with nanopolymers

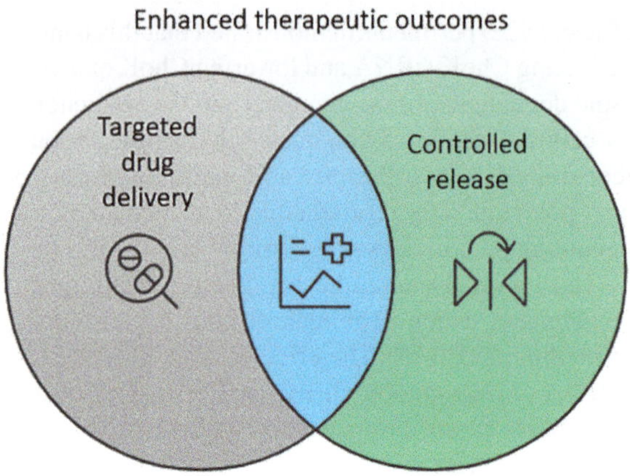

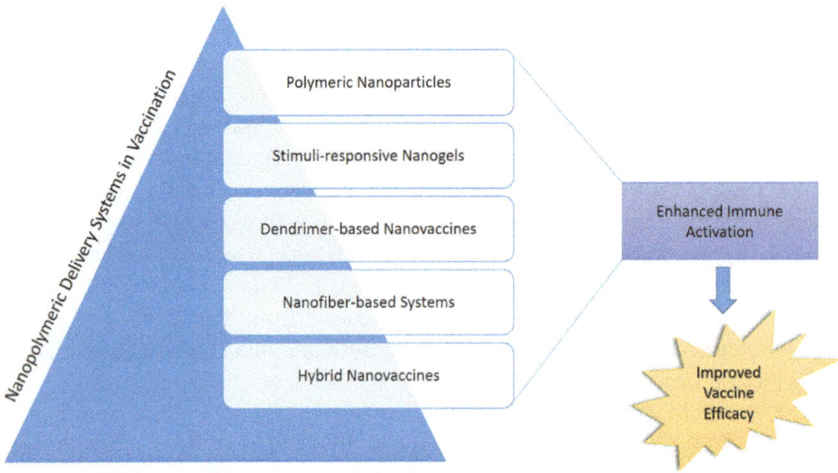

Fig. 6 Nanopolymeric systems for vaccine delivery

adjuvants, enabling the controlled distribution of adjuvants and antigens to stimulate both innate and adaptive immunity while sustaining immune activation over time. This versatility could lead to improved cancer treatments and more effective vaccines [65] (Fig. 6).

2.3 Nanopolymers-Enhanced Sensing

By using nanohybrids, fluorescence chemosensing technology has advanced significantly. These nanohybrids' incorporation with cellulose enhances its overall usefulness, flexibility, and stability. In the field of cellular research, this combination has created new opportunities. Because they provide precise insights at the individual cell level, fluorescence-based technologies have proven essential in deciphering the complexities of cellular processes and dynamics. With the use of these instruments, scientists may perform target-specific imaging inside living cells, giving them a real-time picture of cellular activity. In a variety of cell lines, nanohybrids have shown exceptional efficacy for molecule-specific imaging. Because of their special qualities, they are ideal for identifying and displaying certain molecular targets inside cellular settings, which improves our comprehension of molecular interactions and cellular activities [66]. Luminescent characteristics, specificity, and resistance to photobleaching make fluorescent probes invaluable tools in biotechnology. Their effectiveness in biological detection is enhanced when combined with biomolecules. The phenomenon of fluorescence quenching, which can occur through various mechanisms like molecular rearrangements and energy transfer, proves to be a powerful sensing technique.

An excellent example of this is a glucose detection system that uses thioglycollic acid-coated quantum dots (QDs) in a dual-enzyme configuration. Lifetime measurement-based research has effectively distinguished between static and dynamic quenching mechanisms. These investigations demonstrated that the reduced lifespan seen in CdS QDs upon exposure to oxidized o-phenylenediamine is caused by photo-induced electron transfer.

Additionally, these QDs' interaction with 3,3-diaminobenzid3ine produces QD aggregation and a shift in fluorescence towards longer wavelengths. Interestingly, the content of glucose is correlated with the degree of quenching. The promise of such devices in quantitative glucose sensing applications is highlighted by this relationship [67].

With advancements like nanomachines, tweezers, and imprinted polymers enabling quicker and more sensitive biological tests—which are crucial in clinical laboratories—the incorporation of nanotechnology into diagnostics has revolutionized biomolecular detection. These nanoparticles' tiny size greatly enhances the interactions between analytes and target biomolecules, getting around some of the drawbacks of conventional diagnostics, like laborious procedures and the requirement for specialist knowledge. The development of molecular technologies such as transcriptomics, metabolomics, and genomics has accelerated as a result of this advancement, reaching single-molecule sensitivity [68–72].

Two techniques for binding Alexa Fluor dyes to cellulose nanocrystals for bioimaging applications were created by researchers in 2015. The first method used a controlled reaction with sodium periodate to produce aldehyde groups on the surface of cellulose. Reductive amination was then used to attach Alexa Fluor dyes with terminal amino groups to these aldehyde sites. Adding a chloro-substituted triazine ring to the end of the linker portion of Alexa Fluor dyes was the second technique. Then, at high temperatures, these altered colors interacted with cellulose nanocrystals in acetonitrile. Both methods produced cellulose nanocrystals that were fluorescently tagged while retaining their general structure. In successful proof-of-principle bioimaging tests, these tagged nanocrystals showed promise for monitoring cellulose deposits and their deterioration over time [73].

In bioimaging applications, N-CDs have demonstrated considerable promise, especially for cellular and subcellular imaging. These nanoparticles are perfect for a variety of bioimaging activities because of their exceptional fluorescence qualities, photostability, and biocompatibility. N-CDs have proven to be effective in multicolor imaging of many cell types in cellular imaging. They have been used, for example, to image MCF-7 cancer cells, proving that they can produce distinct fluorescent signals at various excitation wavelengths. N-CDs have shown particular utility for imaging mitochondria at the subcellular level. In contrast to traditional mitochondrial probes, Kaminari et al. created N-CDs that had better photostability and long-term mitochondrial imaging capabilities. Because of their reliance on mitochondrial membrane potential for uptake, these N-CDs preferentially localize in malignant cells, indicating possible uses in medication delivery and cancer-targeted imaging.

Additionally, N-CDs have demonstrated promise in imaging the central nervous system. When given intravenously to mice with gliomas, Li et al.'s LAAM-CDs gradually accumulated in the brain, reaching a peak 8–12 h after injection. This illustrates how N-CDs can be used for drug delivery and imaging in a variety of cancers and disorders affecting the central nervous system [74].

Additionally, N-CDs have been used in animal models for in vivo imaging. They have demonstrated their adaptability in whole-organism imaging studies by being employed for fluorescence imaging in zebrafish. Another aspect of N-CDs' bioimaging capabilities is their capacity to identify particular ions, like $Fe3 +$, using fluorescence quenching mechanisms. Important biological ions can be simultaneously seen and sensed in cellular settings because to this property. With their high sensitivity, selectivity, and biocompatibility across a variety of biological systems and imaging modalities, N-CDs are an all-around intriguing class of fluorescent probes for advanced bioimaging applications [75].

2.4 Nanopolymers in Medical Sensors: Advancing Diagnostic Capabilities

Medical devices are undergoing a revolution thanks to nanoscale materials from inorganic, metallic, and carbon-based sources that improve their stability, sensitivity, selectivity, and affordability. These nanoparticles produce sophisticated nanocomposites with enhanced characteristics, facilitating the downsizing of medical monitoring devices and enabling more accurate sensors. This development opens up new possibilities for patient care by making it easier to track physiological markers in real time using wearable and portable technology. The key components of medical sensor technology are stability and selectivity. Stability ensures stable performance over time and under different conditions, while selectivity ensures exact detection of target analytes while limiting interference. For accurate medical diagnosis and patient monitoring, both must be accomplished. Improved sensitivity and specificity make nanomaterial-based sensors especially promising for disease detection.

Their improved electrical, optical, luminescent, and catalytic qualities open the door to more precise and effective medical equipment. Ensuring biocompatibility and increasing production for broad use, however, continue to present difficulties. Research on medical sensors based on nanomaterials is advancing quickly in spite of these challenges. These sensors have great promise for monitoring infectious infections, diagnosing cancer, and detecting disorders early. More sensitive, selective, and reasonably priced nanomaterial-based sensors are probably in store for the future of medical diagnostics, which could transform patient care and enhance health outcomes [76].

In medical applications, accurate biomolecule detection is essential, especially for the diagnosis of viral and genetic illnesses. Because of their exceptional electrochemical and optical characteristics, graphene-functionalized nanomaterials with certain antibodies provide extremely sensitive biological structure identification, even in trace levels. Biosensors use biological components including enzymes, nucleic acids, and antibodies to recognize analytes and translate these interactions into electrical signals. They are composed of a recognition layer, a transducer, and an electronic component. The capacity of graphene to adsorb aromatic compounds has been used to develop biosensors that detect biomolecules quickly and sensitively using optical and electrochemical processes. Graphene's ability to trap target molecules is greatly improved by functional groups, which allows for in-depth interaction analysis with particular analytes on screen-printed sensors. Further demonstrating graphene's potential for cleverly building printable biosensors and the creation of sophisticated graphene-based sensing tools for healthcare applications, graphene can be functionalized with hydroxyl and carbonyl groups to maximize its usage in medical diagnostics [77]. The conversion of precise biological interactions into digital data is the foundation of aptamer-based biosensing. Target molecules, or analytes, are recognized by particular biological components during this process. A variety of biological components, such as genetic material like DNA and RNA and proteins like enzymes and antibodies, can be detected using biosensors that use graphene as a platform. These sensors are also capable of identifying a variety of pathogenic species [78] (Fig. 7).

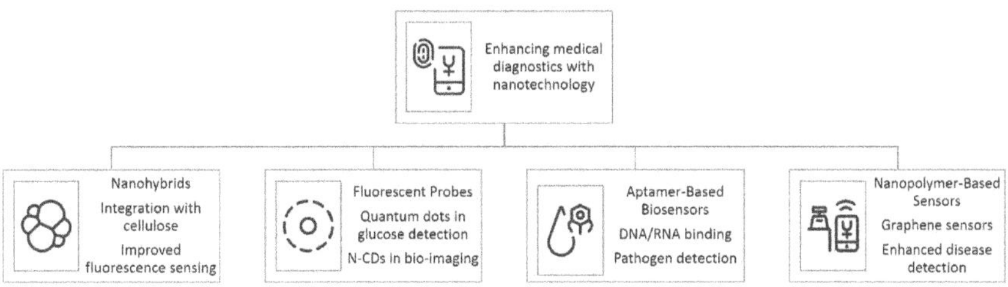

Fig. 7 Advancements in nanotechnology for medical diagnostics

2.5 Nanopolymers in Gene Therapy

Effective and precise delivery of genetic material is made possible by the adaptable carriers known as nanopolymers in gene therapy. Yan et al. created a novel pH-sensitive nanocarrier with chitosan in 2020 for better treatment of liver cancer. Glycyrrhetinic acid, chitosan, polyethyleneimine, and other ingredients made up their sophisticated nanopolymer, which showed strong anti-tumor effectiveness with few side effects. The design of the nanocarrier enables the targeted, controlled release of Bcl-2 siRNA and doxorubicin in the acidic tumor microenvironment. This focused technique improves the combined effectiveness of chemotherapy and gene therapy, providing a more effective liver cancer treatment plan [79].

The nanopolymer has the potential to improve the efficacy of chemotherapy and gene therapy when used together. To increase the catalytic activity of nanoenzymes, scientists created a liposome-TiN system utilizing titanium nitride in 2021. A pH-sensitive glucose oxidase (GOx) chain modified by polyethylene glycol (PEG) was used to functionalize the Liposome-TiN interface. Hydrogen peroxide is produced when the phenylacyl imine link between GOx and PEG breaks in the acidic tumor microenvironment. The nanoparticles catalyze H2O2 when exposed to radiation, generating extremely harmful hydroxyl radicals that kill tumor cells [80]. A nanopolymer called H1 was created by Yao and associates to deliver interleukin-2 (IL-2). H1 created polyplexes with the IL-2 plasmid (H1/pIL-2) that were around 100 nm in size and were made up of low molecular weight PEI, β-cyclodextrin, and folate. When H1/pIL-2 was injected into mice's melanoma tumors, it prevented the growth of the tumor, improved survival, and encouraged the infiltration and activation of immune cells such as natural killer cells, CD4 T, and CD8 T. According to the findings, H1/pIL-2 is a safe and efficient treatment for melanoma that provides an option to rAdv-IL-2 gene therapy [81]. Although MLT, a cytolytic and immunomodulatory peptide, can destroy tumor cells and trigger anti-tumor immunity, its effectiveness is constrained by its quick metabolism and poor absorption. In order to overcome these obstacles, the researchers created LC09-MLT@F127, an OS-targeting nanoparticle that efficiently induces immunogenic cell death (ICD) and sets off a potent immune response against tumors. By protecting MLT against hemolysis and lowering non-specific uptake, the nanoparticle enhanced its safety and targeting. By significantly infiltrating and activating CD4 + and CD8 + T cells in the tumor microenvironment, treatment with LC09-MLT@F127 increased systemic anti-tumor immunity and inhibited tumor growth. Through the activation of both innate and adaptive immune responses, this method exhibits significant potential for overcoming resistance and improving OS treatment [82] (Fig. 8).

By creating a unique biodegradable nanopolymer carrier for AKT1 siRNA delivery, Jere and associates achieved a significant advancement in the treatment of cancer. The AKT1 protein, which is essential for the survival and growth of cancer cells, was successfully silenced by this novel method. Numerous anti-cancer effects, such as decreased cancer cell survival, decreased proliferation, lowered malignant

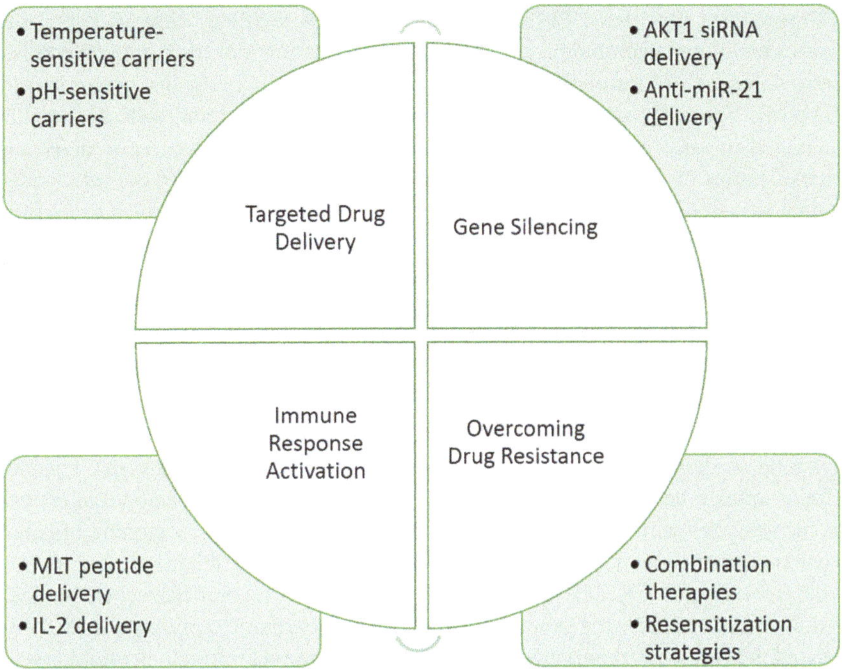

Fig. 8 Enhancing cancer treatment with nanopolymers with improved efficacy and reduced side effects

features, and inhibited metastatic potential, were brought about by the targeted reduction of AKT1. By effectively designing this nanocarrier system, the scientists have created new avenues for targeted cancer treatment, providing a viable strategy to disrupt the AKT1 signaling pathway and maybe enhance cancer patients' prognoses [83]. A novel medication delivery method called PCL-TPE@1 was created by researchers by fusing a fluorescent dye with a PCL-TPE polymer matrix. An in vitro model of diffuse large B-cell lymphoma (DLBCL) demonstrates the promise therapeutic effects of this novel system, which also displays temperature-dependent fluorescence. By reducing PRMT6 expression, PCL-TPE@1 prevents cell invasion, indicating its potential for cancer treatment. Simple synthesis, thermoresponsive fluorescence, adaptability, and biocompatibility are some of the system's main benefits, which make it a flexible platform for a range of biomedical applications. Its potential for combination therapy is further enhanced by its sensitivity to acidic and hyperthermic environments, especially when treating cancers like DLBCL. Future studies will concentrate on improving medication loading, stability, and release control. They will also tackle issues like localized temperature regulation and possible hyperthermia-related side effects. PCL-TPE@1 has the potential to increase its therapeutic applications and enhance the effectiveness of cancer treatment as research progresses [84].

Researchers have used PLGA-PEG nanoparticles coupled with A6 peptide to create a novel medication delivery system for the treatment of triple-negative breast cancer. This novel formulation has shown impressive activity against drug-resistant MDA-MB-231 cells since it contains both doxorubicin and anti-miR-21. Studies conducted in vitro revealed that doxorubicin treatment increased the toxicity and resensitization of resistant cells. Furthermore, the PLGA-PEG-A6 peptide-loaded medications demonstrated accurate tumor targeting in a mouse model, successfully decreasing tumor size while minimizing organ off-target accumulation. This strategy represents a major breakthrough in targeted cancer therapy in addition to providing a possibly safer treatment option. The encouraging data imply that this formulation may address a crucial need in cancer treatment by providing patients with triple-negative breast cancer with more effective and well-tolerated therapeutic outcomes [85].

Nanopolymers play a crucial role in the progress of CRISPR-Cas9 gene editing by acting as delivery vehicles that protect the gene-editing components, enhance cellular uptake, and enable accurate DNA modifications. These nanopolymers can also be designed with additional capabilities like imaging agents or targeting ligands to enhance precision and allow for real-time tracking. The combination of nanopolymeric vectors and CRISPR-Cas9 has the potential to revolutionize personalized medicine and therapies for genetic disorders. Nanopolymeric conjugates effectively deliver CRISPR-Cas9 components to inhibit the laminin-411 protein in glioblastoma multiforme (GBM), leading to tumor growth reduction and improved survival rates. Due to their ability to cross the blood–brain barrier and specifically target GBM cells, these nanopolymers could be utilized on their own or in conjunction with current treatments to develop new, customized therapies for gliomas [86].

Polyethyleneimine (PEI) and various cationic polymers have proven to be effective delivery systems for CRISPR-Cas9. These polymers facilitate successful genome editing by using their positive charge to encapsulate DNA and enhance endosomal escape. Through electrostatic interactions, researchers have discovered that polymers such as PLGA, PEI, chitosan, and poly(amidoamine) can efficiently engage with nucleic acids, allowing for targeted plasmid delivery to cells. By utilizing the proton-sponge effect, these polymers hold significant potential for advancing genetic engineering and therapeutic strategies [87].

Delivering large CRISPR/Cas9 plasmids remains a significant challenge. In HeLa cells, the cationic polymer polyethyleneimine-β-cyclodextrin (PC) forms nanocomplexes with Cas9/sgRNA plasmids, achieving higher editing efficiency at the HBB (19.1%) and RHBDF1 (7.0%) loci compared to PEI (2.8%). Moreover, PC induced the loss of GFP in HeLa-EGFP cells, indicating precise targeting. This study suggests that PC is an effective vehicle for CRISPR/Cas9 delivery in genome editing. To enhance wound healing in the corneas of diabetic patients and to revive stem cell function, scientists developed a distinctive non-toxic nanobiopolymer (NBC) for RNA-based therapies. In HeLa cells, a polyethyleneimine-β-cyclodextrin (PC) delivery method demonstrated greater efficiency in CRISPR/Cas9 genome editing compared to traditional PEI methods. These advancements suggest the possibility of more advanced gene editing and therapeutic strategies [88].

2.6 Cell Activity Regulation

Nanopolymers are essential for controlling cellular processes, especially when it comes to cancer treatment, because they affect autophagy, a carefully regulated mechanism that eliminates damaged cell components [89]. Autophagy is essential for tumor cells to survive, particularly in conditions with low oxygen and nutrition levels. It's interesting to note that overstimulating autophagy can cause cell death, underscoring its potential as a treatment approach. Numerous nanomaterials, particularly those based on metals, have an effect on autophagy through oxidative stress, endoplasmic reticulum stress, mitochondrial damage, and lysosomal dysfunction. Through autophagy, these effects can either increase cell survival or cause cell death. These materials are more appealing for targeted therapies because of their ability to precisely target cancer cells. Nanopolymers are essential for controlling cellular processes, especially when it comes to cancer treatment, because they affect autophagy, a carefully regulated mechanism that eliminates damaged cell components. Autophagy is essential for tumor cells to survive, particularly in conditions with low oxygen and nutrition levels [90]. It's interesting to note that overstimulating autophagy can cause cell death, underscoring its potential as a treatment approach. Numerous nanomaterials, particularly those based on metals, have an effect on autophagy through oxidative stress, endoplasmic reticulum stress, mitochondrial damage, and lysosomal dysfunction. Through autophagy, these effects can either increase cell survival or cause cell death. These materials are more appealing for targeted therapies because of their ability to precisely target cancer cells (Fig. 9).

Chitosan nanopolymers enhance drug stability and efficacy for glioma therapy. Encapsulating eugenol ensures controlled release and bioavailability. In vitro studies show that the formulation induces apoptosis, autophagy, and inhibits metastasis by suppressing NFκB and EMT proteins, highlighting the potential of chitosan as an effective drug carrier for glioma treatment [91]. Ir–Fe nanopolymers for combined photodynamic and chemodynamic cancer treatments are presented in the study. In the acidic lysosome environment, these nanopolymers decompose, generating Ir(III) complexes and Fe ions that cause cell death by ferroptosis, autophagy, and apoptosis. These nanopolymers demonstrate improved tumor targeting and bloodstream circulation when paired with melanoma exosomes (Ir–Fe@EXO). Ir–Fe@EXO is a promising novel cancer treatment approach that almost completely eradicates tumors and prevents lung metastases in a melanoma mouse model with a single injection [92]. The study demonstrates that PLGA-based nanoparticles (NPs) can trigger autophagy, causing their degradation in lysosomes and limiting their drug delivery effectiveness. Cancer cells rely on autophagy to resist chemotherapy, so combining NPs with autophagy inhibitors like 3-MA and CQ enhances their therapeutic effects. Docetaxel-loaded PLGA nanoparticles conjugated with cholic acid showed improved results in both in vitro and in vivo studies, advancing the potential of nanomedicine for clinical applications [93]. In KB cells, PAMAM dendrimers with distinct surface amino groups increased lysosomal pH and had lethal effects. Acetylated dendrimers showed no signs of cytotoxicity. Positively charged dendrimers caused changes in

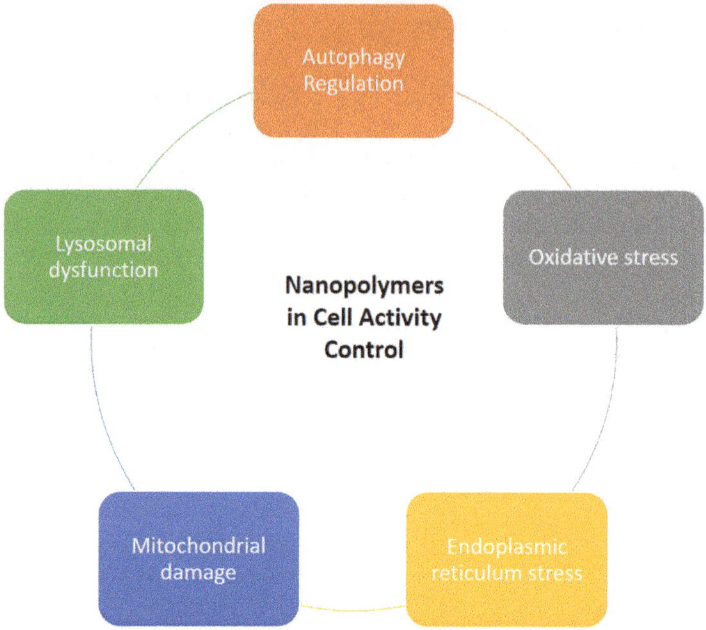

Fig. 9 Nanopolymers in cell activity regulation

mitochondria that led to apoptosis, suggesting a role for both mitochondria-mediated apoptosis and lysosomal endocytosis [94]. Nanoscale PAMAM dendrimers disrupt the mitochondria, most likely by interfering with the mitochondrial membrane, which leads to apoptosis. This process involves interactions with mitochondrial pore complexes and Bcl-2 family proteins, which destabilize membranes and cause cytochrome C to be released, starting the apoptotic process [95]. The research illustrates how biodegradable nanopolymers, particularly PMLA, can modify cellular functions in HER2/neu-positive breast cancer. This nanoplatform targets the HER2/neu receptors present on cancer cells and alters their activity by integrating PMLA with antisense oligonucleotides, trastuzumab, and transferrin receptor antibodies. The ability of nanopolymers to affect cellular behavior in targeted cancer therapies is evidenced by this modulation, which blocks HER2/neu receptor signaling, triggers apoptosis, and reduces tumor growth [96]. A PMLA-based nanopolymer has been developed for the targeted therapy of EGFR-positive triple-negative breast cancer (TNBC). To increase the specificity of medication delivery, this nanopolymer combines transferrin receptor (TfR) antibodies, EGFR antisense oligonucleotides (AON), and the 2C5 anti-tumor monoclonal antibody. In both lab and animal experiments, it effectively inhibits EGFR signaling and tumor growth, which reduces the proliferation of breast cancer cells. The nanopolymer offers a viable approach to treating TNBC by improving treatment efficacy while reducing toxicity and raising specificity [97].

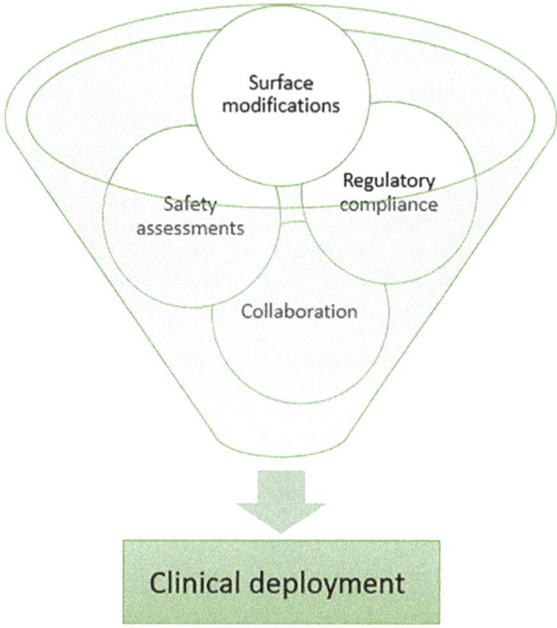

Fig. 10 Pathway to nanopolymer clinical deployment

3 Safety and Regulatory Considerations:

Research indicates that nanopolymers hold significant promise for biomedical applications; however, ensuring their safety and regulatory compliance is crucial for clinical deployment. While short-term studies have not demonstrated acute toxicity, comprehensive long-term assessments are required to determine clearance and safety profiles. Tailored surface modifications may mitigate risks by enhancing bioaccumulation and bioavailability. Since extensive benefit-risk assessments are often necessary for approval processes, regulatory guidelines are vital for standardizing research, ensuring reproducibility, and maintaining product consistency. Collaboration among researchers, industry, and regulatory bodies is essential for the sector to establish clear standards for medicinal products derived from nanopolymers, which include stringent manufacturing protocols and ethical clinical trials. Confirming the safety of these products through biocompatibility assessments, conducting thorough toxicity evaluations, and obtaining informed consent is critical. Crucial steps for verifying these products and ultimately facilitating their secure and effective incorporation into healthcare systems involve confirming biocompatibility, conducting thorough toxicity assessments, and obtaining informed consent (Fig. 10).

References

1. Nanopolymers in medical applications: advances, challenges, and promising innovations for healthcare enhancement. NanoWorld J. **9** (2023)
2. Wang, X., Wang, Y., Chen, Z.G., Shin, D.M.: Advances of cancer therapy by nanotechnology. Cancer Res. Treat. **41**, 1 (2009)
3. Malik, S., Muhammad, K., Waheed, Y.: Emerging applications of nanotechnology in healthcare and medicine. Molecules **28**, 6624 (2023)
4. Sivadasan, D., Sultan, M.H., Madkhali, O., Almoshari, Y., Thangavel, N.: Polymeric lipid hybrid nanoparticles (PLNs) as emerging drug delivery platform—a comprehensive review of their properties, preparation methods, and therapeutic applications. Pharmaceutics **13**, 1291 (2021)
5. Elumalai, K., Srinivasan, S., Shanmugam, A.: Review of the efficacy of nanoparticle-based drug delivery systems for cancer treatment. Biomed. Technol. **5**, 109–122 (2024)
6. Kiriyama, A., Iga, K., Shibata, N.: Availability of polymeric nanoparticles for specific enhanced and targeted drug delivery. Ther. Deliv. **4**, 1261–1278 (2013)
7. Haddad, R., Alrabadi, N., Altaani, B., Li, T.: Paclitaxel drug delivery systems: focus on nanocrystals' surface modifications. Polymers **14**, 658 (2022)
8. Ghosh, B., Biswas, S.: Polymeric micelles in cancer therapy: State of the art. J. Controlled Release **332**, 127–147 (2021)
9. Caminade, A.-M., Turrin, C.-O.: Dendrimers for drug delivery. J Mater Chem B **2**, 4055–4066 (2014)
10. Vitiello, G., et al.: Cationic liposomes as efficient nanocarriers for the drug delivery of an anticancer cholesterol-based ruthenium complex. J. Mater. Chem. B **3**, 3011–3023 (2015)
11. Forssen, E.A., Tökés, Z.A.: Improved therapeutic benefits of doxorubicin by entrapment in anionic liposomes. Cancer Res. **43**, 546–550 (1983)
12. Weiner, N., Martin, F., Riaz, M.: Liposomes as a drug delivery system. Drug Dev. Ind. Pharm. **15**, 1523–1554 (1989)
13. Veronese, F.M., Pasut, G.: PEGylation, successful approach to drug delivery. Drug Discov. Today **10**, 1451–1458 (2005)
14. Singh, V., Sahebkar, A., Kesharwani, P.: Poly (propylene imine) dendrimer as an emerging polymeric nanocarrier for anticancer drug and gene delivery. Eur. Polym. J. **158**, 110683 (2021)
15. Mousavi, S.M., et al.: Bioactive graphene quantum dots based polymer composite for biomedical applications. Polymers **14**, 617 (2022)
16. Mora-Huertas, C.E., Fessi, H., Elaissari, A.: Polymer-based nanocapsules for drug delivery. Int. J. Pharm. **385**, 113–142 (2010)
17. Bahrami, B., et al.: Folate-conjugated nanoparticles as a potent therapeutic approach in targeted cancer therapy. Tumor Biol. **36**, 5727–5742 (2015)
18. Key, J., Park, K.: Multicomponent, tumor-homing chitosan nanoparticles for cancer imaging and therapy. Int. J. Mol. Sci. **18**, 594 (2017)
19. Alford, R., Ogawa, M., Choyke, P.L., Kobayashi, H.: Molecular probes for the in vivo imaging of cancer. Mol. Biosyst. **5**, 1279 (2009)
20. Wickline, S.A., Lanza, G.M.: Nanotechnology for molecular imaging and targeted therapy. Circulation **107**, 1092–1095 (2003)
21. Sim, S., Wong, N.K.: Nanotechnology and its use in imaging and drug delivery (Review). Biomed. Rep. **14**, 42 (2021)
22. Ratner, B.D., Bryant, S.J.: Biomaterials: where we have been and where we are going. Annu. Rev. Biomed. Eng. **6**, 41–75 (2004)
23. Guo, C., Jin, Y., Dai, Z.: Multifunctional ultrasound contrast agents for imaging guided photothermal therapy. Bioconjug. Chem. **25**, 840–854 (2014)
24. Liu, H.-L., Fan, C.-H., Ting, C.-Y., Yeh, C.-K.: Combining microbubbles and ultrasound for drug delivery to brain tumors: current progress and overview. Theranostics **4**, 432–444 (2014)

25. Liu, J., et al.: Low-intensity focused ultrasound (LIFU)-activated nanodroplets as a theranostic agent for noninvasive cancer molecular imaging and drug delivery. Biomater. Sci. **6**, 2838–2849 (2018)
26. Das, S.S., et al.: Stimuli-responsive polymeric nanocarriers for drug delivery, imaging, and theragnosis. Polymers **12**, 1397 (2020)
27. Zhang, X., et al. (2023) NIR light-activated and RGD-conjugated Ultrasmall Fe/PPy nanopolymers for enhanced tumor photothermal ferrotherapy and MR imaging. Chem.—Eur. J. **29**, e202302125 (2023)
28. Kumar, S., Meena, V.K., Hazari, P.P., Sharma, R.K.: PEG coated and doxorubicin loaded multimodal Gadolinium oxide nanoparticles for simultaneous drug delivery and imaging applications. Int. J. Pharm. **527**, 142–150 (2017)
29. Gao, X., et al.: In vivo molecular and cellular imaging with quantum dots. Curr. Opin. Biotechnol. **16**, 63–72 (2005)
30. Lidke, D.S., et al.: Quantum dot ligands provide new insights into erbB/HER receptor–mediated signal transduction. Nat. Biotechnol. **22**, 198–203 (2004)
31. Arias-Ramos, N., et al.: Iron oxide incorporated conjugated polymer nanoparticles for simultaneous use in magnetic resonance and fluorescent imaging of brain tumors. Pharmaceutics **13**, 1258 (2021)
32. Kim, D., Park, S., Lee, J.H., Jeong, Y.Y., Jon, S.: Antibiofouling polymer-coated gold nanoparticles as a contrast agent for in vivo X-ray computed tomography imaging. J. Am. Chem. Soc. **129**, 7661–7665 (2007)
33. Nakagawa, T., et al.: X-ray computed tomography imaging of a tumor with high sensitivity using gold nanoparticles conjugated to a cancer-specific antibody via polyethylene glycol chains on their surface. Sci. Technol. Adv. Mater. **17**, 387–397 (2016)
34. Tian, A., et al.: Polyethylene-glycol-coated gold nanoparticles improve cardiac function after myocardial infarction in mice. Can. J. Physiol. Pharmacol. **96**, 1318–1327 (2018)
35. Namasivayam, S.K.R.: Eco friendly, green route method for the preparation of poly ethylene glycol (PEG) mediated surface modified iron oxide nanoparticles (PEG-IONps) with potential biological activities. Environ. Qual. Manag. **33**, 967–978 (2024)
36. Ma, Q., et al.: Indocyanine green-loaded mesoporous silica nanocomposite for breast cancer imaging. ACS Appl. Nano Mater. **7**, 11377–11385 (2024)
37. Shu, G., et al.: Bismuth drug-inspired ultra-small dextran coated bismuth oxide nanoparticles for targeted computed tomography imaging of inflammatory bowel disease. Biomaterials **311**, 122658 (2024)
38. Yektamanesh, M., et al.: Characterization of multifunctional β-cyclodextrin-coated Bi2O3 nanoparticles conjugated with curcumin for CT imaging-guided synergetic chemo-radiotherapy in breast cancer. Int. J. Pharm. **659**, 124264 (2024)
39. Cen, B., et al.: An efficient bivalent cyclic rgd-pik3cb sirna conjugate for specific targeted therapy against glioblastoma in vitro and in vivo. Mol. Ther. Nucleic Acids **13**, 220–232 (2018)
40. Feng, X., et al.: Conjugated polymer nanoparticles for drug delivery and imaging. ACS Appl. Mater. Interfaces **2**, 2429–2435 (2010)
41. Gulati, S., et al.: Nanobiopolymers in cancer therapeutics: advancing targeted drug delivery through sustainable and controlled release mechanisms. J. Mater. Chem. B **12**, 11887–11915 (2024)
42. Kurul, F., Turkmen, H., Cetin, A.E., Topkaya, S.N.: Nanomedicine: How nanomaterials are transforming drug delivery, bio-imaging, and diagnosis. Nanotechnol. **7**, 100129 (2025)
43. Hillah, I., et al.: Characterization of promising smart nanopolymers that release drugs in response to illness-related signals: a review. Rev. Clin. Pharmacol. Pharmacokinet.—Int. Ed. **38**, 189–192 (2024)
44. Carvalho, A., Fernandes, A.R., Baptista, P.V.: Nanoparticles as delivery systems in cancer therapy. In: Applications of Targeted Nano Drugs and Delivery Systems, pp. 257–295. Elsevier (2019). https://doi.org/10.1016/B978-0-12-814029-1.00010-7
45. Abasian, P., Ghanavati, S., Rahebi, S., Nouri Khorasani, S., Khalili, S.: Polymeric nanocarriers in targeted drug delivery systems: a review. Polym. Adv. Technol. **31**, 2939–2954 (2020)

46. Ahmed, R., Aucamp, M., Samsodien, H.: Surface-modified dendrimer nanoconjugate for targeting delivery of rifampicin. J. Drug Deliv. Sci. Technol. **103**, 106441 (2025)
47. Morgan, M.T., et al.: Dendrimer-encapsulated camptothecins: increased solubility, cellular uptake, and cellular retention affords enhanced anticancer activity in vitro. Cancer Res. **66**, 11913–11921 (2006)
48. Kono, K., et al.: Preparation and cytotoxic activity of poly(ethylene glycol)-modified poly(amidoamine) dendrimers bearing adriamycin. Biomaterials **29**, 1664–1675 (2008)
49. Tafti, M.F., Fayyaz, Z., Aghamollaei, H., Jadidi, K., Faghihi, S.: Drug delivery strategies to improve the treatment of corneal disorders. Heliyon **11**, e41881 (2025)
50. Lin, H., et al.: Subconjunctival dendrimer-drug therapy for the treatment of dry eye in a rabbit model of induced autoimmune dacryoadenitis. Ocul. Surf. **16**, 415–423 (2018)
51. Shi, Z., et al.: Bola-amphiphilic dendrimer enhances imatinib to target metastatic ovarian cancer via β-Catenin-HRP2 signaling axis. ACS Appl. Mater. Interfaces **17**, 2884–2898 (2025)
52. Chau, W.K., Ip, C.K., Mak, A.S.C., Lai, H.-C., Wong, A.S.T.: C-Kit mediates chemoresistance and tumor-initiating capacity of ovarian cancer cells through activation of Wnt/β-catenin-ATP-binding cassette G2 signaling. Oncogene **32**, 2767–2781 (2013)
53. Bi, X., et al.: Polyamidoamine dendrimer-mediated hydrogel for solubility enhancement and anti-cancer drug delivery. J. Biomater. Appl. **38**, 733–742 (2024)
54. Bai, S., Thomas, C., Ahsan, F.: Dendrimers as a carrier for pulmonary delivery of enoxaparin, a low-molecular weight heparin. J. Pharm. Sci. **96**, 2090–2106 (2007)
55. Soltany, P., Miralinaghi, M., Pajoum Shariati, F.: Folic acid conjugated poly (Amidoamine) dendrimer grafted magnetic chitosan as a smart drug delivery platform for doxorubicin: In-vitro drug release and cytotoxicity studies. Int. J. Biol. Macromol. **257**, 127564 (2024)
56. Uram, Ł, et al.: The importance of biotinylation for the suitability of cationic and neutral fourth-generation polyamidoamine dendrimers as targeted drug carriers in the therapy of glioma and liver cancer. Molecules **29**, 4293 (2024)
57. Lee, J.W., et al.: Multifunctional dendrimer-peptide conjugates for MET receptor-specific imaging of cancer cells. Nano Today **59**, 102509 (2024)
58. Cao, Y., Liao, S., Deng, C., Qin, H., Li, Y.: A pH-responsive phase-transition bi-affinity nanopolymer-assisted exosome metabolomics for early screening of osteoarthritis. Talanta **283**, 127144 (2025)
59. Franiak-Pietryga, I., Ziemba, B., Messmer, B., Skowronska-Krawczyk, D.: Dendrimers as drug nanocarriers: the future of gene therapy and targeted therapies in cancer. In: Simonescu, C. M. (ed.) Dendrimers—Fundamentals and Applications. InTech (2018). https://doi.org/10.5772/intechopen.75774
60. Endo, K., et al.: Tumor-targeted chemotherapy with the nanopolymer-based drug NC -6004 for oral squamous cell carcinoma. Cancer Sci. **104**, 369–374 (2013)
61. Chen, Z., Krishnamachary, B., Bhujwalla, Z.: Degradable dextran nanopolymer as a carrier for choline kinase (ChoK) siRNA cancer therapy. Nanomaterials **6**, 34 (2016)
62. Sambi, M., DeCarlo, A., Malardier-Jugroot, C., Szewczuk, M.R.: Next-generation multi-modality of nanomedicine therapy: size and structure dependence of folic acid conjugated copolymers actively target cancer cells in disabling cell division and inducing apoptosis. Cancers **11**, 1698 (2019)
63. Takke, A., Shende, P.: Magnetic-core-based silibinin nanopolymeric carriers for the treatment of renal cell cancer. Life Sci. **275**, 119377 (2021)
64. Mishra, V., Jain, N.K.: Acetazolamide encapsulated dendritic nano-architectures for effective glaucoma management in rabbits. Int. J. Pharm. **461**, 380–390 (2014)
65. Wang, M., Yu, F., Zhang, Y.: Present and future of cancer nano-immunotherapy: opportunities, obstacles and challenges. Mol. Cancer **24**, 26 (2025)
66. Nair, R.R., Hyun, J.H., Kim, J., Jung, K.O., Kim, D.: Recent progress in the development of cellulose-derived organic-nanopolymer and coordination network platforms for application as optical chemosensors. Adv. Compos. Hybrid Mater. **8**, 30 (2025)
67. Malini, S., et al.: Sensing beyond senses: an overview of outstanding strides in architecting nanopolymer-enabled sensors for biomedical applications. Polymers **14**, 601 (2022)

68. Haldavnekar, R., Venkatakrishnan, K., Tan, D.B.: Boosting the sub-cellular biomolecular cancer signals by self-functionalized tag-free nano sensor. Biosens. Bioelectron. **190**, 113407 (2021)
69. Saha, P., Panda, D., Paul, R., Dash, J.: A DNA nanosensor for monitoring ligand-induced i-motif formation. Org. Biomol. Chem. **19**, 1965–1969 (2021)
70. Yarman, A., Kurbanoglu, S., Zebger, I., Scheller, F.W.: Simple and robust: the claims of protein sensing by molecularly imprinted polymers. Sens. Actuators B Chem. **330**, 129369 (2021)
71. Zhang, Y., et al.: A simplified antibody-responsive fluorescence biosensor by proximity ligation-induced quasi-Y-DNA assembly to illuminate emissive DNA looped-Ag nanocluster. Sens. Actuators B Chem. **423**, 136781 (2025)
72. Liang, Q.-H., Cao, B.-P., Xiao, Q., Wei, D.: The application of graphene field-effect transistor biosensors in COVID-19 detection technology: a review. Sensors **23**, 8764 (2023)
73. Grate, J.W., et al.: Alexa fluor-labeled fluorescent cellulose nanocrystals for bioimaging solid cellulose in spatially structured microenvironments. Bioconjug. Chem. **26**, 593–601 (2015)
74. Sangubotla, R., Gubbiyappa, K.S., Devarapogu, R., Kim, J.: Modern insights of nanotheranostics in the glioblastoma: an updated review. Biochim. Biophys. Acta BBA - Mol. Basis Dis. **1871**, 167653 (2025)
75. Liu, Z., et al.: Oxidized nanocellulose facilitates preparing photoluminescent nitrogen-doped fluorescent carbon dots for Fe3+ ions detection and bioimaging. Chem. Eng. J. **384**, 123260 (2020)
76. Saylan, Y., Aliyeva, N., Eroglu, S., Denizli, A.: Nanomaterial-based sensors for coumarin detection. ACS Omega **9**, 30015–30034 (2024)
77. Pal, K., et al.: A critical review on multifunctional smart materials 'nanographene' emerging avenue: nano-imaging and biosensor applications. Crit. Rev. Solid State Mater. Sci. **47**, 691–707 (2022)
78. Luo, M., et al.: Chemiluminescence biosensors for DNA detection using graphene oxide and a horseradish peroxidase-mimicking DNAzyme. Chem. Commun. **48**, 1126–1128 (2012)
79. Yan, T., et al.: Chitosan based pH-responsive polymeric prodrug vector for enhanced tumor targeted co-delivery of doxorubicin and siRNA. Carbohydr. Polym. **250**, 116781 (2020)
80. Liu, J., et al.: A titanium nitride nanozyme for pH-responsive and irradiation-enhanced cascade-catalytic tumor therapy. Angew. Chem. Int. Ed. **60**, 25328–25338 (2021)
81. Yao, H., et al.: Effective melanoma immunotherapy with interleukin-2 delivered by a novel polymeric nanoparticle. Mol. Cancer Ther. **10**, 1082–1092 (2011)
82. Jia, S., et al.: Aptamer-modified melittin micelles efficiently inhibit osteosarcoma deterioration by inducing immunogenic cell death. Colloids Surf. B Biointerfaces **249**, 114512 (2025)
83. Teng, Y., et al.: Molecule interacting with CasL-2 enhances tumor progression and alters radiosensitivity in cervical cancer. J. Transl. Med. **23**, 44 (2025)
84. Lin, Y., Su, R., Huang, Y.: Thermo-responsive fluorescent nanopolymer delivery platform for treatment of diffuse large B-cell lymphoma (DLBCL). J. Fluoresc. (2025). https://doi.org/10.1007/s10895-024-04092-y
85. Gao, L., et al.: Nano-drug delivery system for the treatment of multidrug-resistant breast cancer: current status and future perspectives. Biomed. Pharmacother. **179**, 117327 (2024)
86. Ljubimov, V.A., et al.: Neurosurgery at the crossroads of immunology and nanotechnology: new reality in the COVID-19 pandemic. Adv. Drug Deliv. Rev. **181**, 114033 (2022)
87. İMamoğlu, R., Kaplan, Ö., Gök, M.K., Gökçe, İ.: Polymer-based transfection agents used in CRISPR-CAS9 system. Türk Doğa Ve Fen Derg. **11**, 151–156 (2022)
88. Zhang, Z., et al.: Cationic polymer-mediated CRISPR/Cas9 plasmid delivery for genome editing. Macromol. Rapid Commun. **40**, 1800068 (2019)
89. Saadatpour, F., Mohammadipanah, F., Zohourian, N., Hodjat, M.: Potential of nanoparticles and nanopolymers in treatment of age-associated diseases. Adv. Nat. Sci. Nanosci. Nanotechnol. **12**, 045008 (2021)
90. Singla, R.K., et al.: The role of nanomaterials in enhancing natural product translational potential and modulating endoplasmic reticulum stress in the treatment of ovarian cancer. Front. Pharmacol. **13**, 987088 (2022)

91. Li, Z., et al.: Apoptotic induction and anti-metastatic activity of eugenol encapsulated chitosan nanopolymer on rat glioma C6 cells via alleviating the MMP signaling pathway. J. Photochem. Photobiol. B. **203**, 111773 (2020)
92. Feng, T., et al.: Exosome camouflaged coordination-assembled Iridium(III) photosensitizers for apoptosis-autophagy-ferroptosis induced combination therapy against melanoma. Biomaterials **301**, 122212 (2023)
93. Zhang, X., et al.: The effect of autophagy inhibitors on drug delivery using biodegradable polymer nanoparticles in cancer treatment. Biomaterials **35**, 1932–1943 (2014)
94. Thomas, T.P., et al.: Cationic poly(amidoamine) dendrimer induces lysosomal apoptotic pathway at therapeutically relevant concentrations. Biomacromol **10**, 3207–3214 (2009)
95. Lee, J.-H., et al.: Nanosized polyamidoamine (PAMAM) dendrimer-induced apoptosis mediated by mitochondrial dysfunction. Toxicol. Lett. **190**, 202–207 (2009)
96. Inoue, S., et al.: Polymalic acid-based nanobiopolymer provides efficient systemic breast cancer treatment by inhibiting both HER2/neu receptor synthesis and activity. Cancer Res. **71**, 1454–1464 (2011)
97. Inoue, S., et al.: Nanobiopolymer for direct targeting and inhibition of EGFR expression in triple negative breast cancer. PLoS ONE **7**, e31070 (2012)

Nano Polymers Applications in Food and Beverage Industries

Vijay Chaudhary, Bhasha Sharma, Chanchal Ahlawat, and Shashank Shekhar

Abstract Nanopolymers, characterized by their nanoscale structural features, represent a groundbreaking advancement in material science, offering properties far superior to conventional polymers. Their high surface area, enhanced mechanical strength, and functional versatility make them ideal for addressing complex challenges across various industries, including food and beverages. This chapter explores the innovative use of nanostructured polymers in enhancing the safety, quality, and sustainability of food and beverage systems. It delves into their active and intelligent packaging roles, improving barrier properties, antimicrobial efficacy, and environmental sustainability. Additionally, the chapter highlights nanostructured polymer coatings for storage stability. Furthermore, the chapter highlights the challenges and regulatory considerations that drives the adoption of nanopolymers.

1 Introduction

Nanopolymers, a class of advanced polymeric materials engineered by adding nanofillers at the nanoscale (1–100 nm), represent a transformative innovation in modern material science [1]. Their unique properties, including high surface area, enhanced mechanical strength, tunable functional characteristics, and improved thermal and barrier performance, make them an invaluable resource across various

---

V. Chaudhary (✉)
Department of Mechanical Engineering, Amity School of Engineering and Technology, Amity University Uttar Pradesh, Noida 201313, India
e-mail: vijaychaudhary111@gmail.com

B. Sharma
Department of Chemistry, Deen Dayal Upadhyaya College, University of Delhi, New Delhi, India

C. Ahlawat
School of Computer Science Engineering & Technology (SCSET), Bennett University, Greater Noida 201310, Uttar Pradesh, India
e-mail: chanchalchaudhary827@gmail.com

S. Shekhar
Department of Chemistry, Sri Aurobindo College, University of Delhi, Delhi, India

B. Sharma et al. (eds.), *Nanostructured Polymers*, Engineering Materials,
https://doi.org/10.1007/978-981-95-3581-1_16

Table 1 Classification of nanoparticles based on dimensions [1]

Nanoparticles (nano-nm)	Measurements (nm)		
1-D	Plates/sheets		$t < 100$
	Laminates and shells		$t = 100$
2-D	Nanotubes	$d < 100$	
	Nanofibers	$d < 100$	
3-D	Spherical	$t, d < 100$	
	Silica blends	$t, d < 100$	

t = thickness (nm), d = diameter (nm)

industries [2]. Among these, the food and beverage industry has emerged as a significant beneficiary, leveraging nanopolymers to address challenges related to food safety, quality, shelf life, and sustainability. Nanopolymers are characterized by their nanoscale structural features, which impart distinctive properties that conventional polymers lack [3]. These include improved interaction with environmental factors such as gases, moisture, and microbes, and the ability to be functionalized for specific roles [4]. The incorporation of nanoparticles, nanofibers, or nanocomposites into polymer matrices further enhances their performance, enabling applications that range from active and intelligent packaging to antimicrobial coatings [5].

Nanopolymers are synthesized using a variety of methods, including in situ polymerization, sol–gel techniques, and blending with nanoparticles such as clays, silicates, or metallic oxides [6]. These methods allow for precise control over material properties, tailoring them to meet the diverse demands of the food and beverage industry. Fillers at nanoscale dimensions are capable of producing high surface areas even more than 1000 m^2/g [7]. They significantly improve the properties of the materials. A broad classification of nanoparticles (based on dimensions) and the classification based on chemical nature that are generally used to develop nano polymers were portrayed in Table 1 and Fig. 2 respectively.

The representative synthesis methods of nanoparticles (Fig. 1) is the most important step which in turn helps to control the properties of the nano polymers. A brief information about the methods is highlighted in Table 2. A schematic representation of the overall procedure for the development of packaging materials containing nanoparticles is shown in Fig. 3.

1.1 *Relevance to Food and Beverage Industries*

The food and beverage industry is characterized by stringent requirements for quality, safety, and sustainability, alongside the challenges posed by globalization, extended supply chains, and consumer demands for convenience and functionality. Traditional materials often fall short of meeting these evolving demands. Nanopolymers, with their advanced capabilities, offer solutions that are redefining industry standards.

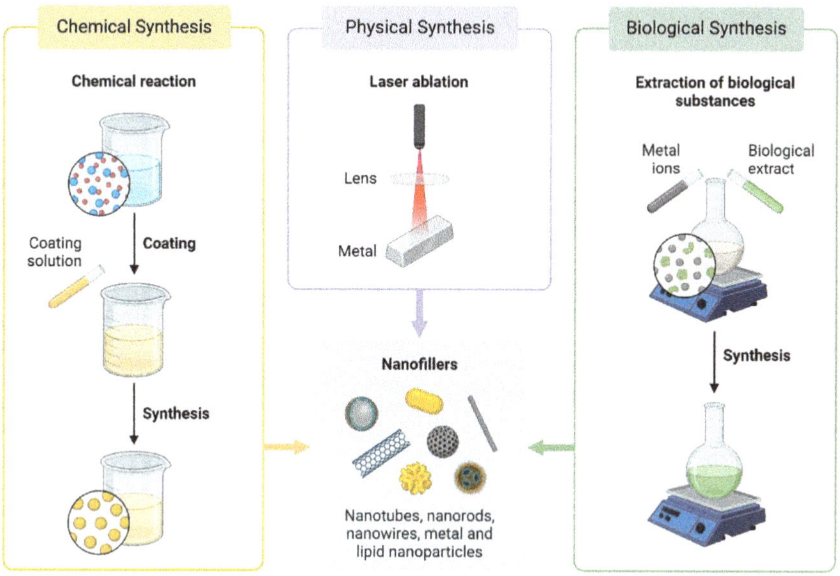

Fig. 1 Synthesis methods of nanoparticles. Created in https://BioRender.com

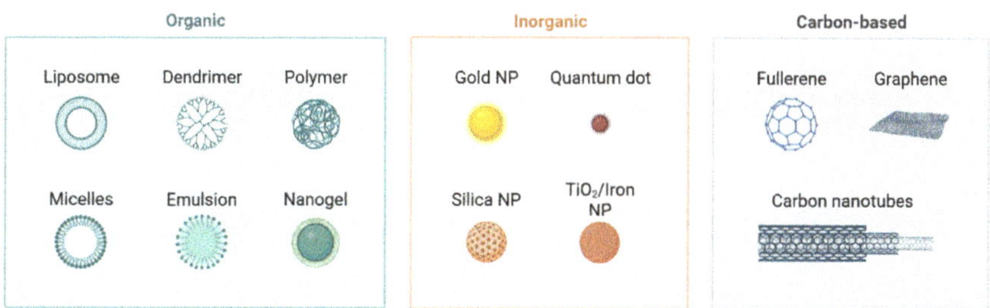

Fig. 2 Classification of nanoparticles. Created in https://BioRender.com

The key aspects on the use of nanopolymers in food and beverage industries are highlighted in Table 3.

1.2 Nano Polymers in Packaging

Packaging is pivotal in the food and beverage industry, acting as the primary barrier to external factors such as oxygen, moisture, light, and microbial contamination [14]. Recent advancements in nanotechnology have enabled the integration of nanopolymers into packaging materials, transforming them from passive barriers to active and intelligent systems capable of enhancing food quality, safety, and shelf life

Table 2 Different synthesis methods [8]

Synthesis method	Steps/process	Outputs	Applications
Chemical synthesis	– Chemical reaction occurs to produce a coating solution – Coating is applied to substrates – Nanoparticles are synthesised through controlled reactions (e.g., sol–gel, precipitation)	– Nanoparticles	– Functional coatings for packaging and electronics – Creation of nanocomposites
Physical synthesis	– Laser ablation involves focusing a high-energy laser on a solid target (e.g., metal) – Laser energy produces nanoparticles in the medium	–Nanotubes, nanorods, nanowires Metal nanoparticles	– High-purity nanomaterials for electronics, optics, and high-strength materials
Biological synthesis	– Biological substances (e.g., enzymes, plant extracts) are combined with metal ions – Biological compounds reduce metal ions into nanoparticles in a sustainable and eco-friendly manner	– Biocompatible nanoparticles	– Medical, pharmaceutical, and eco-friendly applications

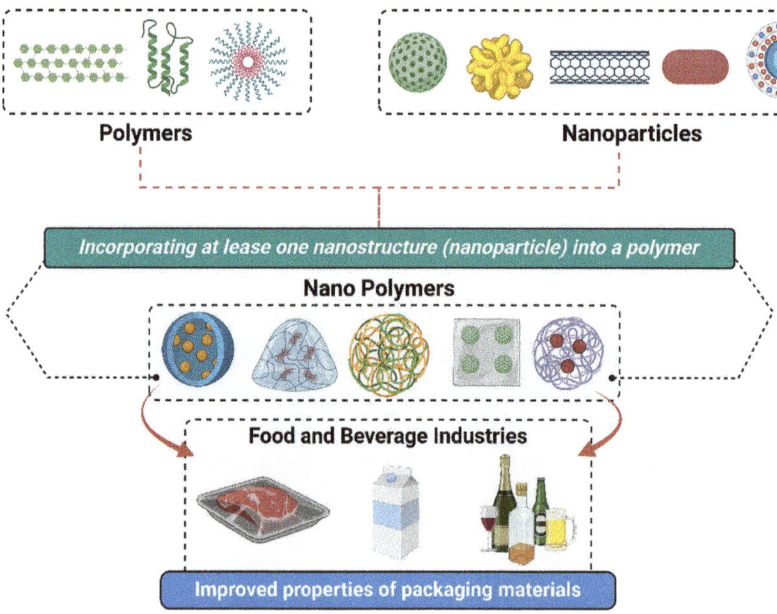

Fig. 3 Schematic representation of the overall procedure for the development of packaging materials containing nanoparticles. Created in https://BioRender.com

Table 3 Key aspects of using nano polymers

Key aspects	Description	References
Enhancing food safety	Nanopolymers play a pivotal role in improving food safety by incorporating antimicrobial agents or oxygen scavengers within packaging materials. These active packaging systems help inhibit microbial growth and oxidation, thereby maintaining food quality and reducing spoilage	[9]
Revolutionizing packaging	Packaging is integral to the preservation and presentation of food products. Nano polymer-based packaging materials offer superior barrier properties against oxygen, moisture, and UV light, extending product shelf life. Additionally, intelligent packaging systems, which integrate sensors or colorimetric indicators into nano polymer matrices, provide real-time monitoring of food freshness and safety	[10]
Minimize food waste	With food waste being a significant global concern, nanopolymers contribute to waste reduction by ensuring prolonged shelf life and quality of perishable products. Their active and intelligent functionalities prevent spoilage during storage and transportation, addressing both environmental and economic challenges	[11]
Sustainability and biodegradable alternatives	The increasing focus on sustainability has driven the development of nanopolymers derived from renewable resources, such as chitosan, polylactic acid (PLA), and starch. These biodegradable materials not only provide an eco-friendly alternative to conventional plastics but also offer comparable or superior performance in packaging applications	[12]
Functional food and beverages	The demand for functional foods and beverages enriched with bioactive compounds has surged in recent years. Nanopolymers facilitate the encapsulation, protection, and controlled release of nutrients, flavours, and antioxidants. These delivery systems enhance the stability and bioavailability of active compounds, providing significant health benefits to consumers	[13]

[15]. Nanopolymers exhibit remarkable properties, including enhanced mechanical strength, barrier efficiency, antimicrobial activity, and responsiveness to environmental stimuli, making them indispensable in modern food packaging systems. A broad classification of packaging systems is shown in Fig. 4.

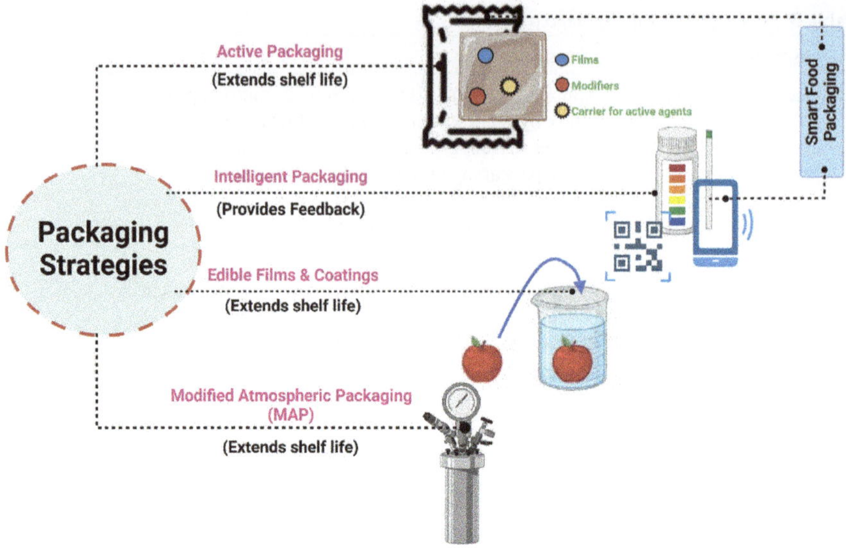

Fig. 4 Classification of packaging systems. Created in https://BioRender.com

1.3 *Active Packaging Technologies*

Active packaging goes beyond the traditional role of being a passive barrier to actively interacting with the packaged food or its environment to extend shelf life and improve quality [16]. Nanostructured polymers have enabled active packaging systems by incorporating functional agents into their matrix, allowing them to release or absorb substances as needed as shown in Fig. 5.

- **Oxygen Scavengers**

  Oxygen is a primary factor in food spoilage, as it promotes microbial growth and oxidative reactions that degrade flavour, colour, and nutrients. Nanopolymers embedded with oxygen-scavenging nanoparticles (e.g., iron oxide or cobalt complexes) can remove residual oxygen from packaging environments [17]. This technology is especially effective for oxygen-sensitive foods, such as processed meats, dairy products, and baked goods. For instance, nanocomposite films with iron oxide have shown remarkable efficiency in maintaining the freshness of snack foods [18].

- **Ethylene Scavengers**

  Ethylene gas, a natural plant hormone, accelerates the ripening and senescence of fruits and vegetables, leading to spoilage during storage and transportation [19]. Polymers incorporating ethylene-scavenging agents, such as potassium permanganate, activated carbon, or zeolites, are highly effective in absorbing ethylene and slowing down the ripening process [19]. For example, polymer films embedded with nano clay

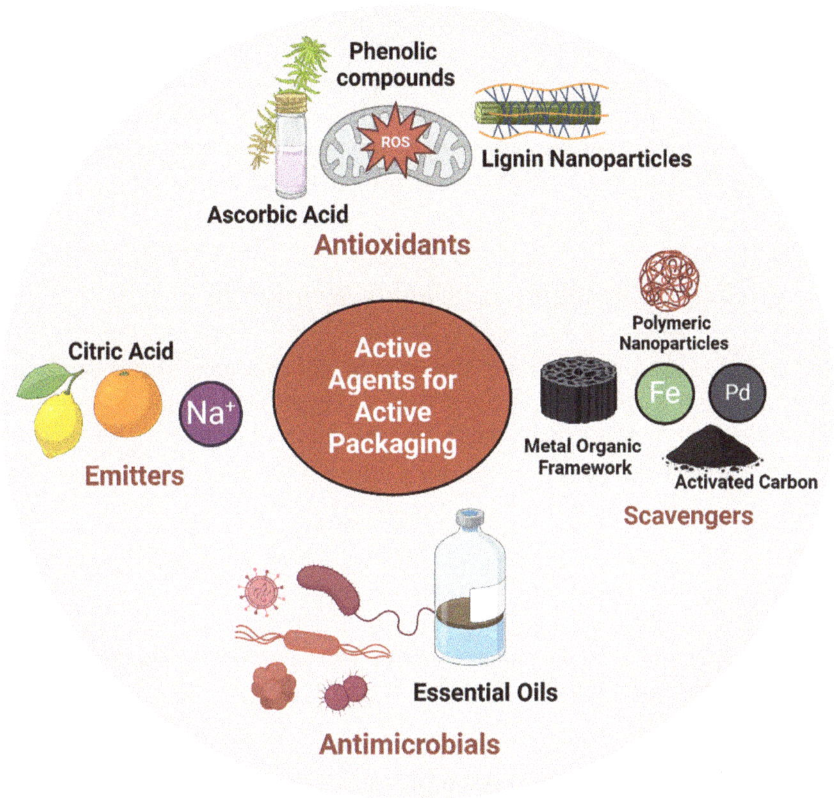

Fig. 5 Active agents used in active packaging. The agents are incorporated with the polymers. Created in https://BioRender.com

and potassium permanganate nanoparticles have been developed for packaging fresh produce, extending shelf life by reducing ethylene concentrations [20]. These ethylene scavengers are particularly useful for perishable items like bananas, avocados, and tomatoes, ensuring better quality and reduced waste along the supply chain [21].

- *Moisture Absorbers*

Humidity can compromise the texture and safety of moisture-sensitive foods, such as cereals and chips. Polymers containing silica nanoparticles or montmorillonite clays act as moisture absorbers, maintaining an optimal humidity level inside the packaging. These materials can absorb excess water vapor, reducing the risk of microbial growth and preserving the crispness of food products [22].

- *Antimicrobial Properties*

The integration of antimicrobial agents, such as silver nanoparticles, zinc oxide, or chitosan, into nanostructured polymers provides effective protection against microbial contamination. These active packaging systems inhibit the growth of bacteria, yeasts, and molds, ensuring food safety and prolonging shelf life. For example, films with silver nanoparticles have demonstrated antimicrobial activity against common foodborne pathogens like *Escherichia coli* and *Salmonella* [23].

1.4 Intelligent Packaging for Quality Monitoring

Intelligent packaging incorporates nanosensors and indicators that monitor and convey information about the condition of the food or the environment in real-time. This innovation provides a more transparent and informed approach to food quality and safety management.

- *Gas Sensors*

Many food products release gases, such as carbon dioxide or ammonia, as they spoil. Nanostructured polymer films integrated with gas sensors can detect these spoilage markers and signal their presence. For example, polyaniline-based nanocomposites are used to monitor carbon dioxide levels in meat packaging, providing an early warning of spoilage [24].

- *Colorimetric Indicators*

These smart packaging systems utilize pH-sensitive nanostructured polymers to indicate freshness visually [25]. For instance, films containing anthocyanin-embedded nano polymers change colour in response to pH changes caused by spoilage [26]. This is particularly useful for monitoring the freshness of fish, poultry, and dairy products.

- *Temperature and Humidity Sensors*

Maintaining optimal storage conditions is critical for food quality. Nanostructured polymers combined with thermochromic materials change colour when exposed to temperature fluctuations beyond the recommended range [27]. Similarly, humidity-sensitive polymers provide real-time feedback on moisture levels inside the packaging. These systems ensure that the food remains within safe storage parameters during transportation and storage [28].

1.5 Barrier Property Enhancements

One of the most significant contributions of nanostructured polymers is their ability to enhance the barrier properties of packaging materials. These materials are designed to reduce the permeability of gases, moisture, and light, thereby extending the shelf life of food products.

- *Oxygen and Moisture Barriers*

Nanocomposites containing clay nanoparticles or graphene oxide create a dense, tortuous pathway that slows the diffusion of oxygen and water vapour. This property is particularly beneficial for cheese, snacks, and ready-to-eat meals, where maintaining low oxygen and moisture levels is critical. For instance, polymer films with montmorillonite clay have demonstrated a 50–70% reduction in oxygen permeability compared to conventional materials [29].

- *UV Light Protection*

Ultraviolet (UV) light can cause oxidation and nutrient degradation in certain foods, such as oils and beverages [19]. Nanostructured polymers doped with titanium dioxide or zinc oxide nanoparticles effectively block UV light, protecting the food from photodegradation. These materials are widely used in packaging for juices, wines, and oils [30].

- *Mechanical Strength and Durability*

Incorporating nanostructures, such as carbon nanotubes or cellulose nanocrystals, into polymer matrices enhances the mechanical properties of the packaging. This results in more robust and puncture-resistant films, which are especially important for protecting fragile or high-value food products during transport and handling [31].

Polymers form the backbone of modern food and beverage packaging due to their versatile properties, such as flexibility, strength, transparency, and barrier performance. Commonly used materials like polyethylene (PE), polypropylene (PP), and polyethylene terephthalate (PET) dominate the industry due to their low cost, durability, and adaptability for various forms of packaging, from bottles to flexible films [20]. Advanced polymers like polylactic acid (PLA) and cellulose-based materials are gaining traction as sustainable alternatives due to their biodegradability, addressing growing environmental concerns [32]. Moreover, speciality polymers like nylon and ethylene vinyl alcohol (EVOH) are critical for applications requiring superior barrier properties, such as vacuum-sealed packaging. Emerging materials like chitosan and other biodegradable polymers bring added functionalities like antimicrobial activity and eco-friendliness, making them suitable for active and intelligent packaging solutions in the industry [22].

Incorporating nanoparticles into these polymers further enhances their functional properties, enabling applications that extend beyond traditional uses. Nanoparticles such as nanoclays, silver, titanium dioxide, and zinc oxide improve mechanical strength, thermal stability, and barrier performance, while reducing gas and moisture permeability [6]. The commonly used polymers in the food and beverage industries are shown in Table 4.

Table 4 Polymers used in food and beverage industries

Polymer	Characteristics	Applications	References
Polyethylene (PE)	LDPE: Flexible, moisture-resistant, good clarity; HDPE: Rigid, high strength-to-density ratio	Flexible films, milk cartons, caps and closures	[33, 34]
Polypropylene (PP)	High melting point, lightweight, good chemical resistance, durable	Food containers, beverage cups, yogurt packaging	
Polyethylene terephthalate (PET)	Excellent clarity, high strength, lightweight, excellent gas and moisture barrier properties	Bottles for water, juices, carbonated beverages	
Polylactic acid (PLA)	Biodegradable, compostable, derived from renewable resources, good clarity	Eco-friendly food packaging, disposable cups and cutlery	
Polyvinyl chloride (PVC)	Excellent clarity, good chemical resistance, can be rigid or flexible	Shrink wraps, cling films, blister packaging	
Polystyrene (PS)	GPPS: Transparent and brittle; HIPS: Impact-resistant; EPS: Lightweight, insulating	Disposable cups, plates, food trays, protective foam	
Polyamide (Nylon)	High mechanical strength, puncture-resistant, good oxygen barrier	Vacuum-sealed bags, meat and cheese packaging	
Ethylene vinyl alcohol (EVOH)	Exceptional oxygen barrier, transparent, used in multilayer films, moisture-sensitive	Multilayer packaging for perishable items	
Polycarbonate (PC)	High impact resistance, excellent clarity, heat-resistant, reusable	Reusable water bottles, baby bottles, beverage dispensers	
Cellulose-based polymers	Biodegradable, moderate barrier properties, natural cellulose base	Food wraps, paperboard coatings, eco-friendly containers	
Acrylic polymers (PMMA)	High clarity, excellent UV resistance, good chemical resistance	Protective coatings for labels and lids	
Chitosan-based polymers	Biodegradable, antimicrobial properties, excellent oxygen barrier	Edible films, active packaging for fresh produce	
Polyvinyl alcohol (PVA)	Water-soluble, excellent oxygen barrier in dry conditions, biodegradable	Multilayer food packaging, beverage capsules	
Polyethylene naphthalate (PEN)	Superior gas and moisture barrier, high thermal stability, premium applications	High-end beverage bottles, films for sensitive foods	

1.6 Coatings Using Nano Polymers in Food and Beverage Industries

Nano polymer coatings represent a transformative approach in the food and beverage industry, providing innovative solutions for enhancing product quality, safety, and shelf life [6]. These coatings, when applied to food surfaces or beverage containers, offer a protective layer that prevents contamination, inhibits spoilage, and maintains freshness. By leveraging the unique properties of nanopolymers, such as their high surface area, tunable functionality, and improved barrier performance, these coatings address critical challenges in modern packaging and storage systems [35].

1.7 Coating Technologies for Shelf-Life Extension

Nano polymer-based coatings are specifically designed to act as active or passive barriers against environmental factors that compromise food and beverage quality. These coatings are applied to food surfaces, packaging materials, or container interiors to protect products from microbial contamination, oxidation, and moisture ingress [36].

- *Protection against contaminants*

Nano polymer coatings with antimicrobial agents, such as silver nanoparticles or chitosan, actively inhibit the growth of pathogens and spoilage microorganisms. These coatings are especially effective for fresh produce, meats, and dairy products, extending their shelf life and reducing foodborne illness risks [37].

- *Oxygen and moisture barrier*

Oxygen and moisture are two major factors that degrade food quality. Nano polymer coatings incorporating nanoclays or graphene oxide create a dense network that reduces oxygen and water vapour permeability, thereby preserving the texture, flavour, and nutritional value of packaged foods. For example, coatings for fresh fruits can delay ripening by minimizing oxygen exposure and ethylene effects [38].

- *Enhancement in functional properties*

Functionalized nano polymer coatings can provide additional benefits, such as UV protection and flavour retention. Titanium dioxide nanoparticles, for instance, block UV light, preventing photodegradation of sensitive food products [39].

1.8 Applications in Beverage Containers

Nano polymer coatings play a crucial role in the beverage industry, where maintaining product quality during storage and transportation is essential. These coatings are specifically tailored to address challenges related to oxygen ingress, carbonation retention, and microbial contamination [40].

- *Reducing oxygen diffusion in wine bottles*

Oxygen ingress in wine bottles can lead to oxidation, altering the taste and aroma of the wine. Nano polymer coatings with enhanced barrier properties prevent oxygen from penetrating through the bottle or closure, preserving the quality and extending the shelf life of wine [41].

- *Improving carbonation retention in soft drinks*

Soft drinks rely on maintaining carbonation levels for taste and quality. Nano polymer coatings applied to the interior of PET bottles or aluminium cans reduce CO_2 permeability, ensuring that beverages retain their fizz over extended periods [42].

- *Protective coatings for glass and metal containers*

Beverage containers made of glass or metal are often treated with nano polymer coatings to enhance scratch resistance, reduce breakage, and prevent corrosion. For example, epoxy-based nano polymer coatings are commonly used in aluminium beverage cans to prevent interaction between the liquid and the metal surface, ensuring product safety and flavour integrity [43].

Nano polymer coatings are an essential innovation in the food and beverage industry, providing robust solutions to extend shelf life, maintain product quality, and enhance consumer safety. By protecting against environmental factors like oxygen, moisture, and UV light, and by preserving the integrity of beverages during storage, these coatings ensure that products reach consumers in optimal condition. As research and development in nanotechnology advance, nano polymer coatings are expected to play an even more significant role in sustainable and intelligent packaging solutions. A wide range of polymers used for coating using nanoparticles is shown in Table 5.

1.9 Engineered Nano Polymers: Towards Sustainability

The integration of nanotechnology into polymer-based packaging systems has brought transformative innovations that address pressing environmental and sustainability challenges. Nano polymers are playing a critical role in reducing plastic waste, improving recyclability, and mitigating food waste, contributing to the development of eco-friendly and resource-efficient packaging solutions.

Table 5 Polymers for coatings using nanoparticles

Polymer	Characteristics	Applications in coatings	References
Polyethylene terephthalate (PET)	High mechanical strength, good barrier properties for gases and moisture	Interior coatings for bottles to improve oxygen and CO_2 barrier properties	[44–47]
Polypropylene (PP)	Lightweight, chemical resistance, and good processability	Thin nano polymer coatings for food containers	
Polylactic acid (PLA)	Biodegradable, compostable, good film-forming capability	Eco-friendly coatings for biodegradable packaging	
Chitosan	Biodegradable, antimicrobial properties, and excellent film-forming ability	Antimicrobial coatings for fruits, vegetables, and other perishable foods	
Cellulose derivatives	Biodegradable, renewable, excellent barrier against oxygen and aroma compounds	Edible coatings and biodegradable barrier films	
Nylon (Polyamide)	High tensile strength, excellent barrier to oxygen and fats	Protective coatings for vacuum-sealed packaging and beverage containers	
Ethylene vinyl alcohol (EVOH)	Exceptional oxygen barrier properties, good transparency	Multilayer coatings to improve gas barrier properties in plastic containers	
Polyurethane (PU)	Durable, flexible, and resistant to abrasion and chemicals	Coatings for beverage cans and glass containers to prevent corrosion and improve longevity	
Acrylic polymers	UV-resistant, glossy finish, and chemical stability	Protective coatings for metal and glass beverage containers	
Silicone-based polymers	Heat-resistant, hydrophobic, and flexible	Anti-sticking and waterproof coatings for food handling equipment and containers	

1.10 Protein and Polysaccharides Based Polymers

Biodegradable polymers are engineered to decompose naturally under environmental conditions, offering a sustainable alternative to conventional plastics. By combining biodegradability with the functional advantages of nanotechnology, these materials are revolutionizing the packaging industry [15]. A broad classification of polymers (Protein and Polysaccharide) that are derived from renewable sources is shown in Figure 6.

Some of the key characteristics of biobased polymers are highlighted in Table 6.

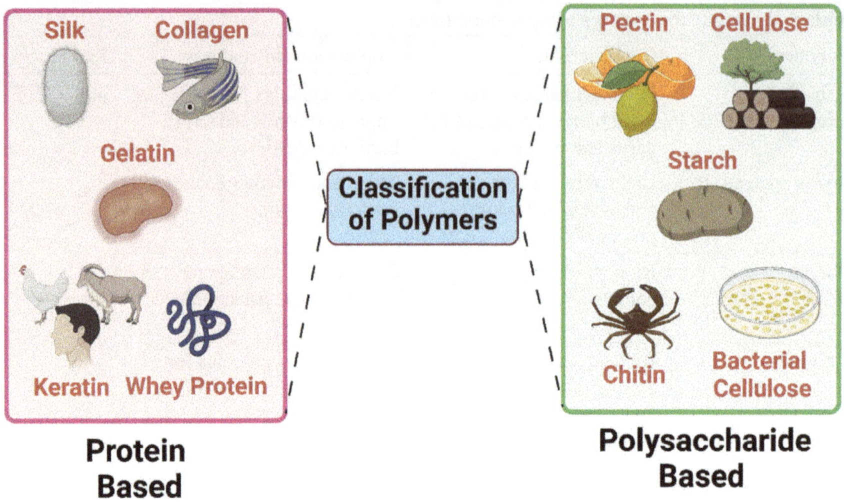

Fig. 6 Protein and polysaccharide based polymers. Created in https://BioRender.com

Table 6 Key characteristics of biobased polymers

Key factors	Description	Reference
Renewable sources	Derived from natural materials such as starch, cellulose, chitosan, and polylactic acid (PLA)	[48]
Environmental benefits	Capable of breaking down into non-toxic residues like carbon dioxide, water, and biomass under composting or natural conditions	[49]
Enhanced properties	Nanoparticles, such as nano clays or cellulose nanocrystals, are incorporated into these materials to improve their mechanical strength, thermal stability, and barrier performance	[50]

1.11 Applications in Packaging and Beverage Industries

- *Edible Coatings*: Biodegradable polymers like chitosan and cellulose derivatives are used as edible coatings for fruits and vegetables to protect against microbial contamination and moisture loss [51].
- *Compostable Packaging Films*: PLA-based nanocomposites are used in flexible packaging, providing excellent performance while being compostable after use [52].

2 Addressing Environmental Concerns

- *Reduction in Plastic Pollution*: Adopting biodegradable polymers helps reduce the accumulation of persistent plastics in landfills and oceans [53].
- *Recyclability*: Many polymer systems are designed to complement existing recycling streams, further minimizing their environmental footprint [54].

3 Reduction of Food Waste

Food waste is a major global challenge, with significant environmental, economic, and social implications. Polymers contribute to reducing food waste by enhancing the preservation of perishable products through improved storage conditions and extended shelf life [12, 15].

3.1 Improving Storage Conditions

- *Active Packaging Solutions*: Polymers with oxygen scavengers or moisture absorbers maintain optimal storage conditions by removing residual oxygen or regulating humidity levels. For example, nano clay composites in films reduce oxygen and moisture permeability, preserving the freshness of baked goods and dairy products [16].
- *Antimicrobial Properties*: Packaging materials embedded with antimicrobial nanoparticles, such as silver or zinc oxide, inhibit microbial growth, reducing spoilage and contamination [44].

3.2 Extending Shelf Life

- *Barrier Enhancements*: Nanocomposite coatings provide superior barrier properties against oxygen, moisture, and UV light, which are key factors in food degradation. For instance, nano-enabled PET bottles for beverages reduce carbonation loss and prevent oxidation, extending product usability [55].
- *Intelligent Packaging Systems*: Nano polymer-based sensors detect spoilage markers, such as gas emissions or pH changes, providing real-time feedback on product quality. These systems allow consumers to use food items before they spoil, reducing unnecessary waste [56].

4 Economic and Environmental Benefits

- *Reduced Food Loss in Supply Chains*: By preventing spoilage during transportation and storage, nano polymers help reduce food loss in supply chains, particularly for fruits, vegetables, and meats [57].
- *Lower Greenhouse Gas Emissions*: By extending shelf life and reducing waste, nano polymers indirectly lower greenhouse gas emissions associated with food production and disposal [58].

5 Challenges and Limitations

The use of nanopolymers in food packaging and related applications offers numerous benefits; however, it is not without challenges and limitations. These concerns primarily stem from potential health risks, environmental impact, regulatory hurdles, and consumer acceptance of nanotechnology.

6 Health and Safety Concerns

- *Migration of Nanoparticles into Food*

One of the most critical safety concerns is the migration of nanoparticles from the polymer matrix into the packaged food or beverage. This can occur due to environmental factors like temperature, mechanical stress, or chemical interactions. Studies indicate that nanoparticles, such as silver, zinc oxide, or titanium dioxide, may leach into food, especially under high temperatures or acidic conditions, posing potential health risks [59].

- *Toxicity and Bioaccumulation*

Nanoparticles have unique properties due to their nanoscale size, which can lead to higher reactivity and potential toxicity. These particles may interact with biological systems, crossing cellular barriers and accumulating in organs, potentially causing cytotoxic, genotoxic, or immunotoxic effects. For example, silver nanoparticles, widely used for their antimicrobial properties, have been linked to oxidative stress and cellular damage in some studies. There is limited long-term data on the chronic effects of exposure to nanoparticles, which raises concerns about their safety in food-related applications [60].

- *Environmental Implications*

Nanoparticles released during manufacturing, use, or disposal can enter the environment, affecting ecosystems and organisms. The persistence and bioaccumulation of nanoparticles in soil, water, or air remain areas of ongoing research [61].

- *Current Research on Safety*

Researchers are focusing on assessing the safety of nanomaterials through toxicological studies and risk assessments. Strategies include the development of inert or biodegradable nanoparticles and coatings that minimize migration and exposure [62].

7 Regulatory and Consumer Perspectives

7.1 Regulatory Challenges

- *Approval Processes*: Regulatory frameworks for nanotechnology in food packaging are still evolving. Agencies like the US FDA, European Food Safety Authority (EFSA), and others require rigorous testing to ensure the safety of nanopolymers before they are approved for use in food applications [63].
- *Labelling Requirements*: In some regions, manufacturers are required to disclose the use of nanotechnology in their products, which adds to the complexity of compliance [64].

7.2 Economic Barriers

- The high cost of developing and commercializing nanotechnology-enabled packaging is another challenge, which may limit its adoption in low-cost food markets.

While nanopolymers hold immense potential to revolutionize food packaging, addressing health, safety, and regulatory concerns is critical for their widespread adoption. Ensuring transparency in communication and establishing robust safety standards can help build consumer trust and confidence. Continued research and collaboration among scientists, regulatory authorities, and industry stakeholders will be essential to overcoming these challenges and unlocking the full potential of nanotechnology in the food and beverage industries.

8 Conclusions

Nanopolymers represent a transformative innovation in the food and beverage packaging industry, offering advanced functionalities such as enhanced barrier properties, antimicrobial activity, and intelligent quality monitoring. These materials address critical challenges such as food spoilage, contamination, and sustainability while aligning with consumer demands for better-quality and longer-lasting products. However, despite their immense potential, the adoption of nanopolymers is not without hurdles. Health and safety concerns, particularly the migration of nanoparticles into food and their potential toxicity remain significant challenges. Moreover, the environmental impact of nanoparticle release and bioaccumulation warrants further investigation to ensure long-term safety.

Regulatory frameworks and consumer acceptance also play a pivotal role in the widespread adoption of nanotechnology in packaging. The lack of standardization and globally harmonized guidelines for evaluating the safety of nanomaterials often delays the commercialization of innovative products. Additionally, public perceptions of nanotechnology, often influenced by misinformation or inadequate communication, highlight the need for transparent labelling and effective science communication. Moving forward, addressing these challenges through interdisciplinary research, stringent regulatory measures, and public engagement will be key to unlocking the full potential of nanopolymers and ensuring their safe, sustainable, and consumer-friendly application in the food and beverage industry.

References

1. Das, P.P., Chaudhary, V., Ahmad, F., Manral, A.: Effect of nanotoxicity and enhancement in performance of polymer composites using nanofillers: a state-of-the-art review. Polym. Compos. **42**(5), 2152–2170 (2021)
2. Arora, B., Bhatia, R., Attri, P.: Bionanocomposites: green materials for a sustainable future. In: New polymer nanocomposites for environmental remediation, pp. 699–712. Elsevier (2018)
3. Bongarde, U.S., Shinde, V.D.: Review on natural fiber reinforcement polymer composites. Int. J. Eng. Sci. Innov. Technol. **3**(2), 431–436 (2014)
4. Alias, A.R., Wan, M.K., Sarbon, N.M.: Emerging materials and technologies of multi-layer film for food packaging application: a review. Food Control 108875 (2022)
5. Afolalu, S.A., Ikumapayi, O.M., Ogedengbe, T.S., Kayode, J.F., Ogundipe, A.T., Jen, T.C.: A review on emerging trends in the scientific application of nano-coatings and nanoparticle synthesis. Mater. Today: Proc. Accessed 21 Feb 2024 (2023). Available: https://www.sciencedirect.com/science/article/pii/S2214785323037719
6. Adeyemi, J.O., Fawole, O.A.: Metal-based nanoparticles in food packaging and coating technologies: a review. Biomolecules **13**(7), 1092 (2023)
7. Daulbayev, C., Sultanov, F., Bakbolat, B., Daulbayev, O.: 0D, 1D and 2D nanomaterials for visible photoelectrochemical water splitting: a Review. Int. J. Hydrog. Energy (2020)
8. Baig, N., Kammakakam, I., Falath, W.: Nanomaterials: a review of synthesis methods, properties, recent progress, and challenges. Mater. Adv. **2**(6), 1821–1871 (2021)
9. Zhou, X., Zhou, X., Zhou, L., Jia, M., Xiong, Y.: Nanofillers in novel food packaging systems and their toxicity issues. Foods **13**(13), 2014 (2024)
10. Hussain, S., Akhter, R., Maktedar, S.S.: Advancements in sustainable food packaging: From eco-friendly materials to innovative technologies. Sustain. Food Technol. **2**(5), 1297–1364 (2024)
11. Moraes, N.V., Lermen, F.H., Echeveste, M.E.S.: A systematic literature review on food waste/loss prevention and minimization methods. J. Environ. Manage. **286**, 112268 (2021)
12. Das, P.P., Kalyani, P., Kumar, R., Khandelwal, M.: Cellulose-based natural nanofibers for fresh produce packaging: current status, sustainability and future outlook. Sustain. Food Technol. **1**(4), 528–544 (2023). https://doi.org/10.1039/D3FB00066D

13. Wang, Z.-C., Yin, Y.-X., Ao, H.-P., Yin, H., Ren, D.-F., Lu, J.: The shelf-life of chestnut rose beverage packaged in PEN/PET bottles under long term storage: a comparison to packaging in ordinary PET bottles. Food Chem. **370**, 131044 (2022)
14. Ahari, H., et al.: Bio-nanocomposites as food packaging materials; the main production techniques and analytical parameters. Adv. Coll. Interface. Sci. **310**, 102806 (2022)
15. Das, P.P., Prathapan, R., Ng, K.W.: Advances in biomaterials based food packaging systems: current status and the way forward. Biomater. Adv. 213988 (2024)
16. Ahmed, I., et al.: A comprehensive review on the application of active packaging technologies to muscle foods. Food Control **82**, 163–178 (2017)
17. Dey, A., Neogi, S.: Oxygen scavengers for food packaging applications: a review. Trends Food Sci. Technol. **90**, 26–34 (2019)
18. Malhotra, S.P.K.: Applications of zinc oxide nanoparticles as an antimicrobial agent in the food packaging industry. In: Zinc-based Nanostructures for Environmental and Agricultural Applications, pp. 125–137, Elsevier. https://www.sciencedirect.com/science/article/pii/B9780128228364000215 (2021). Accessed 24 Jan 2024
19. Gaikwad, K.K., Singh, S., Negi, Y.S.: Ethylene scavengers for active packaging of fresh food produce. Environ. Chem. Lett. **18**, 269–284 (2020)
20. Joung, J., Boonsiriwit, A., Kim, M., Lee, Y.S.: Application of ethylene scavenging nanocomposite film prepared by loading potassium permanganate-impregnated halloysite nanotubes into low-density polyethylene as active packaging material for fresh produce. Lwt **145**, 111309 (2021)
21. Wei, H., Seidi, F., Zhang, T., Jin, Y., Xiao, H.: Ethylene scavengers for the preservation of fruits and vegetables: a review. Food Chem. **337**, 127750 (2021)
22. Ahari, H., Soufiani, S.P.: Smart and active food packaging: Insights in novel food packaging. Front. Microbiol. **12**, 657233 (2021)
23. Ahmad, S.S., Yousuf, O., Islam, R.U., Younis, K.: Silver nanoparticles as an active packaging ingredient and its toxicity. Packag. Technol. Sci. **34**(11–12), 653–663 (2021)
24. Azeredo, H.M., Correa, D.S.: Smart choices: Mechanisms of intelligent food packaging. Curr. Res. Food Sci. **4**, 932–936 (2021)
25. Chi, W., et al.: Developing a highly pH-sensitive κ-carrageenan-based intelligent film incorporating grape skin powder via a cleaner process. J. Clean. Prod. **244**, 118862 (2020). https://doi.org/10.1016/j.jclepro.2019.118862
26. Moradi, M., Tajik, H., Almasi, H., Forough, M., Ezati, P.: A novel pH-sensing indicator based on bacterial cellulose nanofibers and black carrot anthocyanins for monitoring fish freshness. Carbohyd. Polym. **222**, 115030 (2019). https://doi.org/10.1016/j.carbpol.2019.115030
27. Miao, F., Han, Y., Tian, P., Tao, B., Zang, Y., Chu, P.K.: Integrated sensor for humidity, temperature, light, and carbon dioxide based on passive RFID. Sens. Actuators, B Chem. **390**, 133913 (2023)
28. Aeby, X., Bourely, J., Poulin, A., Siqueira, G., Nyström, G., Briand, D.: Printed humidity sensors from renewable and biodegradable materials. Adv. Mater. Technol. **8**(5), 2201302 (2023). https://doi.org/10.1002/admt.202201302
29. Song, H., Choi, I., Lee, J.-S., Chang, Y., Yoon, C.S., Han, J.: Whey protein isolate coating material for high oxygen barrier properties: a scale-up study from laboratory to industrial scale and its application to food packaging. Food Packag. Shelf Life **31**, 100765 (2022)
30. Sani, M.A., Khezerlou, A., Tavassoli, M., Abedini, A.H., McClements, D.J.: Development of sustainable UV-screening food packaging materials: A review of recent advances. Trends Food Sci. Technol. 104366 (2024)
31. Amara, C., El Mahdi, A., Medimagh, R., Khwaldia, K.: Nanocellulose-based composites for packaging applications. Curr. Opin. Green Sustain. Chem. **31**, 100512 (2021)
32. Asim, N., Badiei, M., Mohammad, M.: Recent advances in cellulose-based hydrophobic food packaging. Emergent Mater. **5**(3), 703–718 (2022). https://doi.org/10.1007/s42247-021-00314-2.
33. Sid, S., Mor, R.S., Kishore, A., Sharanagat, V.S.: Bio-sourced polymers as alternatives to conventional food packaging materials: a review. Trends Food Sci. Technol. **115**, 87–104 (2021)

34. Huang, H.-D. et al.: Promising strategies and new opportunities for high barrier polymer packaging films. Prog. Polym. Sci. 101722 (2023)
35. Thirupathi Vasuki, M., Kadirvel, V., Pejavara Narayana, G.: Smart packaging—an overview of concepts and applications in various food industries. Food Bioeng. 2(1), 25–41 (2023). https://doi.org/10.1002/fbe2.12038
36. Abitbol, T., et al.: Nanocellulose, a tiny fiber with huge applications. Curr. Opin. Biotechnol. **39**, 76–88 (2016)
37. Fu, Y., Dudley, E.G.: Antimicrobial-coated films as food packaging: a review. Comp Rev Food Sci Food Safe **20**(4), 3404–3437 (2021). https://doi.org/10.1111/1541-4337.12769
38. Lev, R., Tanninen, P., Lyytikäinen, J., Leminen, V.: Converting and its effects on barrier properties of coated packaging materials: a Review. BioResources **18**(4), 8707 (2023)
39. Chen, M.C., Koh, P.W., Ponnusamy, V.K., Lee, S.L.: Titanium dioxide and other nanomaterials based antimicrobial additives in functional paints and coatings. Prog. Org. Coat. **163**, 106660 (2022)
40. Vallejos, S., et al.: From classical to advanced use of polymers in food and beverage applications. Polymers **14**(22), 4954 (2022)
41. Mercanti, N. et al.: Enhancing wine shelf-life: Insights into factors influencing oxidation and preservation. Heliyon **10**(15). https://www.cell.com/heliyon/fulltext/S2405-8440(24)11719-7 (2024). Accessed 21 Jan 2025
42. Mohammad, Z.H., Ahmad, F., Ibrahim, S.A., Zaidi, S.: Application of nanotechnology in different aspects of the food industry. Discov Food **2**(1), 12 (2022). https://doi.org/10.1007/s44187-022-00013-9
43. Ibrahim, I.D., et al.: Need for sustainable packaging: an overview. Polymers **14**(20), 4430 (2022)
44. Omerović, N., et al.: Antimicrobial nanoparticles and biodegradable polymer composites for active food packaging applications. Comp. Rev. Food Sci. Food Safe. **20**(3), 2428–2454 (2021). https://doi.org/10.1111/1541-4337.12727
45. Dash, K.K., Deka, P., Bangar, S.P., Chaudhary, V., Trif, M., Rusu, A.: Applications of inorganic nanoparticles in food packaging: a comprehensive review. Polymers **14**(3), 521 (2022)
46. de Sousa, M.S., Schlogl, A.E., Estanislau, F.R., Souza, V.G.L., dos Reis Coimbra, J.S., Santos, I.J.B.: Nanotechnology in packaging for food industry: past, present, and future. Coatings **13**(8), 1411 (2023)
47. Khan, A., Ezati, P., Rhim, J.-W.: Chitosan/starch-based active packaging film with N, P-doped carbon dots for meat packaging. ACS Appl. Bio Mater. **6**(3), 1294–1305 (2023). https://doi.org/10.1021/acsabm.3c00039
48. Stark, N.M., Matuana, L.M.: Trends in sustainable biobased packaging materials: a mini review. Mater. Today Sustain. **15**, 100084 (2021)
49. Wandosell, G., Parra-Meroño, M.C., Alcayde, A., Baños, R.: Green packaging from consumer and business perspectives. Sustainability **13**(3), 1356 (2021)
50. Liu, Y., et al.: A review of cellulose and its derivatives in biopolymer-based for food packaging application. Trends Food Sci. Technol. **112**, 532–546 (2021)
51. Díaz-Montes, E., Castro-Muñoz, R.: Edible films and coatings as food-quality preservers: an overview. Foods **10**(2), 249 (2021)
52. Alhanish, A., Abu Ghalia, M.: Developments of biobased plasticizers for compostable polymers in the green packaging applications: a review. Biotechnol. Prog. **37**(6), e3210 (2021). https://doi.org/10.1002/btpr.3210
53. Pandey, P., Dhiman, M., Kansal, A., Subudhi, S.P.: Plastic waste management for sustainable environment: techniques and approaches. Waste Dispos. Sustain. Energy **5**(2), 205–222 (2023). https://doi.org/10.1007/s42768-023-00134-6
54. Bauer, A.-S., Tacker, M., Uysal-Unalan, I., Cruz, R.M., Varzakas, T., Krauter, V.: Recyclability and redesign challenges in multilayer flexible food packaging—a review. Foods **10**(11), 2702 (2021)
55. Chi, K., Wang, H., Catchmark, J.M.: Sustainable starch-based barrier coatings for packaging applications. Food Hydrocoll. **103**, 105696 (2020)

56. Echegaray, N., Guzel, N., Kumar, M., Guzel, M., Hassoun, A., Lorenzo, J.M.: Recent advancements in natural colorants and their application as coloring in food and in intelligent food packaging. Food Chem. **404**, 134453 (2023)
57. Al-Dairi, M., Pathare, P.B., Al-Yahyai, R., Jayasuriya, H., Al-Attabi, Z.: Postharvest quality, technologies, and strategies to reduce losses along the supply chain of banana: a review. Trends Food Sci. Technol. https://www.sciencedirect.com/science/article/pii/S092422442300078X (2023). Accessed 24 Jan 2024
58. Zhao, W., Van Der Voet, E., Zhang, Y., Huppes, G.: Life cycle assessment of municipal solid waste management with regard to greenhouse gas emissions: case study of Tianjin, China. Sci. Total Environ. **407**(5), 1517–1526 (2009)
59. Kaur, G., Sharma, S., Deep, A.: Applications of metal oxide nanoparticles in food safety. In: Nanotechnology Applications for Food Safety and Quality Monitoring, pp. 55–70. Elsevier. https://www.sciencedirect.com/science/article/pii/B9780323857918000082 (2023). Accessed 24 Jan 2024
60. de Oliveira Mallia, J., Galea, R., Nag, R., Cummins, E., Gatt, R., Valdramidis, V.: Nanoparticle food applications and their toxicity: current trends and needs in risk assessment strategies. J. Food Prot. **85**(2), 355–372 (2022)
61. Carroccio, S.C., Scarfato, P., Bruno, E., Aprea, P., Dintcheva, N.T., Filippone, G.: Impact of nanoparticles on the environmental sustainability of polymer nanocomposites based on bioplastics or recycled plastics—a review of life-cycle assessment studies. J. Clean. Prod. **335**, 130322 (2022). https://doi.org/10.1016/j.jclepro.2021.130322
62. del Rosario Herrera-Rivera, M. et al.: Nanotechnology in food packaging materials: Role and application of nanoparticles. RSC Adv. **14**(30), 21832–21858 (2024)
63. EFSA Scientific Committee et al.: Guidance on risk assessment of nanomaterials to be applied in the food and feed chain: human and animal health. EFS2 **19**(8) (2021). https://doi.org/10.2903/j.efsa.2021.6768
64. Guidance, D.: Guidance for industry considering whether an FDA-regulated product involves the application of nanotechnology. Biotechnol. Law Rep **30**(5), 613–616 (2011)

Nanopolymer-Based Agro Nano-chemicals

Michael Ovbare Akharame, Aimuanmwosa Frank Eghomwanre, Fidelis Ifeakachukwu Okolafor, Bamidele Oladapo Fagbayigbo, and Ogheneochuko Utieyin Oputu

Abstract The growing global population poses challenges to conventional agricultural practices, with food scarcity and famine imminent or looming on the horizon. The drive for enhanced agricultural outputs and sustainability is captured in Sustainable Development Goals (SDGs) 2. It encompasses the use of innovative and smart options for sustainable agriculture and increased food production to curb famine and hunger. The use of nanopolymer-based agro nano-chemicals is perceived to hold valuable and sustainable solutions to agricultural challenges faced in conventional systems. These nanopolymers enable precise, controlled, and eco-friendly applications of fertilisers, pesticides, and other agricultural inputs, paving the way for advanced agricultural technologies and sustainable farming practices. In this chapter, we discuss the synthesis methods commonly used to fabricate nanopolymers, their inherent properties, and areas of application in agricultural systems. Salient issues related to environmental and safety considerations and strategies for mitigating the perceived environmental impact of nanopolymer utilization are highlighted. The contribution harped regulatory and ethical considerations and future perspectives for large-scale advanced agricultural and sustainable farming practices.

M. O. Akharame (✉) · A. F. Eghomwanre
Department of Environmental Management and Toxicology, University of Benin, Benin-City, Nigeria
e-mail: michael.akharame@uniben.edu

A. F. Eghomwanre
e-mail: frank.eghomwanre@uniben.edu

F. I. Okolafor
Department of Science Laboratory Technology, University of Benin, Benin City, Nigeria
e-mail: fidelis.okolafor@uniben.edu

B. O. Fagbayigbo
Department of Environmental and Occupational Health, Cape Peninsula University of Technology, Cape Town, South Africa

O. U. Oputu
The Hair and Skin Research Laboratory, Division of Dermatology, Department of Medicine, University of Cape Town, Cape Town, South Africa

B. Sharma et al. (eds.), *Nanostructured Polymers*, Engineering Materials,
https://doi.org/10.1007/978-981-95-3581-1_17

Keywords Nanopolymers · Nanopolymer agrochemicals · Sustainable agriculture · Sustainable development goals

1 Introduction

Conventional agricultural practices are limited by the inefficient use and overuse of traditional agrochemicals, leading to soil degradation, water pollution, and biodiversity loss. The challenge is further aggravated by increasing resistance to conventional pesticides and the need for solutions that minimise non-target effects in their applications. There is a growing need for more robust and innovative practices that ensure precise delivery, reduce input waste, and promote agricultural sustainability. Consequently, agricultural sustainability is a key component of the Sustainable Development Goals (SDGs) 2030. The Sustainable Development Goals 2 covers sustainable and resilient practices to ensure adequate food production; maintain the genetic diversity of seeds, plants, and animals; and technological development and investments in agricultural-based research. SDG 2 is designed to ameliorate the challenges of the growing global population by using sustainable agriculture and increased food production to curb famine and hunger.

The incorporation and utilisation of nanopolymer-based agrochemicals is part of the sustainable practices that are currently defining modern agriculture. Nanopolymers are nanostructured materials with inherent intrinsic properties, such as large surface areas, greater aspect ratios, and functionalised and tuneable surfaces, which allow modifications for several applications [1]. Typically, nanopolymers are formed from repeat units of monomers with nanomaterials dispersed in the matrix. The fabricated nanopolymers may have nanomaterials of different sizes, shapes, or orientations, such as nanorods, nanocrystals, nanosheets, nanowires, nanoshells, nanoparticles, quantum dots, and fullerenes, incorporated into and/or unto the selected polymer matrix, polymer blends or grafts [2]. The intrinsic properties of the formed nanopolymers are strongly influenced by the size scale of the component phases and their degree of interaction [3].

Nanopolymers are deployed as special carrier systems for agrochemicals to curb or mitigate pest attacks and increase crop yields [4]. Agricultural inputs, such as fungicides, herbicides, insecticides, fertilisers or plant growth enhancers, are often embedded in nanopolymer matrices with targeted delivery systems to reduce or minimise the quantity of chemical usage [5]. This practice enhances agrochemical efficiency and efficacy through precise targeting of pests and nutrient delivery. The effectiveness of agrochemicals is prolonged because of their controlled or slow release over extended periods. The minimal agrochemical utilization and use of biodegradable nanopolymers make this technology an environmentally sustainable practice.

The utilisation of nanopolymer-based agro nano-chemicals materials has opened the frontiers of agriculture to collaboration with material scientists. The cross-disciplinary collaboration in materials science and agricultural technology has the potential to ultimately provide solutions to the myriad of challenges in agrochemical usage in conventional agriculture. The enhanced structural properties and functionalities of the hybrid materials aid in the precise targeting of pests and nutrients. Sustainable agricultural practices designed to incorporate nanopolymer-based agrochemicals effectively result in substantial cost savings for farmers due to reduced inputs and improved yields. Moreover, the drive to use biodegradable polymer options will eliminate or reduce long-term residues and alleviate the environmental concerns associated with the use of these nanopolymer-based agrochemicals.

2 Synthesis of Nanopolymers for Agro-Applications

Nanopolymers are synthesised by the incorporation of nanomaterial(s) into appropriately selected polymer matrices [2]. The incorporated nanomaterial is purposely selected to meet the required needs from the plethora of various types, such as metals, mixed metals, metal oxides, carbon-based, and clay/silicate nanomaterials [3]. Numerous polymer matrices are also available to obtain nanopolymers with optimum performance on the basis of the intended usage. Polymers such as polylactic-co-glycolic acid (PLGA), polyethylene glycol (PEG), polycaprolactone (PCL), polyvinyl alcohol (PVA), polyacrylic acid (PAA), alginate polymers, dendrimers, chitosan, and cellulose-based and starch-based polymers provide sustainable and efficient support for the delivery of agro-nanochemicals.

The selection of a suitable synthesis procedure for the fabrication of nanopolymers using the selected nanomaterial(s) and polymer matrix or matrices is highly importance. The synthesis procedure employed should be based on factors such as the area of application, size requirement, particle size distribution, and polymeric system [3]. The common methods utilised for nanopolymers in chemical applications include emulsion polymerisation, electrospinning, and self-assembly techniques.

2.1 Key Methods for Synthesising Nanostructured Polymers

2.1.1 Emulsion Polymerisation

Emulsion polymerisation is one of the key methods for synthesising nanostructured polymers deployed in agro nano-chemicals applications. The synthesis process involves emulsification of the monomers in a continuous phase of an immiscible liquid, leading to an in situ polymerisation reaction with the resulting nanospheres [6]. High-speed mixing or sonication is required in the emulsification process to effectively carry out the polymerisation and ensure optimum particle size distribution. Water-in-oil or oil-in-water-phase systems may be utilised in the polymerisation process, depending on the hydrophobicity of the selected monomer. Generally, fast polymerisation, optimum conversion of monomers, and higher molecular weight polymers can be obtained using emulsion polymerisation techniques, which involve effective heat transfer and easy viscosity control [7, 8].

The loading of the active component for the intended agricultural application is critical in the synthesis phase. The active component can be entrapped on the surface of the nanopolymer, in its core, or in combination with the prevailing nature of the material, depending on the selected polymer matrix or matrices, surfactant, active component, or other additive [9]. For optimum loading and entrapment of the active component, critical selection of the compatibility of the polymer matrix with the active component and nanomaterial is needed [9].

2.1.2 Electrospinning

The use of the electrospinning technique is a veritable way of fabricating nanopolymers deployed in agricultural applications for the controlled release of formulations and soil stabilisers. The technique produces nanofiber-based systems from a top-down approach using different polymer matrix solutions and depends on key variables such as the complex interactions on the material surfaces, rheology, shapes, and electrical charge for the desired morphology [10]. Typically, the setup requires a high-voltage power supply to induce the electrostatic force needed, a syringe with a metal needle for exuding the jet of the polymer, and a metallic collection screen [11]. The induced electric field (electrostatic force) is applied on the jet of the polymer solution, which is deformed, drawn, and elongated into conical shapes called Taylor cones [10, 12]. Electrospinning adopts the principle of high-voltage electrostatic forces used to draw polymer solutions into ultrafine fibres [13]. The resulting structures include nanofibrous mats or scaffolds.

2.1.3 Self-assembly of Block Copolymers

The self-assembly of block copolymers (BCPs) refers to the spontaneous organisation of polymer chains into well-defined, nanoscale structures due to the immiscibility of chemically distinct polymer blocks connected by covalent bonds [14, 15]. This process is driven by the system's attempt to minimise free energy, resulting in ordered morphologies. This technique works on the principle of block copolymers spontaneously organising into nanostructures owing to the segregation of chemically distinct polymer blocks [14]. The key features of this technique include microphase separation, tuneable structures, solvent influence, and thermodynamic/kinetic factors. The common structures include micelles, vesicles, and lamellar phases.

2.2 *Functionalization Techniques for Agrochemical Compatibility*

2.2.1 Surface Modification

One of the most important methods for improving the compatibility of nanopolymers with agrochemicals is surface modification, which guarantees effective delivery and enhanced performance [16]. This procedure involves modifying the surface characteristics of nanoparticles to enable certain interactions with active agricultural agents [17, 18]. Grafting, which involves chemically bonding functional groups to the surface of a nanopolymer, and coating, which involves the application of a thin layer of protective material, are common methods [19]. To satisfy the needs of various agrochemical formulations, these alterations modify the physicochemical characteristics of nanopolymers, such as their hydrophobicity, charge, and stability [16]. For example, hydrophobic coatings are used for herbicides that are resistant to water, guaranteeing that the active components continue to work even when wet [20]. Agrochemical uptake by crops can be increased, and runoff losses can be reduced by engineering surface-modified nanopolymers for improved adherence to plant surfaces [17]. These modifications also enhance dispersibility in systems based on water or oil, which makes nanopolymers adaptable to a variety of agricultural settings [18]. Surface-modified nanopolymers lessen environmental pollution, reduce the application frequency, and decrease chemical waste [21]. These changes address issues such as resource conservation, environmental preservation, and climate adaptation by improving the targeted delivery and retention of agrochemicals, which is in line with the objectives of sustainable agriculture. For this reason, surface modification is essential for the development of nanopolymers for use in contemporary agriculture [22].

2.2.2 Incorporation of Active Ingredients

An innovative technique for enhancing the effectiveness and distribution of agrochemicals is the incorporation of active substances into nanopolymers, such as fertilisers, insecticides, or herbicides [16]. One of the best methods is encapsulation, which embeds the active substance in the nanopolymer matrix [23]. The significant advantages of this approach include prolonged activity, decreased volatility, and regulated pesticide release, which guarantees accurate delivery at the appropriate time and location [21]. Encapsulated fertilisers, for example, are made to release nutrients gradually following the phases of crop growth [23]. By minimising nutrient losses from runoff or leaching, this synchronisation maximises plant uptake and lowers pollution levels in the environment [24]. Similarly, nanoencapsulation lowers pesticide volatility while minimising exposure to nontarget organisms such as beneficial insects and soil microorganisms [25]. This precise delivery technique contributes to a further reduction in environmental pollution by significantly lowering the required quantity of pesticides [26]. Encapsulated herbicides also hold tremendous potential since they may be engineered to release certain active ingredients in response to specific stimuli, such as changes in soil pH or moisture content [24]. This targeted action increases their effectiveness against certain weeds while protecting nontarget vegetation. These advancements in nanopolymer systems show how they are transforming modern agriculture by addressing the shortcomings of traditional pesticide techniques and promoting efficiency, sustainability, and ecological balance [16].

2.2.3 Stimuli-Responsive Functionalisation

A cutting-edge advancement in agricultural nanotechnology, the use of stimuli-responsive nanopolymers, offers a novel method for accurate pesticide distribution [27]. These intelligent systems respond to specific environmental cues, such as pH, moisture, or light, to ensure that agrochemicals are administered at the precise time and place needed [28]. These technologies improve the targeted delivery of inputs, including fertilisers and herbicides, increasing agricultural output while reducing waste and environmental effects [29]. In response to changes in soil acidity, pH-sensitive nanopolymers, for example, can release their active components, ensuring that agrochemicals are delivered under optimal conditions for plant uptake [28]. Similarly, moisture-responsive nanopolymers prevent unnecessary losses in arid conditions by releasing nutrients when there is rain or are irrigated [27]. Light-responsive devices, which are triggered by sunlight, allow for a controlled release mechanism that aligns with plant photosynthesis, thus optimising the timing of pesticide administration for maximum impact [30]. These clever distribution techniques address significant environmental problems by significantly reducing the likelihood of leaching, runoff, and unintentional build-up in nontarget areas. They also increase resource efficiency, which allows farmers to harvest more with fewer inputs [31]. As agriculture faces challenges resulting from climate change and resource scarcity, stimuli-responsive nanopolymers provide a revolutionary approach to sustainability, precision, and ecological stewardship [30].

3 Properties of Nanopolymers Relevant to Agro Nano-chemicals

The nexus between nanopolymers and agriculture is far-reaching, given the recent advancement of nanotechnology as a solution to depleted and dried soil, disease control, and gene editing in plants. Nanopolymers may be classified on the basis of their thermal behaviour, origin, structure, polymerisation, and techniques used for their preparation [32, 33]. Nanopolymers have three major functionalities that are relevant in modern agriculture: nano pesticides, nano fertilisers, and nanosensors. Polymer-based nano-pesticides have been reported to improve the effectiveness of pesticides while reducing their environmental impact [5, 23]. In terms of efficacy, they are known to release pesticides into the environment in a controlled manner, thereby reducing their environmental impact and making them safe for nontargeted organisms. They also have the properties of biocompatibility, biodegradability, and nontoxicity. In the case of polymer-based nanofertilisers, they are known to improve crop yield and reduce possible environmental impacts [34]. Nanofertilisers can be used as polymer coatings, biopolymer matrices, nanogels, and nano encapsulants. On the other hand, polymer-based nanosensors are used as detectors and for monitoring crops for optimisation, crop growth, and waste reduction. Nanosensors monitor parameters such as pathogen presence, nutrient availability, soil moisture, humidity, temperature, soil pH, soil toxicity, and pesticide levels. Examples of some polymers that may significantly contribute to the agricultural use of nanopolymers and their properties and applications are presented in Table 1.

3.1 Controlled Release Mechanisms for Fertilisers and Pesticides

Controlled release of fertilisers and pesticides is a modern innovative approach that is designed to release fertilisers/pesticides in a delayed or controlled manner to achieve optimisation of the amount of fertilisers/pesticides required by plants. This technology allows nutrients and chemicals to be released gradually and consistently over time, aligning with plant needs. Such systems are significant in addressing over-application, reducing waste, and limiting pollution caused by conventional application methods. To achieve a controlled release mechanism for fertilisers and pesticides, green technology using nanopolymer-based products is the future of agriculture. Controlled release by encapsulation is the process whereby a protective coating is added to a water-soluble fertiliser, thereby controlling the dissolution and release of nutrients in a controlled manner [49, 50]. Controlled release can also be achieved by osmosis and diffusion, where dissolved nutrients force themselves out of the nanopolymer-based coatings or through a concentration gradient [51, 52]. Additionally, temperature-dependent release and hydrogels are important ways to achieve the controlled release of fertilisers and pesticides [53] by nutrient or pesticide encapsulation at a temperature sensitive to nanopolymer coatings and by soaking hydrogels in a liquid fertiliser solution, respectively.

3.1.1 Mechanisms of Action

The mechanisms of action of the controlled release of fertilisers and pesticides can be broadly discussed under three major headings: diffusion-based, degradation-based, and osmotic mechanisms.

Table 1 Properties and applications of nanopolymer-based agro nanochemicals

Nanopolymers	Properties	Applications	Example	References
Chitosan-based	Biocompatible, and non-toxic natural polymer derived from chitin, biodegradable a component of crustacean shells. It has excellent film-forming and antimicrobial properties	Encapsulation of pesticides, fertilisers, and plant growth regulators for controlled release Delivery of biopesticides and bio stimulants to enhance plant growth and immunity Use as a bio coating material to protect seeds and fertilisers	Chitosan nanoparticles loaded with urea for slow and sustained nitrogen release in soil	[35–37]
Polylactic-co-glycolic acid (PLGA)	Synthetic, biodegradable polymer approved by the FDA. It degrades into lactic and glycolic acids, which are non-toxic	Controlled release of agrochemicals, such as herbicides and insecticides Delivery of bioactive molecules, including plant hormones Use in precision agriculture to target specific areas or conditions	PLGA nanoparticles encapsulate fungicides for sustained protection against fungal diseases in crops	[9, 38, 39]
Polyethylene glycol (PEG)	Hydrophilic polymer with excellent water solubility and biocompatibility. It is often used as a coating or stabiliser for nanoparticles	Stabilisation of agrochemical formulations to enhance their solubility and bioavailability Coating of nanoparticles to improve their stability and prolong their activity in the field	PEG-coated nanoparticles deliver hydrophobic pesticides to improve their dispersion and effectiveness	[40–42]

(continued)

Table 1 (continued)

Nanopolymers	Properties	Applications	Example	References
Alginate nanopolymers	Naturally occurring polysaccharides extracted from brown algae. It is biodegradable and biocompatible, and forms hydrogels in the presence of calcium ions	Encapsulation of agrochemicals for sustained release Delivery of plant growth-promoting rhizobacteria (PGPR) to enhance soil health and crop productivity	Alginate nanoparticles loaded with bio-fertilisers for gradual nutrient release in soil	[27]
Polycaprolactone (PCL)	Synthetic, biodegradable polyester with good mechanical strength and slow degradation rates	Long-term controlled release of pesticides and herbicides Delivery of hydrophobic agrochemicals to improve their dispersion and uptake	PCL nanoparticles encapsulate hydrophobic insecticides to enhance their efficiency and reduce environmental contamination	[9, 43]
Dendrimers	Highly branched, tree-like synthetic nanopolymers with a high surface area and functional groups for chemical attachment	Delivery of agrochemicals with enhanced solubility and stability Multifunctional carriers capable of loading multiple active ingredients simultaneously	Dendrimer-based systems for codelivery of pesticides and micronutrients to plants	[41, 44]
Polyvinyl alcohol (PVA)	Synthetic, water-soluble polymer that is biodegradable and forms hydrogels	Encapsulation of agrochemicals for slow release Use as a stabiliser for nanoparticles in agrochemical formulations	PVA-coated nanoparticles for the controlled release of herbicides in soil	[45, 46]
Polyacrylic acid (PAA)	Hydrophilic synthetic polymer capable of forming hydrogels and binding metal ions	Carrier for micronutrients such as zinc and iron to improve plant uptake Use as a soil conditioner to enhance water retention and nutrient availability	PAA-based nano gels loaded with micronutrients for slow and efficient nutrient delivery	[47, 48]

Diffusion-based mechanism: In diffusion-based release systems, the fertiliser or pesticide is encapsulated or embedded within a material (polymer, coating, or carrier) that allows controlled molecular diffusion into the soil or plant system [5, 54]. The active ingredient (fertiliser or pesticide) is enclosed in a core surrounded by a semipermeable membrane. Release occurs through the membrane. The rate of release depends on several factors, such as the diffusion coefficient of the active ingredient, the thickness and properties of the coating or matrix material, and environmental factors, such as soil moisture, temperature, and pH. Diffusion-based systems require polymers, which may be biodegradable and nonbiodegradable; hydrogels, which provide diffusion pathways for fertilisers and pesticides; and clays and nanocarriers, which enable precise diffusion and serve as reservoirs for fertilisers or pesticides. Diffusion-based release systems for fertilisers and pesticides offer a promising approach to modern, sustainable agriculture. By providing a steady and controlled delivery of nutrients and agrochemicals, these systems increase crop yields, reduce environmental damage, and support precision farming techniques.

Degradation-based mechanism: Degradation-based release involves the use of biodegradable materials, which ensures minimal environmental impact and reduces the long-term accumulation of non-biodegradable residues in soil and water (Fig. 1). The core advantages of degradation-based release over other methods are improved efficiency, cost savings, and eco-friendly and controlled release rates [5, 55]. By utilising biodegradable materials, these systems offer controlled and extended delivery of nutrients and agrochemicals, reducing environmental impacts and improving resource efficiency. Continued research and innovation are essential to make these technologies affordable and accessible, promoting their adoption for eco-friendly and productive farming.

Osmotic mechanism: The osmotic mechanism for controlled release is an advanced delivery system that relies on osmotic pressure to regulate the release of fertilisers and pesticides (Fig. 2). This system ensures the gradual, sustained, and controlled delivery of nutrients or pesticides to plants, improving resource efficiency and reducing environmental harm. Osmotic systems are particularly beneficial in precision agriculture, where nutrient and pesticide release can be tailored to match plant needs and environmental conditions. The osmotic mechanism works on the principle of osmotic pressure, which occurs due to the movement of water across a semipermeable membrane [56, 57]. This movement is driven by the concentration gradient between the external environment (low solute concentration) and the internal osmotic chamber (high solute concentration). The increase in internal osmotic pressure forces the dissolved active ingredient out through a delivery orifice or microporous membrane. The release rate is controlled by the properties of the semipermeable membrane, the osmotic agent (salt or sugar) used inside the device, and environmental factors such as moisture availability and temperature. The osmotic release of fertilisers such as nitrogen, phosphorus, potassium (NPK), or micronutrients is loaded into osmotic release devices. The gradual release ensures consistent nutrient availability throughout the crop cycle. The major benefits of the osmotic release system are a reduction in nutrient leaching

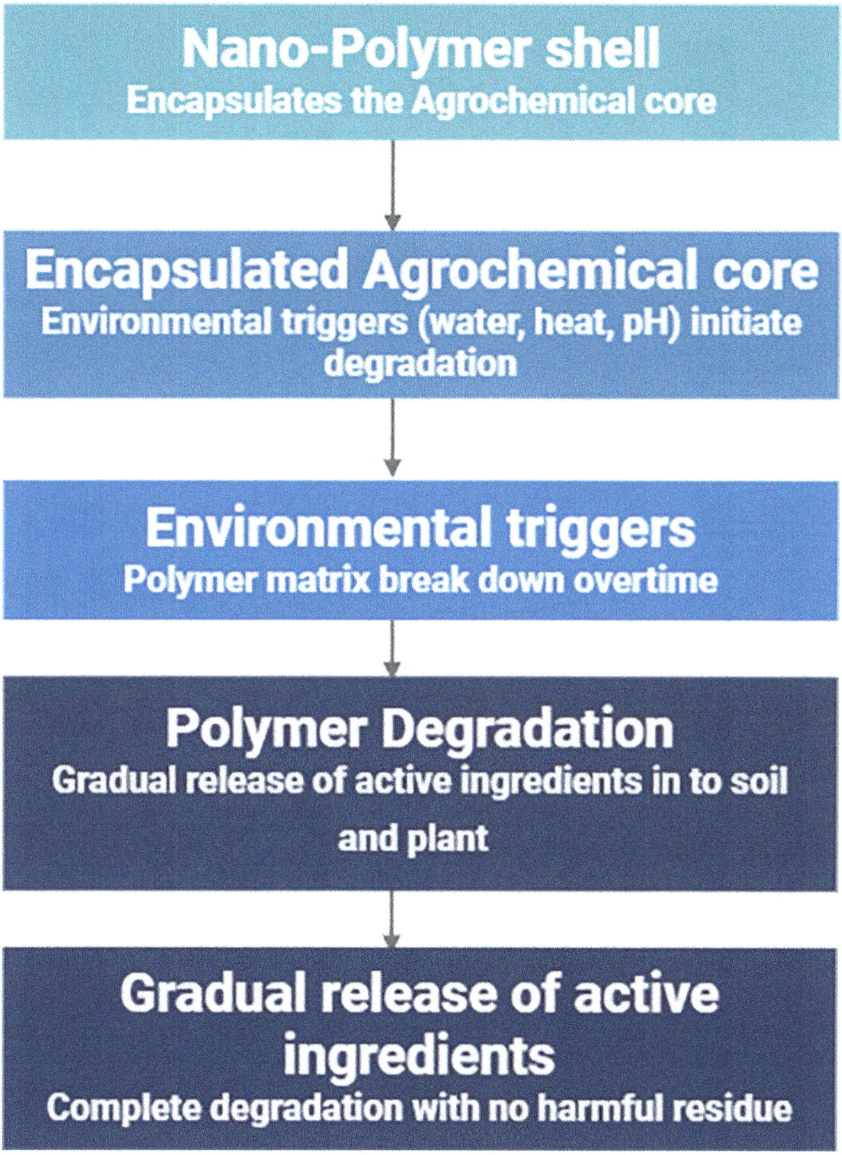

Fig. 1 Chart illustrating the degradation-based release mechanism for agro nano-chemicals

and volatilisation, a matched nutrient supply for plant growth, and increased crop yield [58, 59].

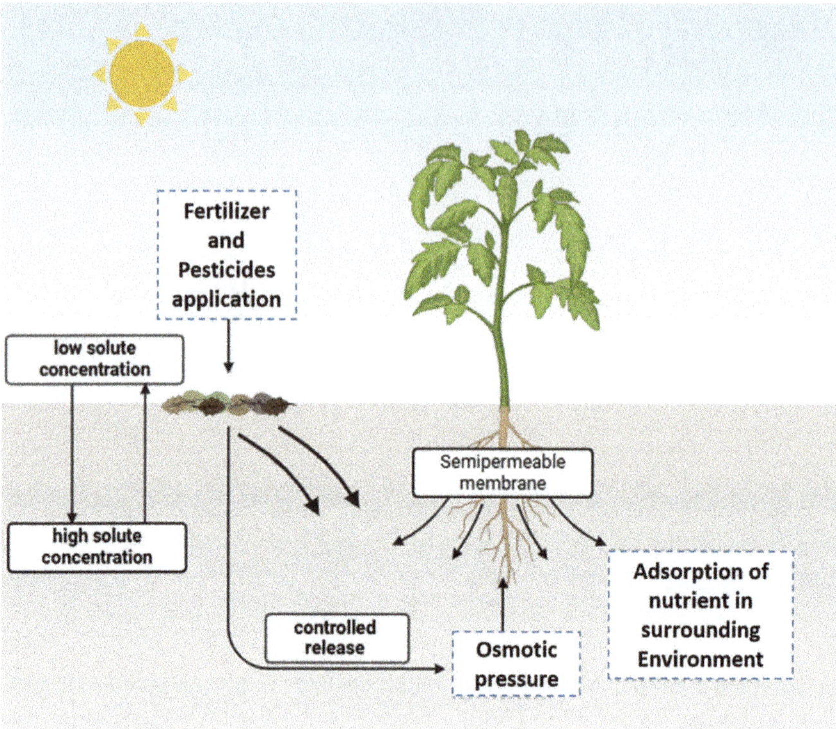

Fig. 2 Osmotic mechanism for agro nano-chemicals release to plants

4 Applications in Agriculture

The application of nanotechnology in agriculture has contributed immensely to increasing global food production and supporting innovation in agrochemical-associated industries to preserve water resources, ecological systems, and the environment [60, 61]. The use of conventional fertilisers such as chemical fertilisers (NPKs) to improve crop productivity and sustainability cannot be overemphasised. However, some notable drawbacks could significantly impact environmental processes and lead to pollution in various environmental compartments, such as water, air, and soil/sediment, while reducing nutrient use efficiency. Studies have shown that less than 50% of applied nutrients in the form of fertiliser application are utilised by plants [62, 63]. The introduction of nanotechnology on a large scale to the agricultural sector is still developing and is rapidly growing across several sectors, with exclusive applications in agriculture and food research [64]. Recently, investigations have focused on the application of nanotechnology for increasing agricultural output, including crop production, livestock management, and preservation and processing of food [65–67]. A study conducted by [68] revealed that increasing productivity and decreasing postharvest expenditure have been accredited

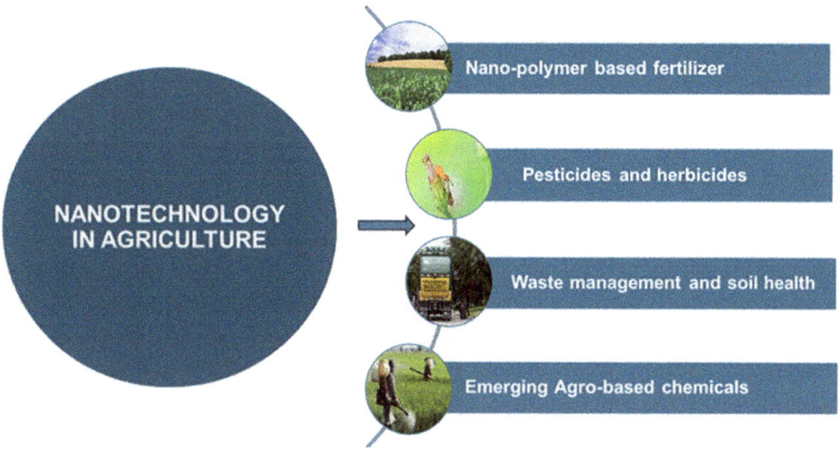

Fig. 3 Areas of applications of nano-polymers-based agro nano-chemicals

via better outcomes, with the support of newer technical investigations with the help of nanotechnology, and that biotechnology in foodstuffs might be the best answer [68]. Novel applications of nanotechnology have been successfully applied to various aspects of agricultural sectors, and their outcomes have contributed immensely to much-needed global food output. Nanotechnology in agriculture is among the emerging innovations that have demonstrated the promise of attaining environmentally sustainable agriculture [67, 69]. The use of nanotechnology in agricultural practices has been explored for different applications, such as providing crop protection and improving cultivation yield, as carriers for existing pesticides [70, 71]. However, the current state of application of nanotechnology in different fields of agriculture globally is under-explored [72]. Nanopolymer-based fertiliser has been frequently applied, as illustrated in Fig. 3.

4.1 Nanopolymer-Based Fertilisers

The application of nanopolymer-based (NPB) fertilisers, otherwise called nano bio fertilisers (NBFs), is pivotal in modern agriculture and represents a revolutionary approach to increase nutrient availability, soil fertility, and overall crop productivity sustainability [73]. The combination of nano fertiliser and biofertiliser technologies offers innovative solutions for targeted nutrient delivery alongside benefits to both crops and ecological components such as flora and fauna [74]. Nano fertilisers are developed by encapsulating nutrients within nanomaterials [75]. The method of encapsulation allows for systematic control of the release of nutrients to the plant and hence enhanced availability where nutrients are needed, which is crucial for plant growth [70, 76]. Sustained-release nanosized encapsulated materials in fertilisers can improve their uptake efficiency by crops. Additionally, it can supply essential nutrients when it can enhance plant resistance to abiotic and biotic stresses, hence promoting sustainable agricultural practices [70]. Examples of materials mostly utilised during the encapsulation process include chitosan and polysaccharides, which increase the bioavailability of enclosed nutrients and facilitate better interactions with soil microorganisms [77]. Several studies have recognised the vast benefits associated with nanotechnology-based agrochemicals; such advantages include the controlled release of encapsulated inputs, rendering them more effective and environmentally benign than conventional formulations [77–79]. Over the years, synthesised nanopolymer-based fertilisers have been developed with a variety of materials sourced from both natural and synthetic sources. It has a unique characteristic that minimises excessive release and uptake of nutrients when mobilised to match the required growth needs [80]. Nanopolymer-based fertilisers are classified into three categories.

4.1.1 Nanosupported Fertilisers

Nanosupport fertilisers are developed when fertiliser is incorporated with nanostructure additives with various preparations and functional groups. This can be described in two forms: (1) entrapment of nano fertiliser, with nutrients doped, encapsulated, or entrapped in nanocarriers with hindered exposure, and (2) adsorption of nano fertiliser, with nutrients incorporated into nanocarriers mainly through physical adsorption. For example, [81] synthesised a hybrid nano fertiliser by doping the synthesised nanoparticles with urea, $Cu(OH)_2$, $Fe(OH)_2$, and $Zn(OH)_2$, which could be continuously released for up to 14 days. Another study by [82] synthesised urea-hydroxyapatite (HA) nanohybrids with a urea ratio of 6:1 by adding H_3PO_4 solution to a suspension containing $Ca(OH)_2$ and urea. The combination of the amine functional groups of urea and the carbonyl groups of HA results in strong N–C–N bonds, thus causing the slow release of urea into the system for one week [73]. Nanosupported fertilisers at the nanoscale have been efficiently used to entrap nutrients in fertilisers.

4.1.2 Nanosised Fertilisers

The development of nanosised fertiliser nutrients has been reported in the literature [83]. Insoluble nutrients are synthesised at the nanoscale to increase their adsorption rate into plant systems. Nanosised fertilisers have been demonstrated to increase uptake efficiency [73]. Owing to their tiny size, nanosised enriched fertilisers are modified with the application of organic functional groups such as organic chemical acids [84, 85]. A study reported by [86] revealed that the synthesis of iron-humic nanosised fertiliser enhances Fe uptake, thus reducing Fe deficiency symptoms in exposed soybean plants in iron-deficient calcareous soil [86]. The small size of nanosised fertilisers is smaller than the pore size on the leaf stoma, which particularly supports foliar fertilisers, which invariably favour direct uptake by crops without drawbacks during application [84].

4.1.3 Nanowrapped Fertilisers

The purpose of wrapping regular-sised fertilisers is to protect against water dissolution and thus reduce nutrient loss. Conventional petroleum-based wrapping materials are usually hydrophobic, which has been effective in reducing the amount of active ingredients in fertiliser. One example [87] described the addition of nanosilica to coatings made of biopolyol and methylene diphenyl diisocyanate (MDI) to reduce porosity through –OH cross-linking, thus extending the release longevity of the coated urea tablet [87]. Nano-supported fertilisers have been extensively investigated for their application as nano agrochemicals, with various preparations and functional concepts with nanostructure additives. Additionally, biopolymer-based nanoparticles can be used to encapsulate and release nutrients slowly, reducing losses [88]. The encapsulation of nutrients in nanopolymer matrices is an effective method to restrict the excessive release of nutrients into farms.

4.2 Pesticides and Herbicides with Targeted Delivery

Pesticides and herbicides play a vital role in modern agricultural production and protection against pest invasion, increasing crop productivity, and promoting sustained and steady agricultural output [89, 90]. Conventional pesticides are generally synthetic organic compounds with high hydrophobicity. Conventional processing and formulation mostly require organic solvents, which can lead to a degree of environmental pollution and associated health risks [91]. The inclusion of nanomaterial pesticides and herbicide formulations has contributed tremendously to crop yield. Studies have shown that nanomaterials, such as nano-TiO_2, nano-Ag, nano-ZnO, nano-Ce_2O_4, and carbon nanotubes (CNTs), possess strong antimicrobial properties that have been explored for water disinfection [92–94]. Nano pesticides can be categorised according to the patterns of release rate regulation (sustained release) and stimulation release. Sustained-release nanopesticide formulations focus on slowing the release rate and thus extending the lifetime of pesticides, whereas stimulation-release nanopesticide formulations are focused on more vigorously releasing pesticides in response to surrounding environmental factors such as temperature, soil pH, and sunlight [85].

The encapsulation or functionalisation of nanopesticides ensures delivery to specific pests or weeds. Polymeric nanocapsules are widely applied, and subsequently, more studies have been conducted on their effective synthesis [84]. The availability of different polymers and their inherent properties have given researchers the option for synthesising nanoencapsulated materials through different methods.

The most developed strategies are nanoprecipitation, emulsion diffusion, solvent evaporation, double emulsification, emulsification coacervation, and layer-by-layer deposition. In an earlier study, [95] encapsulated imidacloprid microcrystals with chitosan and sodium alginate through a layer-by-layer self-assembly technique. This technique offers direct surface modification of nanoengineering particles in a colloidal system. This is a novel strategy in which self-assembly is achieved by electrostatic interactions between polymers and sequential adsorption of oppositely charged polyelectrolytes [95]. Kumar et al. [96] demonstrated that nanofertilisers can reduce nutrient losses by up to 50%, increase crop yields, and improve soil health by reducing the need for chemical inputs. Similarly, nanoformulations can be designed to release active ingredients in response to specific triggers, such as pH changes or the presence of certain enzymes, ensuring that they are activated only in the presence of the target pest or weed [96, 97].

5 Applications in Improving Soil Health

Sustainable agricultural production depends on robust soil quality and access to water resources. Innovations and developments in nanobased polymeric chemicals have significantly improved soil health, increased production and improved crop yield [98]. Recorded breakthroughs in nanotechnology have improved quality and enhanced crop production, which is critically needed to meet the ever-growing global population sustainably without compromising the future of the environment [96]. Some of the benefits of nanopolymer-based chemicals in the management of soil health include the ability to retain soil moisture and soil conditioning.

5.1 Water Retention Capacity

Nanopolymer-based chemicals are designed to increase the ability of soil to retain moisture and are increasingly recognised as vital tools for sustainable agriculture, particularly in arid regions [99]. These advanced materials leverage the unique properties of nanotechnology to improve water retention, reduce irrigation needs, and promote healthier plant growth. Nanopolymers, such as superabsorbent polymers (SAPs), can absorb significant amounts of water, often up to 100 times their weight [99]. When incorporated into cultivated soil, nanopolymers swell and form a gel-like structure that traps moisture and effectively favours the water-holding capacity of the soil [99, 100]. This application is particularly beneficial in arid sandy soils, which typically possess low moisture retention capabilities [100]. Nanopolymers have greatly enhanced the ability of soil to retain available moisture, hence contributing to more efficient water use. This novel breakthrough has reduced the impact of water scarcity by reducing cost and increasing productivity while promoting the conservation of water resources.

5.2 *Soil Conditioning*

Nanopolymer-based formulations have also been explored to improve soil structure and nutrient retention to improve soil conditioning [101]. The application of nanopolymers as soil conditions has the potential to reduce the reliance on chemical fertilisers and pesticides by promoting healthier soil ecosystems. Huang, et al. [102] reported that nanoaqueous adhesives (NAAs) improve soil stability without adversely hindering plant development [102]. Additionally, [103] reviewed edible polymer hydrogels as alternatives to conventional polymers and reported that biocompatible hydrogels could be applicable as soil conditioners, hence regenerating degraded lands. Studies have revealed that cellulose-based hydrogels effectively retain water while being biodegradable [103, 104]. Another example is the application of nanoclay composites for soil conditioning, which are incorporated as nanoparticles such as nanoclays into polymer formulations to enhance structural integrity and further improve water retention capabilities in soils [105]. Some of the available nanopolymers are superabsorbent polymers, which are mostly applicable for improving soil conditioning, and water retention, which reduces the need for irrigation and is biodegradable [105].

6 Potential Risks of Nanopolymers in Agroecosystems

6.1 *Toxicity Concerns*

Nanopolymers are hazardous to organisms that are not their intended target, even though they are designed for certain agricultural uses, such as pesticide and fertiliser delivery systems (Table 2) [27]. According to Nabi [106], beneficial soil microorganisms, such as *Azotobacter* and *Rhizobium*, are essential for nitrogen fixation, organic matter breakdown, and nutrient cycling. Shifts in the types of microbes present and the functions of their enzymes and metabolic processes may result from exposure to nanopolymers [107]. The disruption of microbial cell membranes by nanoparticles has been linked to oxidative stress and decreased microbial diversity, both of which affect plant development and soil health [108].

Insects may unintentionally come into contact with nanopolymers, especially pollinators such as bees, during pollination. If nanoparticles stick to their bodies or are consumed by tainted nectar, their immune system, behaviour, and ability to reproduce can be affected [118]. Similarly, when nanopolymers seep into water systems, aquatic organisms such as fish, algae, and crustaceans are at risk. When these particles build up in the gills or digestive system, they might interfere with physiological functions, including breathing and feeding [27]. For example, *Daphnia magna*, a crucial indicator species in aquatic ecosystems, has demonstrated harmful effects from nanopolymers introduced through agricultural runoff [119].

Table 2 Case studies on the risk assessment of nanopolymers in agro-systems

Application	Nano-polymer	Risk assessment findings	References
Controlled Release Fertilise	Chitosan-based nanoparticles	Reduced nutrient leaching, increased plant uptake, potential for soil accumulation, and toxicity to microorganisms	[109]
Pesticide Delivery	Polymeric micelles	Enhanced pesticide efficacy, reduced environmental impact, potential for human exposure and toxicity	[110]
Seed Coatings	Polymer-based coatings containing embedded nanoparticles (e.g., fungicides, insecticides)	Potential for nanoparticle release into the environment during planting and harvesting. Potential for unintended effects on plant growth and development	[111]
Pesticide Delivery	$Cu(OH)_2$ nanoparticles	$Cu(OH)_2$ Nano pesticides induced oxidative stress in lettuce, triggering dose-dependent metabolic shifts and activating detoxification mechanisms to mitigate the stress	[112]
Pesticide Delivery	Copper nanoparticles (nano-Cu)	Copper nanoparticles (nano-Cu) significantly altered the nutritional supply in cucumber fruit by disrupting key metabolic pathways involved in nutrient synthesis and transport, potentially leading to a reduction in the nutritional quality of cucumber fruit	[113]
Broad Range of Applications	Various Engineered Nanoparticles (ENPs)	ENP significantly impacts their behaviour, fate, toxicity, and changes in environmental conditions (e.g., acidity or alkalinity, salt concentrations, and presence of organic matter)	[114]
Broad Range of Applications in Agriculture	Various Engineered Nanoparticles (ENPs)	ENPs can interact with microbes in various ways, including altering microbial growth, activity, and community composition. ENPs can enhance plant growth by promoting beneficial microbial activities (e.g., nitrogen fixation) or inhibit plant growth by disrupting microbial communities. The interactions between ENPs, microbes, and plants are complex and influenced by factors such as ENP type, size, concentration, and environmental conditions	[115]

(continued)

Table 2 (continued)

Application	Nano-polymer	Risk assessment findings	References
Fertiliser Delivery	Chitosan nanoparticles (derived from shrimp processing waste)	The study found that chitosan nanoparticles can significantly enhance wheat yield and improve certain quality parameters	[116]
Fungicide	Silver nanoparticles (AgNPs)	AgNPs can inhibit microbial activity and alter soil microbial community structure	[117]

Another environmental concern is the persistence of nanopolymers in water and soil systems. In contrast to natural polymers, synthetic nanopolymers breakdown slowly, resulting in build-up over time [120]. Nanopolymers can bind with organic matter in the soil, decreasing its bioavailability while lengthening its residence period. Over time, the presence of nanopolymers can alter the physical and chemical properties of soil, making it more compact and hindering the movement of nutrients [27]. This affects microbial interactions and root formation, both of which are critical to plant health [107].

Nanopolymers can accumulate in sediments in water systems, resulting in ecological hotspots of contamination. Concerns over the longevity of these compounds in aquatic habitats, where they can combine with other pollutants such as heavy metals to increase toxicity, are raised by their sluggish disintegration [121]. Such accumulation can deteriorate water quality over time, endangering aquatic ecosystems and decreasing the amount of clean water available for human and agricultural use [120].

6.2 *Nanomaterial Bioavailability*

The main ways that plants engage with nanopolymers are through surface adherence and root uptake. Depending on their size, charge, and surface coating, nanopolymers added to the soil can either adhere to the surfaces of the roots or pass through the root hair cells [122]. Internalised nanopolymers may disrupt physiological processes in plants, such as increasing their size and complexity, food production from sunlight, and nutrient uptake [107]. Additionally, species that live in soil, such as nematodes and earthworms, unintentionally consume nanopolymers. These particles bioaccumulate within earthworm tissues and are important markers of soil quality. This can cause toxicity, decrease reproduction, and change burrowing habits [123, 124]. Long-term plant production may be impacted by these modifications, which may interfere with soil aeration and nutrient cycling mechanisms [27].

The potential presence of nanopolymers in food chains through unintentional absorption and bioaccumulation poses a threat to the well-being of both humans and animals. These particles may be absorbed by crops cultivated in soil polluted with nanopolymers and accumulate in edible tissues [122]. Concerns regarding food safety have been raised by studies showing that leafy vegetables and cereals absorb nanoparticles. Nanopolymers can increase trophic levels by bioaccumulating and biomagnifying once they reach primary consumers, such as herbivorous insects or tiny aquatic animals. Growth, reproduction, and survival may all be impacted by the cumulative toxicity experienced by predators. Nanopolymer contamination in water or feed for livestock systems can leave residues in meat, milk, and eggs that can be harmful to human health [125].

7 Mitigation Strategies for Environmental Impact

7.1 *Biodegradable Nanopolymers*

By creating biodegradable materials that naturally decompose into innocuous byproducts such as carbon dioxide, water, and organic matter, the environmental hazards of nanopolymers can be reduced [27]. Biodegradable nanopolymers reduce long-term accumulation and their negative impacts on nontarget organisms, in contrast to synthetic alternatives that linger in soil and water systems [30]. The environmentally beneficial qualities of natural polymers such as cellulose derivatives, chitosan, and starch have drawn interest [116]. Agrochemicals such as fertilisers and insecticides can be efficiently transported through starch-based nanopolymers, which are produced from renewable resources such as maize and cassava and are extremely biodegradable [28]. Shrimp shells contain chitosan, a biopolymer that combines antibacterial and biocompatible qualities to ensure plant protection and safe application [116]. Likewise, the natural degradation of cellulose derivatives, such as carboxymethyl cellulose, in soil limits their influence on microbial communities [126]. By striking a balance between environmental safety and performance, these biodegradable materials support the objectives of sustainable agriculture.

7.2 *Targeted Application Techniques*

The precise application of nanopolymers is essential for minimising environmental exposure and off-target effects. Agrochemicals can be released in a regulated and sustained manner via advanced delivery methods, including surface-functionalised particles and nanoencapsulation [30]. By reacting to environmental cues such as soil moisture, pH, or temperature, these formulations ensure that active chemicals are only released where they are required [127]. Targeted application is further enhanced by smart agricultural technologies. Drones and sensor-equipped automated sprayers, for instance, enable site-specific distributions, reducing runoff and abuse [128]. Farmers can reduce the risk of contamination and bioaccumulation by using remote sensing technology and geographic information systems (GISs) to pinpoint exact application sites for nanopolymers [128].

7.3 Monitoring and Feedback Systems

The evaluation and mitigation of the long-term environmental effects of nanopolymers depend on efficient monitoring systems. Understanding the durability, toxicity, and bioaccumulation of nanopolymers in actual agricultural settings is possible through both laboratory and field research [129]. This information can be used to determine safe concentration limits and direct application procedures. By detecting nanopolymer residues in soil and water in real time, technologies such as biosensors and nanosensors allow prompt actions to stop environmental contamination [130]. Monitoring for several growing seasons over an extended period aids in assessing the ecological effects on plants, aquatic life, and soil microbes [129]. Feedback systems establish a useful cycle in which data drive modifications to delivery systems, protocols, and formulations of nanopolymers. Researchers and policymakers can use these data to set concentration restrictions that will protect the environment without reducing agricultural output [131].

8 Regulatory and Ethical Considerations

8.1 Existing Regulations for Agro-Nanomaterials

Although comprehensive international frameworks are still developing, the use of nanomaterials in agriculture is regulated by several laws. If nanomaterials are utilised in agrochemicals, the Environmental Protection Agency (EPA) in the U.S. controls them under the Federal Insecticide, Fungicide, and Rodenticide Act (FIFRA) [132]. The EPA evaluates whether nanoformulations of pesticides present any hazards to the environment or public health. However, agencies find it difficult to keep up with the rapid pace of technological innovation in nanomaterials, which has surpassed legal frameworks [133]. Owing to their small size and special characteristics, nanomaterials react differently from bulk materials, making it challenging for the EPA and other agencies to apply current chemical safety laws to them [134].

The approval of plant protection goods, including nanomaterials, is governed by EU laws such as EC No. 1107/2009. However, the risk evaluation of these materials is constantly being improved [135]. A framework for nanotechnology in agriculture has been established by the EU, emphasising openness, risk assessment, and public involvement. However, compliance is made more difficult by inconsistent regulatory policies in member states [135]. Despite these initiatives, consistent and reliable ways to evaluate nanomaterials are still lacking, and insufficient knowledge about the long-term impacts of developing nanomaterials on the environment and human health presents serious regulatory challenges [107]. The new characteristics of nanoparticles, including their toxicity, bioaccumulation potential, and capacity to cross biological barriers, are currently being adjusted by regulatory agencies. Although this legal loophole allows for innovation, the use of nanomaterials without careful risk assessment could be dangerous.

8.2 Ethical Implications

The use of nanomaterials in agriculture raises several ethical concerns, most notably the harmony between ecological stewardship and technological innovation. The unforeseen repercussions of their extensive use, particularly their toxicity, persistence in the environment, and effects on biodiversity, raise ethical questions [136]. Ensuring that the advantages of nanotechnology do not come at the expense of ecological health is one of the main ethical challenges. Long-term environmental harm is a concern because of the potential for nanomaterials to accumulate in food chains, water, and soil [137]. Therefore, it is crucial from a moral standpoint to thoroughly evaluate potential environmental dangers before any new technology can be implemented. A significant ethical dilemma lies in finding the right approach to reconcile our responsibility to protect the environment with the potential benefits of increased food production [138]. Another significant problem is public concerns regarding the safety of nanomaterials. Owing to the small size and special properties of nanoparticles, which could have the potential for unforeseen negative impacts on both human health and the environment, there is growing awareness and skepticism about the possible concerns linked with them [137]. Establishing public trust in the use of nanotechnology in agriculture requires transparency. Clear communication concerning the safety of nanomaterials, continual observation of their impacts, and public participation in decision-making processes for their use are all examples of ethical behaviours in this field [139].

8.3 Global Cooperation

The difficulty of controlling and regulating nanomaterials in agriculture emphasizes the necessity of worldwide collaboration in the development of guidelines and regulations. National restrictions alone are not enough to address the global concerns that nanomaterials offer because they can easily traverse borders through trade, research, and product movement [138]. To ensure the safe use of nanomaterials in all locations and to standardise standards, international cooperation is essential. By creating standards for the safety assessment of nanomaterials, international such as the Organisation for Economic Cooperation and Development (OECD) have taken action to promote international cooperation [140]. Through programmes that support openness, risk assessment, and public participation, the UN also contributes to the appropriate application of nanotechnology [141]. A worldwide agreement on safety standards for nanomaterials will facilitate the simplification of laws, lower trade barriers, and guarantee the responsible development and application of nanotechnologies. However, different countries have different interests, resources, and regulatory capacities, making it difficult to achieve global collaboration [142].

9 Public Perception and Awareness

9.1 Addressing Concerns

One of the greatest obstacles to the broad use of nanopolymers in agriculture is public apprehension regarding their safety. Transparent communication is essential for addressing these issues. This entails providing the public with a thorough explanation of the advantages and safety precautions of nanopolymers, supported by solid scientific data [137]. Building confidence and overcoming misunderstandings regarding the possible risks of nanomaterials can be achieved by transparency in research findings, risk assessments, and regulatory actions [143]. Before being used in agriculture, regulatory agencies such as the EU and EPA offer recommendations that evaluate the safety of nanomaterials [135]. However, merely putting rules in place is insufficient. Explaining these rules, the procedures used for safety assessments, and how risks are being reduced should be the main goals of public outreach [144]. Concerns regarding health, safety, and environmental hazards can be more successfully addressed, and public trust can be increased by encouraging discussion and making scientific discoveries easily accessible [144]. Furthermore, risk communication should highlight ongoing research to better understand the long-term impacts of nanopolymers while acknowledging the unknowns around their environmental impact. This well-rounded strategy promotes informed decision-making and aids in the management of public concerns [142, 145].

9.2 Educational Campaigns

Targeted instructional programmes are the first step towards effective public participation. Collaborations with important stakeholders, including farmers, legislators, environmental groups, and the general public, should be a part of these programmes [137]. Given that farmers are the main users of nanomaterials in agriculture, educating them is crucial. Educating farmers on the advantages and possible hazards of using nanopolymers responsibly contributes to the safe and effective application of these materials [146]. Additionally, policymakers are crucial in influencing public opinion. Policymakers may promote the responsible use of nanomaterials by conversing with one another and staying up to date on the most recent scientific advancements [147]. They can also have an impact on laws that guarantee that the public is safeguarded and that appropriate safety precautions are taken. Partnerships with scientists, consumer advocacy organisations, and local communities can also be a part of stakeholder engagement to create common knowledge of the possible advantages and difficulties of using nanomaterials in agriculture [148]. The environmental benefits of nanopolymers, such as their capacity to lower the need for fertilisers and pesticides, which can result in reduced pollution, should also be highlighted in educational efforts. These benefits, along with guarantees about safety, can help move the conversation from fear to acceptance.

The acceptability of agricultural advances can result from excellent communication and stakeholder involvement, as shown by many successful public engagement programmes. To fight malnutrition in underdeveloped nations, the Golden Rice Project, which produced genetically engineered rice enriched with vitamin A, is a noteworthy example [149]. Field tests, open communication, and stakeholder participation helped the initiative overcome strong public opposition, particularly from environmental organisations. Engaging local people, outlining safety precautions, and highlighting the advantages of Golden Rice helped the project gain more acceptances and support [150].

Similar concerns about the effects of biological control agents on the environment have initially sparked opposition to their use in pest management. However, biological control agents have become recognised as safe and efficient substitutes for chemical pesticides because of persistent public awareness campaigns spearheaded by environmentalists and agricultural scientists. Campaigns have emphasised the long-term advantages of these methods, such as their negligible negative effects on ecosystems and nontarget species [151, 152].

10 Future Perspectives

The continuous research and development of the use of nanopolymer-based agrochemicals can potentially mitigate several agricultural challenges related to pest control and input delivery. The development of "smart polymers" with programmable release characteristics will further boost agricultural yields and engender sustainability. This can be achieved by embracing the synergy between nanopolymers and cutting-edge technologies such as artificial intelligence (AI) and machine learning for precision farming. The smart farming approach could integrate nanopolymers with environmental sensors for real-time nutrient or pesticide release. The integration of precision agriculture and smart farming technologies will engender data-driven processes for enhanced decision-making in monitoring soil conditions and crop health.

Trends toward the use of biodegradable and renewable nanopolymer materials and an emphasis on carbon–neutral agricultural inputs will eliminate environmental concerns and strengthen the use of nanopolymer-based agrochemicals for agricultural sustainability. Further research into biodegradable materials and affordable manufacturing will be key to broader adoption in global agriculture. Advancing the innovative use of nanopolymer-based agro nano-chemicals requires collaboration among academia, industry, and the government in the development of multifunctional nanopolymers to address diverse agricultural challenges in a single formulation. Innovations in combining plant growth regulators, fertilisers, and pest control agents can be more productive and cost-effective for farmers. Hybrid multifunctional nanopolymer-based agrochemicals could lead to greater sustainability and large-scale applications. Holistic research to profile the health-based and environmental concerns of nanopolymers is critical to address their perceived adverse effects.

In conclusion, the integration of nanopolymers in agro nano-chemicals offers transformative benefits, including enhanced efficiency, reduced environmental impact, and cost effectiveness. These advancements hold immense potential to address the growing global demand for sustainable agricultural practices, ensuring food security while preserving environmental health. By leveraging the unique properties of nanopolymers, the agriculture industry can develop next-generation agrochemicals that are smarter, safer, and more effective.

References

1. Ramu, P., Kumar, C.J., Palanikumar, K.: Mechanical characteristics and terminological behavior study on natural fiber nano reinforced polymer composite—a review. Mater. Today: Proceed. **16**, 1287–1296 (2019)
2. Akharame, M.O., Oputu, O.U., Pereao, O., Fagbayigbo, B.O., Razanamahandry, L.C., Opeolu, B.O., Fatoki, O.S.: Nanostructured polymer composites for water remediation. In: Nanostructured Materials for Treating Aquatic Pollution, pp. 275–306. Springer (2019)
3. Akharame, M.O., Fatoki, O.S., Opeolu, B.O., Olorunfemi, D.I., Oputu, O.U.: Polymeric nanocomposites (PNCs) for wastewater remediation: an overview. Polym.-Plast. Technol. Eng. **57**, 1801–1827 (2018)
4. Pascoli, M., Lopes-Oliveira, P.J., Fraceto, L.F., Seabra, A.B., Oliveira, H.C.: State of the art of polymeric nanoparticles as carrier systems with agricultural applications: a minireview. Energy, Ecol. Environ. **3**, 137–148 (2018)
5. Pavithran, R.K., Reddy, S.G., Kumar, B.S., Kugabalasooriar, S.: Enhancing sustainability in agriculture: natural polymer-based controlled release systems for effective pest management and environmental protection. ES Food Agrofor. **18**, 1276 (2024)
6. Jarai, B.M., Kolewe, E.L., Stillman, Z.S., Raman, N., Fromen, C.A.: Polymeric nanoparticles. In: Nanoparticles for Biomedical Applications, pp. 303–324. Elsevier (2020)
7. Ahmad, Z., Abdullah, M.K., Ali, M.Z., Zawawi, M.A.M.: Polymers in Electronics: Optoelectronic Properties, Design, Fabrication, and Applications. Elsevier (2023)
8. Chern, C.: Emulsion polymerisation mechanisms and kinetics. Prog. Polym. Sci. **31**, 443–486 (2006)
9. Shakiba, S., Astete, C.E., Paudel, S., Sabliov, C.M., Rodrigues, D.F., Louie, S.M.: Emerging investigator series: polymeric nanocarriers for agricultural applications: synthesis, characterization, and environmental and biological interactions. Environ. Sci. Nano **7**, 37–67 (2020)
10. Meraz-Dávila, S., Pérez-García, C., Feregrino-Perez, A.A.: Challenges and advantages of electrospun nanofibers in agriculture: a review. Mater. Res. Express **8**, 042001 (2021)

11. Noruzi, M.: Electrospun nanofibres in agriculture and the food industry: a review. J. Sci. Food Agric. **96**, 4663–4678 (2016)
12. Mercante, L.A., Scagion, V.P., Migliorini, F.L., Mattoso, L.H., Correa, D.S.: Electrospinning-based (bio) sensors for food and agricultural applications: a review. TrAC, Trends Anal. Chem. **91**, 91–103 (2017)
13. Wang, X.-X., Yu, G.-F., Zhang, J., Yu, M., Ramakrishna, S., Long, Y.-Z.: Conductive polymer ultrafine by electrospinning via electrospinning: preparation, physical properties and applications. Prog. Mater Sci. **115**, 100704 (2021)
14. Karayianni, M., Pispas, S.: Block copolymer solution self-assembly: recent advances, emerging trends, and applications. J. Polym. Sci. **59**, 1874–1898 (2021)
15. Feng, H., Lu, X., Wang, W., Kang, N.-G., Mays, J.W.: Block copolymers: synthesis, self-assembly, and applications. Polymers **9**, 494 (2017)
16. Sebastian, A., Nangia, A., Prasad, M.N.V.: Advances in agrochemical remediation using nanoparticles. In: Agrochemicals Detection, Treatment and Remediation, pp. 465–485. Elsevier (2020)
17. Jain, M., Mudhoo, A., Ramasamy, D.L., Najafi, M., Usman, M., Zhu, R., Kumar, G., Shobana, S., Garg, V.K., Sillanpää, M.: Adsorption, degradation, and mineralization of emerging pollutants (pharmaceuticals and agrochemicals) by nanostructures: a comprehensive review. Environ. Sci. Pollut. Res. **27**, 34862–34905 (2020)
18. Rajput, P., Thakur, A., Devi, P.: Emerging agrochemicals contaminants: current status, challenges, and technological solutions. In: Agrochemicals Detection, Treatment and Remediation, pp. 117–142. Elsevier (2020)
19. Singh, J., Singh, B.: Design of innovative agrochemical delivery system for sustainable agriculture and to alleviate environmental and health hazards. Sustain. Chem. Environ. **4**, 100046 (2023)
20. Zheng, S., He, M., Chen, B., Hu, B.: Porous aromatic framework coated stir bar sorptive extraction coupled with high-performance liquid chromatography for the analysis of triazine herbicides in maize samples. J. Chromatogr. A **1614**, 460728 (2020)
21. Singh, H., Sharma, A., Bhardwaj, S.K., Arya, S.K., Bhardwaj, N., Khatri, M.: Recent advances in the applications of nanoagrochemicals for sustainable agricultural development. Environ. Sci. Process. Impacts **23**, 213–239 (2021)
22. Borišev, I., Borišev, M., Jović, D., Župunski, M., Arsenov, D., Pajević, S., Djordjevic, A.: Nanotechnology and remediation of agrochemicals. In: Agrochemicals Detection, Treatment and Remediation, pp. 487–533. Elsevier (2020)
23. Mondéjar-López, M., García-Simarro, M.P., Navarro-Simarro, P., Gómez-Gómez, L., Ahrazem, O., Niza, E.: A review on the encapsulation of "eco-friendly" compounds in natural polymer-based nanoparticles as next generation nanoagrochemicals for sustainable agriculture and crop management. Int. J. Biol. Macromol. 136030 (2024)
24. Machado, T.O., Grabow, J., Sayer, C., de Araújo, P.H., Ehrenhard, M.L., Wurm, F.R.: Biopolymer-based nanocarriers for sustained release of agrochemicals: a review on materials and social science perspectives for a sustainable future of agri-and horticulture. Adv. Coll. Interface. Sci. **303**, 102645 (2022)
25. Kumar, A., Choudhary, A., Kaur, H., Mehta, S., Husen, A.: Smart nanomaterial and nanocomposite with advanced agrochemical activities. Nanoscale Res. Lett. **16**, 1–26 (2021)
26. Rani, T.S., Nadendla, S.R., Bardhan, K., Madhuprakash, J., Podile, A.R.: Chitosan conjugates, microspheres, and nanoparticles with potential agrochemical activity. In: Agrochemicals Detection, Treatment and Remediation, pp. 437–464. Elsevier (2020)
27. Vinzant, K., Rashid, M., Khodakovskaya, M.V.: Advanced applications of sustainable and biological nanopolymers in agricultural production. Front. Plant Sci. **13**, 1081165 (2023)
28. Khodakovskaya, M., Marmiroli, M.: Polymeric nanoparticles for sustainable plant agriculture and food industry. Front. Plant Sci. **14**, 1183938 (2023)
29. Gottardo, S., Mech, A., Drbohlavová, J., Małyska, A., Bøwadt, S., Sintes, J.R., Rauscher, H.: Towards safe and sustainable innovation in nanotechnology: state-of-play for smart nanomaterials. NanoImpact **21**, 100297 (2021)

30. Bhaskar, M., Kumar, A., Rani, R.: Application of nano formulations in agriculture. Biocatal. Agr. Biotechnol. 102934 (2023)
31. Chen, M.-J., Yang, H.-L., Si, Y.-M., Tang, Q., Chow, C.-F., Gong, C.-B.: A hollow visible-light-responsive surface molecularly imprinted polymer for the detection of chlorpyrifos in vegetables and fruits. Food Chem. **355**, 129656 (2021)
32. Sharma, A.K., Priya, B.S. Kaith, Polymer nanocomposite matrices: classification, synthesis methods, and applications. In: Handbook of Polymer and Ceramic Nanotechnology, pp. 403–428. Springer (2021)
33. Khan, Y., Sadia, H., Ali Shah, S.Z., Khan, M.N., Shah, A.A., Ullah, N., Ullah, M.F., Bibi, H., Bafakeeh, O.T., Khedher, N.B.: Classification, synthetic, and characterization approaches to nanoparticles, and their applications in various fields of nanotechnology: a review. Catalysts **12**, 1386 (2022)
34. Chakraborty, R., Mukhopadhyay, A., Paul, S., Sarkar, S., Mukhopadhyay, R.: Nanocomposite-based smart fertilisers: a boon to agricultural and environmental sustainability. Sci. Total Environ. **863**, 160859 (2023)
35. Azmana, M., Mahmood, S., Hilles, A.R., Rahman, A., Arifin, M.A.B., Ahmed, S.: A review on chitosan and chitosan-based bionanocomposites: Promising material for combatting global issues and its applications. Int. J. Biol. Macromol. **185**, 832–848 (2021)
36. Bandara, S., Du, H., Carson, L., Bradford, D., Kommalapati, R.: Agricultural and biomedical applications of chitosan-based nanomaterials. Nanomaterials **10**, 1903 (2020)
37. Chouhan, D., Mandal, P.: Applications of chitosan and chitosan based metallic nanoparticles in agrosciences—a review. Int. J. Biol. Macromol. **166**, 1554–1569 (2021)
38. Kaur, P., Choudhary, R., Pal, A., Mony, C., Adholeya, A.: Polymer-metal nanocomplexes based delivery system: a boon for agriculture revolution. Curr. Top. Med. Chem. **20**, 1009–1028 (2020)
39. Fukamachi, K., Konishi, Y., Nomura, T.: Disease control of *Phytophthora infestans* using cyazofamid encapsulated in poly lactic-co-glycolic acid (PLGA) nanoparticles. Colloids Surf., A **577**, 315–322 (2019)
40. Dhiman, A., Sharma, A.K., Agrawal, G.: Polymer based engineered materials for sustainable agriculture. ACS Agr. Sci. Technol. **2**, 693–711 (2022)
41. Sikder, A., Pearce, A.K., Parkinson, S.J., Napier, R., O'Reilly, R.K.: Recent trends in advanced polymer materials in agriculture related applications. ACS Appl. Polym. Mater. **3**, 1203–1217 (2021)
42. Ditta, A.: Role of nanoclay polymers in agriculture: applications and perspectives. In: Nanohybrids in Environmental & Biomedical Applications, pp. 323–334. CRC Press (2019)
43. Fortună, M.E., Ungureanu, E., Jităreanu, D.C., Ţopa, D.C., Harabagiu, V.: Effects of hybrid polymeric material based on polycaprolactone on the environment. Materials **15**, 4868 (2022)
44. Saied, E., Salem, S.S., Al-Askar, A.A., Elkady, F.M., Arishi, A.A., Hashem, A.H.: Mycosynthesis of hematite (α-Fe2O3) nanoparticles using *Aspergillus niger* and their antimicrobial and photocatalytic activities. Bioengineering **9**, 397 (2022)
45. Soliman, E., Mansour, M.M.: Enhancing soil organic carbon content and water retention using polyvinyl alcohol cross-linked with chitosan and pectin. J. Soil Sci. Plant Nutr. **24**, 791–803 (2024)
46. Nooeaid, P., Chuysinuan, P., Pitakdantham, W., Aryuwananon, D., Techasakul, S., Dechtrirat, D.: Eco-friendly polyvinyl alcohol/polylactic acid core/shell structured fibers as controlled-release fertilisers for sustainable agriculture. J. Polym. Environ. **29**, 552–564 (2021)
47. Pakdel, P.M., Peighambardoust, S.J.: A review on acrylic based hydrogels and their applications in wastewater treatment. J. Environ. Manage. **217**, 123–143 (2018)
48. Lodhi, R.S., Das, S., Das, P.: Recent advances in polymer hydrogels for agricultural applications. Novel Polym. Mater. Environ. Appl. 109–177 (2023)
49. Vejan, P., Khadiran, T., Abdullah, R., Ahmad, N.: Controlled release fertiliser: a review on developments, applications and potential in agriculture. J. Control. Release **339**, 321–334 (2021)

50. Lawrencia, D., Wong, S.K., Low, D.Y.S., Goh, B.H., Goh, J.K., Ruktanonchai, U.R., Soottitantawat, A., Lee, L.H., Tang, S.Y.: Controlled release fertilisers: a review on coating materials and mechanism of release. Plants **10**, 238 (2021)
51. Jariwala, H., Santos, R.M., Lauzon, J.D., Dutta, A., Wai Chiang, Y.: Controlled release fertilisers (CRFs) for climate-smart agriculture practices: a comprehensive review on release mechanism, materials, methods of preparation, and effect on environmental parameters. Environ. Sci. Pollut. Res. **29** 53967–53995 (2022)
52. Biondo, F., Baldassarre, F., Vergaro, V., Ciccarella, G.: Controlled biocide release from smart delivery systems: Materials engineering to tune release rate, biointeractions, and responsiveness. In: Nanotechnology-Based Sustainable Alternatives for the Management of Plant Diseases, pp. 31–147. Elsevier (2022)
53. Shen, M., Liu, S., Jiang, C., Zhang, T., Chen, W.: Recent advances in stimuli-response mechanisms of nanoenabled controlled-release fertilisers and pesticides. Eco-Environ. Health (2023)
54. Lavrentev, F.V., Shilovskikh, V.V., Alabusheva, V.S., Yurova, V.Y., Nikitina, A.A., Ulasevich, S.A., Skorb, E.V.: Diffusion-limited processes in hydrogels with chosen applications from drug delivery to electronic components. Molecules **28**, 5931 (2023)
55. Jain, H.V., Dhiman, S., Ansari, N.G.: Recent trends in techniques, process and sustainability of slow-release formulation for pesticides. Ind. Crops Prod. **216**, 118764 (2024)
56. Manning, G.S., Kay, A.R.: The physical basis of osmosis. J. Gen. Physiol. **155**, e202313332 (2023)
57. Bowler, M.: The physics of osmotic pressure. Eur. J. Phys. **38**, 055102 (2017)
58. Asadu, C.O., Ezema, C.A., Ekwueme, B.N., Onu, C.E., Onoh, I.M., Adejoh, T., Ezeorba, T.P.C., Ogbonna, C.C., Otuh, P.I., Okoye, J.O.: Enhanced efficiency fertilisers: Overview of production methods, materials used, nutrients release mechanisms, benefits and considerations. Environ. Pollut. Manage. (2024)
59. Govil, S., Long, N.V.D., Escribà-Gelonch, M., Hessel, V.: Controlled-release fertiliser: recent developments and perspectives. Ind. Crops Prod. **219**, 119160 (2024)
60. Chhipa, H.: Applications of nanotechnology in agriculture. In: Methods in Microbiology, pp. 115–142. Elsevier (2019)
61. Saberi Riseh, R., Hassanisaadi, M., Vatankhah, M., Varma, R.S., Thakur, V.K.: Nano/microstructural supramolecular biopolymers: innovative networks with the boundless potential in sustainable agriculture. Nano-Micro Lett. **16**, 147 (2024)
62. Duro, J.A., Lauk, C., Kastner, T., Erb, K.-H., Haberl, H.: Global inequalities in food consumption, cropland demand and land-use efficiency: a decomposition analysis. Glob. Environ. Chang. **64**, 102124 (2020)
63. F. FAO, AO The future of food and agriculture–trends and challenges. The Future of Food and Agriculture: Trends and Challenges, vol. 4 (2017)
64. Sadeghi, Z., Fasihi-Ramandi, M., Bouzari, S.: Nanoparticle-based vaccines for brucellosis: calcium phosphate nanoparticles-adsorbed antigens induce cross protective response in mice. Int. J. Nanomed. 3877–3886 (2020)
65. Kaningini, A.G., Nelwamondo, A.M., Azizi, S., Maaza, M., Mohale, K.C.: Metal nanoparticles in agriculture: a review of possible use. Coatings **12**, 1586 (2022)
66. Abobatta, W.F.: Nanotechnology application in agriculture. Acta Sci. Agricult. **2** (2018)
67. Hazarika, A., Yadav, M., Yadav, D.K., Yadav, H.S.: An overview of the role of nanoparticles in sustainable agriculture. Biocatal. Agric. Biotechnol. **43**, 102399 (2022)
68. Yadollahi, M., Namazi, H.: Synthesis and characterization of carboxymethyl cellulose/layered double hydroxide nanocomposites. J. Nanopart. Res. **15**, 1–9 (2013)
69. Singh, R.P., Handa, R., Manchanda, G.: Nanoparticles in sustainable agriculture: an emerging opportunity. J. Control. Release **329**, 1234–1248 (2021)
70. Beig, B., Niazi, M.B.K., Jahan, Z., Zia, M., Shah, G.A., Iqbal, Z.: Biodegradable polymer encapsulated nickel nanoparticles for slow release urea promotes Rhode grass yield and nitrogen recovery. J. Polym. Environ. **31**, 1866–1883 (2023)

71. Elizabath, A., Babychan, M., Mathew, A.M., Syriac, G.M.: Application of nanotechnology in agriculture. Int. J. Pure Appl. Biosci **7**, 131–139 (2019)
72. Prasad, R., Bhattacharyya, A., Nguyen, Q.D.: Nanotechnology in sustainable agriculture: recent developments, challenges, and perspectives. Front. Microbiol. **8**, 1014 (2017)
73. Ansari, M.M., Shin, M., Kim, M., Ghosh, M., Kim, S.-H., Son, Y.-O.: Nanoenabled strategies in sustainable agriculture for enhanced crop productivity: a comprehensive review. J. Environ. Manage. **372**, 123420 (2024)
74. Abdelghany, A.M., El-Banna, A.A., Salama, E.A., Ali, M.M., Al-Huqail, A.A., Ali, H.M., Paszt, L.S., El-Sorady, G.A., Lamlom, S.F.: The individual and combined effect of nanoparticles and biofertilisers on growth, yield, and biochemical attributes of peanuts (Arachis hypogea L.). Agronomy **12**, 398 (2022)
75. DeRosa, M.C., Monreal, C., Schnitzer, M., Walsh, R., Sultan, Y.: Nanotechnology in fertilisers. Nat. Nanotechnol. **5**, 91–91 (2010)
76. Munir, A., Salah, M.A., Ali, M., Ali, B., Saleem, M.H., Samarasinghe, K., De Silva, S., Ercisli, S., Iqbal, N., Anas, M.: Advancing agriculture: harnessing smart nanoparticles for precision fertilization. BioNanoScience **14**, 3846–3863 (2024)
77. Choudhary, R.C., Kumari, S., Kumaraswamy, R., Sharma, G., Kumar, A., Budhwar, S., Pal, A., Raliya, R., Biswas, P., Saharan, V.: Chitosan nanomaterials for smart delivery of bioactive compounds in agriculture. In: Nanoscale Engineering in Agricultural Management, pp. 124–139. CRC Press (2019)
78. Mujtaba, M., Khawar, K.M., Camara, M.C., Carvalho, L.B., Fraceto, L.F., Morsi, R.E., Elsabee, M.Z., Kaya, M., Labidi, J., Ullah, H.: Chitosan-based delivery systems for plants: a brief overview of recent advances and future directions. Int. J. Biol. Macromol. **154**, 683–697 (2020)
79. Ali, M., Cybulska, J., Frąc, M., Zdunek, A.: Application of polysaccharides for the encapsulation of beneficial microorganisms for agricultural purposes: a review. Int. J. Biol. Macromol. **244**, 125366 (2023)
80. Khan, M.R., Rizvi, T.R.: Application of nanofertiliser and nanopesticides for improvements in crop production and protection. Nanosci. Plant Soil Syst. 405–427 (2017)
81. Tarafdar, J., Raliya, R., Mahawar, H., Rathore, I.: Development of zinc nanofertiliser to enhance crop production in pearl millet (Pennisetum americanum). Agric. Res. **3**, 257–262 (2014)
82. Kottegoda, N., Munaweera, I., Madusanka, N., Sandaruwan, C., Sirisena, D., Disanayake, N., Ismail, M., De Alwis, A., Karunaratne, V., Malwana, W.: Plant nutrient nanoparticles encapsulated cellulose matrix for slow and sustained release of nitrogen. Curr. Sci. **101**, 73–78 (2011)
83. Chen, Z.S., Liu, T., Dong, J.F., Chen, G., Li, Z., Zhou, J.L., Chen, Z.: Sustainable application for agriculture using biochar-based slow-release fertilisers: a review. ACS Sustain. Chem. Eng. **11**, 1–12 (2023)
84. Augustine, R., Hasan, A., Primavera, R., Wilson, R.J., Thakor, A.S., Kevadiya, B.D.: Cellular uptake and retention of nanoparticles: insights on particle properties and interaction with cellular components. Mater. Today, Commun. **25**, 101692 (2020)
85. Ma, C., Li, G., Xu, W., Qu, H., Zhang, H., Bahojb Noruzi, E., Li, H.: Recent advances in stimulus-responsive nanocarriers for pesticide delivery. J. Agric. Food Chem. **72** 8906–8927 (2024)
86. Cieschi, M.T., Polyakov, A.Y., Lebedev, V.A., Volkov, D.S., Pankratov, D.A., Veligzhanin, A.A., Perminova, I.V., Lucena, J.J.: Eco-friendly iron-humic nanofertilisers synthesis for the prevention of iron chlorosis in soybean (*Glycine max*) grown in calcareous soil. Front. Plant Sci. **10**, 413 (2019)
87. Yang, G., Zhao, H., Liu, Y., Li, Z., Gao, F., Zhang, Q., Zou, P., Liu, Z., Zhang, M.: Slow release fertilisers based on polyphosphate/montmorillonite nanocomposites for improving crop yield. Arab. J. Chem. **16**, 104871 (2023)
88. Martins, N.C., Avellan, A., Rodrigues, S., Salvador, D., Rodrigues, S.M., Trindade, T.: Composites of biopolymers and ZnO NPs for controlled release of zinc in agricultural soils and timed delivery for maize. ACS Appl. Nano Mater. **3**, 2134–2148 (2020)

89. Beig, B., Niazi, M.B.K., Sher, F., Jahan, Z., Malik, U.S., Khan, M.D., Américo-Pinheiro, J.H.P., Vo, D.-V.N.: Nanotechnology-based controlled release of sustainable fertilisers. Rev. Environ. Chem. Lett. **20**, 2709–2726 (2022)
90. Hussain, N., Sarma, S.: Nanoffertilisers and their potential application in agriculture. In: Nanoparticles Synthesis by Soil Microbes, pp. 271–288. Elsevier (2025)
91. Saroop, S., Tamchos, S.: Impact of pesticide application: positive and negative side. Pesticides Chang. Environ. 155–178 (2024)
92. Biswas, J.K., Rai, M.: Nanotechnology for Environmental Management. CRC Press (2024)
93. Zhao, X., Cui, H., Wang, Y., Sun, C., Cui, B., Zeng, Z.: Development strategies and prospects of nanobased smart pesticide formulation. J. Agric. Food Chem. **66**, 6504–6512 (2017)
94. Yadav, N., Garg, V.K., Chhillar, A.K., Rana, J.S.: Recent advances in nanotechnology for the improvement of conventional agricultural systems: a review. Plant Nano Biol. **4**, 100032 (2023)
95. Guan, R., Li, T., Smagghe, G., Miao, X., Li, H.: dsRNA-based pesticides: production, development, and application technology. In: Frontiers Media SA, pp. 1197666 (2023)
96. Kumar, S., Bhuvaneshwari, R., Jain, S., Nirwan, S., Fatima, Z., Kumar, D., Chhikara, B.S., Rathi, B., Poonam, A.: Systematic review on pesticide-loaded nanocapsules: a sustainable route for pesticide management to enhance crop productivity. Curr. Nanosci. **20**, 280–297 (2024)
97. Lade, B.D., Gogle, D.P., Lade, D.B., Moon, G.M., Nandeshwar, S.B., Kumbhare S.D.: Nanobiopesticide formulations: application strategies today and future perspectives. In: Nanobiopesticides Today and Future Perspectives, pp. 179–206. Elsevier (2019)
98. Javed, Z., Dashora, K., Mishra, M., Fasake, V.D., Srivastva, A.: Effect of accumulation of nanoparticles in soil health-a concern on future. Front. Nanosci. Nanotechnol. **5** (2019)
99. Zaib, M., Raza, I., Zubair, M., Arif, Z., Mumtaz, M.M., Abbas, M.Q., Javed, A., Salman, S., Sikandar, A., Kashif, M.: NanoeNano-enabled soil amendments for improved soil structure and water holding capacity: an in-depth review. Int. Res. J. Educ. Technol. **5**, 344–357 (2023)
100. Venkatachalam, D., Kaliappa, S.: Superabsorbent polymers: a state-of-art review on their classification, synthesis, physicochemical properties, and applications. Rev. Chem. Eng. **39**, 127–171 (2023)
101. Gull, N., Islam, A., Ghaffar, A., Jabeen, S., Iqbal, S.S., Khan, S.M., Khan, R.U., Hussain, N., Bilal, M.: Development and characterization of chitosan and acrylic acid-based novel biodegradable polymeric films for soil conditioning. Int. J. Biol. Macromol. **182**, 950–958 (2021)
102. Huang, W., Du, J., Lai, H., Zhang, Q., Zhou, C., Liu, Z.: Mechanical properties and mechanism of soil treated with nanoaqueous adhesive (NAA). Sci. Rep. **12**, 14711 (2022)
103. Ali, A., Ahmed, S.: Recent advances in edible polymer based hydrogels as a sustainable alternative to conventional polymers. J. Agric. Food Chem. **66**, 6940–6967 (2018)
104. Khan, F., Atif, M., Haseen, M., Kamal, S., Khan, M.S., Shahid, S., Nami, S.A.: Synthesis, classification and properties of hydrogels: their applications in drug delivery and agriculture. J. Mater. Chem. B **10**, 170–203 (2022)
105. Vasile, C., Pamfil, D., Stoleru, E., Baican, M.: New developments in medical applications of hybrid hydrogels containing natural polymers. Molecules **25**, 1539 (2020)
106. Nabi, M.: Role of microorganisms in plant nutrition and soil health. In: Sustainable Plant Nutrition, pp. 263–282. Elsevier (2023)
107. Kumar, N., Samota, S.R., Venkatesh, K., Tripathi, S.: Global trends in use of nanoffertilisers for crop production: advantages and constraints—a review. Soil Tillage Res. **228**, 105645 (2023)
108. Samir, A., Ashour, F.H., Hakim, A.A., Bassyouni, M.: Recent advances in biodegradable polymers for sustainable applications. Npj Mater. Degrad. **6**, 68 (2022)
109. Singh, D.P., Singh, H.B., Prabha, R.: Microbial Inoculants in Sustainable Agricultural Productivity. Springer (2016)
110. Wang, W., Shi, J., Xie, Q., Jiang, Y., Yu, N., Wang, E.: Nutrient exchange and regulation in arbuscular mycorrhizal symbiosis. Mol. Plant **10**, 1147–1158 (2017)

111. Usman, M., Farooq, M., Wakeel, A., Nawaz, A., Cheema, S.A., ur Rehman, H., Ashraf, I., Sanaullah, M.: Nanotechnology in agriculture: current status, challenges and future opportunities. Sci. Total Environ. **721**, 137778 (2020)
112. Zhao, L., Ortiz, C., Adeleye, A.S., Hu, Q., Zhou, H., Huang, Y., Keller, A.A.: Metabolomics to detect response of lettuce (Lactuca sativa) to Cu (OH) 2 nanopesticides: oxidative stress response and detoxification mechanisms. Environ. Sci. Technol. **50**, 9697–9707 (2016)
113. Zhao, L., Hu, J., Huang, Y., Wang, H., Adeleye, A., Ortiz, C., Keller, A.A.: 1H NMR and GC–MS based metabolomics reveal nano-Cu altered cucumber (Cucumis sativus) fruit nutritional supply. Plant Physiol. Biochem. **110**, 138–146 (2017)
114. Markiewicz, M., Kumirska, J., Lynch, I., Matzke, M., Köser, J., Bemowsky, S., Docter, D., Stauber, R., Westmeier, D., Stolte, S.: Changing environments and biomolecule coronas: consequences and challenges for the design of environmentally acceptable engineered nanoparticles. Green Chem. **20**, 4133–4168 (2018)
115. Shende, S.S, Rajput, V.D., Gorovtsov, A.V., Saxena, H.P., Minkina, T.M., Chokheli, V.A., Jatav, H.S., Sushkova, S.N., Kaur, P.: Interaction of nanoparticles with microbes. In: Plant-Microbes-Engineered Nanoparticles (PM-ENPs) Nexus in Agro-Ecosystems: Understanding the Interaction of Plant, Microbes and Engineered Nanoparticles (ENPS), pp. 175–188 (2021)
116. Saad, A.M., Alabdali, A.Y.M., Ebaid, M., Salama, E., El-Saadony, M.T., Selim, S., Safhi, F.A., ALshamrani, S.M., Abdalla, H., Mahdi, A.H.: Impact of green chitosan nanoparticles fabricated from shrimp processing waste as a source of nano nitrogen fertilisers on the yield quantity and quality of wheat (*Triticum aestivum* L.) cultivars. Molecules **27**, 5640 (2022)
117. Xiu, Z.-M., Zhang, Q.-B., Puppala, H.L., Colvin, V.L., Alvarez, P.J.: Negligible particle-specific antibacterial activity of silver nanoparticles. Nano Lett. **12**, 4271–4275 (2012)
118. Naseer, M.M., Rafeeq, H., Afzal, M., BiBi, M., Rehman, A., Fatima, T.: Eco-friendly nanopesticides: a comprehensive review of their applications and environmental impacts. Hamdard Medicus **66** (2023)
119. Wang, M., Wang, W.-X.: Accumulation kinetics and gut microenvironment responses to environmentally relevant doses of micro/nanoplastics by zooplankton Daphnia magna. Environ. Sci. Technol. **57**, 5611–5620 (2023)
120. Gangadoo, S., Owen, S., Rajapaksha, P., Plaisted, K., Cheeseman, S., Haddara, H., Truong, V.K., Ngo, S.T., Vu, V.V., Cozzolino, D.: Nanoplastics and their analytical characterisation and fate in the marine environment: from source to sea. Sci. Total Environ. **732**, 138792 (2020)
121. Venâncio, C., Melnic, I., Tamayo-Belda, M., Oliveira, M., Martins, M., Lopes, I.: Polymethylmethacrylate nanoplastics can cause developmental malformations in early life stages of *Xenopus laevis*. Sci. Total Environ. **806**, 150491 (2022)
122. Arshad, R., Gulshad, L., Haq, I.U., Farooq, M.A., Al-Farga, A., Siddique, R., Manzoor, M.F., Karrar, E.: Nanotechnology: a novel tool to enhance the bioavailability of micronutrients. Food Sci. Nutr. **9**, 3354–3361 (2021)
123. Abbas, Q., Yousaf, B., Ullah, H., Ali, M.U., Ok, Y.S., Rinklebe, J.: Environmental transformation and nanotoxicity of engineered nanoparticles (ENPs) in aquatic and terrestrial organisms. Crit. Rev. Environ. Sci. Technol. **50**, 2523–2581 (2020)
124. Xu, Z., Long, X., Jia, Y., Zhao, D., Pan, X.: Occurrence, transport, and toxicity of nanomaterials in soil ecosystems: a review. Environ. Chem. Lett. **20**, 3943–3969 (2022)
125. Anilakumar, K.: Technological advances in nanoscience for specific food and nutrition delivery. Adv. Process. Technol. 97–121 (2021)
126. Thapliyal, P., Khan, M.: Biopolymer nanocarriers for improved agrochemical performance. In: Biopolymeric Nanoparticles for Agricultural Applications. Springer, pp. 369–391 (2024)
127. Sravani, V., Rajeswari, R., Gopalakrishnan, M., Senthil Kumar, M., Kavitha, M., Bharathi, C.: Nanonutrient Formulations–Impact on Growth and Post-Harvest Quality of Fruits and Vegetables, Communications in Soil Science and Plant Analysis, pp. 1–13 (2024)
128. Rana, A., Thakur, K., Thakur, M.: Smart biosensors for precision agriculture. In: Innovative Methods in Horticultural Crop Improvement: Biosensors and Nanosensors, pp. 237–259. Springer (2024)

129. Dhal, S., Pal, H.: Nanotechnology for climate-resilient agriculture. In: Climate-Resilient Agriculture: Agro-Biotechnological Advancement for Crop Production, vol. 2, pp. 863–880. Springer (2023)
130. Dar, F.A., Qazi, G., Pirzadah, T.B.: Nanobiosensors: NextGen diagnostic tools in agriculture. Nanobiotechnol. Agricul. An Approach Towards Sustain. 129–144 (2020)
131. Haris, M., Hussain, T., Mohamed, H.I., Khan, A., Ansari, M.S., Tauseef, A., Khan, A.A., Akhtar, N.: Nanotechnology—a new frontier of nanofarming in agricultural and food production and its development. Sci. Total Environ. **857**, 159639 (2023)
132. Abd-Elsalam, K.A.: Nanoengineered fertilisers: delivery, impacts, and application methods. In: Nanofertiliser Delivery, Effects and Application Methods, pp. 1–12 (2024)
133. Ghosh, M., Kumar, R.: Regulatory issues in nanotechnology. In: Nanotechnology Theranostics in Livestock Diseases and Management, pp. 765–788. Springer (2024)
134. Suppan, S.: Nanopesticides: introduction to encapsulated products and frameworks for assessing their risks to environmental and human health. Inst. Agric. Trade Policy (2020)
135. Rana, S., Yadav, K.K., Jha, M.: Environmental, legal, regulatory, health, and safety issues of nanoemulsions. In: Industrial Applications of Nanoemulsion, pp. 219–247. Elsevier (2024)
136. Hussain, C.M., Thomas, S.: Handbook of Polymer and Ceramic Nanotechnology. Springer, Berlin/Heidelberg, Germany (2021)
137. Egirani, D.E., Shehata, N., Khedr M.H.: A review of nano materials in agriculture and allied sectors: preparation, characterization, applications, opportunities, and challenges. Mater. Int. **2**, 421–432 (2020)
138. Cherwoo, L., Gupta, I., Bhatia, R., Setia, H.: Improving agricultural practices: application of polymers in agriculture. Energy, Ecol. Environ. **9**, 25–41 (2024)
139. Sahu, M.K., Yadav, R., Tiwari, S.P.: Recent advances in nanotechnology. Int. J. Nanomater. Nanotechnol. Nanomed. **9**, 015–023 (2023)
140. Chávez-Hernández, J.A., Velarde-Salcedo, A.J., Navarro-Tovar, G., Gonzalez, C.: Safe nanomaterials: from their use, application and disposal to regulations. Nanoscale Adv. (2024)
141. Shehata, N., Egirani, D., Olabi, A., Inayat, A., Abdelkareem, M.A., Chae, K.-J., Sayed, E.T.: Membrane-based water and wastewater treatment technologies: Issues, current trends, challenges, and role in achieving sustainable development goals, and circular economy. Chemosphere **320**, 137993 (2023)
142. Rajput, V.D., Singh, A., Minkina, T.M., Shende, S.S., Kumar, P., Verma, K.K., Bauer, T., Gorobtsova, O., Deneva, S., Sindireva, A.: Potential applications of nanobiotechnology in plant nutrition and protection for sustainable agriculture. Nanotechnol. Plant Growth Promot. Protect. Recent Adv. Impacts 79–92 (2021)
143. Wypij, M., Trzcińska-Wencel, J., Golińska, P., Avila-Quezada, G.D., Ingle, A.P., Rai, M.: The strategic applications of natural polymer nanocomposites in food packaging and agriculture: chances, challenges, and consumers' perception. Front. Chem. **10**, 1106230 (2023)
144. Nazeer, A., Ahmad, F., Verma, N., Ahmad, S.: Nanoagro-biotechnology Prospects. In: Targeted Delivery of Nanopesticides and Nanofertilisers in Sustainable Agricultural Farming. Springer, pp. 179–187 (2023)
145. Singh, M., Upadhyay, L., Badekhan, A., Shil, S., Gautam, S.K., Anand, A.: Leveraging remote sensing and nanotechnology to overcome barriers to agroforestry adoption by small holder farmers. J. Sci. Res. Rep. **30**, 113–140 (2024)
146. Okeke, E.S., Ezeorba, T.P.C., Mao, G., Chen, Y., Feng, W., Wu, X.: Nanoenabled agrochemicals/materials: potential human health impact, risk assessment, management strategies and future prospects. Environ. Pollut. **295**, 118722 (2022)
147. Hamad, H.T., Al-Sharify, Z.T., Al-Najjar, S.Z., Gadooa, Z.A.: A review on nanotechnology and its applications on fluid flow in agriculture and water recourses. In: IOP Conference Series: Materials Science and Engineering. IOP Publishing, pp. 012038 (2020)
148. Rahaman, M.T., Pranta, A.D., Repon, M.R., Ahmed, M.S., Islam, T.: Green production and consumption of textiles and apparel: importance, fabrication, challenges and future prospects. J. Open Innov. Technol. Market. Complex. 100280 (2024)

149. Bin Rahman, A.R.M., Zhang, J.: Trends in rice research: 2030 and beyond. Food Energy Secur. **12**, e390 (2023)
150. Mallikarjuna Swamy, B.P., Marundan Jr, S., Samia, M., Ordonio, R.L., Rebong, D.B., Miranda, R., Alibuyog, A., Rebong, A.T., Tabil, M.A., Suralta, R.R., Alfonso, A.A.: Development and characterisation of GR2E Golden rice introgression lines. Sci. Rep. **11**, 2496 (2021)
151. Bonaterra, A., Badosa, E., Daranas, N., Francés, J., Roselló, G., Montesinos, E.: Bacteria as biological control agents of plant diseases. Microorganisms, **10**, 1759 (2022)
152. Elnahal, A.S., El-Saadony, M.T., Saad, A.M., Desoky, E.S.M., El-Tahan, A.M., Rady, M.M., El-Tarabily, K.A.: The use of microbial inoculants for biological control, plant growth promotion, and sustainable agriculture: A review. Eur. J. Plant Pathol. **162**, 759–792 (2022)

Nano Polymers for Clean Energy Solutions

Abdullah Mohammad Sarjish, Md. Wasikur Rahman, Victoria Padilla, Horacio Vasquez, and M. Jasim Uddin

Abstract Nanopolymers are polymers engineered at the nanoscale or composed of nanostructured components with properties such as increased surface area, enhanced conductivity, and optoelectronic performance. These characteristics make nanopolymers highly useful for energy conversion and storage applications, like solar cells, batteries, fuel cells, and other energy related components. Nanopolymers are playing significant roles in the growing demand for more efficient renewable energy sources. In energy storage systems, nanopolymers are greatly improving the performance of batteries and fuel cells by enhancing ion transport and stability of the electrolytes. Due to their small size, nanopolymers possess large surface area, allowing for greater charge storage capacity. They function as efficient hole and electron transfer materials, as well as interface layers, improving carrier separation and minimizing recombination. In organic solar cells, nanopolymers act as donor layers, buffer layers, and polymer-based micro/nanostructures in both binary and ternary devices, enhancing overall device performance. Additionally, the incorporation of nanopolymers in electrocatalysts and proton exchange membranes (PEMs) significantly boosts the efficiency and durability of fuel cells, playing a key role in advancing sustainable energy solutions.

Keywords Solar cells · Fuel cells · Energy storage and conversion · Renewable energy

---

A. M. Sarjish · Md. W. Rahman · M. Jasim Uddin (✉)
Photonics and Energy Research Laboratory, Department of Mechanical Engineering, University of Texas Rio Grande Valley, Edinburg, TX, USA
e-mail: uddinm3@sunypoly.edu

V. Padilla · H. Vasquez · M. Jasim Uddin
Department of Mechanical Engineering, University of Texas Rio Grande Valley, Edinburg, TX, USA

M. Jasim Uddin
Department of Civil and Mechanical Engineering, SUNY Polytechnic Institute, Utica, NY, USA

B. Sharma et al. (eds.), *Nanostructured Polymers*, Engineering Materials,
https://doi.org/10.1007/978-981-95-3581-1_18

1 Introduction to Nanopolymers

The extensive use of fossil fuels leads to greenhouse gas emissions, resulting in rising global temperature and rapid climate change. Investment in clean energy is crucial as renewable energy is becoming more affordable, reliable, and sustainable. Switching to clean energy is not just about technology or government rules, it is about making a more sustainable future for everyone. To make this possible, we need to act on multiple fronts to make some changes. First, we must expand easy access to renewable energy by advocating the policies that stop new fossil fuel projects, expedite the shutdown of coal plants, and ensure energy efficiency initiatives [56, 66]. Second, we should make it easier for people and businesses to choose clean energy, whether through consumer incentives or utility reforms that encourage more sustainable practices. Finally upgrading the power grid is also important which means infrastructure must be developed to incorporate more renewable energy projects. In clean energy applications, nanotechnology is significantly contributing to the acceleration of the clean energy transition by enhancing energy production, storage, and efficiency [6].

Nanopolymers are a class of advanced materials with structural dimensions in the nanometer scale, which range 1–100 nm [51], are gaining importance due to their unique physicochemical properties. These materials integrate nanotechnology with polymer science, to form polymer nanocomposites that significantly enhance their mechanical strength, electrical conductivity, thermal stability, and barrier properties [59]. Nanopolymers play a crucial role in advancing solar technology by optimizing the flow of electric charges, reducing energy loss, and significantly enhancing overall performance, as shown in Fig. 1. Polymers in organic solar cells are often incorporated into many layers to get maximum photoconversion efficiency. Newer nanomaterials like perovskite solar cells and quantum dots are also enhancing the efficiency of components used in renewable energy applications [31].

Nanopolymers are essential to the development of sustainable energy solutions as they maximize energy generation, conversion, and storage. They are essential for developing high-performance batteries, supercapacitors, fuel cells, and solar cells due to their capacity to improve charge transfer, ionic conductivity, and material durability [82]. Incorporating nanomaterials like metal oxides, graphene, and carbon nanotubes into nanopolymer matrices further boosts their efficiency in energy applications [20].

1.1 Application of Nanopolymers in Clean Energy

The sophisticated structural and functional characteristics of nanopolymers make them a valuable addition to renewable energy solutions. In energy storage technology, nanopolymers enable high-capacity performance in lithium-ion batteries and supercapacitors by enhancing electrical conductivity, cycle stability, and charge–discharge

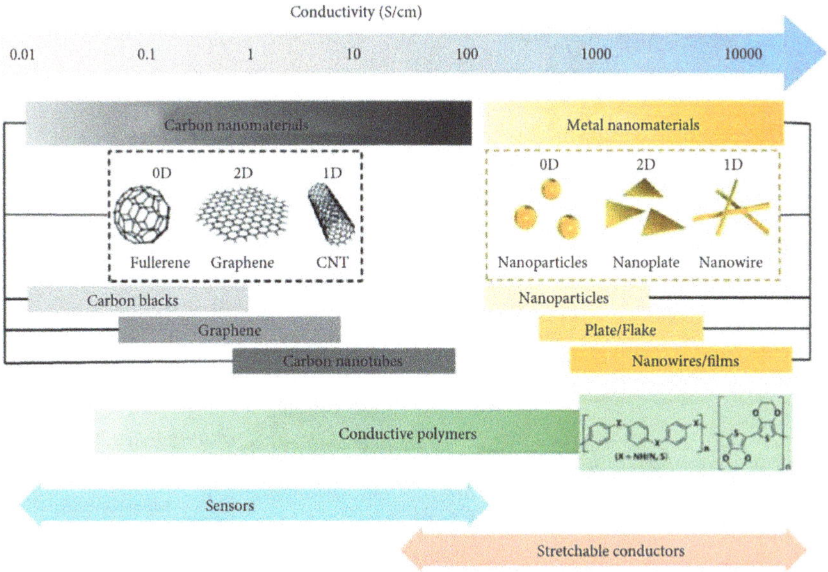

Fig. 1 Advanced uses of conducting nanopolymers incorporated in carbon-based nanomaterials [65]

rates when integrated into electrodes, while issues with efficiency and safety are addressed by solid-state electrolytes that also involve nanopolymers [61].

For solar energy conversion, Poly(3-hexylthiophene) (P3HT) and fullerene derivatives are materials that enhance light absorption and exciton dissociation in organic photovoltaic cells, which greatly increase power conversion efficiencies and make affordable solar technologies [70]. Nanopolymers are crucial for proton exchange membrane in fuel cells due to their nanostructures design. That offers high proton conductivity, mechanical durability and thermal stability which are important for the operational performance of fuel cells [10].

Nanopolymers are responsible for revolutionizing energy sustainability and enhancing energy storage. Besides, they are improving the energy harvesting technologies and reducing the environmental impacts. In triboelectric nanogenerators (TENGs), biodegradable nanopolymers composites, such as POSS-Modified Bone Gelatin (BG-POSS), are boosting the energy output. BG-POSS refers to a hybrid nanopolymer material where Polyhedral Oligomeric Silsesquioxane (POSS) is used to modify bone gelatin (BG) to enhance its properties for applications in sustainable energy systems [21]. In solar thermal fuel (STF) storage, azopyridine based nanopolymers have shown high photothermal energy density around 430 J/g. That enabling the efficient phase changes in energy conversion [43]. Furthermore, nanopolymer composites are increasingly incorporated into electromagnetic energy devices not

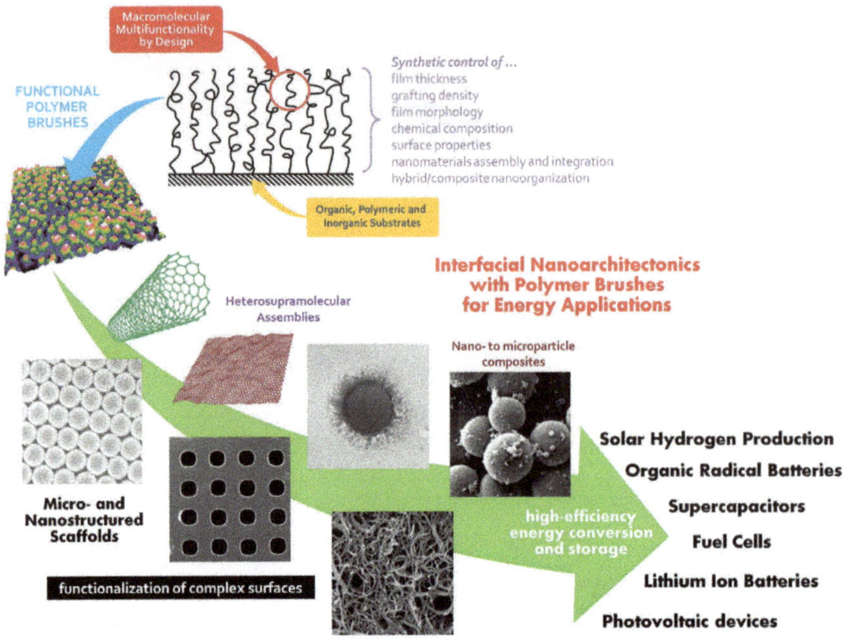

Fig. 2 Combination of polymers and nanomaterials that has the potential to significantly advance energy conversion and storage technologies [24]

only in hybrid energy transport systems but also in electromagnetic shielding materials, which allows usage in high-performance energy components in sophisticated technologies [101].

Polymer based piezoelectric nanogenerator (PENGs) have remarkable energy improvements through nanomaterial integration as it can be used in self-powered application in wearable electronics and IoT (Internet of Things) devices. Recent studies showed that incorporating carbon nanotubes and lead zirconate titanates (PZT) nanofibers into polymer matrices has enhanced the electrical energy output. For example, a CNTs-PZT based nanogenerator achieved a four times higher output which generated 12.5 V and 80 nA/cm^2 with an ultrahigh power density of 91.7 nW/cm^2 [48] (Fig. 2).

1.2 Challenges of Clean Energy

The clean energy sector faces numerous challenges, including enhancing energy conversion efficiency, optimizing storage capacity, and advancing environmental sustainability. Traditional energy materials often suffer from significant drawbacks such as high production costs, low thermal stability, and poor charge mobility, limiting their effectiveness and scalability. These problems prevent renewable energy technologies from being widely adopted [63]. Figure 1 illustrates the conversion of renewable and sustainable clean energy along with its various applications. Renewable energy is often called clean energy which does not produce additional pollution or waste, in contrast to fossil fuel energy sources. It has a minimal carbon footprint and results in lower greenhouse gas emissions. In recent years, clean energy has become increasingly popular as governments and economies work to lessen their dependence on heavily polluting fossil fuels [33].

Nanopolymers provide a transformative solution by resolving these concerns through their adjustable characteristics, increased usefulness, and greater mechanical and chemical stability. Polymer-based nanocomposites increase electrode stability and allow for quicker charge–discharge cycles, hence improving the performance of lithium-ion batteries and supercapacitors. Nanopolymers also boost light absorption and charge transport efficiency in organic photovoltaic cells, whereas dye-sensitized solar cells improve total power conversion efficiencies [49]. Furthermore, their use in fuel cells, particularly as proton exchange membranes not only improves ionic conductivity but also ensures long-term durability.

Beyond the efficiency improvements, nanopolymers are great contributors to environmental sustainability and high technological advancement as they are reducing the materials consumption and improving the recyclability. Furthermore, they are reducing the carbon footprints and minimizing the environmental emission of pollutants. Owing to their lightweight properties, these materials play a crucial role in lowering the energy consumption from transportation to manufacturing sectors. Furthermore, bio based and biodegradable nanopolymers offer a sustainable solution by reducing the plastics waste [34] (Fig. 3).

Fig. 3 Application of bio-based nanomaterials in different sectors [33]

2 Nanopolymers for Energy Storage

Nanopolymers have emerged as transformative materials in energy storage technologies which surpass critical performance, safety, and efficiency barriers in batteries, supercapacitors, and hydrogen storage systems. Their novel nanostructures and customizable features improve ionic conductivity, energy density, and operational stability that result in opening the path for next-generation energy solutions.

2.1 Batteries

Nanopolymer electrolytes are transforming lithium-ion batteries, which were developed to overcome liquid electrolyte problems such as leakage and thermal instability. These sophisticated electrolytes are frequently augmented with nanostructure fillers as resulting in enhancing the ionic conductivity and improving the battery thermal resistance [80]. These improvements not only increase the batteries life but also reduce the potential of thermal issues which are considered as a critical safety design in high energy density applications. In addition, carbon-based nanomaterials like carbon nanofibers (CNFs) and mesoporous carbon perform an important role in LIB electrodes. Their excellent mechanical strength, high surface area and high electrical conductivity enabling rapid lithium-ion transport and reducing internal resistance [93]. CNFs considerably improve the charging and discharging efficiency by generating continuous electron paths, resulting in the important crucial conductive additions in modern electrode architectures [76].

For instance, solid state batteries use polymer nanocomposites to improve the interfacial contact between electrodes and electrolytes, where sodium ion and zinc air batteries use nanostructured polymers to boost ion diffusion kinetics and cycling stability. In recent years some researcher developed newly nanofibers composites membrane that exhibits an ion conductivity of 1.0×10^{-4} S cm^{-1} at 40 °C, that significantly higher than the conventional PEO electrolyte with nanofibers. This improvement was due to the nanofibers' intrinsic lithium-ion conductivity, as well as their favorable impact on the PEO matrix, which decreased both the crystallization temperature and the degree of crystallinity [90] (Fig. 4).

2.2 Supercapacitors

Supercapacitors are recognized for their high-power density and quick charge discharge capabilities, primarily relying on nanopolymer composites to overcome the energy density limitations. When electrode materials such as polyaniline (PANI) and polypyrrole (PPy) are hybridized with carbon nanostructures, they exhibit enhanced

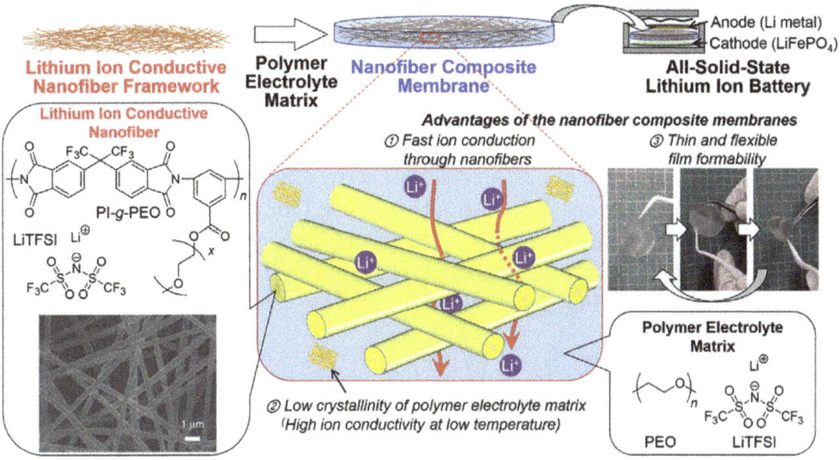

Fig. 4 Diagram of a polymer composite membrane with a lithium-ion conductive nanofiber framework for all-solid-state batteries [90]

surface areas and superior electrochemical activity. These composites enable efficient charge storage through both electric double-layer capacitance and pseudocapacitance, resulting in devices with improved energy and power densities [60]. The function of electrolytes and separators are equally important in supercapacitors. Advanced separators made of nanoporous polymers increase the ion selectivity and thermal stability that ensure the dependable performance in critical functioning situations [84].

2.3 *Hydrogen Storage Technologies for Fuel Cells*

A fuel cell is an electrochemical device that uses chemical energy to generate electricity. It is most typically used as a power source in aerospace applications, such as airplanes, spacecraft, and long-duration space missions. Nanopolymers increase fuel cell efficiency and performance through improved electrode conductivity, catalysis, and overall longevity, making them ideal for use in aerospace and space missions [23].

The use of polymeric nanoparticles as catalyst support matrices has various advantages over their micro- and macro-scale equivalents. These are more surface area, better catalyst particle dispersion, greater metal–metal and metal-support interactions, higher reaction rates, increased electronic conductivity, and improved interfacial behavior [16, 17]. Liao et al. fabricated a tartaric acid-doped polyaniline nanowire network on carbon cloth for use as a dual-chamber microbial fuel cell MFC anode. The modified anode generated approximately four times more power than the unmodified carbon cloth anode and outperformed those doped with inorganic acids [44]. Nanopolymers also enhance hydrogen fuel cells by improving proton exchange membranes (PEMs). For example, Nafion membranes doped with nanoparticles exhibit elevated proton conductivity and chemical stability, reducing reliance on expensive platinum catalysts. Nanostructured polymer-catalyst hybrids further optimize reaction kinetics, enhancing energy efficiency and durability across diverse fuel cell applications [77]. Nanofibers, valued for their high electrochemical surface area, were utilized by Ghosh et al. to radiolytically deposit Pt, Pt–Pd and Pt–Pd–Au nanoparticles on PPy nanofibers. Figure 6 illustrates the synthesis and deposition process [22] (Fig. 5).

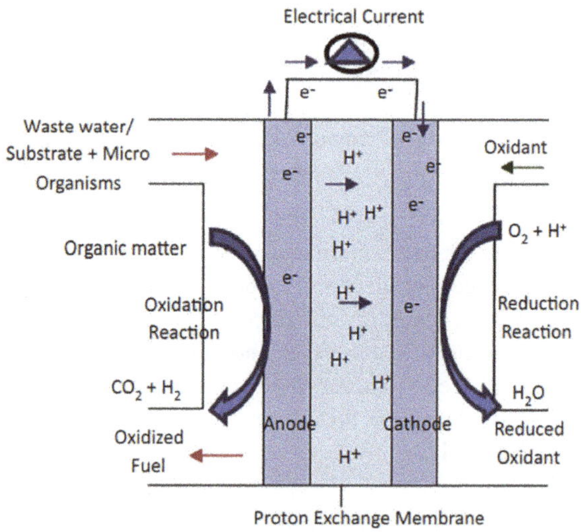

Fig. 5 A schematic representation of a typical microbial fuel cell (MFC) [12]

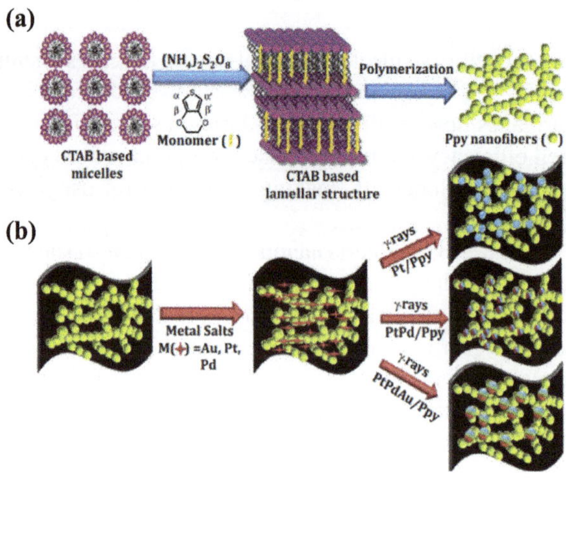

Fig. 6 Schematic representation of **a** radiolytically synthesized polypyrrole (PPy) nanofibers and **b** metallic nanocatalyst deposition on the nanofibers [22]

Despite their promise, scaling nanopolymer technologies for commercial fuel cells faces significant hurdles. Synthesis and scalability remain primary concerns, as techniques like co-precipitation and solid-state reactions are often cost-prohibitive and difficult to replicate at industrial scales [55].

Advances in polymer nanocomposite synthesis and functionalization broaden their potential applications, making them critical to new energy technology and a future based on renewable energy. However, producing in bulk with consistent quality remains a difficulty, as tiny nanostructure morphology and polymer orientation differences might impact performance. Furthermore, typical PEMs such as Nafion are expensive, and nanomaterial inclusion frequently requires unique production techniques, which complicates bulk deployment. Additionally, the long-term stability of nanopolymers under real service conditions, such as humidity, temperature cycling, and mechanical loading, must be assured. Environmental concerns about nanomaterial production, use, and disposal require lifecycle management strategies with a long-term, sustainable outlook. To bridge these gaps, advances in low-cost, high-efficiency synthesis methodologies, such as polymerization in a solventless environment or roll-to-roll processing, and investments in long-term, self-healing polymer architectures will become critical [73]. Academic and industrial collaboration will become critical in balancing performance, scalability, and sustainability in a field in rapid development.

Nanopolymers are also used for clean energy production by improving hydrogen storage systems and fuel cells. In hydrogen storage systems, nanopolymers composites with large surface area and tunable porosity expedite the hydrogen sorption and reduced the desorption temperatures. Besides, nanopolymer-based proton exchange membranes (PEMs) improve fuel cell efficiency by increasing ion conductivity and endurance.

Nanopolymers, notably metal–organic frameworks (MOFs) and covalent organic frameworks (COFs), are changing the way we store hydrogen. These materials have customized nanoporous architectures with ultrahigh surface areas, allowing effective physisorption of hydrogen molecules at moderate temperatures and pressures. The integration of functionalized nanopolymers into MOFs introduces tunable adsorption sites that significantly improve the hydrogen uptake and release of kinetics. Therefore, this advancement addresses the key challenges of traditional storage methods like cryogenic temperatures and high-pressure tanks that enable compact and lightweight solutions [96].

Integrating nanopolymers into hydrogen technologies can promote efficient hydrogen fuel systems and help the shift to sustainable energy [68]. In recent years Rahman et al. research team underscore the efficiency of carbon based nanopolymers such as activated carbon (AC) from jute waste and graphene oxide (GO), in supporting catalytical nanoparticles. Specifically, AC combines with copper oxide (CuO) nanoparticles significantly improved the hydrogen sorption kinetics of magnesium hydride (MgH_2) by increasing the surface area and facilitating the hydrogen desorption at onset temperature from 400 to 250 °C, which underscoring the catalytic synergy between nanopolymers and metal oxide, demonstrates the catalytic potential of nanopolymers in hydrogen storage [68].

Graphene oxide (GO) supported SnO_2 nanoparticles similarly enhance hydrogen storage performance by lowering the desorption temperature of MgH_2 to 274 °C. This effect is attributed to the graphene oxygen-rich functional groups and structural flexibility, which improve the H_2 dissociation and recombination [58]. These nanopolymers can function as both structural frameworks and active matrices, that effectively overcoming the constraints of MgH_2 such as sluggish hydrogen sorption kinetics and strong thermodynamic stability.

The use of nanopolymers not only improves cyclability and storage capacity, but it also works in combination with renewable processes by using renewable precursors like jute. These advancements highlight nanopolymers vital role in the development of efficient, scalable, and environmentally acceptable energy storage technology for automobile fuel cells, grid-scale renewable integration, and a variety of other applications [26, 99]. Furthermore, the physicochemical properties of Nanopolymers, such as high surface area, controlled porosity, and increased ionic conductivity, allow for better charge transport mechanisms [42]. This allows for faster charge–discharge rates, improved interfaces between electrodes and electrolytes, and longer lives, positioning them for future usage in energy storage technologies. As development develops, the synergy between nanotechnology and bio-derived materials will drive future breakthroughs in sustainable energy storage, reducing finite use while increasing device performance and endurance (Fig. 7).

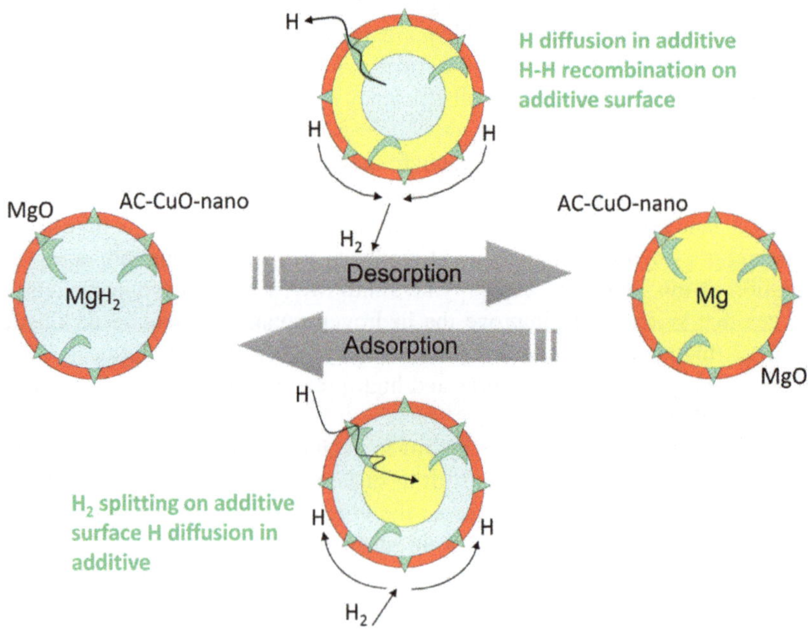

Fig. 7 Schematic illustration of the H_2 adsorption and desorption mechanism using activated carbon-supported CuO nanostructure [68]

3 Nanopolymers in Energy Conversion Techniques

3.1 Solar Energy Conversion

Solar energy is harvested through converting sunlight into direct current (DC) electricity through the photovoltaic (PV) effect. There can be a variety of methods for such a conversion, with crystalline or polycrystalline solar panels being the most efficient and most widespread technology, with a 25% solar-to-electricity efficiency. Solar cells can be classified into three generations based on active materials and power conversion efficiency (PCE).

(a) **First Generation solar cells**: The most prevalent photovoltaic technology is first-generation solar cells, which consist mainly of monocrystalline and polycrystalline silicon. Due to their high purity monocrystalline silicon cells can achieve an efficiency of up to 24.4% that makes them both highly effective and expensive [98].

(b) **Second Generation solar cells**: Thin-film solar cells (TFSCs), often known as second-generation solar cells that represent a significant advancement in photovoltaic technology [64]. They provide a compelling alternative to standard silicon-based solar cells by exploiting tiny layers of semiconductor materials. Semiconductor materials typically utilized in TFSCs include silicon (Si), cadmium telluride (CdTe), and copper indium selenide (CIS) [27].

(c) **Third-generation solar cell**: Third-generation solar cells seek to overcome the efficiency and material limits of first- and second-generation photovoltaic technologies through the use of innovative materials and better light absorption and conversion processes. This class comprises perovskite solar cells, quantum dot solar cells, dye-sensitized solar cells (DSSCs), organic photovoltaics (OPVs), and multi-junction tandem solar cells [25]. Perovskite solar cells have received a lot of attention due to their rapid efficiency increases, which have exceeded 25% in recent years [72]. As well as it is cost-effective manufacturing processes. Quantum dot solar cells use nanoscale semiconductor particles to increase light absorption and give variable bandgaps that allows for more efficient spectral use [37] (Fig. 8).

3.1.1 Photovoltaic Solar Cells

Nanopolymers play a pivotal role in advancing photovoltaic technologies, particularly in polymer solar cells (PSCs) and perovskite solar cells (PSCs). These materials enable the fabrication of lightweight, flexible, and cost-effective solar panels, significantly enhancing charge separation and transport, thereby improving the overall efficiency of the devices.

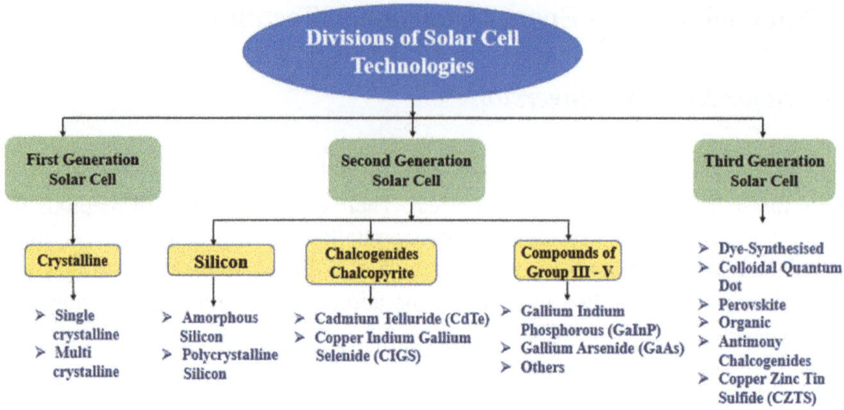

Fig. 8 Classification of solar cell technologies [86]

(a) **Polymer Solar Cells (PSCs)**

Polymer solar cells (PSCs) have gained popularity due to their lightweight, flexible nature, and promise for cost-effective energy conversion. Recent research is focusing on enhancing power conversion efficiency (PCE) and stability through novel material designs and improved device structure. A major milestone has been the invention of non-fullerene acceptors, which offer superior light absorption and enhanced charge transport compared to traditional fullerene-based systems [13, 88].

All polymer solar cells have gained attention due to their mechanical robustness, photostability and thermal stability. The use of polymerized small molecule acceptors (PSMAs) has greatly increased power conversion efficiency (PCE) that makes commercialization and flexible device applications possible. This work constructs and synthesizes two PSMAs, PN-Se and PS-Se, to evaluate the effect of molecular structure on photovoltaic performance [14, 97].

The PN-Se-based all-PSC achieves a high PCE of 16.16%, outperforming the PS-Se-based device (13.83%), due to improved phase separation, faster charge transport, and reduced recombination. This work highlights the significance of SMA selection in PSMA design for high-performance all-PSCs [15] (Fig. 9).

Commercially available solar cells are made from refined silicon crystals, similar to those used in computer chips. However, their high cost and complex production process have driven interest in alternative technologies. Lightweight solar cells are useful for autonomous sensors, offering advantages such as cost-effectiveness, disposability, flexibility, and lower environmental impact. Polymer solar cells, being transparent, have applications in walls, windows, and flexible devices. Four novel random polymers were synthesized using benzothiadiazole and benzotriazole as acceptors and benzodithiophene as the donor unit. In Fig. 10, it provides an introduction to polymer solar cells (PSCs) by illustrating their operating mechanism, spectral response, and morphological structure.

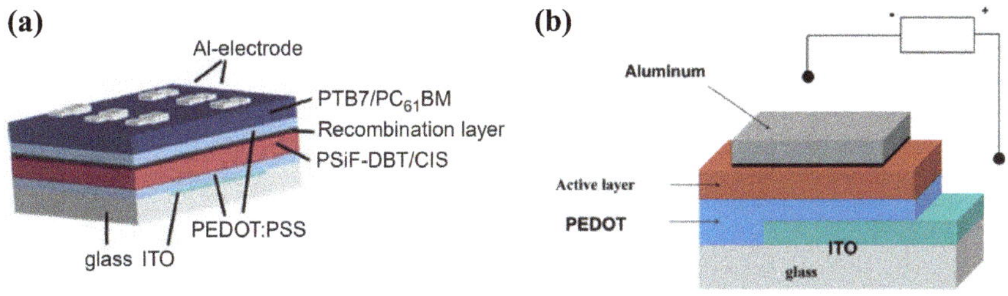

Fig. 9 **a** Schematic diagram of a polymer solar cell [36]. **b** Polymer/fullerene solar cells utilize a bulk heterojunction active layer sandwiched between an indium-tin-oxide/PEDOT:PSS hole transport layer and an aluminum electrode [29]

(b) **Perovskite Solar Cells**

Perovskite solar cells (PSCs) have emerged as revolutionary technology for clean energy production due to their high-power conversion efficiency (PCE) and low-cost fabrication. Over the past decade, the PCE of PSCs has surged from 3.5% to over 26%, positioning them as strong contenders in the photovoltaic market. Their tunable bandgap [62] and superior light absorption properties [81] enable efficient energy harvesting, reducing reliance on fossil fuels. Inverted PSCs (p-i-n architecture) are gaining traction for their improved long-term stability, addressing one of the key challenges in PSC commercialization. Nanostructured designs, including nanorods and nanocones, have been introduced to improve light capture and carrier transport [63]. Despite these advancements, challenges remain, including the need for enhanced durability under real-world conditions. Further innovations in interface engineering and structural optimization are essential for PSCs to achieve widespread adoption. With continued research, PSCs hold immense potential to revolutionize clean energy production and contribute to a sustainable future (Fig. 11).

In the past few years, perovskite solar cells have come a long way. Their efficiency has gone from about 3% in 2009 to over 26% today on small devices (about 0.1 cm^2). It has been shown that perovskite-silicon tandem cells can work at almost 34% efficiency. Perovskite solar cells have become very efficient very quickly, but they are not yet mass-produced, and there are still some problems that need to be fixed before they can be used in industrial PV systems.

For clean energy production, perovskite solar cells (PSCs) are advancing in efficiency, stability, and applications. To be commercially viable, PSCs must exceed 15% efficiency in labs (10–12% for modules). Tandem solar cells, combining multiple sub-cells, enhance photon utilization, with recent solution-processed PSCs reaching 10.6% efficiency. Novel applications include stretchable photovoltaics for wearables and PV-integrated LCDs that harvest wasted light [7]. Stability remains a challenge, requiring at least 7% efficiency and a seven-year lifespan for adoption. Inverted architectures improve durability, with industry efforts like Konarka's flexible PSCs achieving a three-year lifespan. With ongoing advancements, PSCs hold great promise for next-generation solar technology [41].

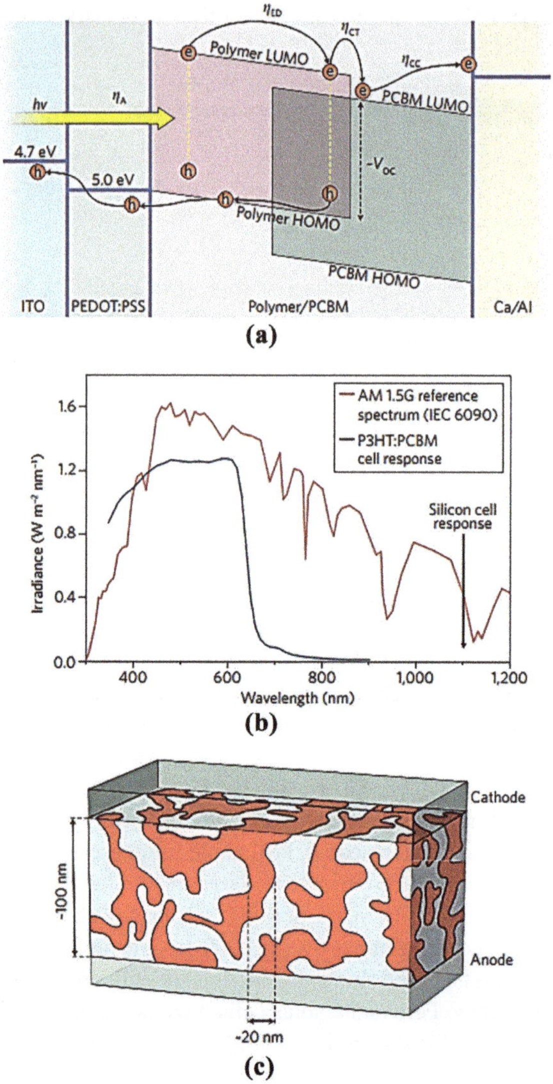

Fig. 10 Overview of PSCs. **a** Operating mechanism. **b** Solar spectrum versus P3HT:PCBM photoresponse. **c** Morphology model illustrating a bicontinuous polymer-acceptor network [41]

(c) **Dye-Sensitized Solar Cells (DSSCs)**

Dye-sensitized solar cells (DSSCs) have emerged as a viable and cost-effective alternative to traditional p-n junction photovoltaic devices. The concept of generating electricity using illuminated organic dyes in electrochemical cells was first explored in the late 1960s. Researchers at the University of California, Berkeley, extracted chlorophyll from spinach to study photosynthesis-related energy conversion. In 1972, the first chlorophyll-sensitized zinc oxide (ZnO) electrode was developed. This innovation marked the first instance where excited dye molecules injected

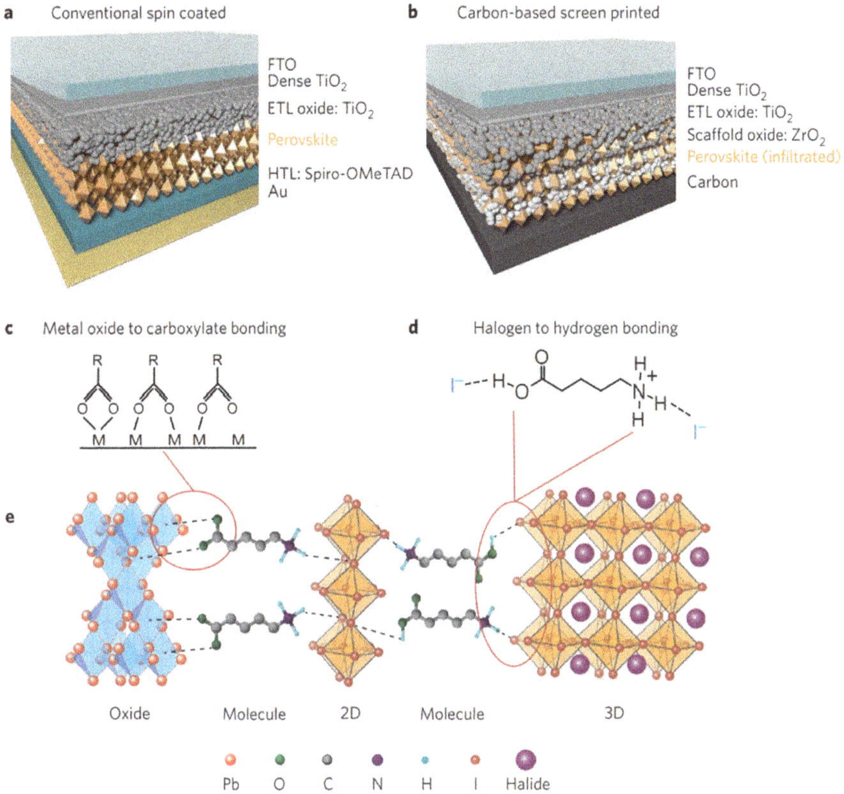

Fig. 11 Interface engineering of perovskite solar cells. **a**, **b** Diagram illustrating halide perovskite solar cells fabricated using the traditional spin-coating technique (**a**) and the carbon-based screen-printing method (**b**). **c–e** The interface modification of interlayers is made through the aminovaleric acid iodide molecule, with carboxyl and amine anchoring groups. Some examples of carboxyl functional groups bonding with semiconducting (TiO_2) or inert scaffold (ZrO_2) metal oxides [53] (**c**) and schematic representation of one possible perovskite/perovskite interaction, the halogen to hydrogen bonding to 2D/3D perovskites 2,4 (**d**). M, metal; FTO, fluorine doped tin oxide; ETL, electrode transport layer; HTL, hole transport layer

electrons into the wide band gap of a semiconductor, enabling the conversion of photons into electricity [79]. The growing energy demand has intensified the depletion of fossil fuels, with global reserves predicted to last approximately 40 years for oil, 60 years for natural gas, and 200 years for coal [40]. This looming resource exhaustion has accelerated the development of renewable energy technologies. In response, the European Union aims to have renewable sources contribute at least 27% of its total energy consumption by 2030, while the United States has allocated over $90 billion toward clean energy advancements through the Recovery Act [11]. Among renewable energy technologies, photovoltaics offers a promising solution for efficiently converting sunlight into electricity. However, the widespread adoption of conventional silicon-based solar cells remains limited due to their high production

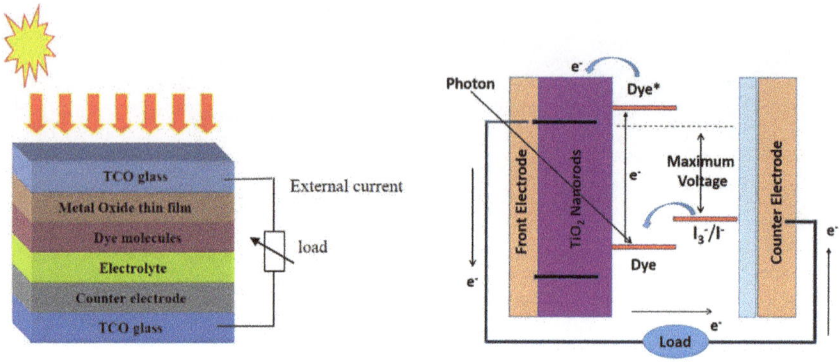

(a) (b)

Fig. 12 **a** Schematic device structure of a typical dye-sensitized solar cell (DSSC) [2] **b** Sensitize solar cell schematic band diagram [69]

costs and environmental concerns, restricting their application primarily to terrestrial markets (Fig. 12).

Dye-sensitized solar cells (DSSCs) are a cost-effective, flexible, and eco-friendly alternative to traditional silicon-based photovoltaics. Developed in the late 1960s, DSSCs employ a photoelectrochemical process by using a dye sensitizer to absorb light and generate charge. Because of their cheap production costs and their compatibility with flexible materials, a significant amount of research has been conducted to improve the efficiency of their power conversion. Therefore, DSSCs continue to be a potential alternative for the generation of sustainable energy. The performance of a DSSC is dependent on its fundamental components, which consist of a photoanode, a dye molecule, an electrolyte, a counter electrode.

3.2 Key Components of DSSC

(a) **Organic Dye (Sensitizer)**

Enhancing dye uptake by the semiconductor layer is critical for increasing DSSC efficiency since it directly affects electron production. A higher dye loading results in more excited electrons and thus a higher current density. To enhance dye absorption, a variety of semiconductor nanostructures have been investigated, including nanorods, nanotubes, aerogels, mesoporous powders, and 3D nanoarchitectures. Surface functionalization, such as co-adsorption with carboxylic or other functional groups, has been investigated as a means of improving semiconductor layer wetting and performance. Additionally, knowing the role of oxygen vacancies in dye loading is critical. To improve dye uptake efficiency in photoanodes and thus device performance, cost-effective solutions should be developed [78].

Enhancing DSSC conversion efficiency can also be achieved by incorporating additional sensitizers. These sensitizers broaden the cell's light absorption and conversion capabilities beyond the range of unsensitized cells. Furthermore, they can enable the injection of multiple electrons upon the absorption of a single high-energy photon and improve electrode properties by providing a structural framework that increases the active surface area. Nanocrystalline quantum dots (QDs), such as CdSe, have demonstrated potential for improving cell characteristics, particularly by reducing electron recombination. The dye molecules absorb incident photons and become excited, initiating the process of electron generation. Common dyes include ruthenium-based complexes, metal-free organic dyes, and quantum dots in advanced DSSC designs.

(b) **Semiconductor Oxide Layer**

Typically composed of TiO_2, this porous nanostructured layer acts as the electron transport medium. The excited dye injects electrons into the conduction band of this oxide, facilitating charge transfer.

(c) **Working Electrode (WE)**

Positioned behind the oxide layer, the WE collect and transports the injected electrons. It is usually made of fluorine-doped tin oxide (FTO) glass, which serves as the transparent conductive electrode.

(d) **Redox Mediator (Electrolyte)**

Often consisting of an iodide/triiodide (I^-/I_3^-) redox couple, the electrolyte regenerates the dye by replacing the lost electrons. Traditional liquid electrolytes face challenges like leakage, degradation over time, and difficulties in flexible DSSC applications. Research is progressing toward solid-state electrolytes to mitigate these issues.

(e) **Counter Electrode (CE)**

The CE facilitates the reduction of the redox mediator, completing the circuit with the working electrode. Typically made from platinum-coated FTO glass, though alternative materials like carbon-based electrodes (e.g., CNTs or graphene) are being explored for cost-effective and flexible designs.

3.2.1 Basic Working Principle of DSSCs

Dye-sensitized solar cells (DSSCs) are a type of photoelectrochemical (PEC) solar technology that uses dye molecules to absorb sunlight and generate charge. DSSCs work differently than traditional p-n junction photovoltaic (PV) devices by exciting dye molecules adsorbed on a semiconductor oxide. DSSCs have a working electrode (WE), a counter electrode (CE), and an electrolyte system mounted on a transparent conductive substrate, such as fluorine-doped tin oxide (FTO) glass [94].

Dye-Sensitized Solar Cells (DSSCs) are solar cells that use organic dye to absorb light and release electrons. These electrons are then introduced into a layer of metal oxide, which contacts a working electrode. A redox mediator is used to retrieve lost electrons, and a counter electrode (CE) ensures continuous charge flow. DSSCs were initially designed with flat-cell structures, requiring a transparent conductive electrode to allow light to flow through and reach the oxide layer. Transparent conductive oxides (TCOs) are commonly used, but they have disadvantages such as high production costs and rigidity, making them unsuitable for flexible solar applications.

However, liquid electrolytes introduce several challenges, such as leakage over time, susceptibility to degradation from temperature fluctuations, and difficulties in manufacturing flexible DSSCs, as the electrolyte must be securely contained. A schematic representation of a modern DSSC, including its primary components—working electrode, counter electrode, electrolyte, and oxide layer—illustrates the cell's structure. Newer DSSC designs increasingly incorporate sensitizers and quantum dots to enhance performance and efficiency.

In a study on dye-sensitized solar cells (DSSCs), researchers created one-dimensional TiO_2 nanofibers using electrospinning. They investigated how the color of the dye and the heating temperature of the cells affect their performance. The results indicated that dye N719 outperforms dye N350 and that lower heating temperatures yield better results. Additionally, the optimization of DSSCs was further enhanced by adjusting the thickness of the TiO_2 nanofibers [35]. Researchers explored the production of electrospun nanofibers associated with Ru(II) complexes, which are utilized as counter electrodes in dye-sensitized solar cells (DSSCs). The study focused on the shape of the fibers, their thermal stability, and electrical conductivity. The finding demonstrates that the fibers associated with Ru have a significant improvement in current density and conductivity, opening up new possibilities for DSSCs [87].

The electrospinning of TiO_2 nanofibers in dye-sensitized solar cells (DSSCs) demonstrates that increased applied voltages produce homogenous nanofibers with improved light absorption and electron transport properties [19]. The ZnO-TiO_2 composite nanofibers have potential as photoanodes in dye-sensitized solar cells (DSSCs), with an ideal ZnO-to-TiO_2 molar ratio enhancing dye loading, electrolyte diffusion, and charge transport, resulting in a 56% increase in efficiency relative to pure TiO_2 nanofibers [9]. These developments in nanofibers exhibits that the essential function of nanofibers uses in DSSC, improve the performance through improved electron transport, increased dye absorption, and diminished charge recombination, positioning them as a viable choice for sustainable solar energy solutions (Figs. 13 and 14).

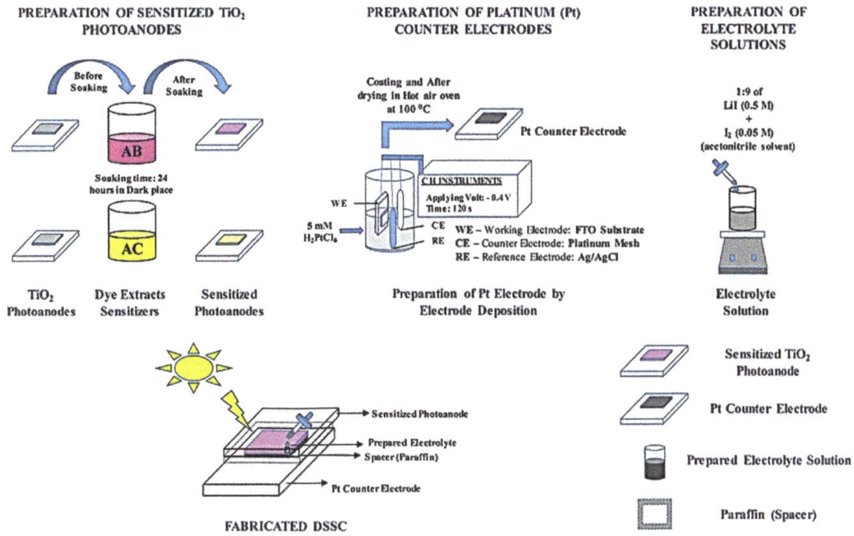

Fig. 13 Schematic representation of the fabrication of DSSC [67]

3.2.2 Mechanism of Dye-Sensitized Solar Cells (DSSC)

The operational process can be divided into several key steps [94]

$$\text{Dye} + h\nu \rightarrow \text{Dye}^{*}\ (\text{Light Absorption})$$

$$\text{Dye}^{*} \rightarrow \text{Dye}^{+} + e^{-}(\text{Electron Injection})$$

$$e^{-}\ (\text{TiO}_2) \rightarrow \text{FTO} \rightarrow \text{External Circuit} \rightarrow \text{Counter Electrode (Electron Transport)}$$

$$\text{Dye}^{+} + 3\text{I}^{-} \rightarrow \text{Dye} + \text{I}_s^{-}\ (\text{Dye Regeneration})$$

$$\text{I}_s^{-} + 2e^{-}\ (\text{Counter Electrode}) \rightarrow 3\text{I}^{-}\ (\text{Electrolyte Regeneration})$$

3.2.3 Electrical Characteristics Parameters:

The performance of dye-sensitized solar cells (DSSCs) is typically evaluated by analyzing their photoresponse, which is represented as a plot of photocurrent density versus voltage. These measurements are conducted under standard illumination conditions using the AM 1.5 G solar spectrum, corresponding to an integrated irradiance of 1000 W/m^2.

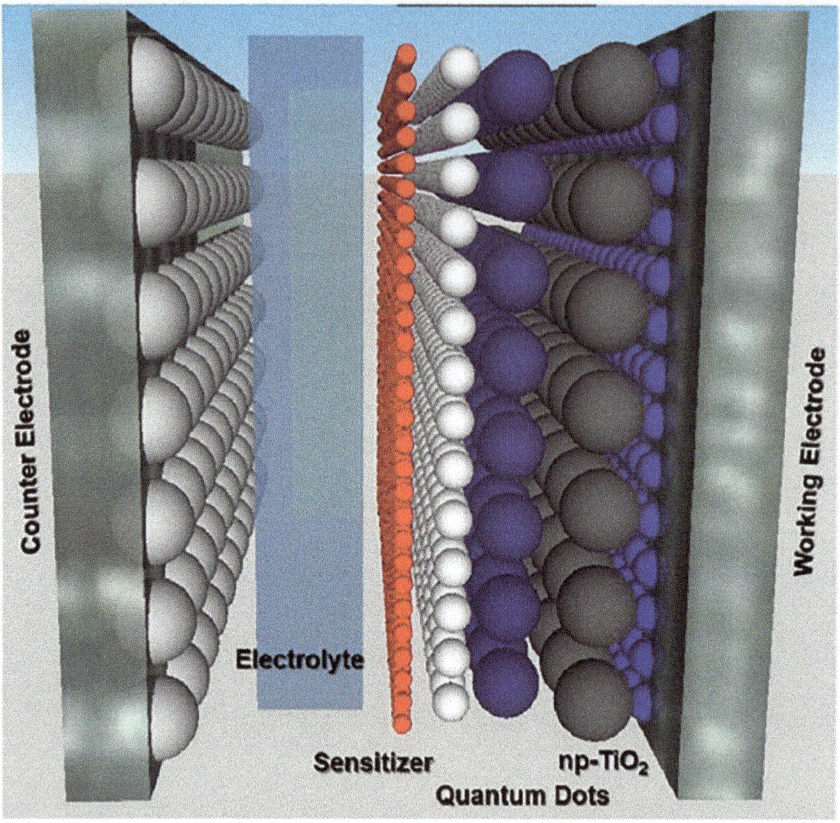

Fig. 14 Schematic and sectional view of quantum dot-based DSSC [85]

The overall power conversion efficiency (η) of a DSSC is determined using the following formula:

$$\eta = \frac{J_{SC} V_{oc}}{P_{in}} \times 100$$

where,

- J_{sc} = represents the short-circuit current density (A/cm^2),
- V_{oc} = is the open-circuit voltage (V),
- P_{in} = denotes the incident light intensity (W/cm^2), and
- FF = is the fill factor, which quantifies the ratio of the maximum power output to the theoretical power output

The fill factor is an essential parameter that characterizes the shape of the current–voltage curve and indicates the efficiency of charge collection within the device. A

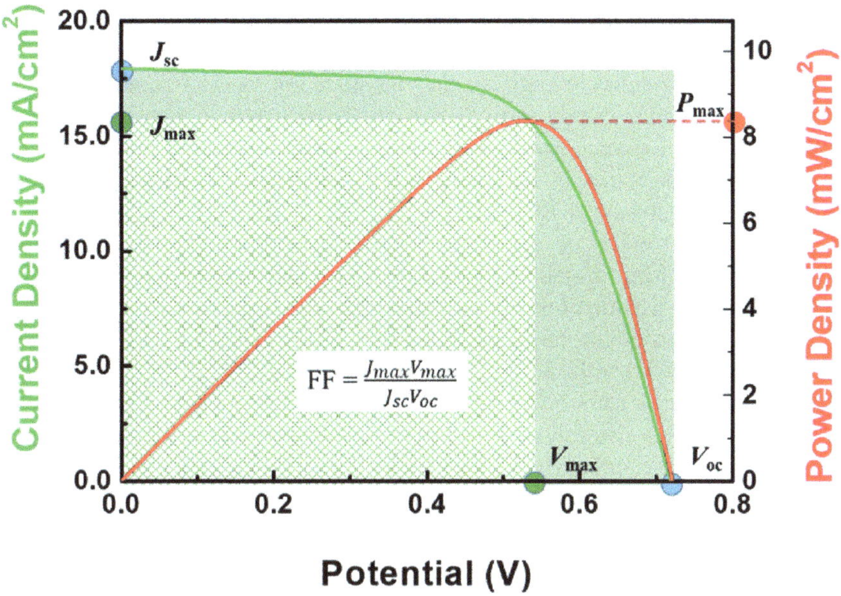

Fig. 15 Steady-state response of current density and power versus voltage of a dye-sensitized solar cell [1]

Table 1 Different photoelectrode of DSSC and their efficiency

Nanopolymer	Preparation methodology	Sensitizer	Electrolyte	Efficiency (%)	References
ZnO nanowire	Hydrothermal method	Ruthenium-based N3 dye	Lithium iodide/ triiodide	0.42	[45]
SnO_2 nanowire	Reactive vapor transport	Tin as source	Ruthenium-based N719 dye	–	[28]
TiO_2 nanoflower	Hydrothermal method	Titanium tetraisopropoxide	Ruthenium-based N719 dye	6.63	[52]
CuO nanowires	Hydrothermal method	Metal-free organic dye	Sulfide-polysulfide electrolyte	1.9	[32]
Fe_2O_3 nanorods	Spray pyrolysis	Organic dye (Perylene diimide)	Alkaline electrolyte	0.85	[8]
$BiVO_4$ nanoplates	Electrodeposition	Organic dye (Eosin-Y)	Sulfite electrolyte	2.4	[38]
TiO_2 nanotubes	Anodization	Ruthenium-based N719 dye	Iodide-based	7.2	[89]

higher fill factor corresponds to a more rectangular shape, reflecting improved device performance (Fig. 15) (Table 1).

3.2.4 Nanoelectrolytes in DSSCs

Electrolyres are essential in DSSC as they facilitate ion transport, enabling charge transfer between photoanode and counter electrolytes. Nanopolymer electrolytes offer a promising alternative to conventional liquid electrolytes by increasing stability and reducing leakage, thereby improving system efficiency and durability [91]. The incorporation of nano-sized fillers into the polymer matrix promotes polymer chain mobility, improves electrode–electrolyte interfacial stability, thermal stability, and long-term electrochemical stability, and creates favorable routes for ionic conduction. As a result, the dispersion of nano-sized fillers into a polymer matrix is proposed as a method of improving the electrolyte characteristics for DSSC manufacturing [39, 71] Liquid electrolytes suffer from corrosion, leakage and volatility problems, on the other hand solid-state electrolytes based on nanopolymers offers a safer alternative. Besides polymer gel electrolytes which is a type of nanopolymers electrolytes that provides high ionic conductivity while maintaining solid like structure. Poly (vinylidene fluoride-co-hexafluoropropylene) (PVDF-HFP), which is widely utilized in DSSCs for its exceptional mechanical stability and ability to retain liquid electrolyte components while minimizing leakage [100]. In another research by [95], they used the application of a polymer gel electrolyte based on poly (methyl methacrylate) (PMMA) in quasi-solid-state DSSCs. Their research highlights the improved interfacial contact with electrodes and enhanced thermal stability achieved by incorporating PMMA-based gel electrolytes [95]. This strategy facilitates efficient charge transport while mitigating the issues associated with liquid electrolytes. Nanopolymers can be used to create these matrices which offer improved mechanical properties, thermal stability, and interfacial contact with the electrodes [5].

3.3 Flexible DSSC Solar Cell

Flexible dye-sensitized solar cells (DSSCs) based on carbon nanofiber electrodes have emerged as a promising technology due to their lightweight, flexibility, and cost effectiveness. Recent research has highlighted several techniques to improve the performance of these cells using novel electrode materials and architectures. Carbon nanofiber (CNF) electrodes, commonly coupled with conductive polymers such as polyaniline (PANI) or polypyrrole (PPY), demonstrate increased electrical conductivity and stability [4, 18]. The addition of PANI into carbon nanotube fibers (CNTF) increased photoelectric conversion efficiency (PCE) from 2.26 to 3.57% [4]. Flexible DSSCs made of TiO_2 nanopapers displayed good mechanical strength and stability in bending tests, even after significant deformation [74] (Figs. 16 and 17).

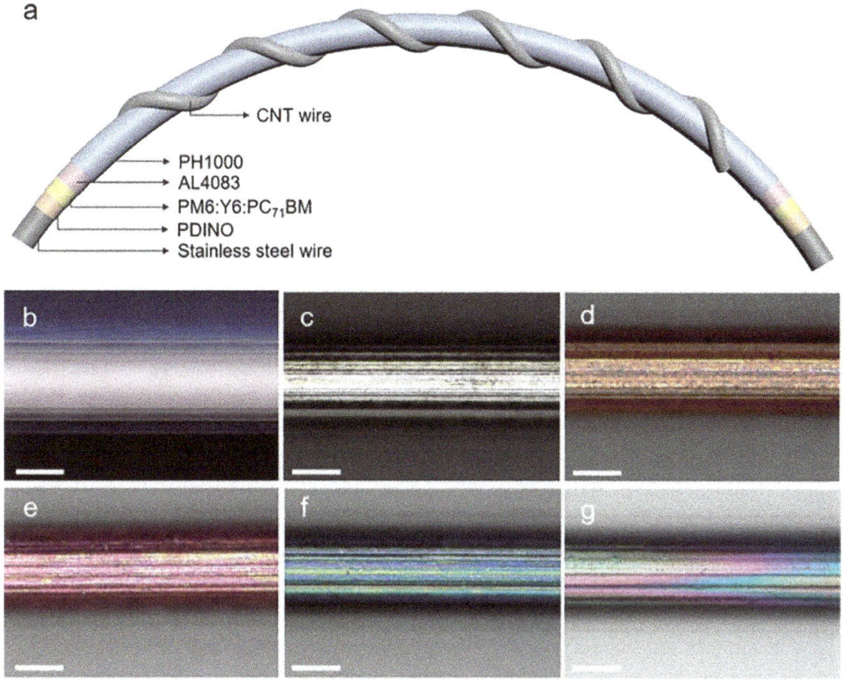

Fig. 16 **a** Representation of a fiber-shaped organic solar cell. **b** Visual microscopy images of lab-grade stainless-steel wire, **c** stainless-steel wire electrode, **d** PDINO ETL, **e** PM6:Y6:PC71BM active layer, **f** A4083 layer **g** PH1000 layer [50]

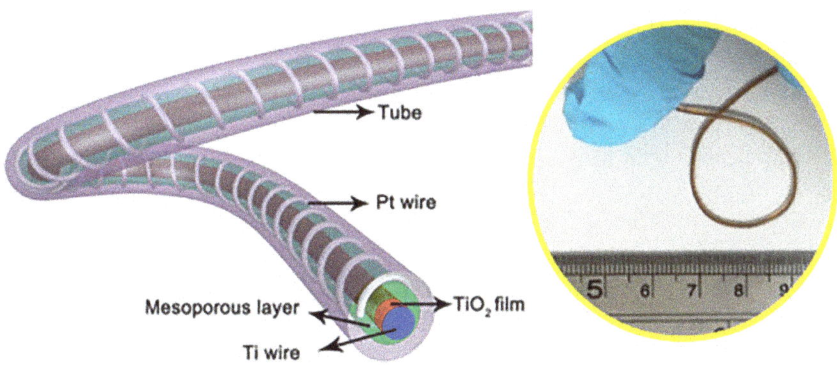

Fig. 17 Flexible DSSC with Ti wire [47]

3.3.1 Preparation of the Working Electrode

The working electrode's carbon nanotube yarn (CNTY) was purchased from Nanocomp Technologies Inc. (USA). To get purity, the polymer coating from the manufacturing process was removed by calcination in air. The CNTY was treated with 35% nitric acid for 20 h to introduce functional groups and improve its surface properties. After this, it was sonicated in ethanol for 30 s to remove residual acid and byproducts. The nanotube was then wrapped around a larger PTFE plate and heated at 290–300 °C for 2 h in tube furnace. Then it sonicated in ethanol for 30 s to eliminate any remaining surface contaminants. The working electrode consists of six twisted CNTYs with an average diameter of 25 μm.

I. **Polyaniline (PANI) Coating**
   In-situ chemical polymerization produced a homogeneous PANI layer on the CNTY surface. The CNTY was immersed in a solution comprising aniline monomer, ammonium persulfate (APS), and 1 M hydrochloric acid (HCl), which allowed polymerization to occur at specified temperatures. Following polymerization, the coated CNTYs were thoroughly washed with Milli-Q water and ethanol before being dried at ambient temperatures.
II. **Deposition of $CuInS_2$ Quantum Dots**
   $CuInS_2$ quantum dots (QDs) were created by a chemical bath deposition (CBD) technique. The CNTY-PANI electrode was immersed in an aqueous solution comprising $CuSO_4$, $In(NO_3)_3$, and thiourea. The process was aided by changing the pH and maintaining a constant temperature. Repeated deposition ensured equal coverage of $CuInS_2$ QDs on the PANI-coated CNTY. The coated electrodes were then cleaned and dried before being characterized.
III. **Fabrication of the Counter Electrode**
   The counter electrode was created by treating CNTY for 3 min with a solution of 1.5 M sulfuric acid and 0.5 M nitric acid, followed by thermal treatment at 300 °C with a constant air flow of 100 ml/min. After cooling, the electrode was cleaned with Milli-Q water, then immersed in Triton-X for 5 min before air-drying. To increase catalytic activity, platinum nanoparticles were sputter-deposited onto the CNTY surface at 8 mA for 40 s, resulting in a homogenous coating of nanoparticles ranging in size from 15–20 nm. The counter electrode was then looped around the working electrode with a 1 mm thread pitch to finish the solar cell assembly.
IV. **Photovoltaic Performance Characterization**
   The photovoltaic performance of the manufactured flexible solar cells was measured using a VeraSTAT3 potentiostat (Princeton Applied Research) with cyclic voltammetry at a scan rate of 50 mV/s. A Honle solar simulator 400, with an AM 1.5G spectrum, was utilized to replicate sunshine for measurements. Prior to characterization, the cells were saturated with an iodine redox couple-based electrolyte consisting of 0.5 M tert-butyl pyridine in 3-methoxy propionitrile (3-Me PRN), which ensured the electrolyte's conductivity and solid-state nature (Fig. 18).

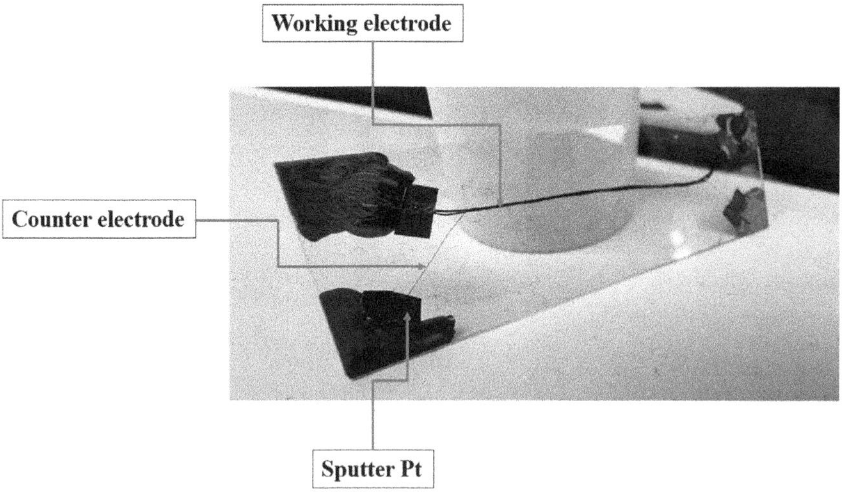

Fig. 18 Schematic view of DSSC mounted on glass slide

3.4 Types of Fabrication Methods (DSSC)

3.4.1 Dip Coating

In the fabrication of DSSC dip coating is a cost effective and simple process that involves immersing a substrate in a precursor solution and withdrawn at a controlled speed. After that, keep drying it to generate a homogeneous thin film. This approach allows for accurate thickness control and is suited for large-area deposition. In recent research dip coating was used to form a thin layer of ZnO on fluorine doped thin oxide. A single cycle yielded a thin coating, while 11 cycles resulted in a 209 nm thick layer that completely covered the FTO surface [83] (Fig. 19).

3.4.2 Hydrothermal Reaction

Hydrothermal synthesis plays an important role in the development of dye-sensitized solar cells (DSSCs) because of their capacity to synthesize high-purity nanomaterials with fine control over size, shape, and crystallinity. This level of precision is crucial for enhancing DSSC performance, as the properties of the constituent materials—particularly the semiconductor photoanode—directly influence light conversion efficiency. In this study, anatase TiO_2 thin films were synthesized using a Ti foil-assisted hydrothermal method. Adjusting the hydrothermal duration (3–12 h) altered the morphology from nanotubes to nanoflowers and increased the film thickness from 2.5 to 4.2 μm. [54] (Fig. 20).

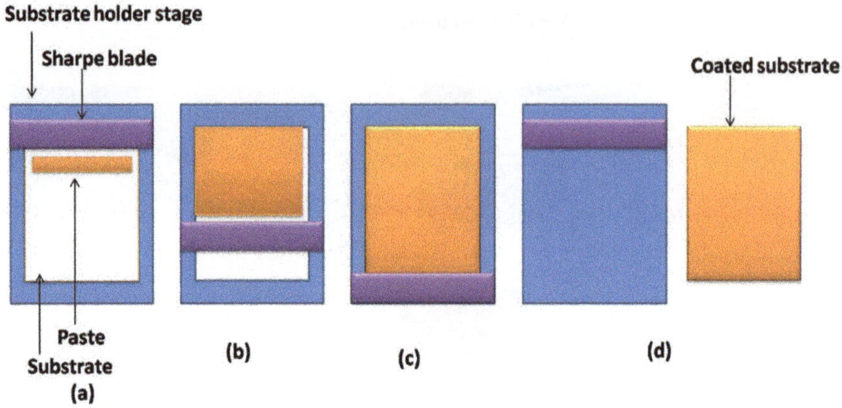

Fig. 19 Doctor blade technique [79]

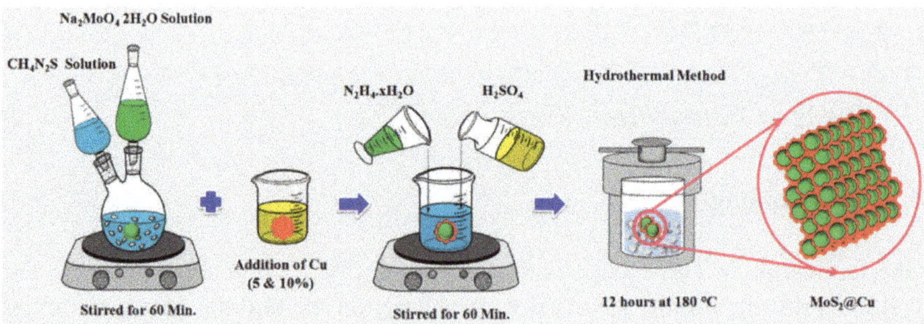

Fig. 20 Synthesis of MoS2@Cu nanoparticles by hydrothermal reactor [75]

3.4.3 Screen Printing Method

Screen printing is a cost-effective and widely used technique for fabricating Dye-Sensitized Solar Cells (DSSCs). J. Liu et al. create a fabrication method which is suitable for textiles that utilize a low-temperature processed TiO_2 paste. Then it is applied through screening and spray painting. This low-temperature method is critical for guaranteeing compatibility with both Kapton, a flexible polymer substrate, and standard 65/35 polyester-cotton woven fabric [46].

3.4.4 Spin Coating

In DSSC fabrication spin coating is a popular technique for producing thin TiO_2 film with a controlled thickeness [92]. This is another type of coating method especially for photoanodes and counter electrodes in DSSC. In a recent study by Al-juaid et al. found that DSSC with ZnO spin-coated TiO_2 electrodes show improved performance

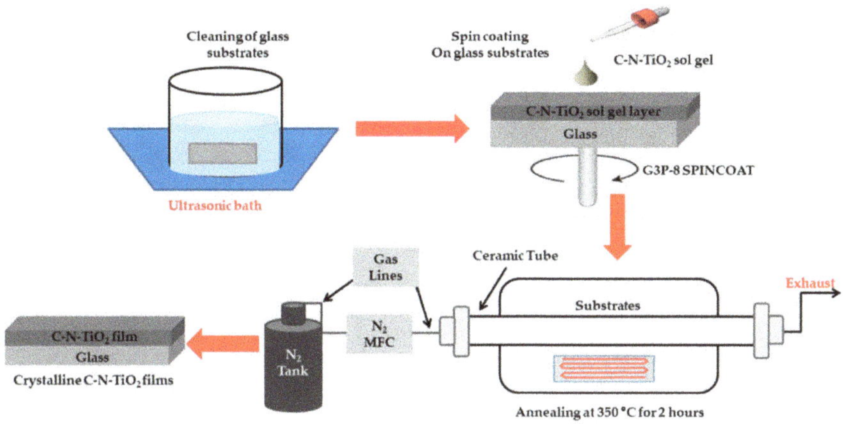

Fig. 21 An experimental set up used for spin coating of C-N-TiO_2 thin films on DSSCs fabrication [57]

compared to bare TiO_2 electrodes. By controlling ZnO deposition parameters, efficiency can be enhanced by over 100%. The highest efficiency was achieved using a single sol drop at a precursor concentration between 0.1 and 0.2 M. The efficiency enhancement is governed by two mechanisms: the formation of an energy barrier at the TiO_2 interface and the decrease in dye adsorption efficiency with excessive ZnO deposition. A thin ZnO layer effectively suppresses electron recombination while preserving dye adsorption further improves the DSSC photovoltaic performance [3] (Fig. 21).

3.4.5 Atomic Layer Deposition (ALD)

Atomic layer deposition (ALD) provides fine control over material production on an atomic scale. Its versatility stems from its ability to fine-tune composition and film thickness while still providing outstanding conformality, low-temperature processing, and real-time monitoring capabilities. A study found that a composite photoanode with ZnO nanobelts and TiO_2 nanoparticles outperformed traditional TiO_2 in terms of DSSC performance. Additional ALD coating of the composite photoanode increased interfacial contact and reduced recombination, resulting in an even larger efficiency boost of 137% [48] (Fig. 22).

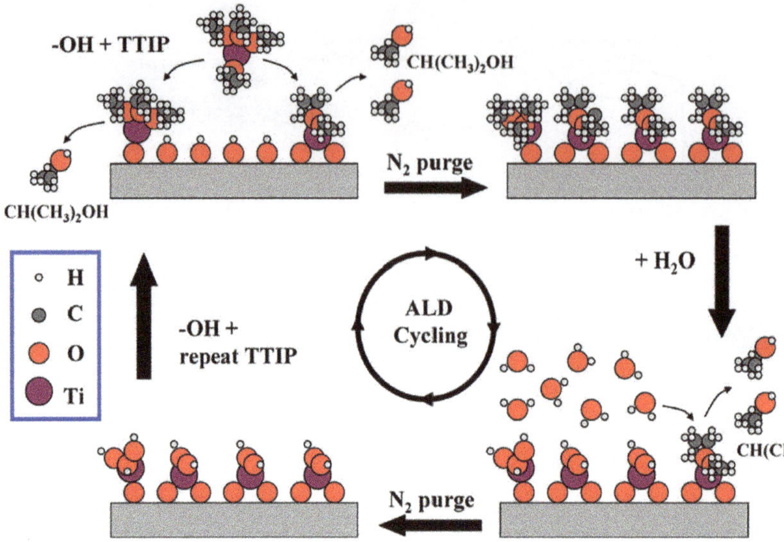

Fig. 22 Illustration of a single cycle of TiO_2 atomic layer deposition (ALD) onto an FTO-coated glass substrate using titanium tetraisopropoxide (TTIP) [30]

4 Vision and Challenges

The use of nanopolymers in renewable energy technology provides a great approach to energy storage, conversion, and sustainability. One of the most potential fields of application is solar energy, where nanopolymer-based photovoltaic (PV) cells such as polymer solar cells (PSCs) and perovskite solar cells (PSCs) have shown improved efficiency, flexibility, and cost-effectiveness. However, establishing large-scale manufacturing with consistent quality is still a big task. Variations in nanostructure morphology or polymer alignment can diminish solar cell performance and stability which will ultimately limit their economic viability. Large-scale production of nanopolymer-based energy sources presents significant challenges including cost and complexity in industrial manufacturing. Nanopolymers must maintain performance under real-world conditions, including humidity, temperature cycling, and mechanical stress. Implementing robust lifecycle management strategies is necessary to address the environmental impact of nanomaterial production, use, and disposal. Addressing these issues not only limited to requiring new material innovations and eco-friendly synthesis methods, but also sustainable energy policies. Collaborations between academia, industry, and government agencies will play a crucial role in overcoming these challenges.

5 Conclusion and Future Outlook

Nanopolymers are becoming increasingly important in the clean energy sector providing enhanced efficiency, durability, and adaptability in various applications such as solar cells, fuel cells, and energy storage systems. Their integration in energy technologies supports a sustainable energy transition by improving device performance and reducing environmental impact. Advancements in polymer nanocomposite synthesis and functionalization have expanded their applicability which bridges the gap between technological innovation and sustainability. However, challenges such as large-scale production, cost-effectiveness, and long-term stability also need to be addressed. Future research should focus on optimizing synthesis techniques, enhancing recyclability, and reducing reliance on expensive materials. The integration of artificial intelligence and machine learning in material design could accelerate the discovery of novel nanopolymer formulations.

References

1. Aftabuzzaman, M., Sarker, S., Lu, C., Kim, H.K.: In-depth understanding of the energy loss and efficiency limit of dye-sensitized solar cells under outdoor and indoor conditions. J. Mater. Chem. A **9**(44), 24830–24848 (2021). https://doi.org/10.1039/D1TA03309C
2. Agarwal, R., Vyas, Y., Ameta, C.: Outdoor Performance and Stability Assessment of Dye-Sensitized Solar Cells (DSSCs) (n.d.). www.intechopen.com
3. Al-juaid, F., Merazga, A., Al-Baradi, A., Abdel-wahab, F.: Effect of sol–gel ZnO spin-coating on the performance of TiO_2-based dye-sensitized solar cell. Solid-State Electron. **87**, 98–103 (2013). https://doi.org/10.1016/j.sse.2013.06.007
4. Alwadai, N., Ali, A., Liaqat, A., Fatima, A., Iqbal, M., Nazir, A., Mnif, W., Algarni, Z., Akyürekli, S., Kaleli, M.: Electrodeposited polyaniline modified CNT fiber as efficient counter electrode in flexible dye-sensitized solar cells. Synth. Met. **307**, 117692 (2024). https://doi.org/10.1016/j.synthmet.2024.117692
5. Aruchamy, K., Ramasundaram, S., Divya, S., Chandran, M., Yun, K., Oh, T.H.: Gel polymer electrolytes: advancing solid-state batteries for high-performance applications. Gels **9**(7), 585 (2023). https://doi.org/10.3390/gels9070585
6. Asamoah, G.A., Korsah, M., Jeyasundar, P.G.S.A., Ahmed, M., Lau, S.Y., Danquah, M.K.: Nanotechnology-based lithium-ion battery energy storage systems. Sustainability **16**(21), 9231 (2024). https://doi.org/10.3390/su16219231
7. Ašmontas, S., Mujahid, M.: Recent progress in perovskite tandem solar cells. Nanomaterials **13**(12), 1886 (2023). https://doi.org/10.3390/nano13121886
8. Bella, F., Gerbaldi, C., Barolo, C., Grätzel, M.: Aqueous dye-sensitized solar cells. Chem. Soc. Rev. **44**(11), 3431–3473 (2015). https://doi.org/10.1039/C4CS00456F
9. Chang, Q., Xu, J., Han, Y., Ehrmann, A., He, T., Zheng, R.: Photoelectric performance optimization of dye-sensitized solar cells based on ZnO-TiO_2 composite nanofibers. J. Nanomate. **2022**(1) (2022). https://doi.org/10.1155/2022/7356943
10. Deng, X., Ma, L., Wang, C., Ye, H., Cao, L., Zhan, X., Tian, J., Tong, X.: Recent progress in materials design and fabrication techniques for membrane electrode assembly in proton exchange membrane fuel cells. Catalysts **15**(1), 74 (2025). https://doi.org/10.3390/catal15010074
11. Department of Energy: Successes of the Recovery Act (2012)

12. Dharmalingam, S., Kugarajah, V., Sugumar, M.: Membranes for microbial fuel cells. Biomass, Biofuels, Biochem. Microbial. Electrochem. Technol. Sustain. Platf. Fuels Chem. Remediat. 143–194 (2019). https://doi.org/10.1016/B978-0-444-64052-9.00007-8
13. Doumon, N.Y., Dryzhov, M.V., Houard, F.V., Le Corre, V.M., Rahimi Chatri, A., Christodoulis, P., Koster, L.J.A.: Photostability of fullerene and non-fullerene polymer solar cells: the role of the acceptor. ACS Appl. Mater. Interfaces. **11**(8), 8310–8318 (2019). https://doi.org/10.1021/acsami.8b20493
14. Du, J., Hu, K., Zhang, J., Meng, L., Yue, J., Angunawela, I., Yan, H., Qin, S., Kong, X., Zhang, Z., Guan, B., Ade, H., Li, Y.: Polymerized small molecular acceptor based all-polymer solar cells with an efficiency of 16.16% via tuning polymer blend morphology by molecular design. Nat. Commun. **12**(1), 5264 (2021). https://doi.org/10.1038/s41467-021-25638-9
15. Du, J., Hu, K., Zhang, J., Meng, L., Yue, J., Angunawela, I., Yan, H., Qin, S., Kong, X., Zhang, Z., Guan, B., Ade, H., Li, Y.: Polymerized small molecular acceptor based all-polymer solar cells with an efficiency of 16.16% via tuning polymer blend morphology by molecular design. Nat. Commun. **12**(1) (2021). https://doi.org/10.1038/s41467-021-25638-9
16. Dutta, K., Das, S., Rana, D., Kundu, P.P.: Enhancements of catalyst distribution and functioning upon utilization of conducting polymers as supporting matrices in DMFCs: a review. Polym. Rev. **55**(1), 1–56 (2015). https://doi.org/10.1080/15583724.2014.958771
17. Dutta, K., Kundu, P.P.: A review on aromatic conducting polymers-based catalyst supporting matrices for application in microbial fuel cells. Polym. Rev. **54**(3), 401–435 (2014). https://doi.org/10.1080/15583724.2014.881372
18. Etefa, H.F., Kumar, V., Dejene, F.B., Efa, M.T., Jule, L.T.: Modification of flexible electrodes for P-type (nickel oxide) dye-sensitized solar cell performance based on the cellulose nanofiber film. ACS Omega **8**(17), 15249–15258 (2023). https://doi.org/10.1021/acsomega.3c00383
19. Fajariah, A.R., Nurazizah, E.S., Faizal, F., Safriani, L.: Synthesis of TiO_2 nanofiber as photoanode of dye sensitized solar cells (DSSC). Key Eng. Mater. **950**, 3–9 (2023). https://doi.org/10.4028/p-1hoXBf
20. Fu, S., Sun, Z., Huang, P., Li, Y., Hu, N.: Some basic aspects of polymer nanocomposites: a critical review. Nano Mater. Sci. **1**(1), 2–30 (2019). https://doi.org/10.1016/j.nanoms.2019.02.006
21. Gao, D., Wu, H., Lyu, B., Zheng, C., Zhou, Y., Ma, J.: Waste-to-energy: sustainable triboelectric stimulating system constructed by bone gelatin based triboelectric nanogenerator for crop growth. Small **21**(3) (2025). https://doi.org/10.1002/smll.202408925
22. Ghosh, S., Bera, S., Bysakh, S., Basu, R.N.: Conducting polymer nanofiber-supported Pt alloys: unprecedented materials for methanol oxidation with enhanced electrocatalytic performance and stability. Sustain. Energy Fuels **1**(5), 1148–1161 (2017). https://doi.org/10.1039/C7SE00126F
23. Ghosh, S., Das, S., Mosquera, M.E.G.: Conducting polymer-based nanohybrids for fuel cell application. Polymers **12**(12), 2993 (2020). https://doi.org/10.3390/polym12122993
24. Giussi, J.M., Cortez, M.L., Marmisollé, W.A., Azzaroni, O.: Practical use of polymer brushes in sustainable energy applications: interfacial nanoarchitectonics for high-efficiency devices. Chem. Soc. Rev. **48**(3), 814–849 (2019). https://doi.org/10.1039/C8CS00705E
25. Gong, J., Darling, S.B., You, F.: Perovskite photovoltaics: life-cycle assessment of energy and environmental impacts. Energy Environ. Sci. **8**(7), 1953–1968 (2015). https://doi.org/10.1039/C5EE00615E
26. Goodenough, J.B., Park, K.-S.: The Li-Ion rechargeable battery: a perspective. J. Am. Chem. Soc. **135**(4), 1167–1176 (2013). https://doi.org/10.1021/ja3091438
27. Grissom, G., Jaksik, J., McEntee, M., Durke, E.M., Aishee, S.T.J., Cua, M., Okoli, O., Touhami, A., Moore, H.J., Uddin, M.J.: Three-dimensional carbon nanotube yarn based solid state solar cells with multiple sensitizers exhibit high energy conversion efficiency. Sol. Energy **171**, 16–22 (2018). https://doi.org/10.1016/j.solener.2018.06.053
28. Gubbala, S., Russell, H.B., Shah, H., Deb, B., Jasinski, J., Rypkema, H., Sunkara, M.K.: Surface properties of SnO_2 nanowires for enhanced performance with dye-sensitized solar cells. Energy Environ. Sci. **2**(12), 1302 (2009). https://doi.org/10.1039/b910174h

29. Hoppe, H., Sariciftci, N.S.: Polymer solar cells. In: Photoresponsive Polymers II, pp. 1–86. Springer Berlin Heidelberg (2007). https://doi.org/10.1007/12_2007_121
30. Hu, H., Dong, B., Hu, H., Chen, F., Kong, M., Zhang, Q., Luo, T., Zhao, L., Guo, Z., Li, J., Xu, Z., Wang, S., Eder, D., Wan, L.: Atomic layer deposition of TiO_2 for a high-efficiency hole-blocking layer in hole-conductor-free perovskite solar cells processed in ambient air. ACS Appl. Mater. Interfaces. **8**(28), 17999–18007 (2016). https://doi.org/10.1021/acsami.6b02701
31. Hussain, I., Tran, H.P., Jaksik, J., Moore, J., Islam, N., Uddin, M.J.: Functional materials, device architecture, and flexibility of perovskite solar cell. Emergent Mater. **1**(3–4), 133–154 (2018). https://doi.org/10.1007/s42247-018-0013-1
32. Ishak, N.N., Ali, M.S.M., Nayan, N., Hasnan, M.M.I.M., Othman, N.A., Ahmad, M.K., Hamed, N.K.A., Fazli, F.I.M.: Investigation of GaN-Aln-Tio_2 nanoflower photoanode for dye-sensitized solar cell application, p. 070002 (2023). https://doi.org/10.1063/5.0124193
33. Jaiswal, K.K., Chowdhury, C.R., Yadav, D., Verma, R., Dutta, S., Jaiswal, K.S., Sangmesh, B., Karuppasamy, K.S.K.: Renewable and sustainable clean energy development and impact on social, economic, and environmental health. Energy Nexus **7**, 100118 (2022). https://doi.org/10.1016/j.nexus.2022.100118
34. Jung, H., Shin, G., Kwak, H., Hao, L.T., Jegal, J., Kim, H.J., Jeon, H., Park, J., Oh, D.X.: Review of polymer technologies for improving the recycling and upcycling efficiency of plastic waste. Chemosphere **320**, 138089 (2023). https://doi.org/10.1016/j.chemosphere.2023.138089
35. Kadam, S.J., More, K.V., Chavan, S.S., Dongale, T.D., Shendokar, S.M.: Electrospun 1D TiO nanofibers for dye-sensitized solar cell application. Mater. Today: Proceed. **65**, 106–110 (2022). https://doi.org/10.1016/j.matpr.2022.05.166
36. Kaltenhauser, V., Rath, T., Edler, M., Reichmann, A., Trimmel, G.: Exploring polymer/nanoparticle hybrid solar cells in tandem architecture. RSC Adv. **3**(40), 18643 (2013). https://doi.org/10.1039/c3ra43842b
37. Kamat, P.V.: Quantum dot solar cells. *The Next Big Thing* in photovoltaics. J. Phys. Chem. Lett. **4**(6), 908–918 (2013). https://doi.org/10.1021/jz400052e
38. Lai, F.-I., Yang, J.-F., Hsu, Y.-C., Lin, K.-J., Kuo, S.-Y.: Enhancing DSSC performance through manipulation of the size of ZnO nanorods. ACS Omega **8**(43), 40206–40211 (2023). https://doi.org/10.1021/acsomega.3c03846
39. Lee, L., Park, S.-J., Kim, S.: Effect of nano-sized barium titanate addition on PEO/PVDF blend-based composite polymer electrolytes. Solid State Ionics **234**, 19–24 (2013). https://doi.org/10.1016/j.ssi.2012.12.011
40. Li, B., Wang, L., Kang, B., Wang, P., Qiu, Y.: Review of recent progress in solid-state dye-sensitized solar cells. In: Solar Energy Materials and Solar Cells, vol. 90, Issue 5, pp. 549–573 (2006). https://doi.org/10.1016/j.solmat.2005.04.039
41. Li, G., Zhu, R., Yang, Y.: Polymer solar cells. Nat. Photon. **6**(3), pp. 153–161 (2012). https://doi.org/10.1038/nphoton.2012.11
42. Li, Z., Zhao, S., Ritchie, R.O., Meyers, M.A.: Mechanical properties of high-entropy alloys with emphasis on face-centered cubic alloys. Prog. Mater Sci. **102**, 296–345 (2019). https://doi.org/10.1016/j.pmatsci.2018.12.003
43. Liang, R., Yuan, B., Zhang, F., Feng, W.: Azopyridine polymers in organic phase change materials for high energy density photothermal storage and controlled release. Angew. Chem. Int. Ed. (2024). https://doi.org/10.1002/anie.202419165
44. Liao, Z.-H., Sun, J.-Z., Sun, D.-Z., Si, R.-W., Yong, Y.-C.: Enhancement of power production with tartaric acid doped polyaniline nanowire network modified anode in microbial fuel cells. Biores. Technol. **192**, 831–834 (2015). https://doi.org/10.1016/j.biortech.2015.05.105
45. Ling, T., Song, J.-G., Chen, X.-Y., Yang, J., Qiao, S.-Z., Du, X.-W.: Comparison of ZnO and TiO_2 nanowires for photoanode of dye-sensitized solar cells. J. Alloy. Compd. **546**, 307–313 (2013). https://doi.org/10.1016/j.jallcom.2012.08.030
46. Liu, J., Li, Y., Arumugam, S., Tudor, J., Beeby, S.: Screen printed dye-sensitized solar cells (DSSCs) on woven polyester cotton fabric for wearable energy harvesting applications. Mater. Today: Proceed. **5**(5), 13753–13758 (2018). https://doi.org/10.1016/j.matpr.2018.02.015

47. Liu, X., Wang, J., Shen, Z., Zhang, H., Li, B., Zhang, Y., Zhang, X.: Enhanced energy harvesting performances of flexible piezoelectric nanocomposite based on CNTs@PZT nanofibers network. J. Alloy. Compd. **927**, 166832 (2022). https://doi.org/10.1016/j.jallcom.2022.166832
48. Lu, H., Tian, W., Guo, J., Li, L.: Interface engineering through atomic layer deposition towards highly improved performance of dye-sensitized solar cells. Sci. Rep. **5**(1), 12765 (2015). https://doi.org/10.1038/srep12765
49. Luhakhra, N., Sharma, D., Abbot, V., Karol, V.: Polymers and Their Composites as Renewable and Flexible Materials for Boosting Energy Storage, pp. 199–221 (2024). https://doi.org/10.1021/bk-2024-1488.ch009
50. Lv, D., Jiang, Q., Shang, Y., Liu, D.: Highly efficient fiber-shaped organic solar cells toward wearable flexible electronics. NPJ Flex. Electron. **6**(1), 38 (2022). https://doi.org/10.1038/s41528-022-00172-w
51. Ma, S., Hou, Y., Hao, J., Lin, C., Zhao, J., Sui, X.: Well-defined nanostructures by block copolymers and mass transport applications in energy conversion. Polymers **14**(21), 4568 (2022). https://doi.org/10.3390/polym14214568
52. Mali, S.S., Betty, C.A., Bhosale, P.N., Patil, P.S., Hong, C.K.: From nanocorals to nanorods to nanoflowers nanoarchitecture for efficient dye-sensitized solar cells at relatively low film thickness: all hydrothermal process. Sci. Rep. **4**(1), 5451 (2014). https://doi.org/10.1038/srep05451
53. Mei, A., Li, X., Liu, L., Ku, Z., Liu, T., Rong, Y., Xu, M., Hu, M., Chen, J., Yang, Y., Grätzel, M., Han, H.: A hole-conductor–free, fully printable mesoscopic perovskite solar cell with high stability. Science **345**(6194), 295–298 (2014). https://doi.org/10.1126/science.1254763
54. Miao, H., Hu, X., Fan, J., Li, C., Sun, Q., Hao, Y., Zhang, G., Bai, J., Hou, X.: Hydrothermal synthesis of TiO_2 nanostructure films and their photoelectrochemical properties. Appl. Surf. Sci. **358**, 418–424 (2015). https://doi.org/10.1016/j.apsusc.2015.08.212
55. Minkiewicz, J., Jones, G.M., Ghanizadeh, S., Bostanchi, S., Wasely, T.J., Yamini, S.A., Nekouie, V.: Large-scale manufacturing of solid-state electrolytes: challenges, progress, and prospects. Open Ceram. **16**, 100497 (2023). https://doi.org/10.1016/j.oceram.2023.100497
56. Mose, P.: The applications of nanotechnology in renewable energy. J. Phys. Sci. **5**(1), 1–12 (2024). https://doi.org/10.47941/jps.1626
57. Mouele, E.S.M., Ngqoloda, S., Pescetelli, S., Di Carlo, A., Dinu, M., Vladescu, A., Parau, A.C., Agresti, A., Braic, M., Arendse, C.J., Petrik, L.F.: Spin coating immobilisation of C-N-TiO_2 Co-doped nano catalyst on glass and application for photocatalysis or as electron transporting layer for perovskite solar cells. Coatings **10**(11), 1029 (2020). https://doi.org/10.3390/coatings10111029
58. Mukta, D.R., Hasan, Md.M., Ali, Md.Y., Haque, K.M.A.U., Khatun, Most. A., Uddin, M.J., Enzo, S., Rahman, Md.W.: Improved in hydrogen storage properties of MgH_2 catalyzed by as-prepared Graphene oxide-supported SnO_2 nanoparticles. Res. Chem. Intermed. **50**(6), 2859–2872 (2024). https://doi.org/10.1007/s11164-024-05294-6
59. Murali, A., Ramesh, R., Sakar, M., Park, S., Han, S.S.: Unveiling the potential of emergent nanoscale composite polymer electrolytes for safe and efficient all solid-state lithium-ion batteries. RSC Adv. **14**(42), 30618–30629 (2024). https://doi.org/10.1039/D4RA05134C
60. Nayak, A., Bhushan, B., Kotnala, S., Kukretee, N., Chaudhary, P., Tripathy, A.R., Ghai, K., Laxmi Mudliar, S.: Nanomaterials for supercapacitors as energy storage application: focus on its characteristics and limitations. Mater. Today: Proceed. **73**, 227–232 (2023). https://doi.org/10.1016/j.matpr.2022.11.259
61. Nayak, R., Harilal, B.S.P., Chandra Ghosh, P.: Polymer Nanocomposite Membrane for Fuel cell Applications, pp. 176–189 (2022). https://doi.org/10.2174/9789815050714122030013
62. Noh, J.H., Im, S.H., Heo, J.H., Mandal, T.N., Seok, S.I.: Chemical management for colorful, efficient, and stable inorganic-organic hybrid nanostructured solar cells. Nano Lett. **13**(4), 1764–1769 (2013). https://doi.org/10.1021/nl400349b
63. Olabi, A.G., Elsaid, K., Obaideen, K., Abdelkareem, M.A., Rezk, H., Wilberforce, T., Maghrabie, H.M., Sayed, E.T.: Renewable energy systems: Comparisons, challenges and

barriers, sustainability indicators, and the contribution to UN sustainable development goals. Int. J. Thermofluids **20**, 100498 (2023). https://doi.org/10.1016/j.ijft.2023.100498
64. Ong, C.B., Ng, L.Y., Mohammad, A.W.: A review of ZnO nanoparticles as solar photocatalysts: synthesis, mechanisms and applications. Renew. Sustain. Energy Rev. **81**, 536–551 (2018). https://doi.org/10.1016/j.rser.2017.08.020
65. Peng, S., Yu, Y., Wu, S., Wang, C.-H.: Conductive polymer nanocomposites for stretchable electronics: material selection, design, and applications. ACS Appl. Mater. Interfaces. **13**(37), 43831–43854 (2021). https://doi.org/10.1021/acsami.1c15014
66. Peter, S., Solomon, N.O., Adenekan, O.A., Obasi, S.C.: Nanotechnology's potential in advancing renewable energy solutions. Eng. Sci. Technol. J. **5**(5), 1695–1710 (2024). https://doi.org/10.51594/estj.v5i5.1137
67. Raguram, T., Rajni, K.S.: Characterization of TiO_2 photoanodes and natural dyes (Allamanda Blanchetti and Allamanda Cathartica) extract as sensitizers for dye-sensitized solar cell applications. J. Sol-Gel. Sci. Technol. **93**(1), 202–213 (2020). https://doi.org/10.1007/s10971-019-05180-3
68. Rahman, Md.W., Ali, Y., Uddin, M.J., Khan, Md.M.R., Enzo, S.: Preparation of jute waste-based activated carbon supported copper oxide nanoparticles for hydrogen storage in MgH2. J. Iran. Chem. Soc. **22**(1), 73–82 (2025). https://doi.org/10.1007/s13738-024-03130-7
69. Rahman, S., Haleem, A., Siddiq, M., Hussain, M. K., Qamar, S., Hameed, S., Waris, M.: Research on dye sensitized solar cells: recent advancement toward the various constituents of dye sensitized solar cells for efficiency enhancement and future prospects. In: RSC Advances, vol. 13, Issue 28, pp. 19508–19529. Royal Society of Chemistry (2023). https://doi.org/10.1039/d3ra00903c
70. Rashid, A.B., Haque, M., Islam, S.M.M., Uddin Labib, K.M.R.: Nanotechnology-enhanced fiber-reinforced polymer composites: recent advancements on processing techniques and applications. Heliyon **10**(2), e24692 (2024). https://doi.org/10.1016/j.heliyon.2024.e24692
71. Rawat, P., Prajapati, R.K., Meena, G.K., Saroj, A.L.: Enhancing the performance of Chitosan/PVP based bio-polymer electrolyte incorporated with SiO_2 nano-particles in dye-sensitized solar cell. Mater. Sci. Eng., B **308**, 117603 (2024). https://doi.org/10.1016/j.mseb.2024.117603
72. Razykov, T.M., Ferekides, C.S., Morel, D., Stefanakos, E., Ullal, H.S., Upadhyaya, H.M.: Solar photovoltaic electricity: current status and future prospects. Sol. Energy **85**(8), 1580–1608 (2011). https://doi.org/10.1016/j.solener.2010.12.002
73. Reddy, K.R., El-Zein, A., Airey, D.W., Alonso-Marroquin, F., Schubel, P., Manalo, A.: Self-healing polymers: synthesis methods and applications. Nano-Struct. Nano-Objects **23**, 100500 (2020). https://doi.org/10.1016/j.nanoso.2020.100500
74. Ri, P., Jong, S., Pae, K., Sin, G., Pae, K., Jon, S.: Flexible dye solar cells with TiO_2 nanopaper and Ti back contact electrodes. J. Saudi Chem. Soc. **27**(6), 101745 (2023). https://doi.org/10.1016/j.jscs.2023.101745
75. Sangeethavanathi, S., Gowthaman, P., Vigneswaran, S.: Facile hydrothermal synthesis of Cu doped MoS2 nanoparticles for enhanced dye-sensitized solar cell performance. Inorg. Chem. Commun. **171**, 113653 (2025). https://doi.org/10.1016/j.inoche.2024.113653
76. Sarfraz, N., Kanwal, N., Ali, M., Ali, K., Hasnain, A., Ashraf, M., Ayaz, M., Ifthikar, J., Ali, S., Hendi, A., Baig, N., Ehsan, M.F., Shah, S.S., Khan, R., Khan, I.: Materials advancements in solid-state inorganic electrolytes for highly anticipated all solid Li-ion batteries. Energy Storage Mater. **71**, 103619 (2024). https://doi.org/10.1016/j.ensm.2024.103619
77. Shah, M., Hakim, N.U.D.: Advances in nanocellulose proton conductivity and applications in polymer electrolyte membrane fuel cells. Next Mater. **6**, 100484 (2025). https://doi.org/10.1016/j.nxmate.2025.100484
78. Shakeel Ahmad, M., Pandey, A.K., Abd Rahim, N.: Advancements in the development of TiO_2 photoanodes and its fabrication methods for dye sensitized solar cell (DSSC) applications. A review. Renew. Sustain. Energy Rev. **77**, 89–108 (2017). https://doi.org/10.1016/j.rser.2017.03.129

79. Sharma, K., Sharma, V., Sharma, S.S.: Dye-sensitized solar cells: fundamentals and current status. Nanoscale Res. Lett. **13**(1), 381 (2018). https://doi.org/10.1186/s11671-018-2760-6
80. Song, Z., Chen, F., Martinez-Ibañez, M., Feng, W., Forsyth, M., Zhou, Z., Armand, M., Zhang, H.: A reflection on polymer electrolytes for solid-state lithium metal batteries. Nat. Commun. **14**(1), 4884 (2023). https://doi.org/10.1038/s41467-023-40609-y
81. Stranks, S.D., Eperon, G.E., Grancini, G., Menelaou, C., Alcocer, M.J.P., Leijtens, T., Herz, L. M., Petrozza, A., Snaith, H.J.: Electron-Hole Diffusion Lengths Exceeding 1 Micrometer in an Organometal Trihalide Perovskite Absorber, vol. 342 (2013). https://www.science.org
82. Swain, S., Samal, H.B., Satpathy, S., Jena, B.R., Pattnaik, G., Bashar, S., Barad, S.: The prospective applications of arising nanostructured dielectric materials in storage of energy: a comprehensive review. Micro Nanosyst. **16**(1), 2–20 (2024). https://doi.org/10.2174/011876402927753223 1231100117
83. Toe, M.Z., Pung, S.-Y., Yaacob, K.A.B., Han, S.S.: Effect of dip-coating cycles on the structural and performance of ZnO thin film-based DSSC. Arab. J. Sci. Eng. **46**(7), 6741–6751 (2021). https://doi.org/10.1007/s13369-021-05418-9
84. Tung, S., Fisher, S.L., Kotov, N.A., Thompson, L.T.: Nanoporous aramid nanofibre separators for nonaqueous redox flow batteries. Nat. Commun. **9**(1), 4193 (2018). https://doi.org/10.1038/s41467-018-05752-x
85. Uddin, M.J., Daramola, D.E., Velasquez, E., Dickens, T.J., Yan, J., Hammel, E., Cesano, F., Okoli, O.I.: A high efficiency 3D photovoltaic microwire with carbon nanotubes (CNT)-quantum dot (QD) hybrid interface. Phys. Status Solidi (RRL)—Rapid Res. Lett. **8**(11), 898–903 (2014). https://doi.org/10.1002/pssr.201409392
86. Vijayan, K., Vijayachamundeeswari, S.P., Sivaperuman, K., Ahsan, N., Logu, T., Okada, Y.: A review on advancements, challenges, and prospective of copper and non-copper based thin-film solar cells using facile spray pyrolysis technique. Sol. Energy **234**, 81–102 (2022). https://doi.org/10.1016/j.solener.2022.01.070
87. Villarreal Gómez, L.J., Iglesias, A.L., Miranda Soto, V., Olivas Sarabia, A., Valdez Castro, R., López Maldonado, E.A., Oropeza Guzmán, M.T., Romero Soto, C.A., Lizarraga Medina, E.G., Vazquez Arce, J.L.: Study of electrospun nanofibers loaded with Ru(ii) phenanthroline complexes as a potential material for use in dye-sensitized solar cells (DSSCs). RSC Adv. **13**(51), 36023–36034 (2023). https://doi.org/10.1039/D3RA07283E
88. Wang, Y., Zhu, Y., Lai, H., Zhu, Y., Lai, X., Tan, P., Shen, X., He, F.: Enhancing efficiency and stability through halogenation of dimerized acceptors in quasiplanar heterojunction organic solar cells. ACS Mater. Lett. **6**(7), 2506–2514 (2024). https://doi.org/10.1021/acsmaterialslett.4c00393
89. Wang, Z.-S., Kawauchi, H., Kashima, T., Arakawa, H.: Significant influence of TiO_2 photoelectrode morphology on the energy conversion efficiency of N719 dye-sensitized solar cell. Coord. Chem. Rev. **248**(13–14), 1381–1389 (2004). https://doi.org/10.1016/j.ccr.2004.03.006
90. Watanabe, T., Inafune, Y., Tanaka, M., Mochizuki, Y., Matsumoto, F., Kawakami, H.: Development of all-solid-state battery based on lithium ion conductive polymer nanofiber framework. J. Power Sources **423**, 255–262 (2019). https://doi.org/10.1016/j.jpowsour.2019.03.066
91. Wu, J., Lan, Z., Lin, J., Huang, M., Huang, Y., Fan, L., Luo, G., Lin, Y., Xie, Y., Wei, Y.: Counter electrodes in dye-sensitized solar cells. Chem. Soc. Rev. **46**(19), 5975–6023 (2017). https://doi.org/10.1039/C6CS00752J
92. Wu, T.-C., Huang, W.-M., Meen, T.-H., Tsai, J.-K.: Performance improvement of dye-sensitized solar cells with pressed TiO_2 nanoparticles layer. Coatings **13**(5), 907 (2023). https://doi.org/10.3390/coatings13050907
93. Wu, Y., Wang, S., Li, H., Chen, L., Wu, F.: Progress in thermal stability of all-solid-state-Li-ion-batteries. InfoMat **3**(8), 827–853 (2021). https://doi.org/10.1002/inf2.12224
94. Yan, J., Uddin, M.J., Dickens, T.J., Okoli, O.I.: Carbon nanotubes (CNTs) enrich the solar cells. Sol. Energy **96**, 239–252 (2013). https://doi.org/10.1016/j.solener.2013.07.027
95. Yang, H., Huang, M., Wu, J., Lan, Z., Hao, S., Lin, J.: The polymer gel electrolyte based on poly(methyl methacrylate) and its application in quasi-solid-state dye-sensitized solar cells. Mater. Chem. Phys. **110**(1), 38–42 (2008). https://doi.org/10.1016/j.matchemphys.2008.01.010

96. Zhang, L., Allendorf, M.D., Balderas-Xicohténcatl, R., Broom, D.P., Fanourgakis, G.S., Froudakis, G.E., Gennett, T., Hurst, K.E., Ling, S., Milanese, C., Parilla, P.A., Pontiroli, D., Riccò, M., Shulda, S., Stavila, V., Steriotis, T.A., Webb, C.J., Witman, M., Hirscher, M.: Fundamentals of hydrogen storage in nanoporous materials. Prog. Energy **4**(4), 042013 (2022). https://doi.org/10.1088/2516-1083/ac8d44
97. Zhang, Z., Li, Y.: Polymerized small-molecule acceptors for high-performance all-polymer solar cells. Angew. Chem. Int. Ed. **60**(9), 4422–4433 (2021). https://doi.org/10.1002/anie.202009666
98. Zhao, J., Wang, A., Green, M. A., & Ferrazza, F.: 19.8% efficient "honeycomb" textured multicrystalline and 24.4% monocrystalline silicon solar cells. Appl. Phys. Lett. **73**(14), 1991–1993 (1998). https://doi.org/10.1063/1.122345
99. Zhao, J., Wu, W., Sun, J., Guo, S.: Triplet photosensitizers: from molecular design to applications. Chem. Soc. Rev. **42**(12), 5323 (2013). https://doi.org/10.1039/c3cs35531d
100. Zhao, X.G., Jin, E.M., Gu, H.-B.: Increased charge transfer of PVDF-HFP based electrolyte by addition of graphite nanofiber and its application in dye-sensitized solar cells. Appl. Surf. Sci. **287**, 8–12 (2013). https://doi.org/10.1016/j.apsusc.2013.09.017
101. Zheng, Q., Wang, J., Cao, W., Zhai, H., Cao, M.: Hetero-dimensional micro-nano architectures toward electromagnetic devices and hybrid energy transport. Adv. Func. Mater. (2024). https://doi.org/10.1002/adfm.202417972

Recent Advances and Prospects of Nano Polymers in Environmental Remediation

Surbhi Singh, Gaurav Mittal, Gaurav Saini, Ravindra Budania, and Ritesh Gupta

Abstract Recent advances in nanotechnology have significantly contributed to the development of novel materials with enhanced properties for environmental remediation. This chapter explores the role of nanopolymers in addressing environmental challenges, particularly in the remediation of pollutants in air, water, and soil. Nanopolymers, owing to their unique characteristics such as high surface area, tunable properties, and biodegradability, have emerged as effective agents in pollutant adsorption, catalytic degradation, and the removal of toxic substances. The chapter delves into the synthesis methods, functionalization strategies, and applications of nanopolymer-based materials, highlighting their potential to improve existing remediation technologies. Furthermore, the chapter examines the environmental impact, sustainability, and future prospects of nanopolymers, offering a comprehensive overview of their potential to revolutionize environmental cleanup efforts. The advancements in this field provide promising avenues for overcoming the limitations of conventional remediation methods, thereby paving the way for more efficient, eco-friendly solutions for a sustainable future.

S. Singh (✉)
School of Applied Sciences and Shodh Lab, Suresh Gyan Vihar University, Jaipur, Rajasthan, India
e-mail: surbhi.singh@mygyanvihar.com

G. Saini
Materials Research Centre, Malaviya National Institute of Technology Jaipur, Jaipur, Rajasthan, India

R. Budania
Department of Civil Engineering, Suresh Gyan Vihar University, Jaipur, Rajasthan, India

G. Mittal · R. Gupta
Department of Metallurgical and Materials Engineering, Malaviya National Institute of Technology Jaipur, Jaipur, Rajasthan, India

B. Sharma et al. (eds.), *Nanostructured Polymers*, Engineering Materials,
https://doi.org/10.1007/978-981-95-3581-1_19

1 Introduction

One of the most urgent problems of the twenty-first century is environmental degradation, as problems like soil contamination, hazardous waste, and air and water pollution get worse [1]. Despite their occasional effectiveness, traditional environmental remediation techniques frequently have drawbacks in terms of expense, effectiveness, and environmental impact [2]. Unquestionably, one of the biggest issues facing society today is environmental pollution [3]. Air, water, and soil pollutants are continuously being remedied through the development of new technology [4]. Among the many alarming contaminants are particulate matter, heavy metals, pesticides, herbicides, fertilisers, oil spills, poisonous gases, industrial effluents, sewage, and organic compounds [5–7]. Since environmental remediation can use a wide range of materials, a large range of strategies can be used to achieve this goal [8]. Recent research has concentrated on the use of nanomaterials for the development of new environmental remediation technologies because the complex mixture of various compounds, high volatility, and low reactivity can make it difficult to capture and degrade environmental pollutants [9–11]. With the advent of nanotechnology, especially the use of nanopolymers, has transformed environmental remediation in recent years, providing creative answers that could potentially solve these problems in more efficient and sustainable ways [12].

The distinct physical characteristics of materials at the nanoscale have drawn a lot of attention to nanotechnology in recent decades [13]. Because of their larger surface-to-volume ratio, nanomaterials exhibit greater reactivity and, thus, greater effectiveness than their bulkier counterparts [14]. Furthermore, in contrast to conventional methods, nanomaterials have the ability to take use of special surface chemistry, which allows them to be functionalised or grafted with functional groups that can target particular molecules of interest (pollutants) for effective remediation [15]. Additionally, by purposefully adjusting the nanomaterials' size, shape, porosity, and chemical makeup, they can acquire other beneficial qualities that have a direct impact on how well they work in contamination cleanup [16]. The nanomaterial's variable physical properties and rich surface modification chemistry provide substantial benefits over traditional approaches to environmental pollution [17]. Therefore, techniques that are created by combining many materials (hybrids or composites) and extracting particular desired qualities from each of its constituents may be more effective, stable, and selective than techniques that rely solely on a single nanoplatform [18]. When compared to using nanoparticles alone, adhering them to a scaffold, for example, can be an alternate method of enhancing the material's durability [19]. Enhancing the material's efficiency and selectivity can be achieved by functionalising it with certain compounds that target relevant contaminant molecules [20].

Materials known as nanopolymers, which are usually between 1 and 100 nm in size, combine the qualities of polymers with nanoparticles [21]. They are particularly well-suited for use in environmental remediation because to their special physicochemical characteristics, which include a large surface area, strong reactivity, and variable chemical functionality [22]. Because of their tiny size, nanopolymers can interact with pollutants at the molecular level, which enhances their capacity to neutralise, degrade, or adsorb dangerous compounds in contaminated environments [23]. Because of their versatility, nanopolymers can be designed to specifically target pollutants such as organic pollutants, heavy metals, and even microbiological infections [24]. More efficiency, improved selectivity, and customisation for particular cleanup jobs are just a few of the many benefits they provide over traditional materials [25].

The high sorption capacity of nanostructured sorbents allows them to be tailored to target particular pollutants [26]. Polymer-nanocomposites and other hybrid sorbents have recently been developed, opening up new possibilities for environmental pollution removal applications [27]. Numerous inorganic and organic pollutants, including dyes, heavy metals, and hydrocarbons, have been eliminated from wastewater using polymer-nanocomposites [28]. This chapter will concentrate on the new idea of employing polymeric nanocomposites for environmental remediation [29].

Environmental pollution is one of the most critical challenges faced by humanity today [30]. It refers to the contamination of air, water, and soil, often as a result of human activities, that causes harm to ecosystems, wildlife, and human health [31]. With the increasing industrialization, urbanization, and agricultural activities, pollution has reached unprecedented levels, impacting the sustainability of natural resources and ecosystems [32]. As the global population grows, finding effective and sustainable ways to address pollution becomes more urgent [33].

2 Significance of Nano Polymers in Environmental Applications

Nanopolymers are a class of materials that combine the properties of polymers with the unique features of nanomaterials [21]. These materials have gained significant attention in environmental applications due to their versatility, strength, and high surface area [22]. In particular, nanopolymers play a crucial role in many areas [23]. Nanopolymers can be engineered to have functional groups that bind and remove contaminants from air, water, and soil [24]. Their high surface area allows them to adsorb large quantities of pollutants, making them highly effective in cleanup processes [25]. Nanopolymer membranes are used in filtration systems to remove nanoparticles, pathogens, and other pollutants from water [26]. These membranes offer high permeability and selectivity, making them more efficient than conventional filters [27]. Nanopolymer-based catalysts can be designed to accelerate the degradation of hazardous pollutants in the environment, such as oil spills or chemical leaks [28]. Nanopolymers can be incorporated into sensors to detect pollutants in real time, enabling quick responses to environmental threats [29].

2.1 *Properties of Nanopolymers*

Nanopolymers, or polymer nanocomposites, are materials that incorporate nanoparticles into a polymer matrix [30]. These materials have unique properties due to their small size (on the nanometer scale) and the interaction between the polymer and nanoparticles [31]. Here are some key properties of nanopolymers:

2.1.1 Enhanced Surface Area

At the nanoscale, the surface area of a polymer increases significantly compared to its volume [32]. Due to their small size and high surface-to-volume ratio, nanopolymers exhibit an increased surface area [33]. This increased surface area can lead to enhanced interactions with other materials and particles [34]. This further increase in surface area results in more reactive sites for interaction with other materials, increasing the material's reactivity and potential for applications in sensors, drug delivery, or surface coatings [35]. The high surface-to-volume ratio also allows for better dispersion of fillers and nanoparticles within the polymer matrix, which can improve mechanical and thermal properties [36].

2.1.2 Enhanced Mechanical Properties

Nanopolymers often exhibit enhanced mechanical properties, such as increased tensile strength, elasticity, and fracture toughness [37]. This is largely due to the ability of nanoparticles or nanoscale fillers to reinforce the polymer matrix, preventing crack propagation and improving the overall stability of the material [38]. The presence of nanoparticles can enhance the stiffness or modulus of the polymer, making it more resistant to deformation under stress [39]. Nanopolymers can be tailored to have superior strength, flexibility, and toughness [40]. By incorporating nanoparticles (such as carbon nanotubes or graphene) into the polymer matrix, the mechanical properties can be enhanced, leading to materials that are both stronger and lighter [41].

Nanopolymers are often stronger and more durable than conventional polymers [42]. This is because the nanostructures can reinforce the polymer matrix, providing improved tensile strength, impact resistance, and overall durability [43]. Despite their increased strength, nanopolymers can retain or even improve their flexibility compared to larger polymers, making them versatile for a wide range of applications, especially in coatings and composites [44].

2.1.3 Unique Chemical Properties

Due to the high surface area and the nanoscale structure, nano-polymers often have higher chemical reactivity compared to bulk polymers [21]. This makes them ideal for applications like drug delivery, where the material can interact more efficiently with biological systems or catalytic processes. Nano-polymers can be tailored by functionalizing the nanoparticles or polymer chains with specific chemical groups [28]. This allows for control over their chemical behavior, making them suitable for various specialized applications, including bio-sensing, advanced coatings, and environmental remediation. The unique surface characteristics of nano-polymers give them high reactivity, making them ideal for applications such as drug delivery or catalysis. The ability to functionalize the surface of nanoparticles can further enhance these properties [28, 42].

At the nanoscale, the surface atoms or molecules of a polymer are more reactive due to their exposed positions [17]. This increased reactivity makes nanopolymers ideal candidates for catalytic applications, such as in chemical reactions or as sensors. The properties of nanopolymers can also be enhanced when combined with catalytic nanoparticles, allowing them to serve as highly efficient catalysts for chemical reactions [11, 23].

2.1.4 Enhanced Thermal Properties

Nano-polymers tend to have improved thermal stability and heat resistance [7]. The inclusion of nanoparticles can enhance the heat resistance of the polymer, making it suitable for high-temperature applications such as in aerospace or automotive industries [23, 35]. Nanopolymers can have improved thermal stability due to the unique arrangements of molecules at the nanometer scale [14]. This makes them more resistant to high temperatures and helps them maintain their properties under harsh conditions [5]. Some nano-polymers, especially those incorporating materials like carbon nanotubes or graphene, exhibit enhanced thermal conductivity, which allows them to efficiently dissipate heat [18, 42]. This property is particularly useful in electronics, where heat management is crucial [9]. Nano-polymers often exhibit better heat resistance and thermal conductivity, which make them ideal for advanced electronics, heat-resistant coatings, and other high-temperature applications [3, 27]. The incorporation of conductive nanoparticles (like carbon nanotubes) can significantly improve the heat management properties of the polymer [31].

2.1.5 Nano Size Effects and Surface Area Enhancement

At the nanoscale, materials exhibit unique characteristics that differ from their bulk counterparts, largely due to quantum effects, surface interactions, and the high surface-to-volume ratio [12, 39]. These effects include nano-polymers having a higher surface area compared to larger particles or bulk materials, which leads to enhanced interactions with other substances [5]. Due to their small size and high surface-to-volume ratio, nanopolymers exhibit an increased surface area [21]. This increased surface area can lead to enhanced interactions with other materials and particles [33]. At the nanoscale, quantum mechanics starts to play a role in the behavior of materials, resulting in changes in optical, electrical, and chemical properties compared to bulk materials [8, 27]. Nanoscale materials tend to have higher surface energy, making them more reactive and useful in applications like coatings or sensors [14, 41].

2.2 *Types of Nanopolymers Used in Environmental Applications*

Nanopolymers have garnered significant attention for their use in environmental applications due to their unique properties at the nanoscale, such as high surface area, enhanced reactivity, and the ability to interact with various environmental contaminants [7, 22]. These polymers can be broadly classified into different types based on their origin, structure, and environmental impact [15, 31, 40]. Here's a detailed explanation of the types of nanopolymers used in environmental applications.

2.2.1 Synthetic Versus Natural Polymers

Synthetic Polymers are human-made polymers, typically derived from petrochemical sources [12]. Examples include polyethylene, polystyrene, and polyvinyl chloride (PVC) [3, 28]. They are engineered for specific functionalities, such as hydrophobicity, resistance to degradation, and high mechanical strength [17]. Their structure can be finely tuned to enhance properties like stability in harsh environmental conditions [45]. Synthetic nanopolymers are used to develop advanced filtration materials that can adsorb or trap pollutants, such as heavy metals, organic compounds, and bacteria [5, 22]. Their ability to form stable composites or coatings helps in removing toxic chemicals and oil spills from water and soil [31, 46]. They are employed in sensors for detecting environmental pollutants due to their high surface area, which increases sensitivity to contaminants [9, 19].

Natural Polymers are polymers derived from natural sources like plants, animals, or microorganisms [14]. For instance, cellulose, chitosan, starch, and alginate [7, 33]. Natural nanopolymers are biodegradable, non-toxic, and often biocompatible, making them more suitable for eco-friendly applications [45]. However, they can be less stable and might degrade under environmental stress unless modified [26]. Natural nanopolymers like chitosan are used in environmental clean-up efforts for their ability to absorb pollutants and support microbial activity in soil and water [2, 46]. Cellulose-based nanopolymers are used in membranes and filters for water purification, especially for the removal of dyes, salts, and heavy metals [18, 37].

Natural polymers can enhance soil properties, improve water retention, and support plant growth, which helps restore degraded ecosystems [11, 29].

2.2.2 Composite Nano-Polymers

Composite nanopolymers are materials that combine nanomaterials (such as nanoparticles, nanotubes, or nanofibers) with polymers, aiming to enhance the performance of the polymer matrix [8, 24]. These composites often combine the desirable properties of both the polymers and the nanomaterials [47]. The incorporation of nanoparticles or nanofillers into the polymer matrix enhances mechanical properties like tensile strength and elasticity [15, 33]. Composite nanopolymers exhibit better resistance to environmental degradation, such as UV radiation, moisture, and chemicals [45]. By modifying the type and concentration of nanoparticles, the polymer composites can be tailored for specific environmental applications (e.g., pollutant adsorption, degradation, or filtration) [12, 46].

Composite nanopolymers are effective in removing a wide range of pollutants from water and soil [21]. For example, composite nanofilters are used for filtering out heavy metals, oils, and other contaminants from wastewater [6, 29]. Nanocomposites can be employed to create high-performance membranes for desalination or to remove toxins like pesticides and herbicides from water bodies [47]. These composites can be used in catalytic processes for the degradation of toxic substances or to enhance the breakdown of organic pollutants in wastewater treatment [18, 45].

2.2.3 Biodegradable Nano-Polymers

Biodegradable nanopolymers are designed to degrade naturally through biological processes, reducing their environmental footprint after use [12, 28]. These polymers are typically derived from renewable sources and can break down into non-toxic byproducts, making them ideal for sustainable environmental applications [48]. Biodegradable nanopolymers degrade into harmless substances like water, carbon dioxide, and biomass when exposed to environmental factors such as moisture, heat, and microorganisms [7, 19]. These materials are non-toxic and do not accumulate in ecosystems, making them safer for wildlife and plant life [49]. Biodegradable nanopolymers can be engineered to degrade at specific rates, providing versatility in environmental applications [15, 33].

Biodegradable nanopolymers are used in the development of biodegradable packaging materials, which reduce plastic waste accumulation [48]. These materials can break down in landfill conditions, offering a solution to the plastic pollution crisis [22, 41]. These polymers are useful in cleaning up oil spills and chemical contaminants [49]. For example, biodegradable nanopolymers can be used to trap and remove toxic substances from soil or water, and their ability to degrade ensures minimal long-term environmental impact [5, 17]. Biodegradable polymers are used to improve soil quality and structure [36]. For example, they can enhance water retention in dry soils, support plant growth, and promote the biodegradation of organic pollutants in contaminated soils [9, 48].

3 Mechanisms of Environmental Remediation Using Nanopolymers

Nanopolymers, with their unique structural and functional properties, have emerged as promising materials for environmental remediation [50]. They possess high surface area, tunable functionalities, and excellent reactivity, which enable them to effectively address a wide range of environmental contaminants [15, 22]. Below, we explore the key mechanisms through which nanopolymers can be utilized in environmental remediation:

3.1 Adsorption and Removal of Pollutants

Nanopolymers play a significant role in the adsorption and removal of various pollutants, particularly organic and inorganic contaminants [51]. Adsorption refers to the process where pollutants adhere to the surface of nano-polymers [28]. Due to their high surface area and porous structure, nano-polymers can adsorb large amounts of pollutants such as heavy metals (e.g., lead, mercury, cadmium) and organic toxins (e.g., pesticides, dyes, pharmaceuticals) from water or soil [7, 33]. Due to their high surface area, nanopolymers provide ample active sites for the adsorption of pollutants such as heavy metals, dyes, and organic compounds [52]. Nanopolymers, especially those with a porous structure, can adsorb large quantities of contaminants [19]. The porous nature of nanopolymers provides a significant surface area for adsorption, allowing them to trap contaminants efficiently [45]. Functional groups on nanopolymers enhance their ability to interact with pollutants through electrostatic interactions, hydrogen bonding, or van der Waals forces, thus improving their adsorption capacity [53]. Many nanopolymers can be regenerated by simple processes (e.g., washing or thermal treatment), making them cost-effective and sustainable for continuous pollutant removal [29]. These properties can be utilised in various applications such as removal of toxic dyes from wastewater and adsorption of heavy metal ions (e.g., lead, arsenic) from contaminated water [54].

Nano-polymers can be functionalized with various groups (amine, carboxyl, thiol, etc.) that can form strong interactions (e.g., ionic bonds, hydrogen bonds, van der Waals forces) with pollutants, enhancing their ability to trap contaminants [12]. Heavy metals can bind to functional groups on the surface of nano-polymers via electrostatic interactions or chelation, while organic pollutants may form hydrogen bonds or π-π interactions with the polymer's surface [55]. Absorption involves the deeper penetration of pollutants into the polymer matrix [48]. Nano-polymers may absorb pollutants into their inner structure, especially if they have a porous or mesoporous nature (nano-polymers with internal cavities or channels) [31]. This increases the overall capacity of nano-polymers for pollutant removal [49].

3.2 Catalysis in Contaminant Degradation

Nanopolymers can function as effective catalysts for the degradation of environmental contaminants [50], particularly in chemical reactions aimed at breaking down harmful compounds such as organic pollutants, into less toxic or benign substances [51]. The nanostructure of the polymeric material increases its surface area and exposes more catalytic sites [52]. The polymer matrix can either be loaded with nano-sized catalytic particles (e.g., metal nanoparticles like platinum, palladium, or copper) or designed to have catalytic activity itself [53]. This can enhance the reaction rates of degradation processes such as oxidation, reduction, and hydrolysis [54]. Nanopolymers can be incorporated with metal nanoparticles (e.g., silver, gold, or iron) or metal oxide nanoparticles (e.g., TiO_2 or ZnO), which have intrinsic catalytic properties [55]. These composites can catalyze the degradation of organic pollutants (e.g., pesticides, herbicides, and plastics) in the environment [22]. Nanopolymers can be used in processes such as photocatalysis, where UV light activates the nanopolymer catalyst to degrade organic contaminants [33]. Some applications are degradation of organic pollutants in wastewater treatment (e.g., removal of pharmaceutical residues) and Catalytic oxidation of volatile organic compounds (VOCs) in air purification [45]. When exposed to light (e.g., UV), nano-polymers with photocatalytic properties can degrade organic pollutants by generating reactive oxygen species (ROS), which oxidize pollutants into harmless byproducts like water and carbon dioxide [29]. Iron-based nano-polymers can catalyze the breakdown of organic contaminants through the generation of hydroxyl radicals in the presence of hydrogen peroxide [48]. Catalytic nano-polymers are used to degrade VOCs, odors, and other hazardous substances in water and air [49]. Catalytic nano-polymers can be employed to break down industrial effluents, including those containing toxic organic solvents or chemicals [50].

3.3 Filtration and Separation Techniques Using Nano-Polymers

Nano-polymers can be utilized in filtration and separation technologies [51], taking advantage of their small size and porous structure to selectively separate pollutants from mixtures [52]. Nano-polymers can be fabricated into membranes that act as filters [53]. These membranes are highly selective, allowing only certain molecules to pass through based on size or charge [54]. They can filter out small contaminants, including heavy metals, microorganisms, and organic pollutants, while allowing water or clean air to pass [55]. The pore size of the polymer matrix can be precisely controlled at the nanoscale, which is crucial for effective filtration [22]. Nano-polymers with tailored pore structures can filter pollutants based on molecular size, charge, or hydrophobicity [33].

In some cases, nano-polymers can be functionalized with magnetic nanoparticles [45], enabling the use of magnetic fields to separate the nano-polymers from the pollutants they have captured [48]. This can be particularly useful for removing heavy metals or other particulate pollutants [49]. Nano-polymer membranes are already being used in advanced water filtration systems, including reverse osmosis and desalination processes, to remove contaminants at trace levels [50]. Nano-polymers can be incorporated into filters for removing fine particulate matter (PM2.5) and gaseous pollutants from industrial emissions or indoor air [51].

4 Role of Nano-Polymers in Microbial Remediation (Bioaugmentation)

Nano-polymers can also play an essential role in enhancing microbial remediation or bioaugmentation [52], where microorganisms are introduced to degrade environmental pollutants [53]. Nano-polymers can be used as carriers or scaffolds for microorganisms, providing a stable microenvironment for their growth and activity [54]. The high surface area of nano-polymers provides a large surface for microbial colonization, improving the effectiveness of bioaugmentation [55]. Certain nano-polymers have antimicrobial or biostimulatory properties that can either enhance the growth of beneficial microbes or inhibit the growth of harmful microbes that compete with pollutant-degrading bacteria [22]. Nano-polymers can be used to deliver essential nutrients or growth factors to microbes over time, supporting the long-term degradation of pollutants in the environment [33]. Some nano-polymers can improve the interaction between microbes and pollutants [45]. For example, the polymers may alter the surface properties of pollutants, making them more accessible to microbial enzymes that break them down [48]. Nano-polymers can assist in the cleanup of contaminated soils by promoting the growth of pollutant-degrading bacteria and fungi [49]. In wastewater treatment, nano-polymers can support the growth of microbial communities that break down organic contaminants, such as petroleum hydrocarbons or agrochemicals [50].

5 Role of Nanopolymers in Oil Spill and Hydrocarbon Degradation

Nanopolymers can significantly contribute to the remediation of oil spills and hydrocarbon contamination in marine and terrestrial environments [51]. Their unique properties allow them to act as efficient sorbents and catalysts for the breakdown of hydrocarbons [52]. Nanopolymers, particularly those with superhydrophobic properties, can efficiently absorb and remove oil and other hydrophobic contaminants from water surfaces [53]. They have high oil absorption capacities due to their porous structure and tailored surface chemistry [54]. Nanopolymers can also be used in the catalysis of hydrocarbon degradation [55]. They can accelerate the breakdown of hydrocarbons into less harmful byproducts through oxidation reactions, often in combination with metal nanoparticles [22]. Biodegradable nanopolymers are being developed as sustainable materials for oil spill cleanup [33], minimizing environmental impact while still effectively remediating the affected area [45].

6 Nano-Polymers in Heavy Metal Ion Removal

Heavy metals are a major environmental threat due to their toxicity, persistence, and bioaccumulation in ecosystems [48]. Nanopolymers have proven to be highly effective in removing heavy metal ions from contaminated water and soil [49]. Certain nanopolymers can chelate metal ions through coordination bonds, trapping them and preventing their toxic effects [50]. For example, nanopolymers functionalized with thiol groups can effectively capture and remove mercury or lead ions [51]. Nanopolymers can also act as ion-exchange resins, swapping harmful metal ions with less toxic ones or water molecules, which further aids in the removal process [52]. Magnetic nanopolymers can be used for heavy metal removal by applying an external magnetic field [53], allowing easy recovery and reuse of the polymer material after the removal process [54].

7 Recent Advances in Nano-Polymer Applications for Environmental Remediation

7.1 Nano-Polymers for Water Treatment

Water contamination, particularly with heavy metals, dyes, and pharmaceuticals, is a growing environmental issue [56]. Traditional methods of water treatment often fall short in efficiently removing these contaminants, necessitating the development of more advanced materials [57]. Nano-polymers have emerged as a promising class of materials for water purification due to their high surface area, tunable properties, and ability to interact with a variety of pollutants [58]. These materials offer significant advantages in terms of efficiency, selectivity, and reusability, making them ideal candidates for use in water treatment processes [59].

Nanopolymers are widely used in water treatment applications due to their high surface area, which allows them to efficiently adsorb pollutants [60]. Nanopolymers can be designed to bind heavy metals (like lead, mercury, cadmium, and arsenic) from contaminated water [61]. This is achieved through functionalization with ligands or chelating agents that selectively bind to metal ions [56]. Once the nanopolymers adsorb the metals, the contaminated water is purified [57]. For instance, a nanopolymer functionalized with thiol groups can effectively remove mercury ions from water [58].

Nanopolymers can absorb hydrophobic substances like oils and organic solvents from water [59]. Nanopolymers with superhydrophobic or oleophilic properties are ideal for removing oil spills and other non-polar pollutants from aquatic environments [60]. Nanopolymers composed of hydrophobic materials like polystyrene or polyethylene glycol can selectively absorb oil from water, aiding in environmental cleanup [61]. The unique properties of nanopolymers also enable them to be used in desalination processes [56]. They can act as efficient filters for removing salts, minerals, and other dissolved ions from seawater, thus providing clean drinking water [57].

Here two main applications of nano-polymers in water treatment: (1) removal of heavy metals, dyes, and pharmaceuticals and (2) the development of smart nanomaterials for selective adsorption have been discussed [58].

7.2 *Removal of Heavy Metals, Dyes, and Pharmaceuticals*

Heavy metals, organic dyes, and pharmaceutical contaminants are persistent and hazardous pollutants that pose significant threats to both human health and the environment [59]. The introduction of nano-polymers into water treatment technologies offers innovative solutions to remove these contaminants effectively [60].

a. **Heavy Metal Removal**

Heavy metals such as lead, mercury, cadmium, and arsenic are toxic and difficult to remove from water using conventional treatment methods [61]. Nano-polymers, with their large surface area and functional groups, can adsorb these metals efficiently [56]. By utilizing chelating agents or incorporating specific nanoparticles (such as magnetic nanoparticles), nano-polymers can form stable complexes with heavy metals, facilitating their removal from water [57]. Polymer-nanoparticle composites can enhance the adsorption capacity for heavy metals through physical interactions (like van der Waals forces) or chemical bonding (such as coordination bonds between the polymer's functional groups and the metal ions) [58]. These composites not only increase the adsorption capacity but also allow for easy recovery and regeneration of the nanomaterials [59]. Magnetic nano-polymers can be synthesized to adsorb heavy metals, and then the contaminated nanomaterials can be easily separated from the water using an external magnetic field [60].

b. **Dye Removal**

Water pollution caused by dyes, particularly from textile industries, is another critical issue [61]. Dyes can be toxic, carcinogenic, and persistent in water bodies [56]. Nano-polymers provide an effective solution due to their high adsorption capacity and ability to be tailored to specific dye molecules [57]. For instance, polymers functionalized with amine, carboxyl, or hydroxyl groups can effectively capture a wide range of synthetic dyes through electrostatic interactions or hydrogen bonding [58]. Nano-polymers with hydrophobic surfaces can adsorb hydrophobic dyes by utilizing van der Waals forces [59]. These materials can be easily regenerated, making them cost-effective and environmentally friendly [60].

c. **Pharmaceutical Contaminant Removal**

Pharmaceuticals, such as antibiotics, hormones, and painkillers, have been detected in various water sources due to their widespread use and inadequate disposal methods [61]. These contaminants can cause significant ecological and human health concerns, as they may affect aquatic ecosystems and contribute to the development of antibiotic-resistant bacteria [56]. Nano-polymers can address this issue by effectively adsorbing pharmaceutical compounds from water [57]. The process generally involves nano-polymers engineered with specific functional groups to selectively adsorb pharmaceutical contaminants, including hydrophilic drugs [58]. Furthermore, the small particle size of nano-polymers enables them to interact more effectively with pharmaceutical molecules, which are often present in trace amounts [59].

7.3 Development of Smart Nanomaterials for Selective Adsorption

One of the most exciting developments in the field of nano-polymers for water treatment is the creation of "smart" nanomaterials [60]. These materials can selectively adsorb pollutants based on specific environmental stimuli (such as pH, temperature, or the presence of certain ions) and offer the potential for highly targeted removal of contaminants [61].

a. **PH-Responsive Nano-Polymers**

Many contaminants in water, such as heavy metals and dyes, undergo changes in their chemical properties under varying pH conditions [56]. Nano-polymers can be engineered to respond to these changes by swelling, changing charge density, or altering their structure, thereby enhancing the adsorption process [57]. For example, pH-sensitive nano-polymers can be designed to swell or contract depending on the pH of the water, increasing or decreasing their surface area for more efficient pollutant removal [58]. These materials can be used for selective removal of contaminants in industrial effluents, where the pH may vary significantly [59].

b. **Temperature-Responsive Nano-Polymers**

Temperature-responsive nano-polymers are another class of smart materials that change their physical properties (such as swelling, charge, or hydrophobicity) in response to temperature fluctuations [60]. These materials can be designed to adsorb pollutants more effectively at certain temperatures and release them when the temperature changes, making them useful for processes requiring regeneration and reuse [61]. For instance, poly(N-isopropylacrylamide) (PNIPAM) is a common temperature-sensitive polymer that can adsorb contaminants at lower temperatures and release them at higher temperatures [56].

c. **Ion-Selective Nano-Polymers**

Certain nano-polymers can be tailored to selectively adsorb specific ions, making them highly effective in removing pollutants that share similar chemical structures [57]. These ion-selective nano-polymers are especially useful for removing heavy metals from water [58]. Ion-imprinted polymers (IIPs) can be synthesized to have cavities that match the shape and charge of a particular ion [59]. When exposed to water containing this ion, the polymer selectively binds to the ion, removing it from the water [60]. These materials can be regenerated by altering the environmental conditions (e.g., pH or temperature) and can provide high selectivity and capacity for pollutant removal [61].

d. **Photocatalytic Nano-Polymers**

Photocatalytic nano-polymers can harness light to trigger chemical reactions that degrade or neutralize pollutants [56]. These materials often include semiconductor nanoparticles like titanium dioxide (TiO_2), which can be combined with polymers to form a composite material capable of both adsorbing and decomposing organic pollutants, such as dyes and pharmaceuticals [57]. UV-activated photocatalytic polymers can degrade organic pollutants in the presence of UV light, breaking them down into less harmful components [58].

7.4 Nano-Polymers for Soil and Groundwater Remediation

Nanopolymers are also effective in soil remediation, where they can help decontaminate polluted soils, removing hazardous chemicals and organic compounds [59]. Polymers can be functionalized with specific groups that interact with toxic organic compounds (such as pesticides, herbicides, and industrial chemicals) in contaminated soils [60]. Nanopolymers act as adsorbents or as carriers for the slow release of cleaning agents [61]. Nanopolymers with functional groups (e.g., amines, hydroxyls) can interact with and trap organic pollutants like dioxins or pesticides, reducing their bioavailability in the soil [56]. Nanopolymers can also be used to encapsulate or deliver nutrients and microorganisms for bioremediation purposes [57]. These nanopolymers act as carriers for the targeted delivery of microorganisms that break down contaminants in soil [58]. Nanopolymers loaded with nutrients can promote the growth of specific microbial strains that degrade petroleum hydrocarbons in soil [59]. Nanopolymers can be used in conjunction with plants for phytoremediation, a process in which plants are used to absorb or degrade pollutants [60]. The nanopolymers may enhance the bioavailability of pollutants to plants or act as scaffolds for the growth of plant roots [61]. Nanopolymers could aid in the transport of nutrients or pollutants to plant roots, enhancing the uptake of heavy metals or other toxic substances [56]. When used in the context of environmental remediation, nano-polymers can play a significant role in bioremediation, phytoremediation, and immobilization of contaminants [57].

7.5 Use in Bioremediation

Bioremediation refers to the use of microorganisms or plants to degrade or neutralize pollutants, especially in contaminated soil and groundwater [58]. Nano-polymers can enhance bioremediation by interacting with the microorganisms or enzymes responsible for breaking down contaminants [59]. Nano-polymers can serve as carriers for microorganisms or nutrients, which improve microbial activity by providing a favorable environment for growth and metabolic activities [60]. For instance, nanoparticles can be used to deliver nutrients or oxygen directly to the contaminated area, stimulating microbial degradation of organic pollutants like petroleum hydrocarbons or pesticides [61]. Nano-polymers can encapsulate biocatalysts such as enzymes or microbial cells, protecting them from harsh environmental conditions (e.g., extreme pH or temperature) [56]. This enhances the stability and effectiveness of the bioremediation process, especially in subsurface environments where harsh conditions may prevail [57].

Nano-polymers can be designed for controlled release of bioremediating agents (e.g., enzymes or microorganisms) over time [58]. By controlling the release rates, nano-polymers ensure that the agents are available in the environment when needed, optimizing the remediation process [59]. The high surface area of nano-polymers provides more reactive sites for microbial attachment, increasing the efficiency of bioremediation [60]. Moreover, the ability of nano-polymers to interact with both contaminants and microbes can accelerate the degradation of organic pollutants, reducing the time required for remediation [61].

7.6 Use in Phytoremediation

Phytoremediation is the process of using plants to remove, degrade, or immobilize contaminants from soil and groundwater [56]. Nano-polymers can be applied to phytoremediation in several ways like nano-polymers can facilitate the controlled and targeted delivery of nutrients or growth regulators to plants, enhancing their growth and, consequently, their ability to absorb and accumulate contaminants [57]. For example, they can deliver micronutrients (e.g., iron, zinc) essential for plant metabolism, improving their ability to tolerate contaminated environments and accelerate contaminant uptake [58]. The properties of nano-polymers can enhance the bioavailability of contaminants, making them more accessible for plant absorption [59]. In the case of heavy metals, for example, nano-polymers can increase the solubility of metals in soil or groundwater, thereby enhancing the uptake by plant roots [60]. Some contaminants can be toxic to plants, inhibiting their growth and reducing the efficiency of phytoremediation [61]. Nano-polymers can help mitigate this issue by forming complexes with pollutants, reducing their toxicity and ensuring that plants can thrive while remediating the environment [56].

Nano-polymers can also be used to create "nano-engineered" plant root systems that have a higher surface area and enhanced interaction with contaminants in the soil [57]. These nano-modified root systems can increase the plant's ability to absorb and filter out harmful substances [58].

7.7 *Immobilization of Contaminants*

Immobilization is the process of rendering contaminants in soil or groundwater less mobile or bioavailable, thereby reducing their potential for spread or toxicity [59]. Nano-polymers are useful for immobilizing contaminants in various ways like nano-polymers can adsorb or trap contaminants (e.g., heavy metals, organic pollutants, radionuclides) onto their surface [60]. The large surface area and functional groups on the polymer surface allow for efficient interaction with contaminants, either through electrostatic attraction, hydrogen bonding, or chemical reactions [61]. This immobilizes the contaminants, preventing their migration into groundwater or surrounding environments [56]. Nano-polymers can be injected into contaminated soils or aquifers to stabilize pollutants in place [57]. For example, nanoparticles can react with contaminants like heavy metals, forming insoluble complexes that are less likely to leach into groundwater [58]. This technique is particularly useful in cases where excavation or removal of the contaminated soil is not feasible [59]. Some nano-polymers can be engineered to react with specific contaminants, such as forming complexes with heavy metals or organic chemicals, thereby immobilizing them [60]. For example, certain functionalized nano-polymers can form stable bonds with heavy metals, preventing them from being bioavailable or mobile in the environment [61]. Nano-polymers can be used as a barrier or coating material to physically contain contaminants [56]. This method is useful when contaminants need to be isolated from surrounding environments or when treatment of large areas is not practical [57]. Nano-polymers can be applied to create coatings or encapsulations around contaminated zones, thus limiting the spread of pollutants [58].

8 Air Pollution Control Using Nano-Polymers

Air pollution is one of the most significant environmental challenges that modern societies face [59]. It stems from a variety of sources, including industrial emissions, vehicle exhaust, and natural sources like dust storms and wildfires [60]. To mitigate the impact of these pollutants, researchers have turned to innovative materials like nano-polymers for controlling air pollution [61]. Nano-polymers are polymers engineered at the nanoscale, offering unique properties that can be leveraged for air purification [56].

Nanopolymers have shown promise in reducing air pollution, particularly in capturing and neutralizing harmful gases and particulate matter [57]. Nanopolymers can be incorporated into filtration materials for air purifiers [58]. Their large surface area and reactive functional groups enable them to capture pollutants such as particulate matter (PM2.5), volatile organic compounds (VOCs), and other gaseous pollutants [59]. Nanoporous polymers can be embedded into filters to adsorb air pollutants, providing an efficient solution for reducing indoor air pollution [60].

Nanopolymers can help capture greenhouse gases like CO_2 from the atmosphere [61]. Some nanopolymers can undergo reversible reactions with CO_2, trapping it in a solid form or turning it into useful products [56]. Polymeric nanomaterials designed to absorb carbon dioxide can be used in carbon capture systems to reduce emissions from industrial processes [57]. Nanopolymers can act as catalysts for the oxidation of harmful air pollutants, such as nitrogen oxides (NOx) and sulfur dioxide (SO_2) [58]. These reactions convert the pollutants into less harmful substances like nitrogen and sulfuric acid [59]. Metal-containing nanopolymers could be used as catalysts in the catalytic converters of vehicles to reduce the emissions of NOx and CO [60].

8.1 Nano-Adsorbents for Air Filtration

Nano-adsorbents are materials that use their vast surface area and high adsorption capacities to remove pollutants from the air [61]. These materials are highly effective in trapping particulate matter (PM), gases, and other toxic compounds from industrial emissions and the atmosphere in general [56]. Nano-polymers are ideal candidates for developing such adsorbents because of their customizable properties at the nanoscale [57]. At the heart of nano-adsorption is the ability of nanoparticles to interact with pollutants through surface forces like van der Waals forces, hydrogen bonding, or electrostatic interactions [58]. By incorporating specific nanomaterials into polymer matrices, scientists can design adsorbents that target particular pollutants, such as volatile organic compounds (VOCs), carbon dioxide (CO_2), sulfur dioxide (SO_2), or particulate matter [59]. Nano-polymers have an exceptionally high surface area-to-volume ratio, which significantly increases their ability to adsorb harmful particles and gases [60]. By modifying the polymer structure or adding functional groups (like amines or hydroxyl groups), nano-polymers can be tuned to selectively adsorb certain pollutants [61]. For example, polymers with amine groups can effectively capture CO_2 from industrial exhausts [56]. The nanoscale structure of these polymers, such as nanoparticles, nanofibers, and nanotubes, can enhance the interaction between the adsorbent and the pollutant, making them more efficient than conventional filtration methods [57].

Nano-polymers can be incorporated into air filters for use in homes and offices [58]. These filters can efficiently remove particulate matter, allergens, and VOCs from the air [59]. Industries like power plants, chemical factories, and refineries can use nano-polymers in filtration systems to reduce the release of harmful gases and particles into the atmosphere [60]. Nano-polymers are being explored as part of cabin air filters and exhaust systems to remove pollutants from car emissions [61].

8.2 Catalysis of Pollutants in Industrial Emissions

Catalysis refers to accelerating a chemical reaction to break down pollutants into harmless substances [56]. Nano-polymers play a significant role in catalysis, particularly in the reduction or conversion of toxic compounds in industrial emissions into less harmful substances [57]. Nano-polymers can be designed to contain nanocatalysts materials that promote chemical reactions without being consumed in the process [58]. These catalysts often consist of noble metals (like platinum, palladium, or gold) or metal oxides (like titanium dioxide) [59]. The high surface area and reactivity of nanomaterials allow these catalysts to work more efficiently and at lower temperatures than traditional catalysts [60].

The nanoscale size and high surface-to-volume ratio of the polymer-based catalysts allow for increased interaction with pollutants [61]. This enhances the efficiency of catalytic reactions [56]. Just as with nano-adsorbents, nano-polymers can be functionalized with specific groups that target particular pollutants for catalytic conversion [57]. For example, polymers with embedded copper or iron nanoparticles can be used to catalytically convert nitrogen oxides (NOx) or sulfur compounds into less harmful products like nitrogen or sulfuric acid [58]. Nano-polymers can also help in lowering the activation energy needed for the reactions, meaning that less energy is required for pollutant conversion [59]. This is particularly beneficial in industrial applications where energy consumption is a key concern [60].

Nano-polymers can be used in catalysts designed to remove nitrogen oxides (NOx) from exhaust gases [61]. These gases are a major source of air pollution and contribute to smog formation and acid rain [56]. Using nano-polymers in catalytic converters can reduce the amount of NOx emitted by vehicles and industrial plants [57]. VOCs released by chemical manufacturing processes and fuel combustion are harmful to human health and the environment [58]. Nano-polymers can catalyze the oxidation of VOCs into non-toxic byproducts such as water and carbon dioxide, reducing their impact on air quality [59]. Another exciting application of nano-polymers in catalysis is their potential to capture and convert CO_2 emissions from power plants and industrial facilities [60]. Through processes like photocatalysis or electrocatalysis, nano-polymers can help reduce CO_2 levels in the atmosphere, contributing to efforts to combat climate change [61]. Nano-polymers can catalyze the removal of sulfur compounds from flue gases, which is essential in preventing acid rain and reducing air pollution caused by sulfur dioxide (SO_2) emissions [56].

9 Wastewater Treatment and Resource Recovery

Nanopolymers play an important role in wastewater treatment, where they can remove a wide variety of pollutants, recover valuable resources, and reduce the environmental impact of wastewater disposal [57]. Nanopolymers can be tailored to selectively remove various contaminants from wastewater, such as industrial chemicals, pharmaceuticals, and personal care products, which are often difficult to remove using traditional methods [58]. Polymers containing functional groups like amides or carboxylates can remove pharmaceutical contaminants such as antibiotics or hormones from wastewater [59].

Nanopolymers can be used to recover valuable resources from wastewater, such as rare metals, phosphorus, and nitrogen compounds, which can be reused or recycled [60]. Nanopolymers can be designed to adsorb phosphates from wastewater, helping to reduce eutrophication in aquatic environments and recover phosphorous for reuse in fertilizers [61].

Waste management and recycling have been major global concerns as the world faces environmental challenges due to the increasing amount of waste and limited resources [56]. Nano-polymers, particularly in the form of nanocomposites, are emerging as innovative solutions in the field of waste management, providing efficient, sustainable, and environmentally friendly methods to treat and recycle various types of waste [57].

9.1 Nanocomposites for Waste Treatment

Nanocomposites are materials made by incorporating nanoscale particles (typically less than 100 nm) into a polymer matrix [58]. These nanoparticles, which can include metals, metal oxides, clay, carbon nanotubes, or other nanoparticles, impart new and often enhanced properties to the polymer [59]. This results in materials with superior mechanical, thermal, electrical, or optical characteristics compared to traditional polymers [60].

Nanocomposites can play a key role in improving various aspects of waste treatment by addressing issues such as degradation, absorption, filtration, and catalysis [61]. Some significant ways in which nanocomposites contribute to waste treatment includes adsorption of toxic substances, including heavy metals (e.g., lead, cadmium, arsenic) from contaminated water [56]. Nanoparticles like nano-clays or graphene oxide have a large surface area, making them highly effective in capturing and immobilizing pollutants [57]. The incorporation of nanocatalysts (e.g., silver, titanium dioxide) into polymer matrices can lead to materials that efficiently degrade organic pollutants such as pesticides, dyes, and pharmaceuticals [58]. Nano-based photocatalysis, where light energy is used to activate nanoparticles for breaking down contaminants, is particularly useful in water and soil treatment [59]. Biodegradable nanocomposites are gaining traction in the field of waste management [60]. For example, polymers can be engineered with nanoparticles of natural sources such as chitosan or cellulose to create materials that degrade naturally, reducing environmental impact compared to traditional non-biodegradable plastics [61]. Nanocomposites are also utilized in converting waste materials into usable energy [56]. Nanocatalysts can enhance processes like pyrolysis, gasification, and bio-digestion, facilitating the conversion of organic waste into biofuels or other valuable energy forms [57].

For instance, Clay-based nanocomposites used for water purification and removal of pollutants from industrial wastewater [58]. Carbon nanotube (CNT) reinforced composites are incorporated into polymer matrices to improve the mechanical strength and adsorption properties for waste containment [59]. Polymer nanocomposites with titanium dioxide (TiO_2): Used for the photocatalytic degradation of organic pollutants in wastewater treatment [60].

10 Integration of Nano-Polymers in Circular Economy Strategies

A circular economy is a system where products and materials are reused, repaired, refurbished, and recycled, extending their lifecycle and reducing waste generation [61]. Nano-polymers can play an essential role in enhancing these strategies, particularly by improving material properties, enabling efficient recycling, and offering innovative solutions for waste reduction [56].

Traditional recycling processes often involve challenges like contamination, degradation of materials, or inefficiency in separating mixed waste [57]. Nano-polymers can help address these issues in the several ways as nanocomposite materials can be designed to selectively interact with certain types of waste (e.g., separating different plastic types) or even specific contaminants, making the recycling process more efficient and precise [58]. Nano-additives in polymers can enhance the properties of recycled materials, improving their strength, thermal stability, and longevity, which makes them more valuable for reuse in manufacturing [59]. Nano-polymers can aid in the design for recycling process by creating materials that are easier to recycle or disassemble [60]. For instance, self-healing polymers or polymers with reversible crosslinking can be used in manufacturing, allowing these materials to be easily reprocessed or repaired, thus reducing waste [61].

10.1 Degradation of Pollutants

Nanopolymers can play a significant role in the degradation of pollutants, breaking down hazardous substances into less toxic or non-toxic compounds [56]. Certain nanopolymers have catalytic properties that allow them to accelerate the breakdown of organic pollutants, such as pesticides, hydrocarbons, and dyes, into simpler, safer substances [57]. Nanopolymers embedded with photocatalysts like titanium dioxide can degrade organic pollutants in the presence of light, breaking down hazardous substances in water or air [58]. Nanopolymers can serve as supports for the immobilization of enzymes that degrade pollutants [59]. The enzymes are protected by the polymer matrix, allowing them to work more effectively and for longer periods [60]. Enzyme-loaded nanopolymers can be used in the degradation of toxic organic compounds in wastewater, facilitating bioremediation processes [61].

10.2 Removal of Radioactive Contaminants

Nanopolymers are also being researched for their ability to capture and neutralize radioactive contaminants, which are particularly challenging to remediate [56]. Nanopolymers can be functionalized to selectively bind radioactive ions like cesium (Cs^{+}) and strontium (Sr^{2+}) in contaminated environments, such as nuclear waste sites or radiological accidents [57]. Functionalized nanopolymers with ligands that bind specifically to radioactive ions can help in the removal of these contaminants from soil or water [58]. Despite their great potential, the application of nanopolymers in environmental remediation faces several challenges [59].

11 Technological Innovations and Novel Approaches

The technological innovations surrounding nano-polymers hold immense promise across various fields, particularly in environmental remediation, energy, and healthcare [61]. The combination of smart, self-healing, and hybrid materials with advances in scalable fabrication techniques ensures that these materials will play a crucial role in addressing global challenges like pollution, resource scarcity, and sustainable development [62]. As fabrication techniques improve and more functionalization strategies are developed, we can expect to see a significant impact on industries ranging from water purification to biomedicine and beyond [63]. Technological innovations and novel approaches in the field of nano-polymers are pushing the boundaries of what can be achieved in environmental remediation, material science, and industrial applications [64]. These key concepts include [65].

11.1 Smart Nano-Polymers for Targeted Remediation

Smart nano-polymers are polymers engineered at the nanoscale that can respond to environmental stimuli (e.g., pH, temperature, light, or specific chemical triggers) to carry out a specific function, such as contaminant removal or toxin degradation [61]. These materials can be used in environmental cleanup, where they target pollutants like heavy metals, pesticides, and organic compounds [62]. By being highly responsive to specific environmental conditions, they can "activate" at the site of contamination, selectively interacting with pollutants without affecting the surrounding environment [63]. Nano-polymers designed to absorb heavy metals from contaminated water or soil when the pH levels drop or increase, making them ideal for targeted and efficient environmental remediation [64].

11.2 Self-healing Nano-Polymers for Long-Term Environmental Impact

Self-healing nano-polymers have the ability to repair damage to themselves without external intervention [65]. This characteristic is particularly valuable for applications where materials are subject to constant wear and tear or exposure to harsh environmental conditions [61]. In environmental remediation, self-healing nano-polymers can be used to repair themselves after they've interacted with contaminants, ensuring prolonged efficacy [62]. This extends the lifespan of the material, reducing the need for replacement and minimizing long-term waste [63]. Self-healing nano-polymers in water treatment systems that maintain their function over extended periods by automatically repairing any structural damage, thus ensuring consistent pollutant removal [64].

11.3 Hybrid Materials: Combining Nano-Polymers with Other Nanomaterials

Hybrid materials involve the integration of nano-polymers with other nanomaterials, such as nanoparticles or nanofibers, to combine the beneficial properties of both materials [65]. This results in materials with enhanced mechanical, chemical, or thermal properties [61]. The combination of nano-polymers with nanoparticles can enhance the material's stability, conductivity, or selectivity, while nanofibers can increase surface area or improve strength [62]. These hybrid materials can be applied in areas like sensor development, environmental sensing, and energy storage [63]. Nano-polymer-based composites combined with carbon nanotubes or metallic nanoparticles for use in water filtration systems that require both selective adsorption of contaminants and mechanical stability [64].

11.4 Functionalization of Nano-Polymers for Enhanced Selectivity and Efficiency

Functionalization refers to the modification of nano-polymers to attach specific chemical groups or molecules to their surface to enhance their ability to interact with particular substances [65]. This can significantly improve the selectivity and efficiency of the material in targeted applications [61]. Functionalized nano-polymers are used in environments where specificity is crucial, such as in biomedical applications (drug delivery), selective pollutant removal in water or soil, and catalysts in industrial processes [62]. Functionalized nano-polymers with surface-bound ligands or antibodies that specifically target heavy metals, pathogens, or toxins in contaminated environments, providing highly selective and efficient remediation [63].

11.5 Advances in Fabrication Techniques for Scalable Production of Nano-Polymers

As nano-polymers offer unique properties at the nanoscale, producing them at a large scale while maintaining these properties has been a challenge [64]. Recent advancements in fabrication techniques are enabling the scalable production of high-quality nano-polymers [65]. Scalable production methods are essential for transitioning nano-polymer technologies from research and development into commercial applications, such as large-scale environmental cleanup or industrial product development [61]. Techniques like electrospinning, roll-to-roll processing, and 3D printing are allowing for the mass production of nano-polymers with tailored properties [62]. These advances are critical for developing large-scale water filtration systems, air purification technologies, and energy storage devices [63].

12 Conclusion

The chapter explored the recent advancements in the development and application of nano-polymers for environmental remediation. It highlighted the unique properties of nano-polymers, such as high surface area, biocompatibility, and the ability to target specific pollutants, making them highly effective in cleaning up various environmental contaminants like heavy metals, pesticides, and organic pollutants. Several innovative techniques, such as nano-polymer-based adsorbents, catalysts, and delivery systems, were discussed, showcasing their potential to efficiently address environmental challenges. The chapter also reviewed various case studies and applications where nano-polymers have demonstrated promising results in water and soil treatment.

Nano-polymers hold transformative potential for environmental remediation due to their remarkable properties that allow for efficient pollutant removal and degradation. Their versatility enables them to be tailored for specific contaminants, whether in water, soil, or air, offering targeted solutions. Additionally, the ability to modify nano-polymers for enhanced reusability and biodegradability makes them a sustainable option, reducing long-term environmental impact. As research continues to advance, nano-polymers could revolutionize current remediation methods by making them more cost-effective, faster, and environmentally friendly.

Looking ahead, nano-polymers are poised to play a critical role in addressing the growing environmental challenges of pollution and climate change. Their ability to integrate with existing remediation technologies and enhance their efficiency could lead to more sustainable practices. As research into nano-polymer synthesis, functionality, and scalability continues to evolve, their widespread application could become a cornerstone of global environmental remediation strategies. However, further investigation into their potential risks, environmental behavior, and long-term effects will be essential to ensure that nano-polymers are implemented safely and effectively. Ultimately, nano-polymers represent a promising avenue for achieving sustainable environmental solutions in the near future.

References

1. Rockström, J., et al.: Planetary boundaries: exploring the safe operating space for humanity (2009). https://www.ecologyandsociety.org/vol14/iss2/art32/
2. Khan, F.I., Husain, T.: Risk-based monitored natural attenuation—a case study (2001). https://www.sciencedirect.com/science/article/pii/S0304389401002369
3. Tchounwou, P.B., et al.: Heavy metal toxicity and the environment (2012). https://www.ncbi.nlm.nih.gov/pmc/articles/PMC4144270/
4. Theron, J., et al.: Nanotechnology and water treatment: applications and emerging opportunities (2008). https://doi.org/10.1080/10408410701710442
5. Qu, X., et al.: Applications of nanotechnology in water and wastewater treatment (2013). https://www.sciencedirect.com/science/article/pii/S0043135413001772
6. Lau, W.J., et al.: Characterization methods of thin film composite nanofiltration membranes (2015). https://doi.org/10.1080/15422119.2014.882355
7. Luckachan, G.E., Pillai, C.K.S.: Biodegradable polymers—a review on recent trends and emerging perspectives (2011). https://doi.org/10.1007/s10924-011-0317-1
8. Sánchez, C., et al.: Applications of hybrid organic-inorganic nanocomposites (2005). https://pubs.rsc.org/en/content/articlehtml/2005/jm/b509097k
9. Fujishima, A., Honda, K.: Electrochemical photolysis of water at a semiconductor electrode (1972). https://www.nature.com/articles/238037a0
10. Stuart, M.A.C., et al.: Emerging applications of stimuli-responsive polymer materials (2010). https://www.nature.com/articles/nmat2614
11. Astruc, D., et al.: Dendrimers designed for functions: from physical, photophysical, and supramolecular properties to applications in sensing, catalysis, molecular electronics, photonics, and nanomedicine (2010). https://doi.org/10.1021/cr900327d
12. Gao, C., & Yan, D.: Hyperbranched polymers: from synthesis to applications (2004). https://www.sciencedirect.com/science/article/pii/S0079670003001345
13. Alivisatos, A.P.: Semiconductor clusters, nanocrystals, and quantum dots (1996). doi/https://doi.org/10.1126/science.271.5251.933

14. Bell, A.T.: The impact of nanoscience on heterogeneous catalysis (2003). https://doi.org/10.1126/science.1083671
15. Babel, S., Kurniawan, T.A.: Low-cost adsorbents for heavy metals uptake from contaminated water (2003). https://www.sciencedirect.com/science/article/pii/S0304389402002637
16. Auffan, M., et al.: Towards a definition of inorganic nanoparticles from an environmental, health and safety perspective (2009). https://www.nature.com/articles/nnano.2009.242
17. Zhang, L., et al.: Multilayer electrospun nanofibrous membranes with antibacterial property for air filtration (2020). https://www.sciencedirect.com/science/article/pii/S0169433220307182
18. De Volder, M.F.L., et al.: Carbon nanotubes: present and future commercial applications (2013). https://doi.org/10.1126/science.1222453
19. Lu, A. H., et al.: Magnetic nanoparticles: synthesis, protection, functionalization, and application (2007). https://doi.org/10.1002/anie.200602866
20. White, S.R., et al.: Autonomic healing of polymer composites (2001). https://www.nature.com/articles/35057232
21. Crini, G.: Recent developments in polysaccharide-based materials (2005). https://www.sciencedirect.com/science/article/pii/S007967000400125X
22. Saleem, J., et al.: Oil sorbents from plastic wastes and polymers: a review (2018). https://www.sciencedirect.com/science/article/pii/S0304389417305885
23. Kim, H., et al.: Thermal conductivity of polymer composites with the geometrical characteristics of graphene nanoplatelets (2010). https://www.nature.com/articles/srep26825
24. Yang, R.T.: Adsorbents: fundamentals and applications (2003). https://doi.org/10.1021/cr0203436
25. Mulligan, C.N., et al.: Remediation technologies for metal-contaminated soils and groundwater (2001). https://www.sciencedirect.com/science/article/pii/S0013795200001010
26. Klemm, D., et al.: Cellulose: fascinating biopolymer and sustainable raw material (2005). https://doi.org/10.1002/anie.200460587
27. Elimelech, M., Phillip, W.A.: The future of seawater desalination: energy, technology, and the environment (2011). https://doi.org/10.1126/science.1200488
28. Siracusa, V., et al.: Biodegradable polymers for food packaging (2008). https://www.sciencedirect.com/science/article/pii/S0924224408002185
29. Sheldon, R.A.: Enzyme immobilization: the quest for optimum performance (2007). https://doi.org/10.1002/adsc.200700082
30. Salt, D.E., et al.: Phytoremediation (1998). https://doi.org/10.1146/annurev.arplant.49.1.643
31. D'Alessandro, D.M., et al.: Carbon dioxide capture (2010). https://doi.org/10.1002/anie.201000431
32. Shannon, M.A., et al.: Science and technology for water purification in the coming decades (2008). https://www.nature.com/articles/nature06599
33. Yarahmadi, A.: Nanotechnology applications in nuclear waste management: challenges and limitations (2025). https://doi.org/10.1007/s10967-025-10015-3
34. Ligon, S.C., et al.: Polymers for 3D printing and customized additive manufacturing (2017). https://doi.org/10.1021/acs.chemrev.7b00074
35. Garcia-Segura, S., et al.: Electrochemical oxidation remediation of real wastewater effluents—a review (2018). https://doi.org/10.1016/j.psep.2017.09.014
36. Elgarahy, A.M., et al.: Recent advances in greenly synthesized nanoengineered materials for water/wastewater remediation: an overview. (2021). https://doi.org/10.1007/s41204-021-00104-5
37. Guesmi, A., et al.: Disinfection of corona and myriad viruses in water by non-thermal plasma: a review. https://doi.org/10.1007/s11356-022-21160-7
38. Badawi, A.K., et al.: Hybrid treatment system for real textile wastewater remediation based on coagulation/flocculation, adsorption and filtration processes (2021). https://doi.org/10.1016/j.jwpe.2021.101963
39. Jawed, A., et al.: Engineered nanomaterials and their surface functionalization for the removal of heavy metals: a review (2020). https://doi.org/10.1016/j.jwpe.2019.101009

40. Suresh, R., et al.: Recent advancements of spinel ferrite based binary nanocomposite photocatalysts in wastewater treatment (2021). https://doi.org/10.1016/j.chemosphere.2021.129734
41. Al-Anazi, A., et al.: Novel franklinite-like synthetic zinc-ferrite redox nanomaterial (2020). https://doi.org/10.1016/j.apcatb.2020.119098
42. Huang, C., et al.: Extremely well-dispersed zinc oxide nanofluids with excellent antibacterial properties (2022). https://doi.org/10.1021/acs.iecr.2c00369
43. Asghar, N., et al.: Application of $MnFe_2O_4$ magnetic silica-covered nanomaterials to the draw solution in forward osmosis (2023). https://doi.org/10.1016/j.chemosphere.2023.138735
44. Wu, Y., et al.: A pH-responsive supramolecular draw solute for arsenic removal (2019). https://doi.org/10.1016/j.watres.2019.114993
45. Sheoran, K., et al.: Recent advances of carbon-based nanomaterials for wastewater treatment (2022). https://doi.org/10.1016/j.chemosphere.2022.134364
46. Akbari, A., et al.: Transition metal oxide nanoparticles as efficient catalysts in oxidation reactions (2018). https://doi.org/10.1016/j.nanoso.2018.01.006
47. Santhosh, C., et al.: Role of nanomaterials in water treatment applications: a review (2016). https://doi.org/10.1016/j.cej.2016.08.053
48. Hojjati-Najafabadi, A., et al.: Magnetic-MXene-based nanocomposites for water and wastewater treatment (2022). https://doi.org/10.1016/j.jwpe.2022.102696
49. Cervantes-Avilés, P., et al.: Incidence of metal-based nanoparticles in conventional wastewater treatment (2021). https://doi.org/10.1016/j.watres.2020.116603
50. Kane, A., et al.: Advanced photocatalytic treatment of wastewater using immobilized titanium dioxide (2022). https://doi.org/10.3390/ma15134547
51. Hussain, F., et al.: E-glass-polypropylene pultruded nanocomposite (2007). https://doi.org/10.1177/0892705707079604
52. Guo, F., et al.: A review of the synthesis and applications of polymer-nanoclay composites (2018). https://doi.org/10.3390/app8091696
53. Yu, J., et al.: Preparation and properties of montmorillonite modified asphalts (2007). https://doi.org/10.1016/j.msea.2006.10.037
54. You, Z., et al.: Nanoclay-modified asphalt materials: preparation and characterization (2011). https://doi.org/10.1016/j.conbuildmat.2010.06.070
55. You, Z., et al.: Nanoclay-modified asphalt materials: characterization studies (2011). https://doi.org/10.1016/j.conbuildmat.2010.06.070
56. Chan, M.L., et al.: Mechanism of reinforcement in a nanoclay/polymer composite (2011). https://doi.org/10.1016/j.compositesb.2011.03.011
57. Rong, K., et al.: Green synthesis of iron nanoparticles for Cr(VI) removal (2020). https://doi.org/10.1016/j.ecoleng.2020.105793
58. Soni, R., et al.: An overview of nanoscale materials for wastewater contaminant removal (2020). https://doi.org/10.1007/s13201-020-01275-3
59. Sarkar, B., et al.: Designer carbon nanotubes for contaminant removal (2018). https://doi.org/10.1016/j.scitotenv.2017.08.132
60. Yu, G., et al.: Applications of nanomaterials for heavy metal removal (2021). https://doi.org/10.3390/SU13020713
61. Pumera, M.: Graphene-based nanomaterials and their electrochemistry (2010). https://doi.org/10.1039/C002690P
62. Perreault, F., et al.: Environmental applications of graphene-based nanomaterials (2015). https://doi.org/10.1039/C5CS00021A
63. Jung, W., et al.: (2018). An efficient reduced graphene-oxide filter for PM2.5 removal. https://doi.org/10.1039/c8ta04587a
64. Shahzad, W., et al.: Enhanced visible light photocatalytic performance (2022). https://doi.org/10.1016/j.ceramint.2022.05.151
65. Guo, M., et al.: Carbon nanotube-grafted chitosan for phenol adsorption (2019). https://doi.org/10.1016/j.scitotenv.2019.05.148

www.ingramcontent.com/pod-product-compliance
Ingram Content Group UK Ltd.
Pitfield, Milton Keynes, MK11 3LW, UK
UKHW020630260926
475829UK00003BA/11
9789819535804